Laser Optics of
Condensed Matter

Laser Optics of Condensed Matter

Edited by

Joseph L. Birman and
Herman Z. Cummins

The City College of the City University of New York
New York, New York

and
A. A. Kaplyanskii

A. F. Ioffe Physico-Technical Institute
Academy of Sciences of the USSR
Leningrad, USSR

Plenum Press • New York and London

Library of Congress Cataloging in Publication Data

Binational USA–USSR (Symposium on Laser Optics of Condensed Matter (3rd: 1987:
  Leningrad, R.S.F.S.R.)
  Laser optics of condensed matter.

  "Proceedings of the Third Binational USA–USSR Symposium on Laser Optics of
Condensed Matter, held June 1–6, in Leningrad, USSR" — T.p. verso.
  Includes bibliographies and index.
  1. Condensed matter — Optical properties — Congresses. 2. Lasers — Congresses. I.
Birman, Joseph Leon, 1927–   . II. Cummins, Herman, Z., 1933–   . III. Kaplian-
skii, A. A. IV. Title.
QC173.4.C65B56   1987   530.4′1   88-2513
ISBN 0-306-42816-4

Proceedings of the Third Binational USA–USSR Symposium on Laser Optics
of Condensed Matter, held June 1–6, 1987, in Leningrad, USSR

© 1988 Plenum Press, New York
A Division of Plenum Publishing Corporation
233 Spring Street, New York, N.Y. 10013

Printed in the United States of America

PREFACE

The Third Binational USA-USSR Symposium titled "Laser Optics of Condensed Matter" was held in Leningrad 1 June - 5 June 1987. This volume contains the full text of 64 papers presented at (or prepared for) the Symposium in both plenary and poster sessions. This Symposium reestablished the very productive series of "Light Scattering" Binational Symposia which were initiated in Moscow in 1975. Unfortunately there was an eight-year hiatus following the Second Symposium in New York (1979). This interval, caused by serious chilling of the climate of USA-USSR collaboration, deprived the active scientists on both sides of the opportunity to meet and interact in the active format of a conference.

During this eight year interval there has been very rapid and intense development of scientific activity in the general area of laser optics phenomena. The development of ultrafast laser sources has permitted rapid advances in time resolved spectroscopy and ultrafast processes; the field of optical bistability and strong nonlinearity became a hot topic; and intense work is now underway to clarify ideas of photon localization. These new developments complement many advances in the study of low dimensional systems such as surfaces, new work on phase transitions, and novel studies of elementary excitations such as polariton-excitons in localized environments such as quantum wells and heterojunctions.

In 1987 the long-overdue Third Binational Symposium occurred. 22 American scientists met with about 50 Soviet colleagues in intense and active discussions at the A. Gorky Scientists' Club on the banks of the Neva river in Leningrad to carry on scientific exchange of ideas. The intensity of the Conference was moderated when delegates visited some local attractions including a trip to Peterhof (after a memorable boat-ride on the river) and the Hermitage Museum which is adjacent to the scientists' meeting hall.

On the scientific and cultural level the Conference was judged by the participants to be very successful and, at the conclusion, there was great enthusiasm for a repeat. It is thus a very pleasant duty to announce that plans are now moving ahead rapidly for the Fourth USA-USSR Symposium in this series which will be held in late 1989, or early 1990 at the University of California, Irvine, organized by Prof. A.A. Maradudin and his colleagues.

We want to express our thanks here to our Soviet colleagues for their help in preparing their manuscripts and sending them to us rapidly. Separate thanks in this regard are due to our co-editor Prof. A.A. Kaplyanskii and his colleagues at the Ioffe Institute and to Prof. A. A. Maradudin for his major contribution to the editing. We also thank Mrs. Rhina Hererra, Ms. Jean Brown, Ms. Tracy Turner and Mr. Spiros Branis for their help in preparing and typing the manuscripts.

New York, September 1987   Joseph L. Birman
                                                           Herman Z. Cummins

CONTENTS

SECTION III
SURFACE OPTICAL PHENOMENA

SECTION IV
OPTICS OF SMALL PARTICLES, DEFECTS

SECTION V
PHOTON LOCALIZATION

SECTION VI
LIGHT SCATTERING

SECTION VII
EXCITONS, POLARITONS

SECTION VIII
PHOTOACTIVE MATERIALS, PHOTOREFRACTION, HOLE-BURNING, HOLOGRAPHY

SECTION IX
OPTICAL BISTABILITY IN SEMICONDUCTORS AND LIQUID CRYSTALS

OPENING REMARKS

Academician V. M. Tuchkevich

A. F. Ioffe Physico-Technical Institute, Leningrad

I am glad to welcome you to the Third Soviet-American Symposium on Laser Optics of Condensed Matter.

Nearly ten years have passed since the Second Symposium was held in New York. I am confident we all regret that this period of ten years was so long. It was the period of mutual distrust, suspicion and misunderstanding which was enhanced by the arms race.

Fortunately for our two countries and for mankind things are changing for the better today. Of course we understand very well that this process takes time and some efforts by both countries. We understand also that a very important factor in improving our relations is increasing contacts between our two nations. I hope that our bilateral Symposium will play an important role in the solution of this problem.

When I looked through the scientific program of the present Third Symposium I became confident that it will be interesting and inspiring. You will discuss here many essential fundamental problems of modern Condensed Matter Optics. I should like to note that some of the problems of this program have not only fundamental meaning for science but also have great application potential.

We are glad to welcome you to our beautiful Leningrad. The city itself is a remarkable monument, a museum, where every stone evokes the events dear to the heart of every Russian. Our city is sacred to Soviet citizens since many glorious events of Russian history, Russian culture and the Russian revolutionary movement are closely associated with Leningrad.

I hope you will find it interesting not only to visit our museums and famous sights but also to meet Leningrad people, to talk to them so as to know us better. The Organizing Committee has also arranged a Social Program for you. You will be shown what is most interesting in Leningrad.

I hope you will enjoy both the scientific and the Social Program. I wish you a pleasant and fruitful stay in our city.

Now, as the chairman of the Soviet Organizing Committee of the Third Symposium of Laser Optics of Condensed Matter, I declare the Third Symposium open.

Thank you.

Joseph L. Birman

City College of the City University of New York

It is a very great pleasure to open the Third Binational USA-USSR Symposium - now titled "Laser Optics of Condensed Matter" - in Leningrad on this 1 June 1987. The American scientists participating in this Symposium are very grateful for all the arrangements which have been made by our colleagues of the Soviet Organizing Committee. We especially express our warmest thanks to Academician V. M. Tuchkevich, Director of the A. F. Ioffe Physico-Technical Institute, USSR Academy of Sciences who is the Chairman of this Committee, and to Prof. A. A. Kaplyanskii of the Ioffe Institute who is the Vice Chairman. On hehalf of all my colleagues I ask them to convey our warmest thanks to all the collaborators of the Soviet Organizing Committee and the Program Committee of this 1987 edition.

Looking at the program and the composition of both the US and USSR scientific delegations, I am pleased to take note of the combination of continuity and novelty in our Symposium. So, for example, in our American group more than half of the scientists are participating in this bi-national Symposium for the first time, and in addition there are a number of "veterans" of our two previous Symposia; the same is true for the Soviet delegation. This mixture of continuity and novelty certainly will give new vitality to our Symposium.

We especially look forward to this 1987 Symposium after the eight year interval since our Second Meeting in New York in 1979. The general field of laser-related optics of condensed matter has undergone an explosive development in recent years. Partly this is due to the development of new and sophisticated experimental techniques for measuring ultrafast phenomena such as on a femtosecond time scale, the development of methods to prepare heterojunctions and quantum wells, as well as other methods to prepare semiconductors and low dimensional - 2-D systems, surfaces, and ultra-small particles. In addition a number of new topics have moved to the forefront such as problems related to localization of optical photons, squeezed optical states with exceptionally low noise, and nonlinear and nonlocal effects such as the predicted phonaritron effects. We look forward in this Symposium to learning much about these aspects of new physics as well as the latest developments in phase transitions, photorefraction and optical bistability.

It is particularly meaningful to anticipate this outburst of new physics here in Leningrad which is the site of so much important physics on optical properties of solids. My own clear recollection is of meeting Prof. E. F. Gross in his laboratory at Ioffe Institute in 1969 and having him point with great excitement to some new data and say to me "You see Dr. Birman, here we have evidence for biexcitons in Cu_2O." I had the privilege of first meeting Prof. Gross at the 1958 Rochester Semiconductor Conference, so experiencing his infectious enthusiasm at our meeting in his laboratory was a double pleasure for me.

In his remarks, in this opening session Prof. Tuchkevich emphasized how important these Symposia are both in the framework of stimulating physics as well as being an element in developing good relations between the US and the Soviet Union. We all know how vital this development is for peace and security in the world. I can underline the importance of building relations and diminishing misunderstandings by recalling a 12 hour train trip I took from Riga to Leningrad in 1986. I had the opportunity then to talk to a number of young, and very sincere Soviet people (they were in their early 20's.) It was very disturbing to me that their picture of the USA was of a country armed to the teeth with missiles and submarines but that they had little understanding of our culture, our science, and our people. No doubt a Soviet physicist taking a train ride from Kansas City to Boulder, Colorado would have been told the corresponding tale about the USSR. Our need to make sure that our children and grandchildren live in a safer world was one of the impelling forces behind my (and I believe everyone else's) desire to realize this 1987 Seminar in spite of many obstacles that had to be overcome.

Now let us look forward. Both US and Soviet physicists have formed Committees to keep continuity of these Symposia. The list of members of the "Permanent Soviet Organizing Committee" and of the "USA Standing Committee for Binational Laser Optics Symposia" is contained in the following pages of the Proceedings. We all anticipate resuming our Binational Symposium on a regular basis by continuing with the Fourth Symposium which we expect will occur in 1989 at the University of California, Irvine.

3

SPECTROSCOPY WITH 6 FEMTOSECOND OPTICAL PULSES

Charles V. Shank

AT&T Bell Laboratories, Holmdel, New Jersey, 07733 - U.S.A.

ABSTRACT

Rapid progress has taken place in the generation and application of femtosecond optical pulses. The impact of these developments is being felt in a broad range of scientific fields including physics, chemistry, and biology. This paper reviews the state of the art and describes the application of femtosecond techniques to the examination of nonequilibrium phenomena.

INTRODUCTION

Progress in gaining new understanding of the world of ultrafast events has dramatically increased with the development of new optical pulse generation and measurement techniques. Recently optical pulses with a duration of 6 femtoseconds (1 femtosecond $= 10^{-15}$ second) have been measured and used for experimental investigations (R. L. Fork and colleagues, 1987). A 6 femtosecond optical pulse contains spectral components that cover a good portion of the visible spectrum and are ideal for obtaining high-resolution time-resolved spectra.

The impact of measurements in the femtosecond time domain covers a broad range of scientific activity. Femtosecond-pulse techniques have contributed significantly to the study of the dynamical properties of molecules and solids. Femtosecond optical pulses have been used to create nonequilibrium, nonthermal population distributions in semiconductors and large molecules. With optical pulses short compared to polarization dephasing times it has been possible to directly observe the process of dynamic spectral hole burning.

PROGRESS IN PULSE GENERATION

The large frequency bandwidth required to generate a short optical pulse has been a major source of difficulty in the generation of ultrashort optical pulses. A fixed-phase relationship must be maintained among all the frequencies oscillating in a mode-locked laser to generate a bandwidth-limited short optical pulse. Until recently, the group velocity dispersion in a laser cavity, due to mirrors, gain medium, and the saturable absorber, limited pulse generation in a dye laser cavity. However, a new device for compensating group velocity dispersion in an optical cavity was invented by Fork and colleagues (1984). The device consists of two pairs of prisms inserted into the optical cavity of a dye laser. Although the prism is typically made of a positively dispersive material, the prism pair arrangement can produce an adjustable amount of negative dispersion. With this cavity configuration, Valdmanis and colleagues were able to generate optical pulses as short as 27 fsec (J. A. Valdmanis and colleagues, 1985).

To generate even shorter pulses the bandwidth of the dye gain medium ultimately becomes limiting. A method of overcoming this limitation was first suggested by Gires and Tournois (1964) and Giordmaine and colleagues (1968) who proposed that optical pulses be shortened by adapting microwave pulse compression techniques to the visible spectrum.

The process of optical pulse compression is accomplished in two steps. In the first step a "chirp" or frequency sweep is impressed on the pulse. Then the pulse is compressed by using a dispersive delay line.

The pulse "chirp" is achieved by passing the optical pulse through a nonlinear medium; an experimentally convenient medium is a short piece of optical fiber. The intense optical field rapidly changes the index of refraction seen by the optical pulse and impresses the frequency sweep on the optical pulse. the chirped pulse is then passed through a "compressor" which provides the appropriate delay required to remove the frequency sweep and produce a compressed optical pulse.

The pulse compressor is usually a pair of gratings (E. B. Treacy, 1969) which form a dispersive delay line. Each frequency component is diffracted by the gratings at a different angle and hence makes a different geometrical path through a set of parallel gratings. This type of compressor is adequate when the pulse frequency bandwidth is a small fraction of the carrier frequency. For very short pulses the grating compressor introduces phase distortions across the pulse spectrum and provides a limitation on the achievable pulse width that can be generated. A pair of prisms can also form a dispersive delay line in much the same way as a pair of gratings. By an accident of nature the unwanted phase distortion induced by the grating pair is of the opposite sign as that for a prism pair. The result is that a grating pair followed by a prism pair can correct phase distortions to the third order in the phase (R. L. Fork and colleagues, 1987). Using such an arrangement an optical pulse of 6 femtoseconds in duration has been generated and measured using the second harmonic autocorrelation technique. An interferometric autocorrelation measurement is shown in Figure 1. For purposes of comparison crosses have been used to indicate the calculated maxima and minima for an interferometric trace of a hyperbolic secant squared pulse having zero phase distortion over the bandwidth of the pulse. The well resolved interference maxima also provide a rigorous calibration of the relative delay.

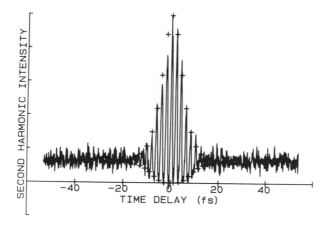

Fig. 1. Plot of the interferometric autocorrelation of a 6 femtosecond optical pulse using the second harmonic technique.

SPECTRAL HOLE BURINING IN LARGE MOLECULES

Ultrashort optical pulse techniques provide a unique means for investigating nonequilibrium energy distribution among vibronic levels in large organic molecules in solution. Using femtosecond pulse techniques it is possible to observe time-resolved hole burning and the process of equilibration to a thermalized population distribution.

The absorption spectrum of a large dye molecule is dominated by vibronic transitions from a thermalized ground state. Typically these large molecules, which have a molecular weight of 400 or more, have a correspondingly large number of degrees of freedom. The optical absorption coefficient may be written as a sum over transitions from occupied vibration levels in the ground state to vibrational levels in the excited state. The absorption coefficient is given by

$$\alpha(\nu) = C\sum_{if}P_iM^2\chi_{if}\nu g(\nu-\nu_{if}) \quad\text{..............................(1)}$$

where C is a constant, P_i is the thermal probability of the initial state, M is the dipole moment of the electronic transition, χ_{if} is the Franck-Condon factor and g is the lineshape profile for each transition. The above expression describes the molecular system in thermal equilibrium. With a short optical pulse it is possible to excite a band of states that are resonant with the pumping energy. Before the molecular system comes into equilibrium, bleaching is observed in a spectral range determined by the convolution of the pump spectrum with the line shape profile of the individual transitions. As time progresses the system relaxes to the thermal equilibrium due to interaction with the thermal bath. The thermal bath couples to the vibronic levels by both intramolecular and intermolecular processes.

Femtosecond pulse techniques make the experimental investigation of such nonequilibrium processes feasible (C. H. Brito Cruz and colleagues, 1986). The experimental apparatus is arranged to pump the sample with a 60 femtosecond optical pulse and a second shorter and weaker pulse of 10 femtoseconds in duration is used to probe the sample absorption. The shorter pulse is used to probe the absorption spectrum of the molecule by passing through the excited sample into a spectrometer and a diode array. The broad spectrum of the short probing pulse allows the measurement a good portion of the S_1 transition of the absorbing molecule.

In Figure 2 we have plotted the absorbance spectrum of the molecule cresyl violet near zero time delay between the pump and the probe before and after excitation. Note that a reduction in the absorbance is observed near the pumping wavelength with additional replica holes above and below the main hole.

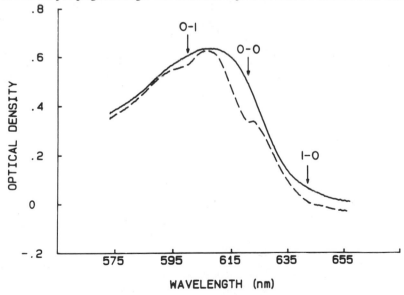

Fig. 2. Plot of the absorbance spectrum of the molecule cresyl violet near zero time delay before (solid line) and after excitation (dashed line) with a 60 femtosecond optical pulse.

The mechanism for the formation of the replica holes is readily understood. Measurements of the Raman spectra of cresyl violet reveal the presence of a strong mode at $590 cm^{-1}$. In a large molecule with a large number of degrees of freedom a correspondingly large number of modes can contribute to the absorption spectrum as was illustrated in equation (1). Usually only a few modes called active or system modes with energies larger than kT change their occupation number during a transition to the excited state. |The strength of the absorption is determined by the Franck-Condon factor, X_{if} The 590 cm^{-1} mode appears to be the dominant mode in the absorption spectrum as evidenced by bleaching both at the 0-0 transition, which is at the excitation energy, and the 0-1 and 1-0 positions of the Franck-Condon progression.

In Figure 3 we plot the differential absorbance spectra as a function of relative time delay between the pump and the probe. The holes broaden and relax to the quasiequilibrium spectrum within the first few hundred femtoseconds.

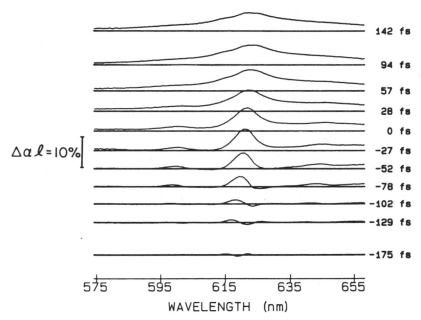

Fig. 3. differential absorbance spectra plotted as a function of relative time delay following excitation with a 60 fsec optical pulse at 618 nm for the molecule cresyl violet.

SPECTRAL HOLE BURNING IN GaAs

Femtosecond optical pulses are nearly ideal for producing nonequilibrium nonthermal carrier distributions in a semiconductor. Nonequilibrium processes in GaAs have been studied using a version of the pump and probe technique. A GaAs sample was prepared with a thickness of 0.5μ. A 50 femtosecond optical pulse at 2.0 eV was used to excite the sample and a second optical pulse with a duration of approximately 6 femtoseconds was used to probe the transmission of the sample at a relative time delay. The frequency spectrum of the 6 femtosecond optical pulse contains spectral components that cover a good portion of the visible spectrum and is ideal for obtaining high-resolution time-resolved spectra.

At high carrier densities ($>10^{17} cm^{-3}$) the thermalization process is dominated by carrier-carrier scattering. In this case at the very earliest relative time delay a spectral hole is observed in the absorption spectrum. As time progress the hole becomes less distinct and by about 100 femtoseconds the carrier distribution becomes thermalized. (See Fig. 4.) In contrast at low excitation densities (Fig. 5.) a second peak in the hole burning spectrum is observed at an energy corresponding to about 1 LO phonon energy below the excitation energy. The accumulation of carriers at this energy is suggestive of

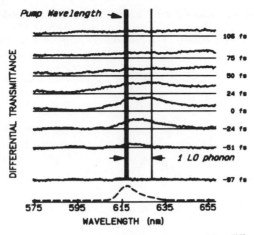

Fig. 4. Time resolved measurements of the differential transmission through a gaAs sample as a function of time for an optically excited carrier density of 10^{19} cm^{-3}.

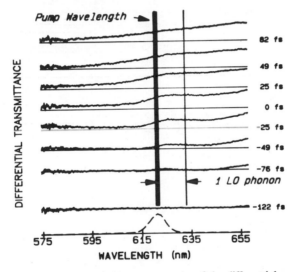

Fig. 5. time resolved measurements of the differential transmission through a GaAs sample as a function of time for an optically excited carrier density of 10^{17} cm^{-3}.

the optical phonon emission. Even at this relatively low excitation density $\approx 10^{17}$ cm^{-3} carrier-carrier scattering is contributing to the thermalization process.

REFERENCES

Cruz, C. H. Brito, Fork, R. L., Knox, W. H., Shank, C. V., Chem Phys Lett 132, 341(1986).
Fork, R. L., Cruz, C. H. Brito, Becker, P. C., Shank, C. V., Optics Lett 12, 483(1987).
Fork, R. L., Martinez, O. E., Gordon, J. P., Optics Lett 9, 150(1984).
Giordmaine, J. A., Duguay, A. A., Hansen, J. W., IEEE J. Quantum Electron. QE-5, 252(1968).
Gires, F., Tournois, P., C. R. Acad. Sci. Paris 258, 6112(1964).
Treacy, E. B., IEEE J. Quantum Electron. QE-5, 454(1969).
Valdmanis, J. A., Fork, R. L., Gordon, J. P., Optics Lett. 10, 131(1985).

HOT ELECTRONS IN SEMICONDUCTORS AND METALS

James G. Fujimoto and Erich P. Ippen

Department of Electrical Engineering and Computer
Science and Research Laboratory of Electronics
Massachusetts Institute of Technology
Cambridge, MA

ABSTRACT

Femtosecond studies have been performed to investigate the scattering and energy relaxation of hot electrons in thin film semiconductors and metals. In both cases nonequilibrium carrier distributions are observed to remain for several picoseconds following ultrashort optical pulse excitation. In metals, during this nonequilibrium period, ultrafast transport has been measured directly.

Absorption Saturation Dynamics in GaAs

Although the dynamics of hot carriers in GaAs and AlGaAs have been the subject of extensive experimental investigation for many years[1], sub-picosecond absorption saturation spectroscopy provided the first direct observations of ultrafast bandgap renormalization and relaxation of a hot carrier distribution[2,3]. Since then, both equal-pulse-correlation[4,5] and non-collinear pump-probe[6,7] methods have extended the study of these materials into the femtosecond time domain. Further, pump and continuum-probe measurements have been used to observe the initial spectral hole burning that accompanies the ultrafast excitation of carriers[8–10]. In this paper we discuss recent results of both identical-pulse pump-probe measurements and multi-wavelength continuum probe studies. The identical pulse measurements monitor absorption saturation at the excitation wavelength and are therefore sensitive to state filling at the excitation energy and the rapid scattering out of this initial distribution. With a continuum probe it becomes possible to study not only dynamics occurring at the excitation wavelength but also the evolution and relaxation of the hot carrier distribution.

The laser source for the identical pulse experiments was a CPM ring dye laser[11] incorporating internal prisms for the control of dispersion[12]. This laser produced transform-limited pulses as short as 35 fs full-width-at-half-maximum. In addition, a pair of prisms was used external to the laser cavity to permit independent adjustment of pulse duration and chirp[13]. Thus, by adjusting both internal and external prisms, we were able to generate either bandwidth-limited pulses of variable duration (35 - 150 fs) or pulses of variable chirp. The control of both parameters was found to be especially valuable for the measurement of responses that occur on the time scale of the pulse. Both the intensity autocorrelation function and the electric field coherence function were measured accurately[14].

Our experimental setup used a conventional delay line geometry for femtosecond pump-probe measurements of transient absorption saturation. Differential detection between the transmitted probe and a reference was used to subtract fluctuations in the incident laser pulse intensity and thereby achieve detection of changes in absorption as small as several parts in 10^6. The zero of delay was determined by interchanging the roles of the pump and probe and observing the point of mirror symmetry.

Our samples were thin layers of GaAs grown by liquid phase epitaxy. Each was approximately 0.5 μm thick and was clad on both sides by transparent layers of $Al_{.85}Ga_{.15}As$ to minimize surface effects. All measurements were made at room temperature and with an optical pumping wavelength of about 625 nm (1.98 eV).

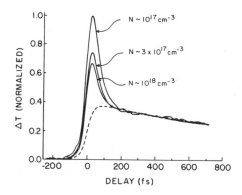

Fig. 1. Absorption saturation response in GaAs measured with perpendicular pump and probe polarizations for three different levels of induced carrier density.

Pump-probe traces taken in GaAs at three different pump intensities (corresponding to the indicated three different excited carrier densities) are shown in Figure 1. All three were taken with orthogonally polarized pump and probe beams. The difference between these traces and those obtained for parallel polarization could be approximated by a coherent coupling contribution equal to one half the amplitude of the parallel case at zero delay. Thus, we infer that induced absorption anisotropy played a negligible role in our experiments. All three curves in Figure 1 demonstrate the presence of an essentially two-component carrier relaxation process. They have been normalized relative to each other by the amplitude of their common, picosecond relaxation component. The dashed line, showing an analytical fit to this slower decay, was obtained by convolving the pulse intensity autocorrelation function with an impulse response having a 1.5 ps exponential time constant. We attribute this decay to the cooling of the carrier distribution to the lattice by LO phonon emission.

The ultrafast partial recovery component evident at the beginning of each of the traces demonstrates that, at these carrier densities, scattering out of the initially excited distribution is very rapid. Deconvolution of the ultrafast component in the $10^{17}/cm^3$ case yields a response time of 30 fs. At the higher densities the response time appears to be considerably shorter than the pulse duration. We have used the decreasing amplitude of this pulsewidth-limited response to estimate times of 17 fs and 13 fs for the 3 x 10^{17} and $10^{18}/cm^3$ cases respectively. Such a carrier density dependence suggests that carrier-carrier scattering plays a strong role in the rapid thermalization of these highly excited carriers, but quantitative analysis is complicated by the nature of the experiment. With our photon energy of 1.98 eV we are exciting and probing carriers with an excess energy of 500 meV in the conduction band, generated by transitions from the light and heavy hole valence bands, as well as carriers with lower energy generated via transitions from the split-off valence band. The relative scattering dynamics of these two different distributions as a function of excitation density are not known. The carriers with excess energy of 500 meV may also scatter into the X and L valleys on this time scale.

In order to monitor more completely the dynamics of the entire carrier distribution, we have also performed experiments using probe pulses derived from a femtosecond white light continuum. For these experiments the output of the CPM ring dye oscillator was amplified using a high-repetition-rate, pulsed copper-vapor laser (CVL)[15]. Femtosecond pulses from the oscillator made six passes through an interferometrically flat dye jet amplifer, pumped by the CVL at a repetition rate of 8 KHz. The net gain of the amplifier was approximately 10^4, resulting in single pulse energies of about 2 μJ. A fraction of the amplifier output was split off for use as a pump beam. The remainder was focussed into a thin jet of ethylene glycol to generate a broadband femtosecond continuum probe pulse. The continuum probe was delayed with respect to the pump by a computer-controlled translation stage before passing through the excitation region in the sample. Both the transmitted probe and a reference beam from the continuum were filtered by a monochromator before differential detection. The probe beam was chopped so that lock-in amplification as well as the real time differential detection played a role in noise suppression.

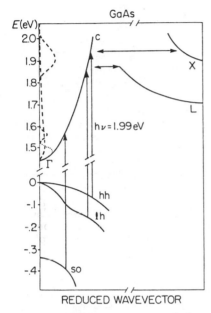

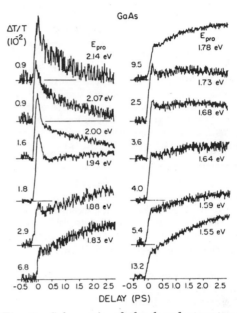

Fig. 2. Absorption saturation date for GaAs obtained with the femtosecond continuum probe. The traces are displayed with arbitrary normalization for comparison of the dynamics occurring at each probe energy.

Fig. 3. Schematic of the band structure for GaAs indicating the optically allowed transitions for 1.98 eV photons and possible intervalley scattering channels.

As in the single-wavelength experiments described above, the pump photon energy of 1.98 eV is sufficient to excite carriers to the conduction band from the split-off valence band as well as from the light and heavy hole bands. This is illustrated in the band structure schematic of Figure 2 which also shows the allowed intervalley scattering channels. The transient absorption saturation and recovery dynamics observed under these conditions for different probe pulse wavelengths are shown in Fig. 3. The excited carrier density is $\sim 10^{18}$ cm^{-3}. The absolute amplitudes of the changes in transmission varied as indicated for different probe photon energy; however, for presentation the traces shown in Fig. 3 are normalized. In all of the traces a two component transient behavior is observed. Measurements performed at energies near the pump photon energy of 1.97 eV indicate, as before, the scattering of the excited carriers out of their initial, optically excited states and are consistent with the single-frequency pump-probe experiments discussed above. The rapid initial transients observed for probe photon energies ranging from 2.14 to 1.88 eV indicate the presence of an absorption saturation hole for photon energies near the incident

13

pump energy. The longer 1-2 ps decay results from a cooling of the thermalized carrier distribution to the lattice temperature.

Measurements performed between 1.73 and 1.68 eV exhibit a similar rapid absorption saturation transient which results from carriers generated by pump transitions from the split-off band to about 150 eV above the conduction band edge. Thus, the saturation hole due to these transitions may be studied separately from that due to the higher energy electrons.

At probe energies throughout the band, ranging from 2.14 to 1.55 eV, data exhibit a rapid onset of the absorption saturation. This indicates that the carriers scatter out of their initial excited states and assume a broad distribution in energy within several tens of femtoseconds. The rapid appearance of carriers throughout the band is further evidence that, at these carrier densities, intercarrier dynamics play an important role in the initial scattering process. The picosecond relaxation of the absorption saturation is a complicated function of energy but is consistent with a cooling of the carrier distribution through carrier-phonon interactions. For photon energies of 2.0 eV, and greater, the decaying absorption saturation results from a decreased occupancy of energy states higher in the band. At intermediate probe energies between 1.76 eV and 1.94 eV the transition from the split-off band probes near the bottom of the conduction band and results in a rising absorption saturation signal. For probe energies less than 1.76 eV, the split-off transition is no longer allowed and only the intermediate energy distribution is probed. For probe energies corresponding to transitions near the band edge, at 1.59 and 1.55 eV, the absorption saturation increases because of an increase in occupancy of states below the Fermi energy as the carrier distribution cools and carriers return from the satellite valleys.

With the above experiments we have observed, for the first time, hole burning due to transitions from the split-off as well as heavy and light hole bands. Our single-frequency experiments indicate that electrons scatter out of their initial excited states in less that 30 fs at carrier densities greater than $10^{17}/cm^3$. Continuum results show that they achieve a broad energy distribution throughout the conduction band also within tens of femtoseconds, consistent with fluorescence studies which indicate that carrier-carrier scattering dominates at these carrier densities[16].

In other recent experiments[17,18] we have investigated the effect of changing material composition on these carrier relaxation dynamics. With thin films of $Al_xGa_{1-x}As$ we observed a dramatic slowing of the initial femtosecond response with increasing mole fraction of Al. For $x > 0.2$, intervalley scattering to the X valley becomes much less probable and transitions from the split-off valence band are no longer permitted. The apparent time constant of the initial rapid recovery increases to 130 fs for $x = 0.3$. Other recent reports[19,20] also indicate that the intervalley scattering can be very fast. Further work with pump and continuum-probe methods or wavelength tunable excitation pulses are still needed to help separate the intravalley and intervalley rates and to better understand their contributions, on both fast and slow timescales, to the relaxation of carriers in these materials.

Nonequilibrium Electron Processes in Metals

Although the possibility of generating transient nonequilibrium electron temperatures in metals was predicted theoretically over three decades ago[21], early experimental attempts using picosecond laser pulses and thermally assisted multiphoton photoemission were unable to demonstrate nonequilibrum electron heating[22,23]. The first evidence for this phenomena was obtained using picosecond thermo-modulation techniques; however, the time resolution was then insufficient to permit a measurement of the electron cooling time[24,25]. Subsequent extension of photoemission studies into the femtosecond regime allowed a measurement of electron cooling but were restricted to high temperatures and suffered from space charge effects[26]. Femtosecond thermomodulation reflectivity[27] and transmissivity[28] measurements now provide the opportunity to investigate nonequilibrium electron temperatures and cooling dynamics in detail. Because metals exhibit small changes in optical properties these investigations have only recently been made possible through the development of new, high-repetition-rate femtosecond sources which provide high sensitivity detection with femtosecond time resolution period

Here we describe an investigation[28] of nonequilibrium electron heating and temperature dynamics in gold using femtosecond transient thermo-reflectivity measurements. We also present femtosecond pump-probe results[29] that reveal ultrafast transport of electron energy through thin films during this nonequilibrium period.

A variety of effects contribute to changes of reflectivity from a metal following excitation. They include smearing of the electronic occupancy near the Fermi level, shifting of the Fermi level, lattice expansion, and electron-phonon interaction. In gold, for photon energies inducing transitions from the d-bands to conduction band states near the Fermi level, the dominant contribution to thermoreflectance is Fermi smearing due to electron heating[30]. This smearing is characterized by increased electronic occupancy above the Fermi level and decreased occupancy below. We have investigated these changes in occupancy by monitoring transient reflectivity over the wavelength range 450 - 580 nm (around the 2.38 eV d-band to Fermi level transition) with probe pulses derived from our femtosecond continuum source. The samples were 1000 Å gold films on sapphire substrates. Nonequilibrium electron temperatures were produced by pulses of 65 fs duration at a wavelength of 630 nm (1.97 eV).

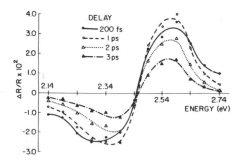

Fig. 4. Transient reflectivity measurements as a function of probe photon energy at various time delays from the heating pulse.

Figure 4 shows the measured transient reflectivity as a function of energy at various time delays. This is a compilation of reflectivity traces taken at fixed time delays with different probe photon energies. The lineshape has an inflection point near the d-band to Fermi level transition energy and a derivative-like structure with minima and maxima above and below the transition energy respectively. The lineshape corresponds closely to results obtained in previous thermomodulation studies[31-33]. The maximum reflectivity change, for an excitation fluence of 4mJ/cm², was a $\Delta R/R$ of about 10^{-2}. $\Delta R/R$ is positive above the Fermi level and negative below. Furthermore, we observe changes of the reflectivity lineshape as the probe pulse is delayed, which indicates a cooling of the electronic distribution. At 200 fs delay, where the electron temperature is near the maximum, we observe the greatest smearing of the lineshape. At longer delays, as the distribution cools, we observe a narrowing of the lineshape and a shifting of the peaks toward the Fermi energy. The temporal evolution of the lineshape is in qualitative agreement with our simple model based on changes in electronic occupancy $\Delta\rho$ and is consistent with previous thermomodulation studies performed as a function of temperature[31-33]. This demonstrates that the reflectivity signal results from changes in electron temperature and that the transient behavior corresponds to a cooling of the electrons to the lattice temperature.

The above observations made use of the fact that the electronic heat capacity C_e is much less than that of the lattice C_l. Transient electron temperatures can therefore be much greater than the lattice temperature would be with the same energy input.

The difference in heat capacity also implies that nonequilibrium electrons have a much larger diffusivity (K/C_e) than that exhibited under equilibrium conditions (K/C_l). In a second set of experiments[29], described below, this difference in ultrafast transport properties was observed directly.

The transit time of heat through a gold film was measured by heating the front surface of the film with one femtosecond pulse and then probing the reflectivity change on the rear surface with another. Pump and probe pulses were obtained directly from our CPM oscillator without amplification. Zero delay between the pump and probe arrival times was determined by reversing their roles and repeating the experiment. The sign of $\Delta R/R$ was negative on both surfaces since the probe photon energy (1.98 eV) was well below the Fermi energy.

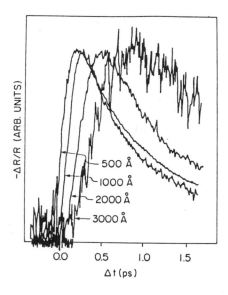

Fig. 5. Rear surface probe data for Au films 500, 1000, 2000, and 3000 Å thick.

Changes in rear surface reflectivity as a function of time after front surface excitation are shown in Figure 5 for films 500, 1000, 2000 and 3000 Å thick. Note that the delay in the rise of the reflectivity change increases with sample thickness. This is a direct consequence of the time needed for heat to propagate through the sample. Furthermore, the measured delays (approximately 100 fs per 1000 Å) are much shorter than would be expected by diffusion of electrons in equilibrium with the lattice (tens of picoseconds). This suggests that heat is transported by the electron gas alone and that the electrons are out of equilibrium with the lattice. It is interesting to note that the delay increases approximately linearly with sample thickness and that the transport velocity one deduces from this is of the same order of magnitude as the Fermi velocity of electrons in Au, 1.4×10^8 cm/sec.

Since the heat moves at a velocity comparable to the Fermi velocity, it is natural to question exactly how the transport takes place. In the diffusion limit, transport is modeled by a pair of coupled non-linear equations[29,34] that describe the evolution of electron temperature under diffusion, with laser irradiation and coupling to the lattice. Without coupling to the lattice, solutions of the diffusion equation exhibit transit times proportional to distance squared ($\Delta t \sim L^2$). When transport and relaxation occur simultaneously, however, this proportionality becomes less clear. Thus it is possible to obtain solutions to the coupled nonlinear diffusion equations that behave qualitatively like the experimental results. Quantitative comparison will require additional measurements.

Simple electron scattering arguments indicate that quasi-ballistic motion may also contribute to heat transport in the time and length regime under study. The electron-electron scattering length in Au has been calculated[35] to vary from 800 Å , for 1 eV electrons, to 350 Å for 2 eV electrons. The electron-phonon scattering length inferred from conductivity data is 420 Å at room temperature[29,36] . Internal photoemission measurements yield "attenuation lengths", a result of electron-electron and electron-phonon effects combined, of about 700 Å [37,38]. None of these lengths is negligible compared with our sample dimensions.

Thus we have seen that, in metals as well as in semiconductors, nonequilibrium electron temperatures can persist for 1-2 ps. During this time electron energy transport can also be different from that under equilibrium conditions. In a metal, because of the small optical excitation and probing depths involved, we have been able to observe such transport directly.

ACKNOWLEDGEMENTS

We gratefully acknowledge the collaboration of W. Z. Lin, R. W. Schoenlein, and S. D. Brorson on various aspects of the research described in this paper. JGF acknowledges support from the National Science Foundation Presidential Young Investigators Program. This work was supported in part by the Air Force Office of Scientific Research and by the Joint Services Electronics Program.

REFERENCES

1. S. A. Lyon, J. of Luminescence **35**, 121 (1986).

2. C.V. Shank, R.L. Fork, R.F. Leheny, and J. Shah, Phys. Rev. Lett. **42**, 112 (1979).

3. R.F. Leheny, J. Shah, R.L. Fork, C.V. Shank, and A. Migus, Solid State Commun. **31**, 809 (1979).

4. C.L. Tang and D.J. Erskine, Phys. Rev. Lett. **51**, 840 (1983).

5. A.J. Taylor, D.J. Erskine and, C.L. Tang, J. Opt. Soc. Am. B **2**, 663 (1985).

6. W.Z. Lin, J.G. Fujimoto, E.P. Ippen, and R.A. Logan, Proceedings of the International Conference on Lasers '86, Orlando, FL, Society for Optical and Quantum Electronics, McLean, VA (1986).

7. W.Z. Lin, J.G. Fujimoto, E.P. Ippen, and R.A. Logan, Appl. Phys. Lett. **50**, 124 (1987).

8. J.L. Oudar, D. Hulin, A. Migus, A. Antonettti, and F. Alexandre, Phys. Rev. Lett. **55**, 2074 (1985).

9. C.H. Brito-Cruz, R.L. Fork, and C.V. Shank, IQEC '87 Technical Digest, Baltimore, MD, April 26-May 1, 1987, pp. 82-83.

10. W.Z. Lin, R.W. Schoenlein, S.D. Brorson, J.G. Fujimoto, and E.P. Ippen, IQEC '87 Technical Digest, Baltimore, MD, April 26-May 1, 1987, pp.82-83.

11. R.L. Fork, B.I. Greene, and C.V. Shank, Appl. Phys. Lett. **38**, 671 (1981).

12. J.A. Valdmanis, R.L. Fork, and J.P. Gordon, Opt. Lett. **10**, 131 (1985).

13. R.L. Fork, O.E. Martinez, and J.P. Gordon, Opt. Lett. **9**, 150 (1984).

14. A.M. Weiner and E.P. Ippen, Opt. Lett. **9**, 53 (1984).

15. W.H. Knox, M.C. Downer, R.L. Fork, and C.V. Shank, Opt. Lett. **9**, 552 (1984).

16. J.C. Tsang and J.A. Kash, Phys. Rev. B **34**, 6003 (1986).

17. W.Z. Lin, J.G. Fujimoto, E.P. Ippen, and R.A. Logan, Appl. Phys. Lett. **51**, 161 (1987).

18. R.W. Schoenlein, W.Z. Lin, E.P. Ippen, and J.G. Fujimoto, Appl. Phys. Lett., in press.

19. F.W. Wise, I.A. Walmsley, and C.L. Tang Appl. Phys. Lett. **51**, 605 (1987).

20. C.H. Brito-Cruz, J. Shah, and C.V. Shank, to be published.

21. M.I. Kaganov, I.M. Lifshitz, and L.V. Tanatarov Zh. Eksp. Teor. Fiz. **31**, 232 (1956) [Sov. Phys. JETP **4**, 173 (1957)].

22. R. Yen, J. Liu, and N. Bloembergen, Opt. Commun. **35**, 277 (1980).

23. R. Yen, J.M. Liu, N. Bloembergen, T.K. Yee, J.G. Fujimoto, and M.M. Salour, Appl. Phys. Lett. **40**, 185 (1982).

24. G.L. Eesley, Phys. Rev. Lett. **51**, 2140 (1983).

25. G.L. Eesley, Phys. Rev. B **33**, 2144 (1986).

26. J.G. Fujimoto, J.M. Liu, E.P. Ippen, and N.Bloembergen, Phys. Rev. Lett. **53**, 1837 (1984).

27. R.W. Schoenlein, W.Z. Lin, J.G. Fujimoto, and G.L. Eesley, Phys. Rev. Lett. **58**, 1680 (1987).

28. H.E. Elsayed-Ali, T.B. Norris, M.A. Pessot, and G.A. Mourou, Phys. Rev. Lett. **58**, 1212 (1987).

29. S.D. Brorson, J.G. Fujimoto, and E.P. Ippen, to be published.

30. M. Cardona, Modulation Spectroscopy, Academic Press, New York, 1969

31. R. Rosei and D.W. Lynch, Phys. Rev. B **5**, 3883 (1972).

32. W.J. Scouler, Phys. Rev. Lett. **18**, 445 (1967).

33. R.Rosei, F. Antonangeli, and U.M. Grassano, Surface Science **37**, 689 (1973).

34. S.I. Anisimov, B.L. Kapeliovich, and T.L. Perelman, Zh. Eksp. Teor. Fiz **66**, 776 (1974) [Sov. Phys. JETP **39**, 375 (1975)].

35. W.F. Krolikowski and W.E. Spicer, Phys. Rev. B **1**, 478 (1970).

36. The relevant parameters were culled from N.W. Ashcroft and N.D. Mermin, Solid State Physics, Saunders College, Philadelphia, 1976.

37. S.M. Sze, J.L. Moll, and T. Sugano, Solid State Electron. **7**, 509 (1964).

38. C.R. Crowell, W.G. Spitzer, L.E. Howarth, and E.E. LaBate, Phys. Rev. **127**, 2006 (1962).

LASER SPECTROSCOPY OF HOT PHOTOLUMINESCENCE IN SEMICONDUCTORS:

ENERGY SPECTRUM AND RELAXATION TIMES

D.N. Mirlin and B.P. Zakharchenya

A.F. Ioffe Physical Technical Institute
Academy of Sciences of the USSR
194021, Leningrad

INTRODUCTION

The methods of hot photoluminescence spectroscopy[1] have been successfully used in recent years to solve a wide range of problems in the physics of semiconductors. After briefly introducing the basic concepts in part 1 we will present some results of recent studies; for more details see surveys.[1-3]

1. In the following discussion the term "hot photoluminescence" (HPL) designates recombination luminescence of photoexcited electrons in the course of their energy relaxation. The distribution function of these electrons differs from Maxwellian. For the electrons to be hot means that their kinetic energy is much higher than the lattice temperature. Because measurements have been performed for p-type crystals (GaAs, InP and $Ga_{1-x}Al_xAs$) at helium temperatures, hot electrons recombine with holes trapped on the acceptor levels. The experiments have been conducted under cw pumping. In spite of this a vast amount of information about ultrafast relaxation processes can be evaluated from the HPL polarization characteristics.

It has been shown that absorption above the bandgap of linearly polarized light in semiconductors with GaAs bandstructure leads to the alignment of the momenta (velocities) of photoexcited electrons. For instance, for electrons created from the subband of heavy holes the momenta are directed preferentially perpendicular to the polarization vector $\vec{e}$ of the exciting light. The recombination radiation of such anisotropically alligned electrons is partly linearly polarized. In a magnetic field the Lorentz force acts on the electrons resulting in the rotation of the distribution function and in a decrease of the degree ρ of linear polarization. So, for the recombination radiation of electrons which are just created $\rho(B) = \rho(0)/(1 + 4\omega_c^2\tau_0^2)$ where ω_c is the cyclotron frequency and τ_0 is the "life time" of the electron in the initial state ε_0. In a general case this time is determined by all relaxation processes (both intra- and interband) which can remove the electron from the $\varepsilon_0(\kappa_0)$ state. The study of a magnetic depolarization spectrum not only allows one to determine the outscattering times but to "keep track" of the whole process of energy as well as momentum relaxation of electrons.

2.   The determination of the intra- and intervalley scattering times in semiconductors is essential for the very active field of ultra-high speed semiconductor devices.   Just these times have been obtained by magnetic depolarization in HPL measurements.   So, in particular, the time of polar optic scattering in GaAs in the electron energy range ε = 100 - 300 meV has been measured.[4,5] Within the accuracy of measurements it was found to be energy independent and equal to 100 fs.   A smaller value for this time - 75 fs at ε = 200 meV - has been found for InP due to the greater magnitude of the Fröhlich electron-phonon coupling constant.[6]

The intervalley transition times in GaAs have also measured.   At ε = 385 meV the time is determined by $\Gamma \to L$ transitions and is equal to 250 fs (at 2K).   This value corresponds to an intervalley coupling constant $D_{\Gamma L}$ = 8×10^8 eV cm^{-1}.   At higher energy the transitions become possible in GaAs not only into the L-valley but into X-one as well.   The outscattering time for the electrons with ε = 570 meV was found to be 18 $\pm$ 2 fs.   It is determined both by intervalley ($\Gamma \to \Gamma$) transitions and by $\Gamma \to L$ and $\Gamma \to X$ ones with the emission of "intervalley" phonons.   For $\Gamma \to X$ channel $\tau_{\Gamma X}$ = (30 $\pm$ 10) fs at this energy corresponding to an intervalley coupling constant $D_{\Gamma X}$ = 1.5×10^9 eV cm^{-1}.

3.   Lately methods of HPL spectroscopy have been successfully used to study band structure away from the centre of the Brillouin zone in a number of A_3B_5 semiconductors.

a) Spin-splitting of the conduction band of GaAs

The degree of circular polarization ρ_C has been studied as a function of initial electron energy.[7]   As ε (and consequently the momentum K) increase the initial value of $\rho^0{}_C$ decreases and differs increasingly from the calculated value $\rho^0{}_C$ = 0.71.   This difference is the result of fast spin depolarization due to the conduction band splitting, which is proportional to K^3.   The magnitude of splitting has been determined for different values of ε.

b) The side valleys of the conduction band

Side valleys are sharply displayed in HLP spectra of GaAs in the form of thresholds.[8-10]   The positions of these valleys could be determined with suficient precision from the energies of these thresholds. The evolution of HPL spectra under uniaxial deformation also made possible an unambiguous determination of the symmetry of the valleys.[10] Although Γ - L-X ordering for GaAs has been generally accepted since Aspnes work,[11] it was based on the analysis of a number of experiments. However, each of these experiments did not exclude an alternative interpretation.   The advantage of experiments[10] is their unambiguity.   The distance from the Γ -bottom to that of L- and X-valleys have been determined at 2K and are $\Delta E_{\Gamma L}$ = 310 $\pm$ 10 meV and $\Delta E_{\Gamma X}$ = 485 $\pm$ 10 meV, respectively.

From the splitting of peaks in HPL spectra the shear deformation potentials of side valleys have been evaluated:

$$\Xi^L{}_u = 14.5 \pm 1.5 \text{ eV}, \quad \Xi^X{}_u = 6.5 \pm 1 \text{ eV}$$

These values differ noticeably from those obtained previously by more indirect methods, particularly for the X-valley.   Let us note, however, that they are very close to the potentials of respective main minima in Ge ($\Xi^L{}_u$ = 14 eV) and GaP ($\Xi^X{}_u$ = 6.2 eV) which are known with good accuracy.

From the ratio of intensities in HPL spectra under deformation the coupling constant D_{LL} between equivalent valleys has been determined for the first time.[10] Thus, the intervalley coupling constants in GaAs determined by the methods of hot photoluminescence spectroscopy are:

$D_{\Gamma L} = (8 \pm 2) \times 10^8$ eV.cm^{-1}, $D_{LL} = (5 \pm 1) \times 10^8$ eV.cm^{-1}, $D_{\Gamma X} = 1.5 \pm 0.4) \times 10^9$eV. cm^{-1}.

c)   Valence band warping

In general, due to valence band warping the degree of linear polarization depends strongly on the orientation of $\vec{e}$ relative to crystallographic axes (this dependence disappears only for direction of excitation and observation along <111>). This phenomenon has been predicted in[12] and experimentally discovered in GaAs crystals.[13] Recently the anisotropy of HPL polarization was studied for a wide range of excitation frequencies.[14] It was shown that the observed increase of anisotropy with energy is due to the change of the wave function of the heavy hole subband and to the decrease of spin-orbit interaction with increasing k. Good agreement with the measured data has been obtained with the Luttinger parameters γ_1, γ_2, γ_3 known from cyclotron resonance and magnetooptic experiments (those settings of γ are practically coincident).[15] A different situation exists for InP: whereas the results for γ_1, and γ_2 are definite, those for γ_3 vary from 1.6 to 2.4.[15] The low limit corresponds to a practically spherical band and the upper one to strong anisotropy. From the comparison of calculated dependence of ρ and its anisotropy on γ_3 with the experimental data γ_3 value has been determined: 2.15 ± 0.05. A close γ_3 value (2.1 ± 0.05) has been obtained from the position of zero-phonon peaks in HPL spectra which is also very sensitive to the magnitude of γ_3.

The dispersion of valence subbands in GaAs has been construced with good accuracy (up to 1 meV) by the Stuttgart team from the zero-phonon peak positions in HPL spectra for a wide set of excitation frequencies.[16] The experimental data obtained for the heavy-hole subband are close to those calculated by the k-p method for $\vec{k} \parallel [110]$ direction (the difference does not exceed 7 meV for $\vec{k}$ up to 10^7 cm^{-1}). This result follows also from the calculation performed in.[14] For the light-hole subband the difference between experimental and calculated dispersion is more pronounced and amounts to 15 meV.

d) HPL spectra in indirect semiconductors

(GaAlAs, GaAsP, AlSb) have been studied.[17-19] New results have been obtained concerning the intervalley scattering times (from the direct minimum into the indirect one), dispersion curves and spin-orbit interaction.

4. The study of acceptor levels in a number of works has been based on the dependence of polarization characteristics of recombination photoluminescence on the shape of the acceptor wave functions. It has been shown that in GaAs crystals under unaxial deformation the linear polarization of the band-to-shallow acceptor luminescence decreases from the value 0.6 at k=0 with increasing electron energy[20] (in this case linear polarization is due to the lifting of the degeneracy of the levels and can be observed under unpolarized excitation too).

The effect discovered in[20] is the direct experimental proof that an admixture of d-functions to s-ones exists in the ground state of a shallow acceptor. The calculation of the selection rules shows that in this case radiative recombination is allowed for electrons in "s" and "d" states. Low energy electrons recombine only from the "s-state due to the small overlap of electron and hole d-functions at small k. As k increases the relative contribution into recombination increases for the electrons with orbital momentum equal to two ("d" electrons). It is precisely this fact that explains the decrease of the polarization.

One should note that the polarization dependence studied in[20] is not connected with the allignment of electron momenta.

When circular polarization is studied due to acceptor level splitting in a magnetic field, the study of hot electron recombination is advantageous, because it is free from complications connected with the magnetization of electrons by the external magnetic field. In this way the antiferromagnetic interaction of holes in the ground state of Mn-acceptor with Mn-ions has been established.[20] In this case the polarization is opposite in sign as compared with that of nonmagnetic acceptors (Zn,Ge). The results of[20] were confirmed, when recombination into the excited state of Mn-acceptor in quantum wells has been studied.[21] The exchange interaction for the excited state turned out to be not so strong and could be destroyed by a moderate magnetic field of the order of 3T. In this case the sign of the polarization changed (it became "usual", as for nonmagnetic acceptors).

Recently the great possibilities of hot photoluminescence spectroscopy began to be used by a number of teams in the USA, France and FRG. Some of this work was already cited. One may mention interesting work by a French team on HPL polarization in heavy doped crystals,[22,23] and HPL study in nipi superlattices.[24] A more detailed survey of this and other work is beyond the scope of this paper.

The authors thank V.I. Perel for many useful discussions.

References

1. B. P. Zakharchenya, D. N. Mirlin, V. I. Perel and I. I. Reshina, Sov. Phys.-Uspekhi, 25:143 (1982).
2. D. N. Mirlin in: "Optical Orientation," F. Meier and B. Zakharchenya, eds., North Holland, Amsterdam (1984).
3. S. A. Lyon, Journ. of Lumin., 35:121 (1986).
4. D. N. Mirlin, I. Ya Karlik, L. P. Nikitin, I. I. Reshina and V. F. Sapega, Sol. State Commun. 37:757 (1981).
5. I. Ya Karlik, D. N. Mirlin and V. F. Sapega, Sov. Phys. Semicond., 21:No6 (1987).
6. I. Ya Karlik, D. N. Mirlin and V. F. Sapega, Sov. Phys. Solid State, 27:1326 (1985).
7. M. A. Alekseev, I. Ya Karlik, I. A. Merkulov, D. N. Mirlin, L. P. Nikitin and V. F. Sapega, Sov. Phys. Solid State, 26:2025 (1984).
8. D. N. Mirlin, I. Ya Karlik, L. P. Kikitin, I. I. Reshina and V. F. Sapega, JETP Lett., 32:31 (1980).
9. E. A. Imhof, M. I. Bell, and R. A. Forman, Sol. State Commun., 54:845 (1985).
10. I. Ya Karlik, R. Katilius, D. N. Mirlin and V. F. Sapega, JETP Lett., 43:319 (1986); D. N. Mirlin, V. F. Sapega, I. Ya Karlik and R. Katilius, Solid State Commun., 61:799 (1987).
11. D. E. Aspnes, Phys. Rev., B14:5331 (1976).
12. V. D. Dymnikov, Sov. Phys. Semicond., 11:868 (1977).
13. D. N. Mirlin and I. I. Reshina, Sov. Phys. JETP, 46:451 (1977).
14. M. A. Alekseev, I. A. Karlik, I. A. Merkulov, D. N. Mirlin, Yu. T. Rebane and V. Sapega, Sov. Phys. Solid State, 27:1589 (1985).

15. Landoldt-Bornstein, Vol. 17a, Semiconductors, O. Madelung., ed., Springer (1982).
16. G. Fasol and H. P. Hughes, Phys. Rev., B33:2953 (1986).
17. I. Ya Karlik, D. N. Mirlin, V. F. Sapega and Yu. P. Yakovlev, Sov. Phys. Solid State, 28:1039 (1986).
18. F. F. Charfi, M. Zouaghi, R. Planel and C. Benoit a la Guillaume, Phys. Rev., B33:5623 (1986).
19. M. Maaref, F. F. Charfi, M. Zouaghi, C. Benoit a la Guillaume and A. Joullie, Phys. Rev., B34:8650 (1986).
20. I. Ya Karlik, I. A. Merkulov, D. N. Mirlin, L. P. Nikitin, V. I. Perel and V. F. Sapega, Sov. Phys. Solid State, 24:2022 (1982).
21. A. Petrou, M. C. Smith, C. H. Perry, J. M. Worlock, J. Warnock and L. L. Aggarwal, Sol. State Commun., 55:865 (1985).
22. A. Twardowski and C. Hermann, Phys. Rev., B32:8253 (1985).
23. Bo. E. Sennelius, Phys. Rev., B34:8696 (1986).
24. G. Fasol, K. Ploog and E. Bauser, Sol. State Commun., 54:383 (1985).

STATIC AND DYNAMIC COOLING RATE OF

PHOTOCREATED PLASMAS IN SEMICONDUCTORS

S. E. Kumekov and V. I. Perel'

A.F. Ioffe Physico-Technical Institute
USSR Academy of Sciences, 194021 Leningrad

The cooling of optically created semiconductor plasmas has been intensely investigated over the past several years (see the References in Ref. 1).[1] The main result is that the cooling rate has generally been found to decrease with increasing carrier density. This means that the rate of optical phonon emission by carriers slows down at a high plasma density. This was attributed either to the screening of electron-phonon interaction[2] or to the heating of optical phonons. In the present work we discuss conditions when screening is not essential. The phonon emission reduction is related to the fact that in a certain wave vector interval Δq the phonons reach equilibrium with the carriers, so that further emission of such phonons occurs only to compensate for their anharmonic decay.

We consider a simple model (Fig. 1) in which a plasma interacts with LO phonons only. These phonons in turn may decay into acoustic phonons

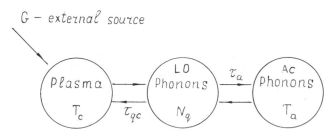

Fig. 1.   The model of plasma cooling.

due to anharmonicity. (Hole interactions with TO phonons may be considered analogously.)   The carriers with density n have temperature T_c, the temperature of acoustic phonons is T_a ("lattice" temperature.)   The purpose of the present work is to obtain analytical expressions for the cooling rate for such a model in the steady state as well as in the dynamical regime and to study its dependence on the parameters of the plasma and of the crystal.

Two characteristic times are significant: the lattice-dynamical lifetime τ_a, during which LO phonons reach equilibrium with the lattice and the time τ_{qc} during which equilibrium between LO phonons and the carriers is established.

The time dependence of the phonon occupation numbers N_q and the plasma energy density E is described by the following equations

$$\frac{dN_q}{dt} = - \frac{1}{\tau_a} (N_q - N_a) - \frac{1}{\tau_{qc}} (N_q - N_c) \tag{1}$$

$$\frac{dE}{dt} = \int \frac{d^3q}{\tau_{qc}} (N_q - N_c) \frac{\hbar\omega}{(2\pi)^3} + G \tag{2}$$

where N_c and N_a are the equilibrium occupation numbers for temperatures T_c and T_a respectively, ω is the LO phonon frequency (the q-dependence of ω is neglected), G is the power input to the plasma from an external source.

Eqs. (1) and (2) are evident and well known,[1,4] however their physical consequences were not fully elucidated.   For a two-component plasma $\tau_{qc}^{-1} = \tau_{qh}^{-1} + \tau_{qe}^{-1}$.   The main contribution to the plasma energy loss rate is given by holes, therefore the q dependence of τ_{qh} will be discussed below (τ_{qe} behaves similarly).   For non-degenerate holes and any type of interaction τ_{qh} may be given in the form[5]

$$\tau_{qh} = \tau_{qh}^* \exp\left[\frac{m_h\omega^2}{2T_cq^2} + \frac{\hbar^2q^2}{8m_hT_c}\right] \tag{3}$$

where τ_{qh}^* is proportional to n^{-1} and has a relatively weak q dependence. For the Fröhlich interaction

$$\tau_{qh}^* = q^3/n\nu_h,$$

$$\nu_h = \frac{2(2\pi)^{3/2}n\omega e^2\sqrt{m_h}}{\varepsilon_\infty\hbar\sqrt{T_c}} \; sh \; (\frac{\hbar\omega}{2T_c}), \quad \eta = 1 - \frac{\omega^2_t}{\omega^2}$$

Here ε_∞ is the high-frequency dielectric constant, ω and ω_t are the LO and TO phonon frequencies.   The exponent in Eq. (3) is a consequence of conservation laws for the process of phonon absorption by a hole and it does not depend on the type of interaction.   The exponential factor makes the plasma-phonon interaction ineffective for significantly small q as well as for large q.

For the steady state it follows from Eqs. (1) and (2) that

$$G = n \cdot J_{st} \qquad J_{st} = \frac{(N_c - N_a)\hbar\omega}{n} \int \frac{d^3q}{(2\pi)^3} \frac{1}{\tau_a + \tau_{qc}} \qquad (4)$$

Here the steady-state plasma energy loss rate per electron-hole pair J_{st} is introduced. At low concentration $\tau_{qc} > \tau_a$ for all q, so the quantity τ_a in Eq. (4) may be neglected and the usual concentration independent Frohlich expression J_F for J_{st} is valid. With increasing concentration τ_{qc} decreases and becomes less than τ_a in a certain interval Δq. LO phonons in this interval reach the temperature of plasma. Thus J_{st} begins to reduce. This decrease may be described analytically for high ($T_c > \hbar\omega$) and for low ($T_c < \hbar\omega$) temperatures.

If $T_c > \hbar\omega$ a wide interval exists $q_{min} < q < q_{max}$ where $t_{qc} \approx \tau_{qc}{}^*$. Here $q_{min} = \sqrt{2T_c/m_n\omega^2}$, $q_{max} = \sqrt{8m_nT_c/\hbar^2}$. The minimum value of τ_{qc} is reached close to the lower boundary of this interval

$$\tau_{qhmin} = 0,05 \frac{\varepsilon_\infty m_n\hbar\omega}{ne^2n} \qquad (5)$$

For GaAs $\tau_{qhmin} = 4.10^{16}$ ps/n(cm^{-3}). The energy loss rate J_{st} begins to depend on n for $\tau_{qhmin} < \tau_a$. An estimate of this dependence may be obtained by restricting the integration range in Eq. (4) between the values q_{min}, q_{max} and putting $\tau_{qc} = \tau_{qh}{}^*$

$$J_{st} = (N_c - N_a) \frac{\hbar\omega\nu_h}{6\pi^2} \ell n \frac{q^3{}_{max} + n\nu_h\tau_a}{q^3 + n\nu\tau} \qquad (6)$$

In this case the screening of the relevant phonons is static. Calculations for GaAs show that it is not important for $n < 5 \times 10^{18}$ cm^{-3}. Expressions (4) and (6) give similar results, and agree with experimental data[6] the interpretation of which was given in.[7]

If $T_c < \hbar\omega$ the exponential factor in Eq. (3) has a sharp minimum at $q_m = \sqrt{2m_h\omega\,\hbar}$. In the vicinity of this minimum

$$\tau_{qh} = \tau_{qhmin} \exp[\hbar^2(q-q_m)^2/2m_hT_c] \qquad (7)$$

$$\tau_{qhmin} = 0,2 \frac{\varepsilon_\infty m_h\omega}{ne^2n} \sqrt{\frac{T_c}{\hbar\omega}} \qquad (8)$$

For $\tau_{qhmin} < \tau_a$ the integral in Eq. (4) may be estimated by taking the product of the integrand at $\tau_{qc} = 0$ and the volume of spherical layer of radius q_m and thickness $\Delta q = 2\sqrt{m_h(T_c/\hbar^2)} \ell n \, (\tau_a/\tau_{qhmin})$. Thus we obtain

$$J_{st} = \frac{1}{\tau_a} \frac{2m_h\omega^2}{\hbar\pi^2} \frac{(N_c - N_a)}{n} \sqrt{m_hT_c \ell n(\tau_a/\tau_{qhmin})} \qquad (9)$$

One can see that at low temperatures J_{st} is approximately proportional to n^{-1} and only weakly depends on the form of the plasma-phonon interaction. The screening in this case is of a dynamical phonon interaction. The screening in this case is of a dynamical nature and is of little significance for $n < 5 \times 10^{17}$ cm^{-3}. Another important conclusion, which may be drawn from Eq. (9), is the strong dependence of J_{st} on the effective mass of carriers ($\sim m^{3/2}$). This agrees with the results[8] where a much lower energy loss rate was observed for electrons than for holes in GaAs quantum wells.

Consider now the cooling rate in the dynamical regime after the pumping is switched off ($G = 0$). If the heating is small, one may linearize Eqs. (1) and (2) and look for the time dependence of their solutions in the form $\exp(-\lambda t)$, λ^{-1} being the dynamical relaxation time. Then one obtains the following equation for λ:

$$\lambda = \frac{\hbar\omega\Delta N_c}{\Delta E} \int \frac{d^3q}{(2\pi)^3} \left[\frac{\tau_a}{1-\lambda\tau_a} + \tau_{qc} \right]^{-1} \qquad (10)$$

Eq. (10) has two real solutions: $\lambda_1 > \Delta E/nJ_F$ (fast process) and $\lambda_2 < \tau_a^{-1}$ (slow process). If initially the plasma was heated by a short optical pulse, while the phonons were cool, then the relaxation process consists of two stages. At first fast cooling of plasma and heating of phonons occurs during a short time λ_1^{-1}, and after that the plasma and phonons together slowly cool down with a characteristic time λ_2^{-1}.

For the case of strong heating, interesting results arise in the low temperature region ($T_c < \hbar\omega$). In the q-interval where the phonons have the plasma temperature we may insert in Eq. (1) $dN_q/dt = N_c(\hbar\omega/T_c^2)$ dT_c/dt. For $N_a = 0$, using the same approximations that led to Eq. (9), we obtain

$$C \frac{dT_c}{dt} = n.J_{dyn} \qquad J_{dyn} = \frac{J_{st}}{1+A} = \frac{C}{n\tau_a} \frac{T_c^2}{\hbar\omega} \frac{A}{1+A} \qquad (11)$$

$$A = \frac{2}{C\pi^2} \left(\frac{\hbar\omega}{T_c}\right)^{3/2} \left(\frac{n_h\omega}{\hbar}\right)^{3/2} \sqrt{\ln\left(\frac{\tau_a}{\tau_{qhmin}}\right)} \exp\left(-\frac{\hbar\omega}{T_c}\right)$$

where C is the plasma heat capacity. One can see that for $A \gg 1$ (if, however, the concentration is sufficiently large for the condition $\tau_{qcmin} < \tau_a$ to be satisfied) the dynamical cooling rate in much less than the static one. The value J_{dyn} (for $A \gg 1$) is practically concentration independent, just as for small n, but much lower. Such behavior was observed in a number of experiments.[1]

The calculated temperature dependence of J_{st} and J_{dyn} in GaAs is depicted in Fig. 2 for $\tau_a = 21 ps$, $C = 3n/2$, $T_a = 0$. These dependences are quite similar to those observed.[9] One should bear in mind that at temperatures $T_c < 50$ K the interaction of carriers with acoustic phonons becomes essential for the cooling process. This interaction was not taken into account in the present work.

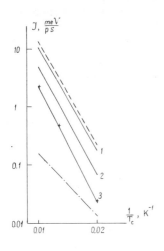

Fig. 2. The comparison of steady state and dynamic cooling rates. Dashed lines - J_F, solid curves - J_{st} (Eq. (14)): 1 - n = 10^{16} cm^{-3}, 2 - n = 2×10^{17} cm^{-3}, n = 4×10^{17} cm^{-3}. Crosses are calculated from Eq. (9) for n = 4×10^{17} cm^{-3}. Dotted line - J_{dyn} (Eq. (10)), n = 4×10^{17} cm^{-3}.

References

1. S. A. Lyon, Journal of Luminescence 35:121 (1986).
2. E. J. Yoffa, Phys. Rev. B 23:1909 (1981); M. Pugnet, J. Collet, A. Cornet, Solid St. Commun. 38:531 (1981); A. R. Vasconsellos, R. Luzzi, Solid St. Commun. 49:587 (1983).
3. H. M. Van Driel, Phys. Rev. B 19:5928 (1979).
4. W. Potz, P. Kocevar, Phys. Rev. B 28:7048 (1983).
5. V. L. Bonch-Bruevich, A. G. Mironov, Fiz. Tv. Tela 2:489 (1960); V. L. Gurevich, Yu. A. Phyrsov, Zh. Eksp. Teor. Fiz. 40:199 (1961).
6. I. L. Bronevoy, R. A. Gadonas, V. V. Krasauskas, T. M. Lifshits, A. S. Piskarskas, M. A. Sinitsyn, B. S. Yavich, Zh. Eksp. Teor. Fiz. Pis'ma 42:322 (1985).
7. I. L. Bronevoy, S. E. Kumekov, V. I. Perel', Zh. Eksp. Teor. Fiz. Pis'ma 43:368 (1986).
8. J. Shah, A. Pinczuk, A. G. Gossard, W. Wiegmann, Phys. Rev. Lett. 54:2045 (1985).
9. K. Kash, J. Shah, D. Block, A. C. Gossard, W. Wiegmann, Physica 134B:189 (1985).

GaSe NONLINEAR SUSCEPTIBILITY AND RELAXATION PROCESS RATES

V. M. Petnikova, M. A. Kharchenko, and V. V. Shuvalov

Physics Department, Moscow State University, Lenin Hills
Moscow 119899, USSR

ABSTRACT

An effective complex investigation of semiconductors by means of nonlinear spectroscopy is reported. The main nonlinear physical mechanisms are investigated, and characteristic times of some relaxation processes are measured in GaSe at 293 K.

1. Relaxation times are determined by comparing experimental data with theoretical models - the choice of the latter determines the authors´ conclusions. Usually the simplest models are used, but the construction of a complete theory is difficult because experimental data are obtained under different conditions. The physics can be understood by the use of a maximum number of possible spectroscopic methods under the same conditions.

We report the analysis of experimental results of cubic nonlinearity dispersion in GaSe in the region of interband and exciton transitions at 293K. A complex of nonlinear spectroscopy (NS) methods - biharmonic pumping (BP), probe pulse (PP), saturation spectroscopy (SS), and degenerate four-photon spectroscopy (DFPS) - was utilized. All these methods are based on the generation of a nonlinear polarization with wave vector $\vec{k}_4 = \vec{k}_1 - \vec{k}_2 + \vec{k}_3$ and frequency $\omega_4 = \omega_1 - \omega_2 + \omega_3$ [1-4]. The independent control of the time delay, polarizations, frequencies, and intensities of all pulses was exercised (Table 1).

2. The computer controlled setup is shown in Fig. 1. Spectral-limited pico- (25 ps duration, 100 kW power) and nanosecond (10 MHz bandwidth, 1 kW) dye laser pulses tuned in the 605-645 nm range were used. The pulse lengths were intermediate between two characteristic groups of fast (< 1 ps) and slow (> 100 ps) semiconductor relaxation times. The fast (polarization) relaxation was investigated by the spectral analysis method (BP), and the slow, nano- and subnanosecond (interband transitions, spatial diffusion of carriers, etc.) - by analysis in the time domain (PP).

3. By the SS method the PP transmission spectrum $T(\lambda_1)$ (Fig. 2) and its relative deviation $\delta T(\lambda_1)$ (Figs. 3, 4) with variations in intensity I_2, polarization (with respect to probe), and frequency (λ_2 = 609, 617.5 nm), have been investigated. Under 20 kW of power, 20 μm GaSe

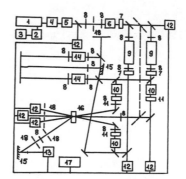

Fig. 1. Experimental set-up: 1 Nd:YAG laser; 2-photodiode; 3-oscilloscope; 4-decoupling system; 5-amplifier; 6-doubler; 7-filter; 8-lens; 9-dye-laser; 10-delay line; 11-polarization governance system; 12-energy detector; 13-photomultiplier; 14-Fabry-Perot etalon; 15- gratings; 16-sample; 17-computer; 18-diagram.

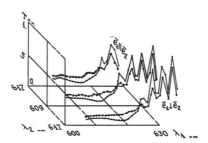

Fig. 2. SS Method. Transmission spectrum $T(\lambda_1)$, $I_2 \neq 0$ (o), $I_2 = 0$ (•).

Table 1. Classification of Nonlinear Spectroscopy Methods Used.

NS method	$\vec{K}_4 =$	Main Variable	Varied Parameters	Measured Quantity	Fig.
	$\underline{\vec{K}_1 - \vec{K}_1 + \vec{K}_1}$		Frequency	Transmission	
SS	$\vec{K}_1 - \vec{K}_2 + \vec{K}_2$	Pulses intensities	Frequencies, polarizations	$T(\lambda_1)$, T deviation	2-4
	$\underline{\vec{K}_1 - \vec{K}_2 + \vec{K}_2}$		Frequency	T deviation	3-5
PP	$\vec{K}_1 - \vec{K}_2 + \vec{K}_3$	Pulses time delay	detuning, polarizations	Diffraction efficiency	6
DFPS	$2\vec{K}_1 - 2\vec{K}_2$	Pulses frequencies	Polarizations of waves	Self-Diffraction efficiency	7
BP	$2\vec{K}_1 - \vec{K}_2$	Frequencies detuning	Central frequency	Self-Diffraction efficiency	8

thickness, and a 0.2-0.5 mm excitation diameter, the concentration of electron-hole pairs n was estimated as $> 10^{16}$ cm^{-3}. The SS method evolves into a variant of PP absorption when a time delay between the interacting pulses is introduced (Fig. 3-4). In the PP method the efficiency of diffraction from transient gratings, formed by simultaneously acting pump pulses, was measured as a function of the PP time delay τ was measured (Fig. 6). The detuning of the frequencies of the acting waves Δ and their relative polarizations were varied. The evolution of PP transmission determined the dynamics of carrier population relaxation $n(\tau)$ (Fig. 5).

In the DFPS method the self-diffraction (SD) efficiency η was measured as a function of the wavelengths of the interacting beams $\lambda = \lambda_1 = \lambda_2 = \lambda_3$ (Fig. 7) for orthogonal and parallel polarizations of the pump waves. The nonlinear susceptibility $|\chi|^2$ was calculated with the help of $T(\lambda)$. In some experiments additional illumination by Nd: YAG laser second harmonic pulses with peak intensity I_0 was used (Fig. 7).

In the BP method the SD efficiency η was measured as a function of the frequency detuning $\Delta = \omega_2 - \omega_1$ (Fig. 8). The central frequency – frequencies coincidence point in GaSe absorption spectrum $\lambda_2 = 620.4$ (the exciton line wing); 617.5 (exciton line); 614; 609 nm (interband transitions) – was varied. The orthogonal pump wave polarizations were chosen to eliminate scattering from acoustic gratings[1].

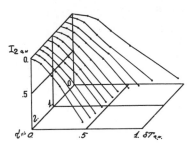

Fig. 3.　SS, PP method.　Deviation of $T(I_2,\tau)$, λ_2 = 617.5 nm, $\vec{e}_1 \| \vec{e}_2$.

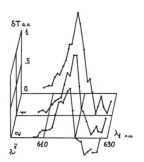

Fig.4.　SS, PP methods.　Deviation of $T(\lambda_1,\tau)$, λ_2 = 617.5 nm, $\vec{e}_1 \perp \vec{e}_2$.

4. The analysis of the experimental results obtained enables one to draw the following conclusions.

1) In the region investigated the nonlinear response of GaSe is formed by exciton, electron, and polariton mechanisms. The corresponding frequencies were identified. The peaks of the transmission coefficient deviation δT (SS, Fig. 4) and the SD nonlinear susceptibility $|\chi|^2$ under parallel polarizations (DFPS, Fig. 7) lay near the exciton resonance. DFPS under orthogonal polarizations showed significant increase with frequency tuning to interband transitions (Fig. 7). In the region of transparency subsidiary peaks in the SD process efficiency η were observed (BP, Fig. 8). Their frequency positions corresponded to known LO and TO phonon frequencies. Exciton line saturation (SS, Figs. 2, 4), exciton screening by free carriers (DFPS with illumination by Nd: YAG laser second harmonic pulses, Fig. 7), band renormalization (SS and PP absorption variant, Figs. 2, 4) were observed.

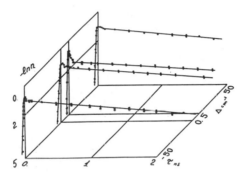

Fig. 5. PP methods. Deviation of $T(\lambda_1,\tau)$, λ_2 = 617.5 nm, $e_1 \| e_2$.

2) Under our experimental conditions several types of relaxation processes played the main role. The subpicosecond polarization relaxation dynamics was determined by the electron-phonon interaction. It was displayed most clearly in the BP experiments (Fig. 8). The central symmetrical peak of SD process efficiency η is connected with low frequency phonons. The interaction with high frequency phonons, unexcited in equilibrium, broadens the $\eta(\Delta)$ Stokes wing. Under tuning to the region of interband transitions polariton subordinate peaks are becoming smooth, and under λ_1 = 609 nm (Fig. 8) $\eta(\Delta)$ the blue wing may be approximated by the Lorentzian $(1 + \Delta^2/\gamma^2)^{-2}$. Its effective width γ = 150 + 20 cm^{-1} corresponds to a transverse relaxation time T_2 = $(\gamma)^{-1}$ = 36 ± 5 fs (two-level model). But the

asymmetry of the dispersion curve points to the inapplicability of such models: the polarization dynamics may be more complicated.

The decay dynamics of picosecond transient gratings (42 ± 20 ps) is due to the spatial diffusion of the carriers. It was recorded by the PP method with picosecond pulses and by BP with nanosecond pulses[4]. The nanosecond relaxation dynamics of the population n with a characteristic time (5±2 ns) was recorded by the PP method (Fig. 5).

3) Under excitation of electron (exciton) transitions in semiconductors the phonon subsystem as well as molecular media vibrational one is effectively excited by BP (Fig. 8) or subpicosecond pulses spectral components in PP, SS, DFPS and cannot be described as a Markovian bath. Polarization decay must be described by dynamical models that take into account the coherence of the electron-phonon interaction.

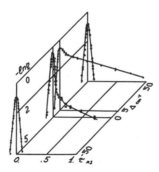

Fig. 6. PP method. Diffraction efficiency of transient gratings $\eta(\tau,\Delta)$.

5. Use of a complex of NS methods under the same conditions enables significantly more general physical conclusions to be drawn. Thus, tuning the BP central frequency through the exciton resonance did not show any features in the behavior of $\eta(\Delta)$ (Fig. 8). Comparison of these results with those obtained by the DFPS method (Fig. 7) permits one to make the conclusion that under orthogonal pump polarizations the excitonic nonlinearity is significantly smaller than the electronic one.

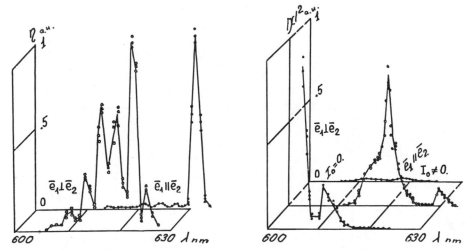

Fig. 7. DFPS method. Self-diffraction efficiency $\eta(\lambda)$ (a) and nonlinear susceptibility $|\chi|^2$ (λ) (b).

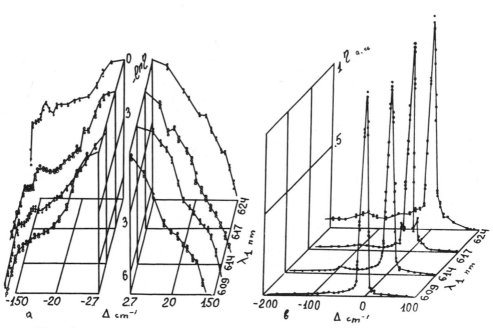

Fig. 8. BP method. Self-diffraction efficiency $\eta(\Delta)$, tuning of λ_2: logarithmic (a) and linear (b) scale.

REFERENCES

1. V. M. Petnikova, S. A. Pleshanov, V. V. Shuvalov, Sov. Phys. JETP,
   61, 211 (1985); Opt. Spectr., 57, 965 (1984); 59, 288 (1985).
2. S. A. Pleshanov, V. V. Shuvalov, Opt. Spectr. 60, 998 (1986).
3. K. E. Asatrian, V. M. Petnikova, V. V. Shuvalov, Opt. Spectr. 63, 123
   (1987).
4. V. M. Petnikova, M. A. Kharchenko, V. V. Shuvalov, Opt. Spectr. 63,
   296 (1987).

PICOSECOND LASER STUDIES OF COLLISIONAL EFFECTS ON ROTATIONAL

PROCESSES IN LIQUIDS

Robin M. Hochstrasser

Department of Chemistry
University of Pennsylvania
Philadelphia, PA   19104-6323

Introduction

Recent advances in laser technology are at last making it possible for direct measurements of liquid state phenomena to be carried out on the timescale of the collisions. The complimentarity of these experimental studies to large scale computer simulations of molecular dynamics in liquids signals the beginning of an exciting period of chemical physics during which there are likely to be significant advances in our understanding of chemical processes in solutions.

In this paper I will summarize some recent work from our laboratory concerning the role of frictional forces in determining motion and chemical reactivity in liquids. This study begins with a summary of our work aimed at following the geometric isomerism of trans-stilbene from the inertial through the diffusive regime (1). In this case the inertial motion is along the isomerization coordinate. This leads naturally to a consideration of the relation between the overall rotation of the molecule and the internal rotation along the reaction coordinate (2). We then explore the question of whether quasi-free rotation of a solute, in this case aniline (3), could be observed directly in fluorescence polarization experiments. The application of some of these ideas to a chemical reaction is presented. The example involves proton transfer from one part of a molecule to another and we show that this process is mediated by the overall rotational motion of the molecule (4). Finally, the effects of rotational reorientation on Raman polarization is evaluated (19).

Friction Dependence of Rotation and Internal Rotation over a Barrier

The simplest of the motions is free rotation. At some instant of time a particular solute molecule might be regarded as having a definite rotational angular momentum. This molecule is therefore embarked on a definite trajectory which will not be allowed to fully develop because of interruption by collisions that alter the orientations and magnitude of the angular momentum. This picture of angular motion in liquids was first formulated by Gordon (5). A free rotor in a definite angular momentum state has an infinite rotational correlation time, τ_R, and the introduction of a weak frictional force decreases τ_R (see Figure 1). On the contrary, for rotational diffusion in which the mean time between effective collisions is very small in comparison with τ_R, the molecule goes from one angular configuration to another by small step random walk. In this case small steps can be regarded as the rotational trajectories or inertial parts of the overall motion.

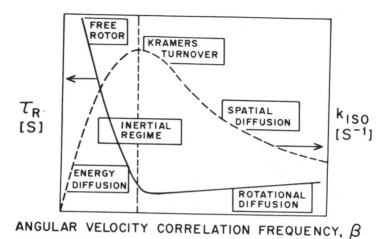

ANGULAR VELOCITY CORRELATION FREQUENCY, β

Figure 1. Variation of Free and Hindered Rotation with Friction

The angular velocity correlation frequency defined as $\beta = \int \langle \omega(o)\omega(t) \rangle dt$ where $\langle \omega(o)\omega(t) \rangle = (kT/I) \exp\{-(\zeta/I)|t|\}$, is sketched as a cartoon versus rotational correlation time τ_R (solid line) and barrier crossing rate for an internal rotation k_{iso} (dashed line). The fluctuations in the angular velocity are induced by the collisions.

The anisotropy r(t), is given by:

$$r(t) = \frac{2}{5} \langle P_2[\hat{\mu}(o) \cdot \hat{\mu}(t)] \rangle \qquad (1)$$

This quantity or its square is measured in numerous nonlinear optical experiments including time resolved CARS, polarization methods, transient birefringence, transient gratings or time resolved conjugate four-wave mixing (6). It is also measured from the decay of the fluorescence polarization (7). The anisotropy is quite different for free and diffusing rotors (8) as shown in Figure 2, so we expect that in real fluids the anisotropy will be some combination of the two motions (see Figure 3). At short times we may expect the nearly Gaussian form for inertial motion

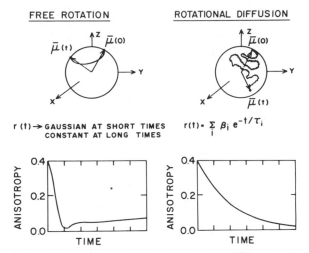

Figure 2. Comparison of Inertial and Diffusive Rotation

The free rotation of a dipole corresponds to the motion of a point in a smooth trajectory on the surface of a sphere (Top,left). Even after t→∞ the whole surface is not explored so the correlation time is infinite. In rotational diffusion the point moves in random walk eventually exploring the whole surface of the sphere (Top,right). The anisotropy for the free rotor is a superposition of many oscillatory components which reinforce only near t = 0 (and at regular intervals corresponding to the reciprocal of the rotational B-value, not shown on the Figure) (Bottom,left). The rotational diffusion anisotropy is a sum of exponentials (Bottom,right).

while at long times the system anisotropy will exhibit the exponential decays of a rotational diffuser. Steele (9), Chandler (10) and others (8) have addressed theoretically the problem of the intermediate case where the motion is neither inertial nor diffusive. Since characteristic timescales for free rotation of molecules at 300K are $(I/kT)^{1/2}$ ~1ps (I is the moment of inertia) and effective collision frequencies are in the range 0.1 ps for simple liquids, we can expect that subpicosecond experiments will expose any inertial effects that occur in the angular motion.

The internal rotations of molecules involve additionally the accelerating and decelerating forces deriving from the intramolecular potential function V. The stochastic Langevin equation for such motions about an angle θ in the presence of a fluctuating torque F(t) has the form:

$$\ddot{\theta} + \beta \dot{\theta} + \frac{dV}{d\theta} = F(t) \tag{2}$$

For the case that V is a piecewise parabolic potential Kramers (11) calculated the barrier crossing dynamics using the interpretation of θ from equation (2). The time between collisions that randomize the angular velocity is characterized by $\beta^{-1} = 2I/\zeta$, where ζ is the zero frequency friction. The rate of barrier crossing in the low friction, or inertial, regime was found to be directly proportional to β. In this regime

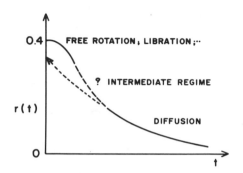

Figure 3.   Anisotropy in the Time Regime of the Collisional Period (Cartoon)

The anisotropy resembles at early time that of an inertial rotor (Figure 2, bottom left) while at later time that of a diffuser (Figure 2, bottom right). The extrapolation of the exponential part back to t = 0 does not yield the correct r(0). At t≈0 the value of dr(t)/dt approaches zero. Theories for the intermediate regime are addressed in references (9) and (10).

collisions are needed to energize molecules sufficiently for them to pass over the barrier (see Figure 4) just as in conventional gas phase kinetics. At very large friction the rate of isomerism is inversely proportional to β. For the high friction solution, the maximum rate occurs for $\beta \rightarrow 0$ in which case the barrier crossing is determined by the inertial oscillations at frequency ω_a: This represents the transition state limit.

In attempting to relate theory and experiment some measure of the effective friction is required. The rotational friction is defined as βI which in turn is found to be proportional to the shear viscosity by means of hydrodynamics (Stokes Law). Since the hydrodynamic relations are essentially continuum approximations, it is unsatisfactory to compare observed rates with macroscopic viscosities in order to obtain tests of the theory. However, in the high friction regime and where the fluctuations in angular velocity produce a linear response in the overall molecular orientation (see part of curve labelled `rotational diffusion' in Figure 1) it is found that $\beta \propto \tau_R$ (12). It follows that β could be obtained independently of hydrodynamic approximations if the values of τ_R were available.

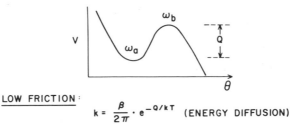

LOW FRICTION:

$$k = \frac{\beta}{2\pi} \cdot e^{-Q/kT} \quad \text{(ENERGY DIFFUSION)}$$

HIGH FRICTION:

$$k = \frac{\omega_a \omega_b}{\beta} \cdot e^{-Q/k_B T} \quad \text{(SPATIAL DIFFUSION)}$$

VANISHING 'HIGH' FRICTION:

$$k = \omega_a e^{-Q/k_B T} \quad \text{(TRANSITION STATE DESCRIPTION)}$$

Figure 4.   Kramers Theory: Limiting Solutions

Upper: The potential function $V(\theta)$ appropriate to describe the internal forces acting on the stilbene molecule during its isomerization is characterized by well and barrier frequencies ω_a and ω_b, and barrier height Q.

Lower: The three limiting solutions of the barrier crossing problem:
Low Friction: The rate is proportional to β and collisions energize molecules. This corresponds to the left hand side of the "inertial line" in Figure 1.
High Friction: This is the Smoluchowski limit relevant to the extreme right hand side of Figure 1.
Vanishing High Friction: Corresponds to the region at the Kramers turnover approached from the high friction side.

Methods:

Rotational correlation times of a few picoseconds or less are readily measured for neat liquids using a nonlinear response such as the electronically non-resonant Kerr effect (13) but the signal strength in such experiments is barely large enough to study solutes in dilute solution. In such cases some resonance enhanced method is desired such as fluorescence polarization or the population grating response. We have used both fluorescence polarization at a time resolution of ca. 1 ps (2) and transient birefringence methods (3) to obtain τ_R.

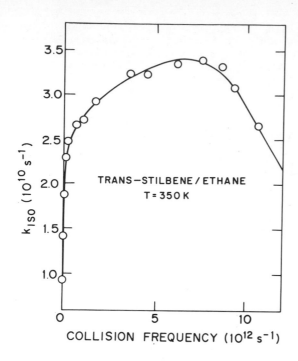

Figure 5. Variation of the Barrier Crossing Rate of Stilbene with Collision Frequency [After ref. (1)]

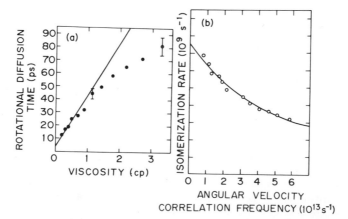

Figure 6. Barrier Crossing in the Spatial Diffusion Regime [after ref. (2)]

(a) Measurements of τ_R for stilbene in hydrocarbon solvents; the solid line corresponds to a Stokes–Einstein–Debye fit in which $\zeta \alpha \eta$.

(b) Solid line is fit to barrier crossing rates using Kramers theory and values of β from data in (a).

Barrier crossing processes can be studied by measuring the population in the initial potential well by means of transient absorption or fluorescence methods. In the case of trans–stilbene (14,15) the initial well at the trans configuration in the electronically excited state is a strongly fluorescent configuration while the product states reached after crossing the barrier are not fluorescent.

Inertial Effects in Barrier Crossing of Electronically Excited States (1).

Figure 5 shows the results for trans to cis isomerism of stilbene in supercritical fluid ethane at various pressures. There is an evident turnover regime in the rate versus collision frequency. The conventional gas phase kinetics (Lindeman–Hinshelwood theory) explains quite well the form of the initial rise in rate. The turnover occurs at a collision frequency of 6×10^{12} s^{-1}. The inertial regime approaching the maximum rate of isomerization, does not exhibit smooth variations of rate vs. collision frequency, suggesting that more molecular detail of the solvent–solute structure is needed to explain the behavior.

Spatial Diffusion Regime in Barrier Crossing of Excited Stilbene (2).

Values of β for stilbene were estimated from measurements of τ_R in different solvents covering a range of viscosities (see Figure 6a). The marked deviation of τ_R from linearity with viscosity dramatizes that the attempts to use hydrodynamic approximations for β (such as $\beta \propto \eta$) are simply not justified. The Kramers equation fit to the isomerism data using the derived values of β is shown in Figure 6b. These results suggest that the Kramers approach has predictive value for isomerism processes.

Inertial Motion in the Overall Rotation of Molecules (3).

We have been seeking manifestations of nondiffusive (inertial or "free") rotation in a direct time domain experiment. As seen in Figure 7, the rotational relaxation of aniline in isopentane occurs with an effective time constant of 1.2 ps which is only about a factor of 2 slower than that for rotation of the isolated molecule. This suggests that the reorientation involves unhindered rotation through significant fractions of a radian between collisions that change the angular momentum and interrupt the inertial motion. A more convincing demonstration of an inertial component to the reorientation will require the identification of a near Gaussian anisotropy decay near time zero as indicated in Figure 3.

The orientational motion of aniline in hydrocarbon solvents is reasonably well described by slip hydrodynamics modified to account for the free rotation contribution. In alcohols, on the other hand, rotation is slower than predicted by slip hydrodynamics, but faster than stick. Presumably the reason for the large difference in rotational dynamics between hydrocarbon and alcoholic solvents involves aniline's ability to hydrogen bond with alcohols. Evidence for the role of hydrogen bonding comes from the reorientation times observed for N,N–dimethylaniline (see Figure 8). In the hydrocarbon solvents this substituted aniline rotates more slowly than aniline itself, consistent with its larger size. In methanol (which has 2.5 times the viscosity of isopentane) aniline rotates just a little faster (16 ps) than predicted by stick hydrodynamic boundary conditions whereas the addition of the methyl groups causes it to rotate about three times faster (3.5 ps) and closer to the rate expected for slip boundary conditions. Apparently, a slight chemical change of this nature can radically alter the reorientation dynamics. In this case, because we know that aniline will hydrogen bond to the solvent, a proper assessment of the friction determining the rotational motion should incorporate the specific chemical interactions which are obviously not included in the hydrodynamics.

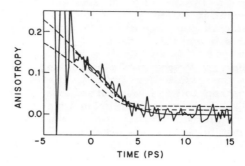

Figure 7.   Fluorescence Anisotropy of Aniline in Isopentane: [After
      ref. (3)]

Comparison of experimental and free rotor anisotropy decays.
Solid Curves: Experimental data and best exponential fit.
Upper dashed curve: Free rotor anisotropy of aniline as prolate
top and short–axis polarization.   Lower dashed curve: Free rotor
anisotropy for aniline as prolate   top and mixed short and long
axis transition.

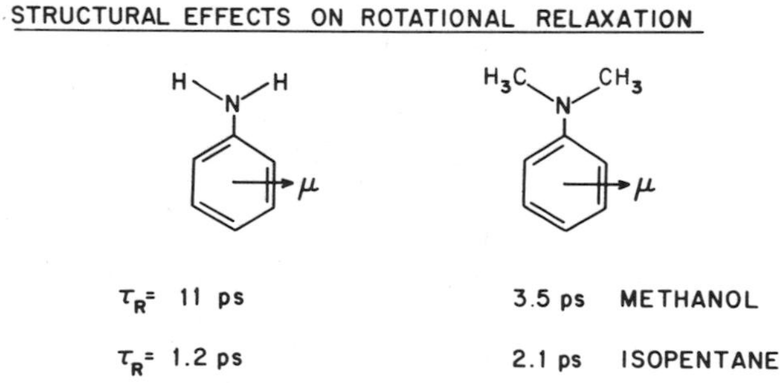

STRUCTURAL EFFECTS ON ROTATIONAL RELAXATION

τ_R= 11 ps   3.5 ps   METHANOL

τ_R= 1.2 ps   2.1 ps   ISOPENTANE

Figure 8.   Solvent Effects on Rotational Relaxation:

The values of τ_R for aniline (left) and dimethylaniline (right)
vary dramatically with the nature of the solvent.

48

Rotationally Mediated Chemical Reactions (4)

A common way of describing reaction kinetics in solution is to use the concept of an encounter complex. For example, for the bimolecular reaction, $A + B \longrightarrow C$ to occur, the reacting molecules, A and B, must first come sufficiently close together. According to a theory originally formulated by Smoluchowski, the rate constant for this step is diffusion-controlled and $k = 4\pi rD$, where r is the sum of the radii of A and B and D is the sum of translational diffusion coefficients of A and B (16). When the two molecules are in close contact, the encounter complex is still required to achieve the structure and energy appropriate for reaction (17,18). It is therefore expected that rotational diffusion of solvent and solute molecules will contribute to the overall rate of the reaction.

Systems which could be used to explore cage dynamics are those that undergo intramolecular proton transfer. As the proton moves from one part of the molecule to another, the molecule itself may require some rearrangement to accommodate the new structure, in which case the rotational dynamics of both solvent and solute will control the reaction rate. This is in a sense the rotational analogue of the Smoluchowski concept of bimolecular reaction rates. However, for the rotational case it is also interesting to explore the importance of inertial as well as diffusive reorientations.

In some recent experiments with a benzotriazole (see Figure 9), in which the proton shifts from the oxygen to a neighboring nitrogen when the molecule is electronically excited, we have found that the overall rotational reorientation rate is approximately the same as the proton transfer rate in a series of alcohol solvents varying in viscosity over a factor of more than 20 (4). The results for the proton transfer rate (measured by picosecond time resolved fluorescence methods) are shown in Figure 10. The curved line through the points corresponds to the sum of the rotational relaxation rate of the solute $(1/\tau_R)$ and the Debye relaxation rate $(1/\tau_D)$ of the solvent. This fit suggests that the reaction rate is in this case controlled by the local rotational motions of the solute and the solvent. Our picture is that the tunneling is very fast (less than 1 ps) but that a configuration optimized for tunneling can only be achieved after significant solvent mediated motions of other atoms have taken place.

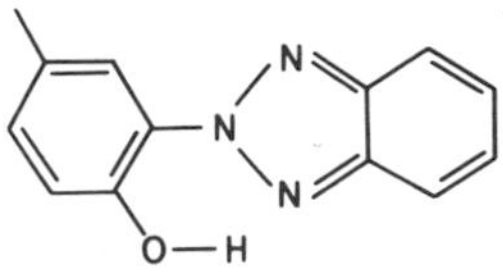

Figure 9. Molecular Structure of the Triazole studied in Figure 10

Rotational Effects on Raman Scattering in Liquids (19)

As a final example of how studies of molecular reorientation can provide novel information about solution phase processes, the case of Raman scattering is now considered. The collisions that establish the frictional forces of the previous examples interrupt the coherent evolution of the ensemble of molecules interacting with a radiation field. The

Raman scattering, which corresponds to the spontaneous emission from the coherently excited ensemble, is therefore damped by these dephasing collisions. The excitation is redistributed and then reirradiated at new frequencies (fluorescence). It is therefore clear that studies of the time evolution of the Raman scattering will provide information about the dephasing. One approach to probing the Raman dephasing is to study the polarization of the scattered light in a steady state experiment.

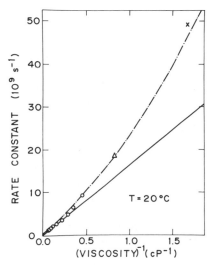

Figure 10. Variation of Proton Transfer Rate with Viscosity: [After ref. (4)]

Experimental proton transfer rates (circles) are compared with measured values of $1/\tau_R$ (solid line is best fit through the high viscosity data, not shown) and $1/\tau_D$ (dashed-dot line is the sum: $1/\tau_R + 1/\tau_D$). The data points Δ and X are respectively taken from: A. L. Huston, G. W. Scott and A. Gupta, J. Chem. Phys. 76, 4978 (1982); and K. P. Ghiggino, A. D. Scully and J. H. Leaver, J. Phys. Chem. 90, 5089 (1986)

The polarization of Raman emission depends on the elapsed time between the annihilation of the incident photon and the creation of the scattered photon. When a molecule is excited on resonance, this time is governed by the excited state lifetime. In the presence of collisions the coherence time is reduced to the order of the collisional period. Now it is apparent that the inertial motion of the molecule must be considered, since the polarization will be influenced by how much the molecule rotates during the scattering process. A simple analysis shows that the steady state anisotropy r(∞) for a transition onto a particular final state |f⟩ is given by:

$$r(\infty) = \frac{\int dt\ e^{-2\Gamma t}\ |\langle f|i(t)\rangle|^2\ r(t)}{\int dt\ e^{-2\Gamma t}\ |\langle f|i(t)\rangle|^2} \tag{3}$$

where Γ is the dephasing rate of interest and r(t) is given by eqn. (1). The quantity ⟨f|i(t)⟩ is the time dependent projection of the final state in the Raman process onto the initial state propagated on the excited state potential surface. The result is more obvious if the detector senses all possible final states, in which case:

$$R(t) = 2\Gamma \int dt\ e^{-2\Gamma t}\ r(t) \tag{4}$$

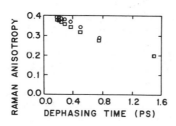

Figure 11. Calculated Raman Anisotropy for Aniline versus Dephasing Time [After ref. (19)]

The circles and squares are respectively the anisotropies predicted for Raman scattering detected on a single vibrational transition at 1307 cm⁻¹ and for the signal summed over all Rayleigh and Raman transitions.

Both these results represent experimentally realizable situations. Figure 11 shows the theoretical predictions for the anisotropy of the resonance Raman of aniline as a function of dephasing time (1/Γ) for the two experimental approaches assuming the motion defining r(t) is purely inertial free rotation.

Conclusion

Measurements of r(t) by various experimental techniques have proved useful in probing fast responses in liquid solutions. The importance of nondiffusive motions, typified by free rotation, is brought out through studies of barrier crossing processes and ultrafast rotational reorientational dynamics. These measurements are readily extended to other systems including solids, polymers and proteins where librations cause the decay of r(t) at short times. Furthermore, it was shown that such inertial motions also influence light scattering in a fundamental way.

Acknowledgements

This research was supported by grants from NSF, NIH and AROD. I am indebted to colleagues cited in references (1) to (4) and (19) upon whose work this lecture was based.

References

1. M. Lee, G.R. Holtom and R.M. Hochstrasser, Chem. Phys. Lett. 118:359 (1985).
2. M. Lee, A.J. Bain, P.J. McCarthy, C. Han, J.N. Haseltine, A.B. Smith III and R.M. Hochstrasser, J. Chem. Phys. 85:4341 (1986).
3. A.B. Myers, P.L. Holt, M. Pereira and R.M. Hochstrasser, J. Chem. Phys. 86:5146 (1987).
4. M. Lee, J.T. Yardley and R.M. Hochstrasser, J. Phys. Chem. in press.
5. R.G. Gordon, J. Chem. Phys. 44:1830 (1966).
6. A.B. Myers and R.M. Hochstrasser, IEEE J. Quant. Elect. QE-22: 1 482 (1986).
7. A. Szabo, J. Chem. Phys. 34:1 (1976).
8. B.J. Berne and R. Pecora, Dynamic Light Scattering, Wiley, New York (1976).
9. W.A. Steele, Adv. Chem. Phys. 34:1 (1976).
10. D. Chandler, J. Chem. Phys. 60:3508 (1974).
11. H.A. Kramers, Phyica 7:284 (1940); S. Chandrasekhar, Rev. Mod. Phys 15:1 (1943).
12. P.S. Hubbard, Phys. Rev. 131:1155 (1963).
13. J.M. Halbout and C.L. Tang, Appl. Phys. Lett. 40:765 (1982); B.I. Greene and R.C. Farron in "Picosecond Phenomena III" Eisenthal, Hochstrasser, Kaiser and Laubereau Springer 1982; p 209
14. R.M. Hochstrasser, Pure and App. Chem. 52:2892 (1980)
15. G. Rothenburger, D.K. Negus and R.M. Hochstrasser, J. Chem. Phys. 79:5360 (1983).
16. M.V. Smoluchowski, Z. Phys. Chem. 92:129 (1917)
17. See, for example, A.M. North, Collision Theory of Chemical Reactions in Liquids Methuen, London, 1961.
18. (a) R.M. Noyes, Progr. React. Kinet. 1:733 (1961); (b) K. Solc and W.H. Stockmayer, J. Chem. Phys. 54:2981 (1971); Int. J. Chem. Kinetics 5:733 (1973); (c) S. Lee and M. Karplus, J. Chem. Phys. 86: 1883; 1904 (1987), and references therein.
19. A.B. Myers and R.M. Hochstrasser, "Resonance Raman Depolarization Ratios as a Probe of Dephasing Times in Liquids," J. Chem. Phys., in press.

OPTICAL MANIFESTATIONS OF ENERGY AND PHASE RELAXATION

IN VIBRONIC SYSTEMS

V. V. Hizhnyakov

Institute of Physics of the Estonian SSR Acad. Sci.
Riia 142, Tartu, 202400, USSR

1. Introduction

In this work, the vibrational relaxation of impurity centers of crystals in excited electronic states is examined. A dynamical theory of this relaxation is proposed which is based on a model Hamiltonian system. The model includes a localized (optical) electron and a set of vibrational harmonic oscillators whose number N is very large. $N \geq 10^{20}$, with a quasi-continuous spectrum of frequencies. After an electronic transition the equilibrium positions and elastic forces of atoms of the center and its nearest environment are changed. The motion of these few atoms described by symmetrized shifts (s.c. configurational coordinates) determines the vibronic (optical) transitions in the center. This motion is considered by using the method of the statistical operator. This consideration allows quantum fluctuations to be taken into account exactly at all stages of relaxation. In particular, a description is given of the temporal behavior of the states whose fluctuations change in the course of relaxation. It is such states that are created upon optical excitation of centers when the elastic constants change after electronic transition; at T = 0 these can be considered as a generalization of squeezed states for a multimode system. A generalization of the Callen-Welton formula (s.c. fluctuation-dissipation theorem) for a nonstationary case is presented and a simple model of the squeezed pulse is proposed. We also describe the relaxation of the energy of the electronic transition and its quadratic fluctuations which determine, in particular, the hot luminescence spectrum.

Besides the relaxation of energies which is rather fast, as a rule considerably slower processes of phase relaxation of the electronic excitation also take place. The latter manifest themselves directly in the width and shape of the zero-phonon line. The method proposed for the consideration of the quadratic vibronic interaction in optical spectra allows one to formulate a new nonperturbative theory of this relaxation. The theory predicts a difference of the widths of zero-phonon lines in absorption and luminescence spectra and an asymmetric shape of these lines.

2. Nonequilibrium Statistical Operator in the Configurational Coordinate Representation

The statistical operator of a multicoordinate system in the representation of the configurational coordinate q_1 (which may be a vector in a space of some dimensionality) equals[1,2]

$$\rho(Q,Q';t) = \frac{1}{2\pi} \int_{-\infty}^{\infty} d\nu \; e^{i\nu(Q+Q')/2} \, \chi(\mu,\nu;t), \tag{1}$$

where $\mu = Q'-Q$, Q and Q' are instantaneous values of the coordinate q_1,

$$\chi(\mu,\nu;t) = \langle \psi_t | e^{i\mu p_1 + i\nu q_1} | \psi_t \rangle \tag{2}$$

is the characteristic Wigner function, $p_1 = i\partial/\partial q_1$.

Before excitation let the impurity center be in the vibrational state $|i\rangle$ of the ground electronic level (E_i, the energy of the state $|i\rangle$). With the probability $n_i \sim \exp(-E_i/kT)$. As a result of optical excitation at the time moment $t = 0$ and of the subsequent relaxation, at time t this state gives the value $|\psi_t\rangle = \exp(-itH_2)|i\rangle$. Here and below H_1 and H_2 are the vibrational Hamiltonians of the ground (1) and excited (2) electronic level. In the harmonic approximation,

$$H_1 = \sum_i \omega_{1i} a^+{}_{1i} a_{1i},$$

$$H_2 = \sum_j \omega_{2j} a^+{}_{2j} a_{2j} + \omega_0,$$

where $a^+{}_{1i}$, $a^+{}_{2j}$ and a_{1io}, a_{2j} are the creation and annihilation operators in the ground and excited electronic states, ω_{1i} and ω_{2j} are the phonon frequencies, ω_0 is the frequency of purely electronic transition, $\hbar = 1$. In this approximation,

$$\chi(\mu,\nu;t) = \langle e^{itH_2} e^{i\mu p_1 + i\nu q_1} e^{-itH_2} \rangle_1$$
$$= \exp[i\langle L\mu\nu(t)_2\rangle_1 - \frac{1}{2} \langle (L\mu\nu(t)_2 - \langle L\mu\nu(t)_2\rangle^2{}_1], \tag{3}$$

$\langle ... \rangle_m \equiv Sp[\exp(-H_m/kT)...]/Sp[\exp(-H_m/kT)]$ is the sign of the averaging over the vibrations of the m-th electronic level; $m = 1,2$,

$$L\mu\nu(t)_2 = e^{itH_2}(\mu p_1 + \nu q_1)e^{-itH_2}$$

If the equilibrium positions of nuclei and the elastic constants are changed up on electronic transition, then

$$V \equiv H_2 - H_1 = V_0 + (aq) + \frac{1}{2} (bq^2),$$

where a and q are vectors and b is the tensor, as a rule, of some (not large) dimension, the vector of the configurational coordinates q can be represented in the forms

$$q = \sum_i e_{1i} \chi_i + Q^{(0)} = \sum_j e_{2j} y_j .$$

Here $\chi_i = (a^+_{1i} + a_{1i})(2\omega_{1i})^{-1/2}$ and $y_j = (a^+_{2j} + a_{2j})(2\omega_{2j})^{-1/2}$ are the normal coordinates of the ground and excited electronic levels, related to each other by the Duschinsky transformation

$$\chi_i = \chi_{oi} + \sum_j c_{ij} y_j ,$$

where

$$c_{ij} = (e_{1i} b e_{2j})/(\omega^2_{2j} - \omega^2_{1i}$$

is the mode-mixing matrix found in,[3] $\chi_{oi} = - (a e_{1i})\omega_1^{-2}{}_i$, $Q^{(0)} = \sum_i e_{1i}\chi_{oi}$,

$e_{1i} = \sum_j c_{ij} e_{2j}$. The first component of the vector q gives the configurational coordinate q_1.

In the model under consideration the phonon operators in different electronic states are related by the multimode-squeezing-type relaxation[4]

$$a^+_{1i} = - \xi_{1i} + \frac{1}{2} \sum_j \frac{c_{ij}}{\sqrt{\omega_{1i}\omega_{2j}}} [(\omega_{1i}+\omega_{2j})a^+_{2j} + (\omega_{1i}-\omega_{2j})a_{2j}], \qquad (4a)$$

$$a^+_{2j} = \xi_{2j} + \frac{1}{2} \sum_i \frac{c_{ij}}{\sqrt{\omega_{1i}\omega_{2j}}} [(\omega_{1i}+\omega_{2j})a^+_{1i} + (\omega_{2j}-\omega_{1i})a_{1i}], \qquad (4b)$$

where $\xi_{1i} = (a e_{1i})(2\omega_{1i}{}^3)^{-1/2}$, $\xi_{2j} = ((a-Q^{(0)}b)e_{2j})(2\omega_{2j}{}^3)^{-1/2}$. Expressing q_1 and p_1 via a^+_{2j} and a_{2j}, using the relations

$$a^+_{2j}(t)_2 = e^{it\omega_{2j}}a^+_{2j}, \quad a_{2j}(t)_2 = e^{-it\omega_{2j}}a_{2j}$$

and going over to the operators a^+_{1i} and a_{1i}, one can find

$$\chi(\mu,\nu,t) = \exp[i(\nu Q_t + \mu \dot{Q}_t) - \frac{1}{2}\mu^2 \Delta_t - \frac{1}{2}\mu\nu \dot{\Delta}_t - \frac{1}{2}\nu^2 \delta_t \qquad (5)$$

and

$$\rho(Q,Q';t) = (2\pi\Delta_t)^{-1/2} \exp[-\frac{1}{2}\Delta^{-1}{}_t (\frac{Q+Q'}{2} + i\dot{\Delta}_t \frac{Q-Q'}{2} - Q_t)^2 \qquad (6)$$

$$- \frac{1}{2}\delta_t(Q-Q')^2 + i\dot{Q}_t(Q'-Q)]$$

Here

$$Q_t = 2\pi^{-1} \int_0^\infty d\omega\, \omega^{-1} \cos\omega t\, (a\,\mathrm{Im}\bar{G}(\omega)), \tag{7}$$

$$\Delta_t = \pi^{-1} \int_0^\infty d\omega(2n_\omega+1)\, [\mathrm{Im}G(\omega)|D(\omega,t)|^2]_{11} \tag{8}$$

and

$$\delta_t = \pi^{-1} \int_0^\infty d\omega(2n_\omega+1)\, [\mathrm{Im}G(\omega)|\omega D(\omega,t) + bv(t)|^2]_{11} \tag{9}$$

describe respectively the classical time dependence of configurational coordinate, its quadratic quantum fluctuation (Δ_t) and quadratic quantum fluctuation of the corresponding momentum (δ_t); the subscripts 11 denote the first diagonal matrix element; and

$$v(t) = 2\pi^{-1} \int_0^\infty d\omega\, \sin\omega t\, \mathrm{Im}\bar{G}(\omega), \tag{10}$$

$$D(\omega,t) = 1 + b\,\bar{G}(\omega) + \frac{b}{2} \int_t^\infty d\tau\, e^{-i\omega\tau-\varepsilon\tau} v(\tau)$$

($\varepsilon \to +0$), the dynamical Green's functions

$$G(\omega) = \sum_i e^2{}_{1i}\, (\omega^2-\omega^2{}_{1i}-i\varepsilon\omega)^{-1}$$

and

$$\bar{G}(\omega) = \sum_j e^2{}_{2j}(\omega^2-\omega^2{}_{2j}-i\varepsilon\omega) = (1-G(\omega)b)^{-1}\, G(\omega)$$

describe the vibrations of the center in the ground and excited electronic states, (in each particular case, these functions can be calculated by standard methods of lattice local dynamics [5]); finally $n_\omega = [\exp(\omega/kT)-1]^{-1}$.

Formulas (8) and (9), connecting the quadratic fluctuations of the relaxing system with its spectrum Im $G(\omega)$, can be considered as a generalization of the Callen-Welton formula [6] (fluctuation-dissipation theorem) for a nonstationary case. At small and large time formulas (8) and (9) give the equilibrium values for the ground ($t \to 0$) or the excited ($t \to \infty$) state, corresponding to the ordinary Callen-Welton formula.

Using the formulas obtained, one can find the time dependence of the energy of the electronic transition V_t and its quadratic fluctuation $\varepsilon^2{}_t$ in the course of the relaxation. For example, in the case of a one-dimensional b

$$V_t = V_0 + aQ_t + \frac{1}{2} b (Q^2_t + \Delta_t),$$

$$\varepsilon^2_t = (a + bQ_t)^2 \Delta_t + \frac{1}{2} b^2 \Delta^2_t. \tag{11}$$

These functions, in particular, determine the envelope function of the time-dependent hot luminescence spectrum:

$$I(\omega,t) \sim \varepsilon^{-1}_t \exp[-2\gamma_0 t - (\omega-V_t)^2/2 \, \varepsilon^2_t]. \tag{12}$$

Here γ_0 is the radiative decay constant of the excited state. The calculations of hot luminescence for concrete systems have been made.[7]

3. Exponentially-Decaying Mode; Squeezed Pulse

Let us examine the properties of the state (6) by a simple model

$$G(\omega) = (\omega^2 - \omega^2_1 - 2i\omega\Gamma)^{-1} \tag{13}$$

describing the exponentially decaying mode

$$Q_t = Q^{(0)} \cos\omega_2 t \; e^{-\Gamma t}.$$

Here ω_1 and $\omega_2 = (\omega_1^2+b)^{1/2}$ are the frequencies of the mode in the ground and excited electronic states. In the case of slow damping ($\Gamma \ll \omega_1,\omega_2$, $|\omega_1-\omega_2|$) the quadratic fluctuations equal

$$\Delta_t = \Delta_1 e^{-2\Gamma t} [1+(\frac{\omega_1^2}{\omega_2^2} - 1) \sin\omega_2 t] + \Delta_2 (1-e^{-2\Gamma t}),$$

$$\delta_t = \delta_1 e^{-2\Gamma t} [1+(\frac{\omega_2^2}{\omega_1^2} - 1) \sin\omega_2 t] + \delta_2 (1-e^{-2\Gamma t}), \tag{14}$$

where $\delta_1 = \omega^2_1 \Delta_1 = (\bar{n}_1+\frac{1}{2})\omega_1$ and $\delta_2 = \omega^2_2 \Delta_2 = (\bar{n}_2+\frac{1}{2})\omega_2$ give the equilibrium values of fluctuations in the ground and excited electronic states, $\bar{n} = [\exp(\omega m/kT)-1]^{-1}$. In the absence of damping ($\Gamma = 0$) formulas (6) and (12)-(14) describe the squeezed state[8] (more exactly, its generalization for $T = 0$), characterized by diminishing fluctuations of one quadrature in comparison with its value for zero vibrations (the condition $\Delta_t < \Lambda_0 = (2\omega_1)^{-1}$ or $\delta_t < \delta_0 = \omega_1/2$ may be fulfilled). Moreover, state (6) has the same property in general (with parameters given by general expressions (7) - (11)), for a sufficiently large parameter $|b|$. Consequently, the formulas obtained above describe the squeezed state of a multimode system. In particular, if one replaces t by $|t|$ in (12)-(14) then a simple model of the squeezed pulse is obtained. In this model, Q_t describes (up to a normalization factor) the temporal behavior of the average value of the field strength in the pulse, while Δ_t and δ_t characterize the time dependence of quadratic fluctuations of two quadratures of the field.

Let us note that the squeezed pulse, unlike the coherent one, cannot be represented as a simple product of the squeezed states of the oscillators of the continuum: at a finite energy of the pulse the degree of squeezing of a state of each oscillator is infinitely small ($\sim N^{-1}$). As a result, no squeezing of the fluctuations of the locally-measured quadratures of the field arises. Only states with the mixing of continuum modes (namely such states are considered here) give a finite squeezing for quadrature fluctuations.

4. Width and Asymmetry of the Zero-Phonon Line

The absorption (luminescence) spectrum in the Condon approximation is determined[9] by:

$$I(\omega) = \frac{1}{2\pi} \int_{-\infty}^{\infty} d\tau \; e^{i\omega\tau - \gamma_0 |\tau|} \; F(\tau),$$

where

$$F(\tau) = \langle f_\tau \rangle_{1(2)}; \qquad f_\tau = e^{-i\tau H_1} e^{i\tau H_2}$$

The derivative $\dot{F}(\tau) = i\langle f_\tau V\rangle_1$ in the case of quadratic vibronic interaction V is determined by the normalized Fourier amplitudes of resonance Raman scattering of the first order $(\langle f_\tau a^+_{1i}\rangle_1)$ and second $(\langle f_\tau a^+_{1i} a^+_{1i'}\rangle_1$ and $\langle f_\tau a^+_{1i} a_{1i'}\rangle_1)$ orders. Replacing in (4a) the formulas for amplitudes a^+_{1i} by a^+_{2j} and a_{2j}, rearranging the latter with $\exp(i\tau H_2)$, and, using (4b), going back to the operators a^+_{1k} and a_{1k}, one gets linear equations[4] after the rearrangement with $\exp[-H_1(i\tau+1/kT)]$ and cyclic permutation. In the case of a finite number of degrees of freedom N, these equations can be solved numerically. For large N it is possible to proceed to equivalent integral equations which can also be solved numerically.

Let us examine the asymptotic behavior of the Fourier transform. The decay of $F(\tau)$ at $\tau \to \infty$ is described by the Fourier amplitudes $\sigma_{ii'} = \langle f_\tau(a^+_{1i} a_{1i'} - \bar{n}_i \delta_{ii'})\rangle_1$, which are determined at $|\tau| \to \infty$ by the equation[4]

$$\sigma_{ii'} = F\,(\bar{n}_i+1) \sum_k d_{ik}\,(\bar{n}_{i'}\delta_{i'k} + e^{-\frac{\omega_{1k}}{kT}}\sigma_{ki'}), \tag{15}$$

where

$$d_{ik} = \sum_j \frac{c_{ij}(e_{1k} b e_{2j})\omega_{1i}^{1/2}}{2\omega_{2j}\omega_{1k}^{1/2}(\omega_{2j}-\omega_{1k})}\,(e^{i\tau(\omega_{2j}-\omega_{1k})} -1).$$

At low temperature and an arbitrary b (as well as for arbitrary temperature and small $|b|$ the items $\sim \exp(-\omega_{1k}/kT)\sigma_{ki'}$ can be neglected. Then we get

$$F(\tau) \sim \exp[-\gamma|\tau| + i\alpha\,\text{sign}\,\tau], \tag{16}$$

where

$$\gamma = \text{Re} \int_0^{\infty} d\tau \; (bg_2(\tau)\,bg_1(-\tau)), \tag{17}$$

$$\alpha = \text{Im} \int \int_0^{\infty} d\tau d\tau' \; (bg_2(\tau+\tau')\,bg_1(-\tau-\tau')), \tag{18}$$

$$g_1(\tau) = \pi^{-1} \int_0^\infty d\omega \, e^{i\omega\tau}(n_\omega+1) \, \text{Im} \, G(\omega),$$

$$g_2(\tau) = \pi^{-1} \int_0^\infty d\omega \, e^{i\omega\tau}(n_\omega+1) \, \text{Im} \, \overline{G}(\omega).$$

The parameter γ determines the width of zero-phonon line and the rate of the pure dephasing of electronic excitation, and the parameter α, the asymmetry of zero-phonon line. These parameters are, in general, different for absorption and luminescence. The larger the difference is, the more the weighted phonon densities $\text{Im}G(\omega)$ and $\text{Im}\overline{G}(\omega)$ in different electronic states differ; the difference decreases if $|b| \to 0$ and $T \to 0$.

The conclusion of this theory about the difference of dephasing rates (and the widths of zero-phonon lines) in absorption and luminescence is opposite to the conclusion of the nonperturbative theory by Osad'ko[10] and Skinner and Hsu.[11]

References

1. E. Wigner, Phys. Rev. 40:749 (1932).
2. V. I. Tatarskii, Upsk. Fiz. Nauk 139:587 (1984).
3. V. Hizhnyakov, Phys. Stat. Sol. 114:72 (1984).
4. V. V. Hizhnyakov, Trudy Instituta Fiziki AN Estonskoj SSR 59:55 (1986).
5. A. A. Maradudin, Z. W. Montroll and G. H. Weiss, Theory of Lattice Dynamics in the Harmonic Approximation, New York, 1963.
6. H. B. Callen and T. A. Welton, Phys. Rev. 83:34 (1951).
7. G. S. Zavt, V. G. Plekhanov, V. V. Hizhnyakov and V. V. Shepelev, J. Phys. C 17:2839 (1984).
8. D. F. Walls, Nature 280:451 (1979).
9. M. Lax, J. Chem. Phys. 20:1752 (1952).
10. I. S. Osad'ko, Uspekhi Fizicheskikh Nauk 128:31 (1979); Sov. Phys. Usp. 22:311 (1979).
11. J. L. Skinner and D. Hsu, Phys. Chem. 90:4931 (1986).

THE OSCILLATIONS OF 2D-ELECTRON DENSITY OF STATES

IN A TRANSVERSE MAGNETIC FIELD

I.V. Kukushkin and V.B. Timofeev

Institute of Solid State Physics, Academy of Sciences of the
USSR, 142432 Chernogolovka

INTRODUCTION

The problem of dealing with the electron density of states (DOS) in the presence of disorder associated with a random potential due to defects[1] is the focus of discussion of the energy spectrum of 2D-systems in a transverse magnetic field. The importance of this problem is caused by the necessity of a microscopic description of the magnetotransport properties of the 2D-space charge layers for a wide range of variations of the filling of the quantum states, including the regimes of integral and fractional quantum Hall effect (QHE). In order to construct a microscopic theory one has to have detailed information on disorder in the system, viz., on the random potential of scatterers. Here comes the question of screening of random potential fluctuations as a function of the filling factor of quantum states. These problems can be solved experimentally in terms of a spectroscopic method employed to study the energy distribution of the density of one-particle electronic states - D(E). The previously employed methods for investigations of 2D-DOS at the Fermi level - dn_S/dE_F were based on the measurements of magnetization,[2] electron heat capacity,[3] magnetocapacitance,[4] contact potential difference[5] and thermaly activated conductivity.[6] With account taken of the electron-electron interaction and the related effects of random potential screening, the thermodynamic DOS - dn_S/dE_F and $D(E_F)$ are different. An advantage of the spectroscopic method is that it enables one to detect how the energy distribution of the DOS varies with the filling factor ν of the quantum states ($\nu = n_S h/eB$, n_S being the electron density) and, also, with the amplitude and linear scale of long-range random potential fluctuations, the magnetic field and the electron mobility. Finally, the spectroscopic method may be used to determine the gap values in one-electron energy spectrum of 2D-electrons in a transverse magnetic field and, what is especially important, the Coulomb gap values of incompressible Fermi-liquids in the fractional QHE regimes.

1. Two-dimensional DOS and screening of random potential fluctuations

The spectroscopic method is based on the measurements of luminescence spectra related to the radiative recombination (RR) of 2D-electrons with photoexcited holes in silicon MOS-structures (p-Si (001) - MOS-structure.[7,8]

Under the conditions of excitation of nonequilibrium electron-hole pairs near the Si-SiO$_2$ interface and at positive voltage across the gate $V_g > V_T$ (V_T is the threshold voltage), there arises a 2D-electron accumulation channel completely screening the gate field. Therefore the energy bands as far as 10^2 A behind the channel appear actually flat in the depth of the semiconductor (the deplotion layer disappears).[8] The RR spectrum is a convolution of distribution functions of 2D-electrons and noneequilibrium holes on acceptors. It is significant that the width of the energy distribution of holes, participating in the recombination, is not large in the case in question ($\approx$ 0,8 meV).[8]

Fig. 1 demonstrates that in the absence of the magnetic field the 2D-electron luminescence spectrum is a step-like energy function (Spectrum 2) in accord with the constance of the 2D-DOS for H = 0. The structure related to Landau's quantization arises in a transverse magnetic field (spectrum 3). The energy values corresponding to the dimensional quantization band bottom, E_0, and the Fermi level, E_F, are determined, at a preset n_s, by means of constructing the Landau-level fans (see the top of the figure 1). The 2D-character of the electron subsystem is proved by a regular alteration of the quantization range $\hbar\omega_c$ with varying angle of the interface inclination with respect to the H direction that is kept constant (spectrum 4).

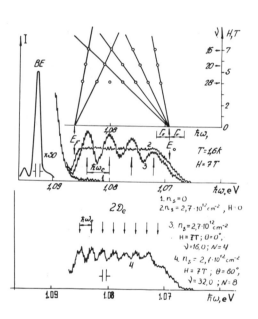

Fig. 1. Spectra of radiative recombination of 2D-electrons with photoexcited holes, measured at W=10^{-3} W/cm^2 T=1.6 K, n_s=2.7 10^{12} under the conditions when the magnetic field H=0 (Spectrum 2), H=7T and is normal to the 2D layer (Spectrum 3). H=7T and the field is inclined by 60° to the normal (Spectrum 4). Spectrum 1 is obtained for n_s=0 ($v_g < v_t$). The line BE corresponds to the radiation of electrons bonds on boron atoms. The top of the figure demonstrates the Landau-level fans (spectral position in the lines) constructed at n_s=2.7 10^{12}cm^{-2} under a complete filling of four, five and seven layers. Extrapolations to H → 0 determine the positions for the dimensionally quantized band E_0 and the Fermi energy E_F.

Fig. 2 illustrates the behaviour of the luminescence spectra at an alteration of the filling factor of quantum states when one Landau level is operative. It is seen that the shape of the luminescence spectra that directly reproduces the energy distribution of the density of one-particle states on the Landau level, D(E), is markedly dependent on the filling factor. The luminescence peak width is maximal under the condition of complete filling of the Landau level, and it is minimal when the quantum states are half-filled.

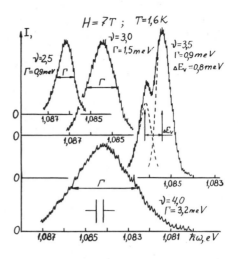

Fig. 2. Spectra of the 2D-electron radiative recombination, measured for H=7, T=1.6 K, W=10⁻³W/cm² for different filling of one Landau-level values.

An oscillating behaviour of the luminescence line width as a function of ν (with one or several Landau levels being operative) is illustrated in Fig. 3. These oscillations have greatest amplitude when the recombination is related to only one (lowest) Landau level. When several Landau levels are filled, the oscillations exhibit the same character but their depth is the smaller the deeper is the quantum state under the Fermi level (the smaller quantum number N is). The observed oscillations of the line widths in the recombination spectra are a direct consequence of the oscillating behaviour of the DOS peak widths on the Landau levels as a function of the electron concentration. The nature of this phenomenon is undoubtedly related to random potential fluctuations screening,[9] namely, to its long-range components. In fact, under the condition of a complete filling of the Landau level the Fermi

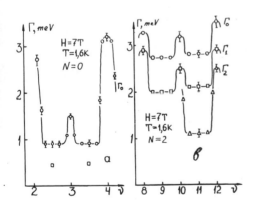

Fig. 3. Landau levels widths as a function of ν measured for H=7 T T=1.61 K for the Landau levels with N=0.

level is strictly in the centre of the energy spectrum gap, therefore screening in this case is rather weak. In the absence of screening the Landau levels along the interface reproduce the potential profile caused by long-range random potential fluctuations. It should be concluded that under a complete filling the luminescence line width and os of the DOS peak is determined by the amplitude of these fluctuations. In this case the DOS peak width, Γ_{max}, is described by the equation:[10]

$$\Gamma_{max} = Q(1+\ell H^2(4N + 2)/d^2)^{-1/2} \qquad (1)$$

where Q and d are the amplitude and the linear scale of the potential fluctuations, ℓH is the magnetic length. Investigating the $\ell H/d$ - sensitive luminescence peak width on the completely filled Landau level as a function of the magnetic field H and the quantum number N, one can easily determine the linear scale of fluctuations d. Subsequently, when the dimension of the cyclotron orbit $\ell_H (2N+1)^{1/2}$ exceeds the linear scale of the long range fluctuations, d, the luminescence peak width will no longer oscillate with ν. In other words, when the condition $\ell_H(2N+1)^{1/2}>d$ is fulfilled the dependence $\Gamma_{max}(N)$ with H = const are no longer different. Hence there comes an independent method, although less accurate, for the determination of d.

Under the condition of half-filling of quantum states the long-range random potential fluctuations are completely screened, and the DOS-peak width is narrowed and becomes minimal, Γ_{min}. The electrons in this case are scattered only by short-range scatterers that remain unscreened. According to the theory[11] Γ_{min} here must depend on the sole parameter H/μ (μ is the electron mobility).

2. Coulomb gaps in the energy spectrum of an incompressible Fermi-fluid under fractional Quantum Hall Effect

The spectroscopic method is effective for the solution of other problems associated with investigation of the 2D-electron energy spectrum in Si MOS-structure. We shall now discuss how, using this method, one can measure the values of Coulomb gaps in the energy spectrum of an incompressible quantum liquid whose ground state is observable in the fractional QHE regime.[12] According to the theory[13] elementary excitations in an incompressible Fermi-liquid are quasiparticles with fractional charges $e^*=e/q$ (q=3,5,... is odd integer). In this model introducing an extra electron into the sysem (with $\nu=p/q$, p being integer) is equivalent to the creation of q excitations, viz., quasielectrons, and diminishing the number of electrons by unity is equivalent to the generation of q quasiholes. Quasiparticle excitations are separated from the ground state by an energy gap of Coulomb origin. The gap values for quasielectrons, Δ_e, and quasiholes, Δ_h, are in principal different.[13] To date the gaps in the incompressible Fermi-liquid spectrum have been measured by means of the temperature dependence of the magnetotransport coefficients - σ_{xx} or ρ_{xx} in the corresponding minima, with $\nu=p/q$.[14] Fig. 4a presents as an example the spectra of magnetoconductivity $\sigma_{xx}(\nu)$ measured in the magnetic field H=8T and at T=0.35 and 1.5 K for (001)Si - Corbino MOSFET. The σ_{xx} minma at fractional values of $\nu=p/3$ due to the electron gas condensation are clearly seen. When the excitations are thermally activated to the mobility edges $\sigma_{xx} \sim exp(-W/kT)$, that enables one to find, from the corresponding temperatue dependences, the activation energy W and to estimate the total gap in the spectrum: $\Delta=\Delta_e + \Delta_h \approx 2W$. The temperature dependence $\sigma_{xx}(T)$ measured for $\nu=4/3$ and H=8T is plotted as in σ_{xx}/T^{-1} in Fig. 4b.

It is seen that in the low temperature region the dependence $\sigma_{xx}(T)$ does not fit a simple Arrhenius law since here the activation processes are strongly masked by a variable-range-hopping conductivity. It is quite evident that the limitations to the accuracy of the scale of Coulomb gap determination so arising in this method are matters of principle. In view of this, other independent methods for the Δ value measurement are needed. Our spectroscopic method for the determination of the value of

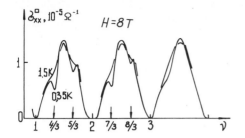

Fig. 4a. Conductivity σ_{xx} for transverse magnetic field H=8 T as a function of ν at T=1.5 K and T=0.35 K.

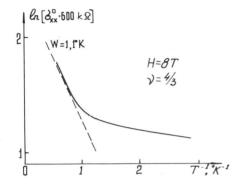

Fig. 4b. Temperature dependence of the conductivity measured at ν = 4/3 and H=8 T and plotted as in σ_{xx}/T^{-1}.

Coulomb gaps of quasiparticle excitations of incompressible quantum liquid is based on that fact that the spectral position of the luminescence line, measured in the fractional QHE regime, is closely related to the chemical potential of the interacting electrons, ξ. It is known that the chemical potential measurable as a function of the filling factor - $\xi(\nu)$ must exhibit a discontinuity $\Delta\xi = q\Delta$ at fractional value of the filling factor ν = 1/q, that is when a fractional QHE is observed.[15] The discontinuity corresponds to the generation (absorption) of an appropriate number of quasiparticle excitations. In full agreement with a discontinuous behaviour of $\xi(\nu)$ at fractional ν, one has to expect a nonmonotonic behavior of the spectral position of the luminescence line. Fig. 5 demonstrates that with varying ν from 2.27 to 2.4 (the filling factor region ν = 7/3) the luminescence line shape (2De) is actually unaltered, but a nonmonotonic dependence of its spectral position is observed.[16] In the direct vicinity of the filling factor ν = 7/3 (from ν - 2.32 to ν = 2.35) the luminescence line has a doublet character (see Fig. 6). As the filling factor is growing in this region, the intensities of the doublet component get redistributed so that, initially, a low-energy component of the doublet arises and, subsequently, as its intensity begins to decrease, the component on the greater energies side grows. The observed doublet character of the luminescence spectrum indicates directly the presence of the gap in the energy spectrum of an incompressible Fermi-liquid arising at a fractional ν. The spacing in the energy scale between the doublet components determines the total magnitude of this gap. In order to establish how the effect of 2D-electron condensation manifests itself on the energy position of the $2D_e$-line, the observed alterations in the spectral position of the lines in Figures 6 and 7 have to be compared with the spectra measured under the conditions when the Fermi-liquid does not arise, namely at T > Δ (the corresponding spectra are shown by dashed lines in Fig. 5 and 6).

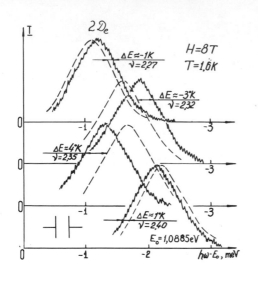

Fig. 5. 2D-electron recombination spectra measured at T=1.6 K and T=4 K (dashed line) for different values of ν: 2.27; 2.32; 2.35; 2.40 (H=8T) ΔE is the difference in the spectral position of the lines, measured at T=1.6 K and T=4 K, E_0=1.0885 eV.

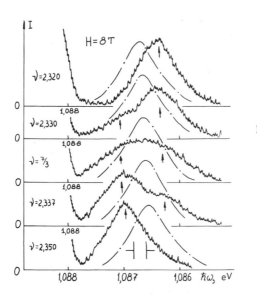

Fig.6. 2D-electron recombination spectra measured in the vicinity of ν = 7/3 at T=1.5 K and T=2.1 K (dashed line) for different values of ν: 2.320; 2.330; 7/3; 2.337; 2.350.

The results of the comparison of the $2D_e$-line spectral positions in the vicinity of ν = 7/3 (for T < Δ) are illustrated in Fig. 7a. The difference in ΔE in the energy position of the $2D_e$-lines, determined for T = 1.5 K and T = 2.1 K > Δ, characterizes the effects of interaction of 2D-electrons upon their condensation to a quantum liquid. Fig. 7b illustrates the dependences ΔE(ν) measured for two Si MOS-structure (the mobilities being μ(T = 0.35 K) = 4x10⁴ cm²/v sec and μ(T = 0.35 K) = 3x10⁴ cm²/v sec with H = 8T in the vicinity of ν = 7/3 and 8/3. It is seen that ΔE is other than zero only near to ν = 7/3 and 8/3, therefore the anomalous behaviour of the $2D_e$-line spectral position is related precisely to the 2D-electron gas condensation. It is important that the ΔE(ν) value behaves in a nonmonotonic manner: it is negative and minimal for ν < 7/3 (8/3), then it reverses its sign and reaches a maximum at ν > 7/3 (8/3). This dependence can be attributed to the fact that during the recombination act the the number of electrons is decreased by unity. In terms of the incompressible Fermi-liquid model this is equivalent, for example for $\nu \leq 7/3$, to the generation of three quasi-hole one-particle excitations. Analogous considerations, with account taken of the electron-hole symmetry, hold for the region $\nu \leq 8/3$ and $\nu > 8/3$ as well. At the absorption of three quasielectrons the energy of the emitted photon is increased by $3\Delta_e$, and at the generation of three quasi-holes it is decreased by $3\Delta_h$. Fig. 7b suggests that with H = 8T and $\nu = 7/3$ $3\Delta_e =$ (4±0.3) K. The measured magnitudes of $\Delta = \Delta_e + \Delta_h$ at $\nu = 7/3$, 8/3 obtained by means of optical spectroscopy method are close to activation gaps found at the same fractions with the use of thermoactivated conductivity, namely $\Delta(\nu) \approx 2W(\nu)$. It should be interesting to know how the magnitude of the Coulomb gap is changed with temperature. In the framework of the thermally activated conductivity method this problem cannot be solved because of some matters of principal.

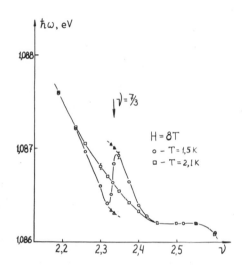

Fig. 7a. Dependence of the spectral positions of the $2D_e$-line maximum, measured at T=1.5 K and T=2.1 K as a function of ν in the vicinity of $\nu = 7/3$ for H=8 T. Symbols (*) correspond to spectral positions of doublet line close to $\nu = 7/3$.

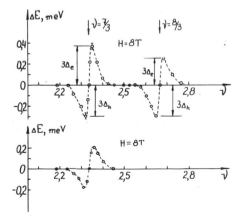

Fig. 7b. Difference ΔE of the spectral positions of the lines $2D_e$ measured at T=1.5 K and T=2.1 K as a function of ν in the vicinity of $\nu = 7/3$ and $\nu = 8/3$ for two MOSFET, having mobility at T=0.35 K μ=4 10^4 cm²/v s - (curve b) and μ=3 10^4 cm²/v s - (curve c).

By means of the spectroscopy method, however, this problem is experimentally solvable. To this end, one should detect how the $2D_e$-line spectrum, corresponding to fractional filling fator, $\nu = 7/3$ for example, alters with varying temperature. The measurements were conducted for two spectral regions of the $2D_e$-line, indicated by arrows in Fig. 8. It is quite evident that with varying temperature the spectrum alters in a discontinuous manner, and the critical temperature, corresponding to this discontinuity is $T_C = 1.96$ K. This experiment directly demonstrates that the phenomenon of condensation to an incompressible Fermi-liquid is a phase transition and it is characterized by a critical temperature which is probably dependent on the magnitude of the gap.

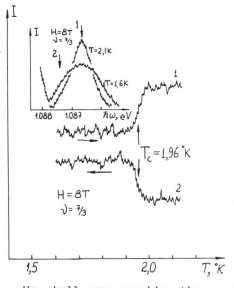

Fig. 8. Temperature dependences of $2D_e$-line intensities measured at spectra positions (1) and (2) indicated by arrows.

We shall now consider the question of how the value of the gap, measured spectroscopically, changes due to disorder, related to the random potential fluctuations.[17,18] To this end we measured the values of the Coulomb gaps in the region of $\nu = 7/3$, varying the mobility value that is the measure of disorder. It is seen qualitatively from Fig. 7b,c that a decrease of the electron mobility results in a decrease of the measurable value of the gap. The mobility dependence of the gap value for $\nu = 7/3$ (in Coulomb units), measured by means of themoactivated conductivity, $\tilde{W}$, and spectroscopically, $\tilde{\Delta}$, is presented in Fig. 9 in the coordinates $\tilde{W}$, $\tilde{\Delta}/\mu^{-1}$. As should be expected for fractions with $q=3$ $\Delta\xi=3(\Delta_e+\Delta_h) = 6W$ throughout the region of the measurement. Fig. 9 also presents an analogous dependence for the critical temperature $\tilde{T}_c/\mu^{-1}$. This dependence is well approximated by a linear law, namely

$$\tilde{\Delta} = \tilde{\Delta}_0 \ (1 - \mu_0/\mu), \tag{2}$$

here $\tilde{\Delta}_0$, is the Coulomb gap at $\mu \to \infty$ (the gap in the absence of disorder). μ_0 is a minimal mobility (when the Coulomb gap gets collapsed due to disorder). The dependences $\tilde{W}(\mu)$, $\tilde{T}_c(\mu)$ and $\tilde{\Delta}(\mu)$ are seen to yield approximately the same value for a minimal mobility coinciding with a graphical estimate by the formula $\mu_0 \approx (h/e)^3 \ (\varepsilon/m).$[18] So, spectroscopy methods offer new possibilities for the investigation of condensation phenomena in an incompressible Fermi-liquid.

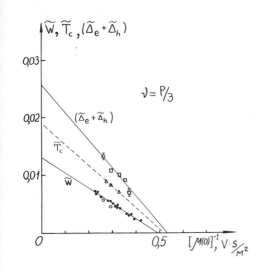

Fig. 9. Activation mobility ga-$\widetilde{W}$, Coulomb gap-$\widetilde{\Delta}$ and critical temperature -$\widetilde{T}_c$ made dimensionless in Coulomb energy units and measured as a function of reciprocal mobility for $\nu = p/3$.

References

1. T. Ando, A. B. Fowler, F. Stern, Rev. Mod. Phys., 54:437 (1982).
2. J. P. Eisenstein, H. L. Stormer, V. Narayanamurti, A. Y. Cho, A. C. Gossard, Phys. Rev. Lett., 55:875 (1985).
3. E. Gornik, P. Lassnig, G. Strasser, H. L. Stormer, A. C. Gossard, W. Weigmann, Phys. Rev. Lett., 54:1820 (1985).
4. T. P. Smith, B. B. Goldberg, P. J. Stiles, M. Heiblum, Phys. Rev. B 32:2696 (1985).
5. V. M. Pudalov, S. G. Semenchinskii, V. S. Edelman, Pis'ma Zh. Eksp. Teor. Fiz. (JETP Lett.), 41:265 (1985).
6. D. Weiss, K.von Klitzing, V. Mooser, Solid State Sciences, Springer-Verlag Series, 67:204 (1986).
7. I. V. Kukushkin, V. B. Timofeev, Pis'ma Zh. Eksp. Teor. Fiz. (JETP Lett.), 40:413 (1984); 43:387 (1986).
8. I. V. Kukushkin, V. B. Timofeev, Zh. Eksp. Teor. Fiz. (JETP) 92:258 (1987); 93: (1987).
9. B. I. Shklovskii, A. L. Efros, Pis'ma Zh. Eksp. Teor. Phys. (JETP Lett.), 44:520 (1986).
10. S. Hikami, E. Brezin, Surf. Sci., 170:262 (1986).
11. T. Ando, J. Phys. Soc. Japan, 52:1740 (1983).
12. D. C. Tsui, H. L. Stormer, A. C. Gossard, Phys. Rev. Lett., 48:1559 (1982).
13. R. B. Laughlin, Phys. Rev. Lett., 50:1395 (1983); Surf. Science, 142:163 (1984).
14. D. C. Tsui, Proc. XVII Int. Conf. Phys. Semic., San Francisco, 247 (1984).
15. H. Halperin, Helv. Phys. Acta, 56:75 (1983).
16. I. V. Kukushkin, V. B. Timofeev, Pis'ma Zh. Eksp. Teor. Fiz. (JETP Lett.), 44:179 (1986).
17. R. B. Laughlin, Yamada Conf. XIII EP2DS, Kyoto, 251 (1985).
18. I. V. Kukushkin, V. B. Timofeev, Zh. Eksp. Teor. Fiz. (JETP), 89:1692 (1985).

EXCITONIC OPTICAL NONLINEARITIES IN SEMICONDUCTOR QUANTUM WELLS

D.S. Chemla

AT&T Bell Laboratories
Holmdel, NJ 07733

ABSTRACT

In this article we review our experimental and theoretical investigations of the nonlinear optical processes seen at room temperature in GaAs quantum well structures photoexcited close to the fundamental absorption edge.

INTRODUCTION

High intensity optical excitation of semiconductor near the fundamental absorption edge produces several distinct nonlinear optical effects. Excitation below the gap, where only virtual electron-hole (e-h) pairs are generated, induces processes which are qualitatively different from those seen when real populations are created by above gap excitation. Furthermore, in the latter case it is important to distinguish the processes associated with thermalized e-h plasma from those due to nonthermal e-h plasma or to exciton populations. These nonlinear optical processes are also observed in semiconductor quantum wells (QW), however in these materials they are qualitatively and quantitatively different from those seen in bulk material. The difference originates from the reduced dimensionality of bound and free electron-hole (e-h) pairs which is induced by the confinement in the ultrathin layers. In this article we review our experimental and theoretical investigations of the nonlinear optical processes associated with quasi-two dimensional (2D) excitons in GaAs quantum well structures (QWS) excited below, at and above the lowest energy exciton resonances.

EXCITONIC RESONANCES IN QWS

For any physical process quantum size effects are expected to be seen whenever the dimensions of the sample becomes smaller or of the order of the characteristic length governing the quantum mechanics of the process. In the case of absorption in semiconductors, the length measuring the electron hole correlation is the exciton Bohr radius. Therefore in ultrathin layer whose thickness L_z is of the order of the bulk exciton Bohr diameter (e.g. $\approx 300\,A$ for GaAs) the exciton structure and hence of the absorption resonances are strongly modified.

The linear optical properties of III-V QWS have been reviewed recently [1,2] and indeed experiments show that the excitonic behavior is strongly modified in these structures. The excitonic binding and oscillator strength are enhanced because of the artificial reduction of the average distance between the electron and the hole. At low temperature, sharp and well resolved exciton resonances are observed at the onset of each intersubband transition [1,2]. Because of the reduced symmetry of the QWS the heavy hole (hh) and the light hole (lh) band give distinct excitons resonances. A surprising result of the increased exciton binding energy, which has important consequences in nonlinear optics, is the observation of well resolved excitonic resonances at room temperature (RT) in III-V QWS [3,4,5]. The first curve of Fig. (1) gives an example of the low excitation absorption spectrum of a $L_z = 96\,A$ GaAs QWS at RT.

The binding energy of excitons in GaAs QW approximately $100\,A$ thick is of the order of 10 meV. It is much smaller than the LO-phonon energy (36 meV in GaAs) and is also smaller than $k_BT = 25meV$ at RT. Thus at this temperature, the e-h bound states which are unstable against collision with LO-phonons are quickly ionized due to the relatively large population of thermal phonons. The life time of excitons at RT can be estimated from the temperature broadening of the resonance. It is found that in GaAs QW excitons live less than half a picosecond before releasing free e and h pairs with a substantial excess energy [3]. Theses free e-h pairs then thermalize very quickly among themselves and with the lattice.

NONLINEAR OPTICAL EFFECTS DUE TO REAL POPULATIONS

From the discussion of the previous paragraph we can conclude that when excitations long compared to the ionization time are used to probe the nonlinear optical response of QW, the nonlinearities depend only on the density of free e-h pairs created directly or indirectly by thermal-phonon ionization of selectively generated bound pairs. Experiments show that this is indeed verified and demonstrates that the nonlinear response is insensitive to the wavelength and to the duration of the excitation [1,6,7,8]. An example of this behavior is shown in Fig. (1) where the absorption spectra of a sample containing a density of e-h pairs increasing from $N = 0$ to $N = 2 \times 10^{12}\ cm^{-2}$ are presented. A smooth transition from a excitonic absorption profile to a 2D continuum absorption is seen. This particular set of spectra were taken one picosecond after excitation with a 100 fs pulse with photon energy above the resonances. Similar results are obtained at long delays (> 1ps) after ultrashort pulse resonant excitation or with long duration pulse excitation [1,3,7]. Even the very weak absorption in the band tail many linewidths below the hh-exciton peak produces the similar effects, once phonon-assisted transfer promote e-h pairs into the bands [9,10].

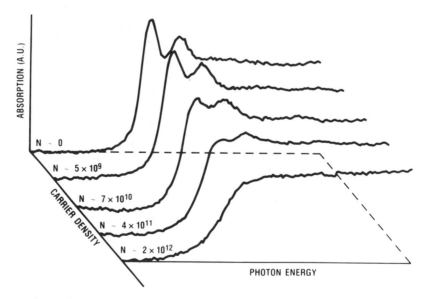

Figure 1 - Changes in the room temtperature absorption spectrum of a 65 periods of 96A GaAs and 98A Al.3Ga.7As layers QWS for increasing e-h plasma density as shown on the figure.

The optical nonlinearity associated with the generation of thermalized plasma is huge ($\chi \approx 0.06esu$) it corresponds to very large changes in refractive index and in absorption coefficient [1,3,7]. However these effects saturate easily and remain as long as the e-h plasma is present in the sample i.e. several nanoseconds. The theoretical explanation of the magnitude of these nonlinearities will be discussed hereafter.

The extreme sensitivity of the exciton to the presence of carriers was utilized to study the Urbach absorption tail and the phonon-sidebands [9,10]. Residual absorption as low as 0.01 cm^{-1} can be measured in 1 μm thick samples using a very weak probe centered at the excitons to monitor the transmission as a tunable and intense pump excites the sample well below the gap. It was found that at high temperature the absorption below the exciton peaks has indeed an exponential dependence on the pump photon detuning in agreement with the Urbach rule. However, as the temperature is lowered structures appear on the low energy side of the exciton resonance. They where identified as due to phonon assisted absorption with the one-phonon sideband threshold appearing clearly. The line shape is well described by a second order perturbation theory adapted to quasi-2D excitons [10].

The generation of e-h pairs (bound or unbound) modify the optical transitions through several mechanisms [phase-space filling (PSF), long range Direct Screening (DS) and short range Exchange Interaction (EI)]. PSF is the blocking mechanism which is the direct consequence of the Pauli exclusion principle, DS can be interpreted as a "classical" screening due to the Coulomb interaction and EI is the modification of DS due to the exclusion principle i.e. the antisymmetry of the wavefunctions. The effectiveness of these three mechanisms is different if the e-h pairs are free (plasma) or bounded (exciton gas) [11]. In three-dimensions (3D) the dominant effect of e-h pair photogeneration is the DS of free e-h pairs, EI is only important in very dense plasmas [11]. Screening by the neutral excitons is much weaker [12]. Excitonic PSF although properly included in correct manybody theories [11] has not been investigated experimentally in details. Because the consequences of generating e-h plasma and exciton gas are so different, it should be possible to time resolve the exciton ionization by measuring the time evolution of the absorption spectrum of a QWS after selective generated of exciton by resonant excitation. If QWS behave as bulk material, one should see first a very small bleaching of the exciton resonances when only exciton are present, followed by a strong increase of the bleaching as the exciton ionize into a charged e-h plasma. At RT to be able to excite and probe the QW samples in times short compared to the exciton ionization time it is necessary to utilize the recently developed femtosecond spectroscopy techniques [13].

The experiments give results absolutely opposite to those expected [14,15], as shown on Fig. 2 where are presented the differential absorptions spectra measured with a wide band 80fs continuum on a sample under excitation resonant with the hh-exciton by a 80 fs pump pulse [15]. A fast and extremely efficient transient beaching of hh-exciton resonance is seen for the first half ps. Then the resonance recovers and after about 1 ps the changes in the absorption spectra stabilize to the magnitude and

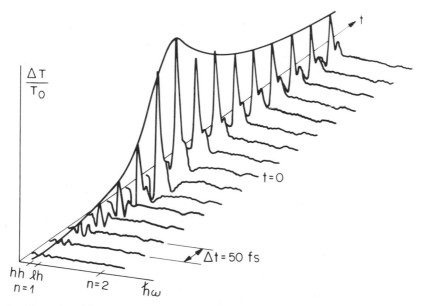

Figure 2 - Absorption differential spectra measured with a broad band (1.45-1.6 eV) 80fs continuum at various delays after the excitation with a 80fs pump pulse resonant with the hh- exciton. First the absorption at the peak of the exciton is very efficiently reduced then after the excitons are ionized by collisions with thermal phonons the spectrum evolve toward that measured with long duration excitations.

shape observed when e-h plasma are generated and remain constant for tenths of ps. The ionization time of the hh-excitons at RT deduced from these measurements is $\tau_i \approx 300$ fs. The most surprising results of these experiments is that the selective generation of excitons produce a much larger reduction of the excitonic absorption than the generation of free carriers. As mentioned above this behavior is in contradiction with theories [11] and experimental results [12] in bulk semiconductors.

The origin of the transient response of excitons in QW resides in the very special conditions in which excitons are in QW and at RT. They are generated with ultrashort optical pulses and observed before they can interact with the thermal reservoir are not in thermodynamic equilibrium with the lattice. They have not yet a "temperature" and occupy the lowest energy state of the crystal. Then their first interaction with the thermal LO-phonons destroys then and releases a warm plasma, which then thermalizes in less than a ps. However, the relative magnitude of the "excitonic" and "e-h plasma" bleaching imply that the mechanisms responsible are different in 2D and in 3D. The PSF and EI are governed by the occupancy of the states that enter in the exciton wavefunction as measured by U(k); the exciton relative motion wavefunction in k-space [11]. The overlap between the wavefunction of the real populations and U(k) is slightly different in 2D as compared to 3D, but the change in overlap in the two cases is too small to explain the experimental result. On the other hand there are indications from that DS is strongly reduced in 2D [16].

The relative strength of DS as compared to PSF and EI in quasi-2D systems was measured directly in the following manner [15]. Since PSF and EI are effective only when the wavefunction of the photo excited e-h pairs overlap with that of the excitons, a nonthermal carrier population generated in the continuum between the $n_z = 1$ and the $n_z = 2$ exciton resonances will first interact with these excitons through DS. Then as the carrier thermalize to the bottom of the bands the effects of the Pauli principle turn on at the $n_z = 1$ resonances. However, for the $n_z = 2$ excitons which are built up out of plane waves belonging to the $n_z = 2$ subband (almost orthogonal to the $n_z = 1$ subband) and are much higher in energy, the overlap should remain small. The switching of the PSF and EI has indeed been observed [15]. This is shown on Fig. 3 where are presented the differential absorption spectra seen at 50fs interval during and after the generation of a non-thermal carrier distribution in the $n_z = 1$ continuum less than one LO-phonon energy above the gap and at a density $N \approx 2 \times 10^{10} cm^{-2}$. The turning on of the absorption at the $n_z = 1$ resonance as the carrier thermalize is clearly seen. As expected the effects at the $n_z = 2$ excitons, which are only sensitive to the DS, do not vary in time showing that DS remain essentially constant as the carriers thermalize. These results put an upper limit on the magnitude of DS which is at least six time smaller than the sum (EI+PSF). They also provide direct informations on the carrier-carrier scattering in these important heterostructures. The thermalization time is $\tau_{th} \approx 150$ fs and the carriers leave the states in which they are created in time less than (but of the order of) 80fs. Recent investigations of the thermalization at much higher densities show that τ_{th} is a rather slow varying function of the concentration. It decreases to $\tau_{th} \approx 30$ fs i.e. about a factor 5 larger when the density of photocarriers is increased to $N \approx \times 10^{12} cm^{-2}$ [17].

A theory explaining the saturation of 2D-excitons by e-h plasma and by exciton gas has been developed using the many body formalism [18]. Assuming that the interband transition matrix element does not depend on the wave vector ($p_{cv}(k) \approx p_{cv}$), we can write the oscillator strength of the excitonic transition as,

$$f_x \approx |p_{cv}|^2 \sum_{k,k'} [1 - f_e - f_h] U(k) U^*(k') \tag{1}$$

The reduction of the transition probability due to PSF is described by the term $[1 - f_e - f_h]$. Whereas screening (DS+EI) modify the envelope wavefunction, U(k), because they change the way the electron and the hole interact. Since the effects of DS are small we will neglect them and only consider PSF and EI. Furthermore in the small signal regime we can add the effects of the two perturbations and write;

$$\frac{\delta f_x}{f_x} \approx -N \times [\frac{1}{N_s^{PSF}} + \frac{1}{N_s^{EI}}] \tag{2}$$

The PSF saturation density, N_s^{PSF} has a very simple physical interpretation. It measures the average occupation in k-space of the exciton envelope wavefunction and it has a very simple form that relates directly to its physical meaning [18],

$$\frac{1}{N_s^{PSF}} = \frac{1}{U(r = 0)} \times \sum_k [f_e + f_h] U(k) \tag{3}$$

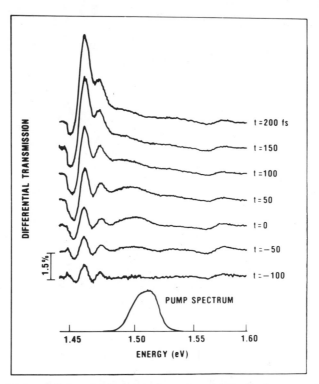

Figure 3 - Differential absorption spectra for a non-thermal carrier distribution generated by a 80fs pump centered at 1.509eV and observed at 50fs interval with a broad 80fs continuum. This resolves the effects of the exclusion principle from that of the direct Coulomb screening.

The EI saturation density N_s^{EI} describes the reduction in the e-h attraction due to the fact that like-particles with parallel spin avoid each other at short distances. This result in a change of the exciton envelope wavefunction, however the functional form of N_s^{EI} is not simple [18]. In any case both of these quantities originate from the exclusion principle and not surprisingly their effects relate directly to the overlap in k-space between the state occupied by the photoexcited e h pairs and the states out of which the excitons are formed. For a thermalized e-h plasma the occupation of the phase-space is given by the Fermi or the Boltzmann distributions according to the plasma temperature. For excitons it can be shown, by analyzing the multi-exciton Wannier equation [11], that the exclusion principle effects produced by N_x bound pairs are the same as those due to the special distributions of electrons and holes: $f_e(k) = f_h(k) = (N_x/2) \times |U(k)|^2$. The results of the theory are shown in Fig. 4. Here the saturation density is plotted as a function of the temperature in normalized units both for an exciton gas (dotted) line and an e-h plasma (solid line). They explain the most salient features seen experimentally i.e. that at low temperature the plasma is more efficient than the excitons gas, but that as the temperature increases the plasma efficiency decreases as $E_{1S}/k_B T$ and drops below that of the excitons around $k_B T \approx 2.5 \times E_{1S}$. The model also give an absolute value of the nonlinear cross section for thermalized e-h plasma and for exciton gas generation in very good agreement with experiments [14,15,18].

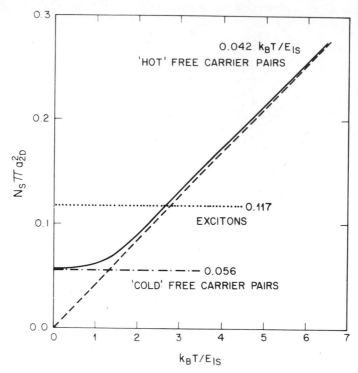

Figure 4 - Plot of the theoretical saturation density (normalized to the exciton area) for exciton gas (dotted line) and for a plasma (solid line) versus the temperature (normalized to the exciton binding energy). This show that the a cold plasma is more efficient than an exciton gas which is more efficient than a warm plasma.

NONLINEAR OPTICAL EFFECTS DUE TO VIRTUAL POPULATIONS

At low temperature and for excitation far enough below the resonance, so that only very few or no real transition occur, one observes a strong and transient change in the transmission of a weak probe beam tuned at the exciton resonance [19]. As shown in Fig. 5, its temporal as well as spectral behavior are very distinct from that of the weak background of e-h plasma effects that are induced by phonon assisted transitions. In fact for larger detuning of the pump photon energy the plasma effect disappear and only the instantaneous transient is observed [19]. This effect is also seen with femtosecond excitation[20]. Both for ps and fs it last only as long as the pump pulse is applied to the sample. The cause of this transient nonlinear effect has been identified as the Dynamical Stark Shift (DSS) of the ground state lh and hh-exciton resonances [19,20]. It is found experimentally [19] that the transmission change, which is proportional to the DSS of the resonances, is inversely proportional to the the pump detuning and proportional to the pump intensity.

A theory treating simultaneously the Coulomb interaction responsible for exciton formation and the field induced coherent interband polarization that corresponds to the absorption-emission of photons involved in virtual transitions, gives and excellent description of all the experimental results [21]. This theory can be interpreted in terms of a gas of non-ideal and weakly interacting of Bosons formed by the

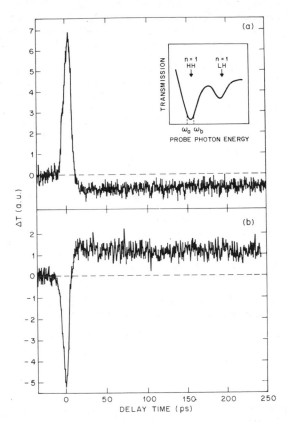

Figure 5 - Time resolved change of the probe beam transmission 1 meV below and 1 meV above the hh-exciton peak for an excitation 25 meV below the resonance. The fast response is due to the exciton AC Stark shift, the slow response to the effects of the e-h plasma created by phonon assisted absorption.

virtually created excitons. These Bosons, under the external symmetry breaking pump field, form a condensate with however significant anharmonic exciton-exciton and exciton-photon interactions [21,22]. The latter being due to the PSF discussed in the previous paragraph. The weak test beam probes the collective excitations of the condensate thus measuring "renormalized" resonances shifted from those of nonexcited sample by the DSS. The theory also predicts that the test beam stimulates a depletion of the condensate i.e. that a gain without population inversion should be seen at a photon energy symmetric from the exciton peak with respect of the pump photon energy. It includes as limiting cases Elliot's theory of excitons (when no pump is present) and Mollow's theory of two level systems (when the Coulomb interaction is neglected).

In the dilute limit the DSS takes a simple and physically meaningful form;

$$\delta E_{1S} \approx 2 \frac{|er_{cv}E_p|^2}{E_{1S} - \hbar\omega_p} \frac{|U_{1S}(r=0)|^2}{N_s^{PSF}} \tag{4}$$

The first term describes the "atomic" DSS associated with the valence to conduction band matrix element er_{cv}. In the second term, $|U_{1S}(r=0)|^2$ describes the enhancement of the oscillator strength due to the "pairing" of electrons and holes whereas N_s^{PSF} measures the maximum density of excitons allowed by the PSF. This functional form agrees with the behavior of the DSS measured experimentally. The magnitude of the DSS is also correctly given by Equ. (4) without any adjustable parameter. For example at a pump detuning of 31 meV and for a pump intensity of $I_p = 8MW/cm^2$ the experimental DSS of the hh-exciton is [19]; $\delta E_{1S} \approx \, '0.2mev$ as compared to the theoretical value [21] $\delta E_{1S} \approx \, '0.15 \, mev$. Furthermore, the ratio of the lh-exciton DSS to the hh-exciton DSS, which is independent of the pump intensity is found to be 4 experimentally as well as theoretically.

CONCLUSION

The nonlinear optical processes seen in semiconductor QWS present some specific features not seen in bulk material. The magnitude of the mechanisms responsible for exciton resonance bleaching are reversed by the quantum confinement and at RT the optical response exhibit unusual transient due to exciton instability to collision with the large density of thermal phonons. For below gap excitation a dynamical Stark shift of the ground state exciton is observed. This latter effect is well describes by a theory that treats the virtual excitons in close analogy with a non-ideal Bose condensate.

Acknowledgments: This work was performed in close collaboration with D.A.B. Miller and S. Schmitt-Rink.

We have extremely benefited from valuable contributions of W. H. Knox, A. Von Lehmen, J. S. Weiner, J. E. Zucker, J. P. Heritage, C. Hirlimann, J. Shah, C. V. Shank and H. Haug.

The high quality samples we have studied were expertly grown by A. C. Gossard.

REFERENCES

[1]   D.S. Chemla, D.A.B. Miller J. Opt. Soc. Am. **B2**, 1155 (1985).
[2]   R.C. Miller, D.A. Kleinman J. Lum. **30**, 144 (1985) and D.S. Chemla J. Lumin. **30**, 502 (1985).
[3]   D.A.B. Miller, D.S. Chemla, P.W. Smith, A.C. Gossard, W. Wiegmann Appl. Phys. **B28**, 96 (1982).
[4]   J.S. Weiner, D.S. Chemla, D.A.B. Miller, T.H. Wood, D. Sivco, A.Y. Cho Appl. Phys. Lett. **46**, 619 (1985).
[5]   H. Temkin, M.B. Panish, P.M. Petroff, R.A. Hamm, J.M. Vandenberg, S. Sunski, Appl. Phys. Lettl. **47**, 394 (1985).
[6]   D.A.B. Miller, D.S. Chemla, P.W. Smith, A.C. Gossard, W.T. Tsang, Appl. Phys. Lett. **41**, 679 (1982).
[7]   D.S. Chemla, D.A.B. Miller, P.W. Smith, IEEE J. Quant. Electron. **QE-20**, 265 (1984).
[8]   J.S. Weiner, D.S. Chemla, D.A.B. Miller, H. Haus, A.C. Gossard, W. Weigmann, C.A. Burrus Appl. Phys. Lett. **47**, 664 (1985).
[9]   A. Von Lehnen, J. E. Zucker, J.P. Heritage, D.S. Chemla, Appl. Phys. Lett. **48**, 1479 (1986).
[10]   A. Von Lehnen, J. E. Zucker, J.P. Heritage, D.S. Chemla, Phys. Rev. **B35**, 6479 (1987).
[11]   H. Haug and S. Schmitt-Rink, Prog.Quant.Electron. **9**, 3 (1984).
[12]   G.W. Fehrenbach, W. Schafer, J. Treusch, R.G. Ulbrich, Phys. Rev. Lett. **49**, 1281 (1982) and G.W. Fehrenbach, W. Schafer, R.G. Ulbrich, J. Lumin. **30**, 154 (1985)
[13]   See for example C.V. Shank and E. Ippen in this volume.
[14]   W.H. Knox, R.L. Fork, M.C. Downer, D.A.B. Miller, D.S. Chemla, C.V. Shank, A.C. Gossard, W. Wiegmann, Phys. Rev. Lett. **54**, 1306 (1985).
[15]   W.H. Knox, C. Hirlimann, D.A.B. Miller, J. Shah, D.S. Chemla, C.V. Shank, Phys. Rev. Lett. **56**, 1191 (1986).
[16]   T. Ando, A.B. Fowler, F. Stern, Rev. Mod. Phys. **54**, 437 (1982).
[17]   W.H. Knox and D.S. Chemla to be published.
[18]   S. Schmitt-Rink, D.S. Chemla, D..B. Miller Phys. Rev. **B32**, 6601 (1985).
[19]   A. Von Lehnen, D.S. Chemla, J. E. Zucker, J.P. Heritage, Optics Lett. **11**, 609, (1986).
[20]   A. Mysyriwicz, D.Hulin, A.Antonetti, A.Migus, W.T. Masselink, H. Morkoc, Phys. Rev. Lett. **56**, 2748, (1986).
[21]   S. Schmitt-Rink, D.S. Chemla, Phys. Rev. Lett. **57**, 2752 (1986).
[22]   S. Schmitt-Rink, D.S. Chemla, H. Haug to be published.

EXCITONS IN QUANTUM WELLS AND SHORT PERIOD SUPERLATTICES*

M.D.Sturge

Physics Department
Dartmouth College
Hanover NH 03755

ABSTRACT

Layer width fluctuations in quantum wells and superlattices produce a random potential which has a pronounced effect on the dynamics of direct and indirect excitons. The potential fluctuations broaden the direct exciton line into a band in which excitons of sufficiently low energy are localized, being separated from the delocalized ones by an effective mobility edge. Four-wave mixing, resonant Rayleigh scattering, hole-burning and and photon echoes have been used to detect the localization of direct excitons in quantum wells. The mobility edge is found to be at the position predicted by classical percolation theory. In the case of indirect excitons, as in alloys, the most pronounced effect of the disorder is on the decay kinetics, which are strongly affected by localization.

INTRODUCTION

New techniques of epitaxial crystal growth have made it possible to fabricate layered hetero-structures in which the interfaces are atomically flat. We will consider two such structures here: "quantum wells" (QW), and "superlattices" (SL). QW consist of a layer, typically ~100A thick, of a narrow gap semiconductor such as GaAs, sandwiched between "barrier" layers of a wider gap semiconductor such as $Al_xGa_{1-x}As$. The barrier layers in a QW structure are sufficiently thick that electrons and holes cannot tunnel through and are essentially isolated in the well; to increase the sensitivity of optical experiments one usually works with "multi-QW" structures in which there are many such wells, but for most purposes they can be treated as independent single wells.(Note, however, that they are coupled through the electromagnetic field, so that energy can be trans-ferred from one well to another). In a QW, motion normal to the layer is quantized and the motion of electrons and holes is two-dimensional (2D). Excitons formed from these electrons and holes are also 2D, both in their

* Work done at Bell Communications Research, Red Bank NJ,and at Bell Labs, Murray Hill NJ, in collaboration with J.Hegarty, L.Schultheis, E.Finkman, M.-H.Meynadier, L. Goldner, M.C.Tamargo, A.C.Gossard and W.Wiegmann.

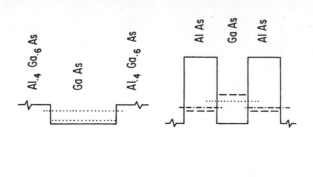

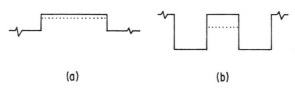

(a)   (b)

Figure 1.(a) Potential energy (full line) and confined
k=0 states (dotted lines) in a GaAs/Al$_x$Ga$_{1-x}$As quantum well.
The n=1 and n=2 Γ electron states and the n=1 hole state
are shown. The light-heavy hole splitting is too small to
see on this scale.

(b) Potential and k=0 states in a GaAs/AlAs short period superlattice. The full line is the potential at the Γ point; n=1 Γ
electron and hole states are shown dotted. The dashed lines show
the potential for electrons at the X point; the confined X electron
state is shown dot-dashed. The horizontal lengths of the lines
indicate the approximate spatial extent of the wave-functions.

internal structure and in their motion. In SL the barrier is sufficiently
thin that tunneling is possible, and motion is three-dimensional (3D) but
highly anisotropic. The distinction between a multi-QW structure and a SL
may not always be clearcut (since a particular structure might behave as
a QW for heavy holes but as an SL for electrons) but is convenient.

The general features of the optical spectra of QW are well understood
in terms of the effective mass approximation (EMA), in which electrons and
holes are assumed to have the same masses as in the bulk but are subject
to a one dimensional potential due to the difference in bandgap[1]. Band
mixing due to the high spatial frequency of this potential is neglected.
This model is illustrated for the usual case of abrupt interfaces (giving
a square well potential) in Fig.1a. Motion parallel to the layer is assumed
to be unaffected by the presence of the barrier. It has proven rather
difficult to determine what fraction of the band-gap difference is to be
assigned to the valence band (VB) discontinuity ("offset") and what to
the conduction band (CB); recent work[2] on the GaAs/Al$_x$Ga$_{1-x}$As interface has
shown that the VB offset is 0.55x eV, where x is the molar fraction of AlAs.
However, there are significant differences between the VB offsets obtained
by different techniques; these discrepancies may reflect failure of the
EMA at a fundamental level. The energy levels of QW are not sensitive to
the exact value of the band offset. On the other hand, those of short
period SL, to be considered below, are very sensitive; furthermore, it
appears to be necessary to go beyond the EMA to understand them.

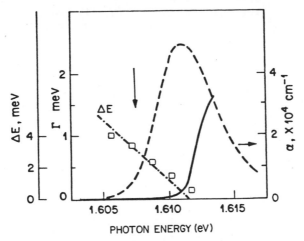

Figure 2. Absorption (dashed line) and homogeneous
linewidth Γ at 5K (full line) for excitons in a
GaAs/Al$_x$Ga$_{1-x}$As multi-quantum well sample with GaAs
layer thickness 51A. The points are the activation
energies ΔE for $\Gamma(T)$ at the exciton energies indicated.
The arrow indicates the position of the luminescence peak.

EXCITON DYNAMICS IN QUANTUM WELLS

While the energy levels of excitons in QW are well described on the
assumption that the interfaces are perfectly flat and abrupt, this is not
true of the exciton dynamics. Spatial fluctuations in the layer width
produce corresponding fluctuations in the exciton energy, so that the
exciton moves in a 2D random potential and the exciton band-edge is
broadened out. According to classical percolation theory, the exciton
states below the unperturbed band-edge are localized[3,4]. Those above it
are delocalized and are spread in k-space by scattering, so that their
wave functions have some k=0 character and they can be created in an
optical transition. The energy at which changeover from localized to
delocalized behavior occurs is called the "mobility edge". According to
current ideas[5] the ground state of a 2D system is always localized at
0K; however, delocalized excitons are never in their ground state, even if
radiative decay is neglected, since they can always decay to a localized
state by phonon emission.

We have studied the dynamics of excitons in GaAs/Al$_x$Ga$_{1-x}$As QW in a
series of experiments, whose conclusions will be summarized here. Four wave
mixing (transient grating) experiments[4,6] have shown that excitons with
energies above the center of the exciton absorption line propagate diffusively over distances of order microns within their subnanosecond lifetime,
while below the line center diffusion is much slower. Unfortunately such
direct measurements of exciton motion are difficult to make quantitative. A
more precise, if less direct, probe of exciton localization is the homogeneous linewidth Γ. Delocalized excitons are strongly scattered by the
potential fluctuations and have a temperature-independent $\Gamma \sim 1$ meV, while

for localized excitons Γ is limited only by radiative decay and by phonon scattering, so that Γ is strongly temperature dependent and tends to less than 10^{-2} meV at 0 K. We have measured Γ down to 1.5K, by hole-burning[7], photon echoes[8], and resonant Rayleigh scattering[9,10], for a number of QW with layer widths in the range 50 to 100 A. Of these techniques, only the last can be used at exciton densities sufficiently low ($< \sim 10^7$ cm^{-2}) that the results are unaffected by exciton-exciton scattering. On the other hand, it is less direct, the measurements of Γ are relative, not absolute, and there is no satisfactory theory for the low temperature, low energy region where $\Gamma < \omega_{LT}$, the polariton splitting.

In spite of these difficulties, the overall picture which emerges from these experiments is clearcut, except possibly in a narrow region around the mobility edge, for which experiments at lower temperatures are needed. Results for a 51A layer are shown in Figure 2. The prediction of classical theory is followed: the mobility edge is close to the center of the exciton absorption line. Above it $\Gamma > \sim$ 1 meV, while below it, over a certain temperature range, Γ at any particular energy shows Arrhenius behavior with an activation energy roughly equal to the exciton energy measured downwards from the line center. This suggests that the dominant dephasing mechanism of localized excitons is phonon-assisted activation to the mobility edge.

INDIRECT EXCITONS IN SHORT PERIOD SUPERLATTICES

We now turn to experiments on SL. To reduce the number of variables we confine our attention to "50:50" GaAs/AlAs SL, in which the well and barrier have (nominally) equal thicknesses. For SL periods >60A (i.e. layer thicknesses >30A) the EMA is found to give an accurate picture of the lowest exciton states; the exciton is direct, with both electron and hole predominantly in the GaAs layer. However, AlAs has an indirect gap, the lowest CB edge being at the X point (or near it), and for periods <60A the GaAs Γ electron state lies higher than the AlAs X state, as shown in Figure 1b[11]. (Note that the ordering of the levels is very sensitive to the VB offset). The lowest exciton is now indirect both in momentum and in real space, although for these thin layers penetration of the wavefunctions into the barrier layer (AlAs for holes, GaAs for electrons) is substantial. The absorption spectrum of such SL is still dominated by the direct Γ exciton, but this decays rapidly by phonon emission to the indirect X exciton, whose radiative decay, made allowed by various Γ-X mixing processes, dominates the luminescence at low temperatures.

In a cubic crystal there are three equivalent X minima, but in an SL the symmetry is reduced so that the "X_z" minimum, with its momentum parallel to the growth direction, is no longer degenerate with the "X_x" and "X_y" minima, whose momenta lie in the plane of the layer. The superlattice potential has a (001) component (momenta are expressed in units of $2\pi/a_0$) which mixes $X_z = (001)$, but not $X_x (100)$ or X_y (010), with Γ. It also has the effect of mixing X_x and X_y with each other, since it connects (100) to (101), which is equivalent to (010). Hence the degeneracy of X_x and X_y is split and the symmetric combination "X_{xy}" is lowest[12]. While the EMA predicts[13] that X_z will always be lower than X_{xy}, because the longitudinal

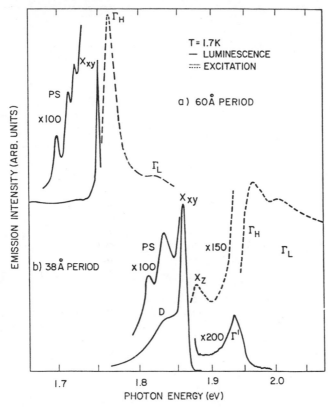

Figure 3. Luminescence (full line) and excitation
(dashed line) spectra for GaAs/AlAs superlattices
with periods (a) 60A and (b) 38A.

mass at the X-point is much larger than the transverse mass, we will see
that experimentally X_{xy} is lowest.

Figure 3 shows the excitation (equivalent to absorption) and
luminescence spectra of a SL with period 38A. The shoulder in the luminescence labelled D is sample dependent and we attribute it to defects or
impurities. At much higher gain two or three very weak lines can be seen,
labelled "PS"; these are phonon sidebands of the no-phonon line labelled
X_{xy}. In order to see them it is necessary to wait ~200nsec to allow the "D"
luminescence to decay. The main features in the excitation spectrum are
the light and heavy hole Γ excitons, Γ_L and Γ_H , while the weak line at
lower energy is most reasonably assigned to the X_z exciton, made allowed
by the superlattice potential. There is a very weak luminescence line
labelled Γ' associated with the Γ_H exciton, which is very short-lived
(<100psec), as is to be expected since this state rapidly decays to the
lower indirect state. The X_{xy} luminescence has a relatively slow decay,
shown by the points in Figure 4, as expected for an indirect transition.

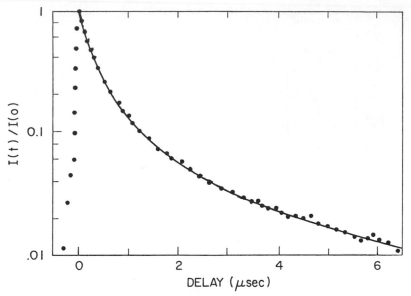

Figure 4. Time dependence of the decay of the X_{xy} luminescence after short pulse excitation (points). The line shows the theoretical prediction for purely random Γ-X mixing, with fitted mean rate w = 1.5×10^6 sec^{-1}.

The decay is non-exponential, indicating decay rates that vary over at least two orders of magnitude. This is to be expected for localized indirect excitons whose radiative decay is made allowed by random scattering, which in this case must come from disorder at the AlAs/GaAs interface. The theory of such decay has been worked out previously[14]; for excitons exactly at the X-point, with no non-random decay process occurring, the theory predicts the full line in Figure 4. Agreement is excellent and the non-random contribution to the decay must be less than 3×10^4 sec^{-1}, corresponding to a Γ-X mixing of less than 10^{-5}; this is quite inconsistent with the observed strength of the X_z absorption, so the luminescence must be from X_{xy}.

The average decay rate w varies as N_s/Δ^2, where Δ is the Γ-X separation and N_s the density of scattering centers[15]. Comparison of a series of SL with different periods but grown under similar conditions shows that N_s varies inversely as the period[11], confirming the hypothesis that scattering is at the interfaces. As growth techniques are improved and interfaces become more perfect it should be possible to achieve lifetimes at low temperatures of the order of milliseconds, ultimately limited by phonon-assisted decay.

Note that the excitons must be localized for the decay to be non-exponential. If they move an appreciable distance in their lifetime, their decay rates, which vary because of spatial fluctuations in the potential, will be averaged out and if the decay is still radiative exponential decay with rate w should occur. This has been observed in 3-D ternary alloys[15]. It has not so far been seen in SL, since delocalized indirect excitons readily migrate to non-radiative centers. As the temperature is raised, and the excitons are thermally excited to delocalized states, the intensity of the exciton luminescence decreases and the lifetime shortens due to non-radiative decay.

REFERENCES

1. R.C.Miller and D.A.Kleinman, J.Lumin. **30,** 520 (1985).
2. Reviewed by G.Duggan, J.Vac. Sci. Technol. **B3,** 1224 (1985).
3. See, for example, J.M.Ziman *Models of Disorder* (Cambridge University Press, 1979) p.485.
4. J.Hegarty and M.D.Sturge, J.Opt. Soc. Amer. **B2,** 1143 (1985).
5. P.W.Anderson, Physica **117/118B,** 30 (1983).
6. J.Hegarty, M.D.Sturge, A.C.Gossard and W.Wiegmann, Appl. Phys. Lett. **40,** 132 (1982).
7. J.Hegarty and M.D.Sturge, J.Lumin. **31/32,** 494 (1984).
8. L.Schultheis, M.D.Sturge and J.Hegarty, Appl. Phys. Lett. **47,** 995 (1985); L.Schultheis and J.Hegarty, J. de Physique **46,** C7-167 (1985).
9. J.Hegarty, M.D.Sturge, C.Weisbuch, A.C.Gossard and W.Wiegmann, Phys. Rev. Lett. **49,** 930 (1982).
10. J.Hegarty, L.Goldner and M.D.Sturge, Phys. Rev. **B30,** 7346 (1984).
11. E.Finkman, M.D.Sturge and M.C.Tamargo, Appl. Phys. Lett. **49,** 1299 (1986).
12. J.Ihm, Appl. Phys. Lett. **50,** 1068 (1987).
13. G.Duggan and H.I.Ralph , private communication (1987); M.-H.Meynadier, private communication (1987).
14. M.V.Klein, M.D.Sturge and E.Cohen, Phys. Rev. **B25,** 4331 (1982).
15. M.D.Sturge, E.Cohen and R.A.Logan, Phys. Rev. **B27,** 2362 (1983).

RECOMBINATION PROCESSES IN GaAs/AlGaAs MULTI-QUANTUM

WELL STRUCTURES

P.S. Kop'ev, V.P. Kochereshko, I.N. Uraltsev, and
D.R. Yakovlev

A.F. Ioffe Physico-Technical Institute
USSR Academy of Sciences
194021, Leningrad, USSR

INTRODUCTION

Using steady state photoluminescence (PL) techniques, we have studied recombination processes via excitons, impurities, and traps in undoped MQW structures grown by MBE.

I.   Exciton Localization

Layer width fluctuations in QW produce a random potential which exercises a pronounced effect on the exciton recombination at low power densities.[1]   The line width and Stokes shift analysis of the heavy hole exciton PL and excitation spectra of investigated MQW samples has given evidence for an island-like structure of the QW interface with 1 or 2 monolayer height and lateral extent higher than an exciton diameter.[2]

Magnetoluminescence techniques provides a precise probe of exciton localization. When the exciton wave function shrinks in the presence of a magnetic field normal to the QW layers, the exciton can be trapped by the islands of lower lateral extent. This effect gives us the opportunity of examining the distribution function of the extent of islands.

We found that the localized exciton PL line exhibits an additional shift in comparison to the diamagnetic shift of the delocalized exciton PL line which is shown in Fig. 1 by closed and open circles, respectively. The Stokes shift decrease induced by magnetic field has been analyzed using the exponential distribution function of island lateral extent.[3]   The lateral extent of the effective exciton trapping by islands has been evaluated for the first time by fitting with the exciton translation mass $M = 0.2m_0$ (Fig. 1, dashed line).

The shrinking of the wave function usually results in deeper localization, as is the case in alloys. Since the height of QW interface fluctuations is limited by one monolayer, the exciton localization energy decreases in the presence of a magnetic field. This permits development of a quantative method for characterizing the QW interface microstructure.[3]

II.   Exciton Formation

The effect extrinsic barrier properties on the exciton formation process has been studied at low excitation densities, where the strong competition between exciton and impurity related PL and nonradiative recombination occurs. Exciton recombination processes in QW are found to be suppressed by the influence of impurities spaced in barrier domains.[4]

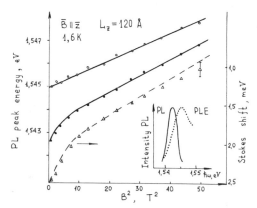

Fig. 1.   The energy positions of the heavy hole exciton PL line (solid curves) taken at excitation densities $W < 10$ W·cm^{-2} (full circles) and $W > 30$ W·cm^{-2} (open circles), when the PL line is related to localized and delocalized excitons, respectively, are shown as a function of square magnetic field B^2 applied normally to the QW layers. Stokes shift dependence on magnetic field (dashed curve) is calculated with $M = 0.2m_0$, a_{ex} ($B = 0$) = 120 Å and the island lateral extent $R = 300$ Å. The Stokes shifted PL line and excitation peak PLE spectral positions at $B = 0$ are shown on inset.

The efficiency of exciton recombination increases drastically under an additional pulse illumination by extremely low intensity light with $\hbar\omega_i > E_g$ (Fig. 2), where E_g is the bandgap of AlGaAs barriers. The threshold-like excitation spectrum of the exciton PL (Fig. 3) also gives evidence that the effect is associated with the influence of barrier impurities. The exciton PL is practically absent for selective photoexcitation of QW, i.e. at $\hbar\omega < E_g$, and enhances strongly when photoexcitation is absorbed into barrier domains, $\hbar\omega > E_g$. When the illumination pulse duration and/or intensity increases, the saturation of the effect is observed due to the limited number of residual barrier impurities. The effect of the illumination is then only revealed on the exciton line. It was not observed on the electron-acceptor PL line.

Using time-resolved spectroscopy we were able to study the decay curve of the effect shown in Fig. 4. The time decay values of the illumination effect were evaluated as $\tau_0 \sim 1$ ms, as expected for the tunnel relaxation time from deep impurities spaced in barrier domains.

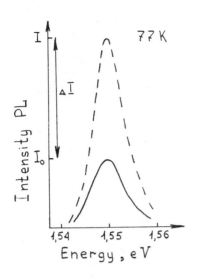

Fig. 2. Illumination-induced inten-
sity increase ΔI ($\hbar\omega_i$ = 2.41
eV) of the PL exciton line
(I_o) taken at excitation $\hbar\omega_o$
= 1.647 eV L_z = 100 Å.

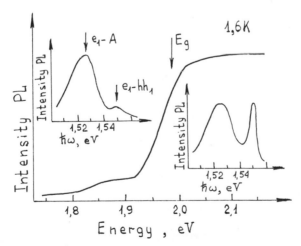

Fig. 3. Excitation spectrum of the e_1-hh_1 heavy hole exciton line at W =
100 µW·cm^{-2}. Insets show PL spectra taken at $\hbar\omega \gtrless$ Eg.

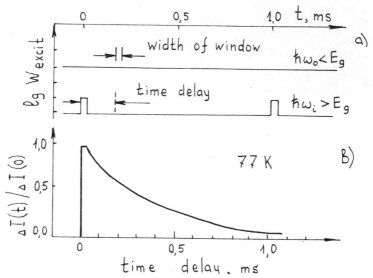

Fig. 4. Excitation power densities for the probe and illumination beams as a function of time (a) and decay curve of the illumination effect with $\tau_0 = 1$ ms (b).

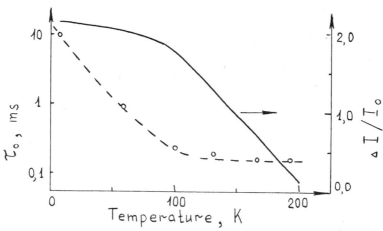

Fig. 5. Temperature dependences of the illumination effect quenching and τ_0 decay time are shown by solid and dashed lines, respectively.

The temperature dependence of the decay kinetics (Fig. 5, dashed line) is found to be in rough agreement with a deep level ionization process. The quenching of the illumination effect at T ~ 200 K (solid line) is accompanied by the appearance of the free carrier recombination.

To understand the effect of the potential due to barrier impurities on the QW exciton recombination we have determined the electron lifetime by an optical orientation technique. The increase of the Hanle shape half width from 0.38 to 0.80 T and of polarization degree from 0.36 to 0.46 have been observed under illumination. The electron lifetime decreases by 2.5 times and the exciton PL intensity increases by the same amount are clear evidence of the enhancement of the rate of exciton formation.

The barrier impurities produce a random potential in the QW which has a pronounced effect on the exciton formation rate. When filling of the deep impurity states occurs due to photoexcitation into barrier domains, the potential fluctuations disappear resulting in a domination of the exciton recombination process in QW.

III. Impurity Recombination

In a quantum well the binding energy of a carrier to an impurity markedly depends on whether the impurity is located in the center of the well or in the heterointerface.[5,6] As a result the recombination rate is governed by the impurity distribution function in the heterostructure, and by the optical response (overlap matrix element) of each individual impurity. Analysis of the impurity PL polarization is found to be a tool for non-destructive characterization of impurity profiles in heterostructure.[7]

Impurity PL, which is related to confined electron recombination with residual carbon acceptors, is observed at low excitation densities. Two maxima associated with recombination at the well acceptor A_c and the heterointerface acceptor A_i are revealed in PL intensity dependence in both temperature and excitation density.[8]

The magnetic field induced circular polarization degree is found to depend on acceptor position in QW (Fig. 6a). The values of polarization degree at the energy positions A_i and A_c are shown as a function of magnetic field B in Fig. 6b. To analyze these dependences we used the equation for bulk e-A recombination in magnetic field.[9]

$$\rho(B) = \frac{5}{4} \frac{\tau_h}{\tau_h + \tau_s} \, th \, \frac{\beta_0 g_A B}{kT}$$

where g_A - acceptor g-factor, β_0 = $eh/2m_0C$, τ - hole lifetime and τ_s - hole spin relaxation time on acceptor. The hole lifetime, as determined from the saturation of the polarization degree with B, turnes out to be 2 times shorter for an acceptor in the center of QW than for an interface acceptor.[7] When measuring hole lifetimes at PL intensity saturated regimes, we can obtain the electron recombination probabilities on the A_i and A_c. The dependence of the g-factor on the acceptor position in QW has been evaluated by a fitting procedure at low magnetic fields. The value g_{A_c} = 0.55 ± 0,05 coincides with that for the bulk acceptor,[10] as predicted in.[11] The value g_{A_i} = 0.30 ± 0,05 has been obtained for the first time.

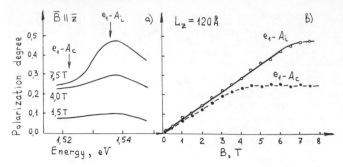

Fig. 6. The circular polarization degree as a function of (a) spectral position at B = 1.5, 4, 7.5 T and (b) magnetic field at energies of the e_1-A_i and e_1-A_c transitions shown by solid and dashed lines, respectively.

The dependence of the electron recombination rate on the acceptor position is found in optical orientation spectra. When the intensity of the e_1-A_i transition is saturated by the increase of excitation density and with e_1-A_i transition dominating, the degree of circular polarization becomes higher (Fig. 7), due to shortening of the electron lifetime. To determine the electron lifetime dependence on an impurity position we measure the half width of the Hanle shape under the same conditions. When a_1-A_i recombination dominates, the electron lifetime is found to be 4 times shorter, than in the case of e-A_c recombination. Both the acceptor distribution function through QW and the electron-acceptor transition probabilities govern the electron lifetime dependence so evaluated.

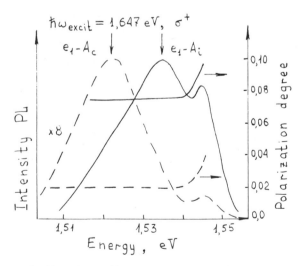

Fig. 7. Spectra of the impurity related PL lines and of the circular polarization degrees excited by σ^+ circular polarized light at $\hbar\omega$ = 1.647 eV with W ≤ 0.05 W·cm^{-2} (dashed) and W = 0.25 W·cm^{-2} (solid line).

Using the calculations of the acceptor binding energy as a function of impurity position in QW,[5,11] we can compare the spectral dependences of the e-A transition probability and the electron lifetime obtained by polarized PL and optical orientation techniques, respectively. So we can obtain the distribution function through the QW. We find the evaluated acceptor (carbon) profile, as shown in Fig. 8.

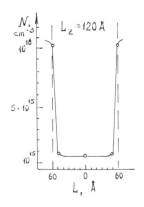

Fig. 8. Shallow acceptor (carbon) distribution profile through QW is nor-malized to residual impurity background N = 1.5 x 10^{15} cm^{-3} as measured on epitaxy layers grown under the same conditions. Heterointerfaces GaAs/Al$_{0.4}$Ga$_{0.6}$ as are shown by dashed line.

References

1. C. Weisbuch, R. Dingle, A. C. Gossard, W. Wiegmann, Sol. St. Com. 38:709 (1981).
2. P. S. Kop'ev, B. Ya. Meltser, I. N. Uraltse, Al. L. Efros, D. R. Yakovlev, JETP Lett. 42:402 (1985); in: Proc. 18th Int. Conf. Phys. Semicond., Stockholm, p. 219 (1986).
3. P. S. Kop'ev, I. N. Uraltsev, Al. L. Efros, D. R. Yakovlev, A. N. Vinokurova, Sov. Phys. Semicond. 22:N2 (1988).
4. P. S. Kop'ev, V. P. Kochereshko, I. N. Uraltsev, D. R. Yakovlev, JETP Lett. 46:N2 (1987).
5. G. Bastard, Phys. Rev. B 24:4717 (1981).
6. R. C. Miller, A. C. Gossard, W. T. Tsang, O. Munteanu, Phys. Rev. B 25:3871 (1982).
7. P. S. Kop'ev, V. P. Kochereshko, I. N. Uraltsev, D. R. Yakovlev, Sov. Phys. Semicond. 22:N2 (1988).?
8. Zh. I. Alferov et al., Sov. Phys. Semicond. 19:434 (1985).
9. M. I. Dyakonov and V. I. Perel, Fiz. Tverd. Tela 14:1452 (1972).
10. D. Bimberg, Phys. Rev. B 18:1794 (1978).
11. W. T. Masselink, Y.-C. Chang, H. Morkoc, Phys. Rev. B 32:5190 (1985).

PICOSECOND RAMAN AND LUMINESCENCE STUDIES OF CARRIER RELAXATION

IN QUANTUM WELLS

D. Y. Oberli, D. R. Wake, and M. V. Klein

University of Illinois at Urbana-Champaign (UIUC)
Materials Research Laboratory
104 South Goodwin Avenue
Urbana, Illinois 61801 USA

T.Henderson and H. Morkoç

UIUC: Coordinated Science Laboratory
Urbana IL 61801

ABSTRACT

Using two optical techniques, we have probed the dynamics of hot electrons in GaAs/(GaAl)As quantum wells. We have studied the relaxation of photoexcited electrons in undoped multiple quantum well samples using time-resolved luminescence and Raman techniques. By use of a synchroscan streak camera we have observed the cooling of the photoexcited plasma and its transformation into an exciton gas. A more precise probe of intersubband relaxation is provided by resonance Raman techniques. By tuning the photon energy of a time-delayed weak probe, we study the n=1 to n=2 or n=2 to n=3 intersubband transitions by electronic Raman scattering. The lifetime and the intersubband scattering time of the electrons photoexcited on the lowest electronic subbands are determined separately. Calculation of the rate of intersubband scattering by longitudinal acoustical phonons succesfully accounts for the observation of relatively long lived electrons on the second subband of a 215Å MQWS. When the well is narrower, so that the n=2 to n=1 transition can occur with emission of a longitudinal optic phonon, we find that the lifetime of carriers in the n=2 band is too short to measure with our technique.

INTRODUCTION

The photoexcitation of an electron-hole plasma in semiconductors is a powerful tool to probe the interactions of the carriers between themselves and with the lattice. Previous studies[1,2,3,4,5] of the dynamics of photoexcited carriers have shown that on a very short time scale (less than one picosecond), carrier-carrier scattering effectively randomizes the carrier momentum and establishes a quasi-equilibrium electron-hole plasma which subsequently cools towards the lattice temperature through the emission of phonons. While the mechanisms involved in the cooling of hot carriers are relatively well understood in bulk semiconductors,[6] some aspects of the relaxation of carriers in two-dimensional systems remain unclear. In semiconductor quantum well structures the motion of the carriers is confined in the direction normal to the layers by the higher potential of the barriers effecting a ladder of energy subbands. Earlier work has examined the effect of carrier confinement on the interaction of electrons and holes with phonons.[7,8] Time resolved absorption [9,10] and photoluminescence[11,12,13] have been the major optical techniques to study the dynamics of the carriers following photoexcitation and to deduce the cooling rates and the temperature of the plasma. However, these experiments do not directly determine the carrier densities of the plasma and they have not distinguished between the inter- and intrasubband acoustic phonon scattering of carriers. Acoustic phonon scattering is expected to be the dominant mechanism for intersubband electronic transitions if the electronic energy loss is less than the energy of a longitudinal optical (LO) phonon.

PHOTOLUMINESCENCE

In a sample with 220Å wells we pumped into the n=2 subband and have observed the intersubband relaxation from n=2 to n=1 take place inthe order of 300ps by correlating the observed shift in the luminescence peak with known positions of n=1 conduction band (C) to n=1 heavy hole band (HH) and n=2 C to n=2 HH transitions. Timing was provided by a tunable 5ps dye laser pulse and a Hamamatsu Synchroscan streak camera. The results are similar to those reported earlier.[12] After the carriers were essentially all in the n=1 subband, we could observe the luminescence linewidth narrow as they cooled and formed excitons.

TIME-RESOLVED RAMAN TECHNIQUE

In the remainder of this paper, we report on use of time-resolved Raman scattering to examine the mechanism of intersubband relaxation of electrons within a GaAs quantum well via longitudinal phonon emission. In order to determine the dynamics of the photoexcited carriers, we have developed a resonant-Raman scattering version of the pump and probe technique. In our approach we probe the inelastic light scattering of the electronic intersubband excitations of a two-dimensional plasma to look directly at the electrons.[14,15] We have observed both single-particle (SP) and collective intersubband excitations of the 2D-plasma. The collective excitations provide an independent and absolute calibration of the excitation density. Each SP intersubband excitation is characterized by a peak in the Raman spectra whose energy shift from the laser frequency is equal to the subband energy separation. The peak originates from a probe photon inelastically scattering an electron from the lower to the upper energy subband. The intensity of the SP peak is proportional to the difference of the carrier densities on the two subbands.

We used pulses of less than 5 ps duration from two independently tunable dye lasers synchronously pumped by a mode-locked Argon-ion laser at a repetition rate of 76 MHz. The temporal resolution was limited by jitter of the dye lasersto be about 8 ps. The probe-pulse energywas tuned in resonance to the optical gap separating one of the conduction subbands and the spin-orbit-split-off valence band (the $E_0+\Delta_0$ gap). Both beams were sent colinearly onto the sample at Brewster's angle, and the back-scattered light was collected normal to the sample and analyzed in a Spex Triplemate spectrograph. We measure both polarized $z(x',x')\bar{z}$ and depolarized $z(x',y')\bar{z}$ spectra. In this notation, x'=[110], while y'=[1$\bar{1}$0], and z=[001]. A key element of our technique is the large sensitivity and parallel detection provided by a charge coupled device detector.[16] The sample was held in a cryostat at 5K by a flow of cold helium gas. Pump fluence was typically $2 \mu J \cdot cm^{-2}$ per pulse, and the ratio of pump to probe laser fluences is kept close to 10. Pump and probe laser beams are respectively focused to a spot size of 50 and 25 μm and carefully centered with pin holes.

The GaAs-Al$_{.3}$Ga$_{.7}$As MQWS were grown by molecular beam epitaxy on a <100> GaAs substrate and consist of 30 periods of 100 Å of Al$_{.7}$Ga$_{.3}$As and either 215 Å or 116 Å of GaAs. The electron-hole plasma is generated by a near infrared laser pulse whose penetration depth varies from 1 to 2 μm.

RESULTS

Figure 1 shows how tuning the energy of the probe photon affects the spectra of the electronic SP-intersubband excitations. At a laser energy of 1.905 eV, the incident photon is in resonance with a transition between the split-off valence band and the second conduction subband. One observes one peak at 26.8 meV (216 cm^{-1}) in the depolarized spectrum which is attributed to SP-intersubband scattering involving vertical transitions of an electron between the first and the second subbands. In the second spectrum, an additional intersubband excitation, whose energy shift is 41.4 meV (334 cm^{-1}), is seen when the incident photon energy is tuned in resonance with the third electronic subband. Thus by selecting the resonance, one may preferentially enhance the sensitivity to transititions associated with particular subbands. A Krönig-Penney calculation based on the growth parameters allowed us to estimate the energy separation of the lowest electronic subbands. These results are shown in Table 1 for the two samples studied here.

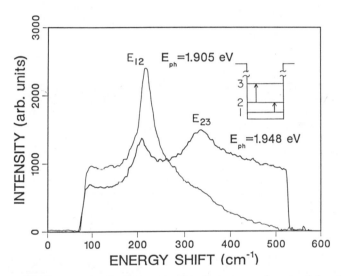

Fig. 1. Raman scattering spectra for two resonant photon energies at a delay time of 50ps. The scattering geometry is z(x',y')z̄.

Table 1. Comparison between experimental and calculated values of the energy separations of the lowest electronic subbands (n=1 refers to the ground subband; energies in meV).

Well Width	E_{12} (cal.)	E_{23} (cal.)	E_{12} (exp.)	E_{23} (exp.)	E_{13} (exp.)
215 Å	26.9	44.4	26.8	41.4	70.7
			(216 cm^{-1})	(334 cm^{-1})	(570 cm^{-1})
116 Å	72.3	107.9	64.2		
			(518 cm^{-1})		

The 41.4 mev spectral band is present only if the photon energy of the pump laser exceeds the threshold for exciting electrons directly into the second subband. This indicates that the electronic distribution probed by the second pulse is a nonequilibrium one photoexcited by the pump pulse.

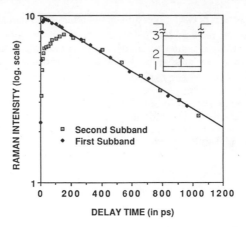

Fig. 2   Raman intensity of the SP-intersubband excitation at various times for two
exciting conditions (first subband: energy of the pump photon is 1.530 eV,
second subband: E_{ph}= 1.577 eV). The least square fit results in a lifetime of 750
± 25 ps in this 215 Å MQWS.

We now show how the intersubband electronic transitions can reveal the presence of
excited carriers on the second subband. Figure 2 presents the time evolution of the n=1 to
n=2 intersubband peak following excitation of electrons onto the two lowest subbands as we
change the time delay between the pump and probe pulses. Solid points correspond to the
excitation of electrons to the bottom of the first subband. After an initial fast rise, the
intensity decreases exponentially with a time constant of 750 ps for the 215 Å MQWS. This
decay time is the result of radiative and non-radiative recombination of the electron-hole
plasma. Any role of the plasma expansion is believed to be minimized by the choice of a
ratio of the focal spot areas of the pump and probe pulses equal to four. When electrons are
initially excited on both the first and the second subband the overall intensity is significantly
reduced during the first 300 ps (open squares). We attribute this reduction to the blocking
effect of the electronic distribution on the second subband. This indicates a relatively long
lifetime for the electrons on the second subband. As the electrons scatter down from the
second to the first subband the signal continues to rise. In a 215 Å quantum well, the
subband energy separation E_{12} (26.8 meV) is far less than the energy of an LO phonon (36.7
meV). For this reason, we do not expect the thermalized electrons of the second subband to
scatter down to the first subband with the emission of LO phonons. However, this would
not be the case for a narrower quantum well. We carried out a similar experiment in a 116 Å
MQWS, and tuned the probe laser photon into resonance with the second electronic subband.
In this narrow well, the intensity of the SP-intersubband excitation is not reduced when the
pump photon energy is raised above the excitation threshold of the second subband (see Fig.
3). We conclude that the electrons on the second subband initially excited by the pump pulse
are rapidly scattered to the lower subband by the emission of an LO phonon. The
spontaneous lifetime of the electrons in the 116 Å MQWS is also shorter: 400 ps. This
supports the idea of an enhancement of the radiative lifetime caused by the increased overlap
of the electron and hole wavefunctions when the size of the well is decreased.[12]

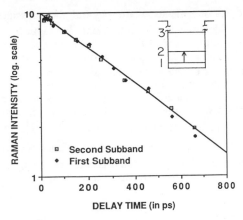

Fig. 3. Raman intensity of the SP-intersubband excitation at various times for two exciting conditions (first subband: energy of the pump photon is 1.556 eV, second subband: $E_{ph} = 1.656$ eV). The least square fit results in a lifetime of 400 ±10 ps in this 116 Å MQWS.

To further test our interpretation of the 300 ps time constant, we have probed the electron resonance at $E_{photon}=1.948$ eV in the 215 Å well sample (see Fig. 1). The intersubband excitation peak E_{23} is observable at this resonance if the energy of the pump photon is tuned to excite above the second subband. Since no electrons are excited onto the third subband, the intensity of this peak is simply proportional to the carrier density of the second subband. Fig. 4 indicates the time dependence of the peak intensity. A least squares fit to this data confirms a lifetime of 325 ps for the lifetime of the electron plasma on the second subband.

INTERPRETATION

The decay rate found in the plot of Fig. 4 is interpreted as the true lifetime of the electrons excited on the second subband of the 215 Å QW. This lifetime is related to the average intersubband scattering time τ_{12} as follows:

$$\frac{1}{\tau} = \frac{1}{\tau_{r2}} + \frac{1}{\tau_{12}}$$

where τ_{r2} is the radiative recombination time on the second subband. Assuming the recombination lifetime to be identical on both subbands, we deduce a value of τ_{12} equal to 570 ± 70 ps. Because of the larger effective mass of the holes, the relaxation time of the holes on the second subband is much faster. This will effectively increase the radiative lifetime of the electrons on the second subband and reduce the value of τ_{12}.

The transition rates for both electon scattering and phonon scattering have been calculated using Fermi's golden rule and the appropriate interaction matrix elements[17]. Because of the large momentum transfer associated with intersubband scattering, the strength of the electron-electron scattering is at least one order of magnitude smaller than the deformation potential scattering by the longitudinal acoustic phonons[18]. The acoustic phonon contribution to the intersubband scattering rate is inversely proportional to the square of the well width and in first approximation independent of the initial momentum of the electron (except for electrons whose phase velocity is less than the sound velocity of the LA phonons). Taking the deformation potential energy to be 7 eV[19], we calculated a value of τ_{12} equal to 490 ps. Considering the experimental uncertainty of the measured lifetime, the agreement with this value is excellent. This determination of the intersubband scattering time is unique, since it cannot be derived from transport measurements.

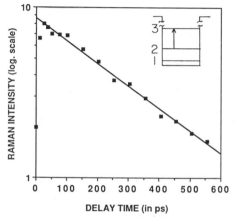

Fig. 4. Raman intensity of the SP-intersubband excitation E_{23} at various delay times. The least square fit results in a lifetime of 325 ±25 ps for the electron gas excited on the second subband of this 215 Å MQWS.

In conclusion, we have demonstrated that the carrier density of an electron plasma can be directly determined for arbitrary delay times following its excitation by using a Raman scattering version of the pump and probe technique. These measurements have revealed a relatively long life time of the electrons that are photoexcited on the second subband of a 215 Å MQWS. Intersubband scattering by the deformation potential of the acoustic phonons substantiates this result in the wide-well limit, while intersubband scattering by longitudinal optical phonons becomes dominant in the narrow-well limit.

One of us (D. Y. Oberli) would like to acknowledge fruitful discussions with J. P. Leburton. This work was supported by the National Science Foundation under DMR-82-03523 and 83-16981 and by AFOSR and JSEP.

REFERENCES

1. C. V. Shank, R. L. Fork, R. Yen, J. Shah, B. I. Greene, A. C. Gossard, and C. Weisbuch, Solid State. Commun. **47**,981 (1983).
2. C. V. Shank, R. L. Fork, B. I. Greene, C. Weisbuch, and A. C. Gossard, Surf. Science **113**, 108 (1982).
3. J. L. Oudar, A. Migus, D. Hulin, G. Grillon, J. Etchepare and A. Antonetti, Phys. Rev. Lett. **53**, 384 (1984).
4. D. J. Erskine, A. J. Taylor, and C. L. Tang, Appl. Phys. Lett. **45**, 54 (1984).
5. C. L. Tang, and D. J. Erskine, Phys. Rev. Lett. **51**, 840 (1983).
6. Jagdeep Shah, Journal de Physique, Colloque C7, 445 (1981).
7. C. H. Yang, J. M. Carlson-Swindle, S. A. Lyon, and J. M. Worlock, Phys. Rev. Lett. **55**, 2359 (1985).
8. J. Shah, A. Pinczuk, A. C. Gossard, and W. Wiegmann, Phys. Rev. Lett. **54**, 2045 (1985).
9. R. F. Leheny, J. Shah, R. L. Fork, C. V. Shank, and A. Migus, Solid State Commun. **31**, 809 (1979).
10. C. V. Shank, R. L. Fork, R. Yen, J. Shah, B. I. Greene, A. C. Gossard, and C. Weisbuch, Solid State Commun. **47**, 981 (1983).
11. J. F. Ryan, R. A. Taylor, A. J. Turberfield, Angela Maciel, J. M. Worlock, A. C. Gossard, and W. Wiegmann, Phys. Rev. Lett. **53**, 1841 (1984).
12. E. O Göbel, H. Jung, J. Kuhl, and K. Ploog, Phys. Rev. Lett. **51**, 1588 (1983). See also: R. Höger, E. O. Göbel, J. Kuhl, K. Ploog, and G. Weinmann, Proceedings of the 17th International Conference on the Physics of Semiconductors edited by J. D. Chadi and W. A. Harrison (Springer-Verlag, New York, Berlin,Heidelberg, Tokyo, 1985), p. 575.
13. Z. Y. Xu, and C.L. Tang, Appl. Phys. Lett. **44**, 692 (1984)
14. A. Pinczuk, J. Shah, A. C. Gossard, and W. Wiegmann, Phys. Rev. Lett. **46**, 1341 (1981).
15. A. Pinczuk, and J. M. Worlock, Surf. Science **113**, 69 (1982).
16. C. A. Murray, and S. B. Dierker, J. Opt. Soc. Am. A **3**, 2151 (1986).
17. P. J. Price, Annals of Physics **133**, 217 (1981).
18. Unlike for the case of intraband processes, piezoelectric scattering by the acoustic phonons is much weaker than the deformation potential scattering (at very low temperature).
19. D. L. Rode, Phys. Rev. B **2**, 1012 (1970). The 7eV value of the deformation potential has been recently questioned; see e.g. P. J. Price, Phys. Rev. B **32**, 2643 (1985).

GAIN SATURATION AND FEATURES IN THE THRESHOLD BEHAVIOR OF $A^{III}B^{V}$ HETEROSTRUCTURE LASERS

D. Z. Garbuzov and V. B. Khalfin

A. F. Ioffe Physico-Technical Institute, USSR Academy of Sciences, 194021 Leningrad, Polytechnicheskaya 26, USSR

ABSTRACT

Properties of semiconductor diode lasers have been improved drastically due to the development of $A^{III}B^{V}$ double heterostructures (DH) with very thin quantum well (QW) active regions. The research in this field now is in progress. Recently[1,2] it has been established that in distinction to conventional DH lasers the threshold current densities (I_{th}) of QW DH lasers with active region thicknesses d_a smaller than 200Å start to increase sharply with decrease of their cavity lengths (L) in the region $L < L_o$, where $L_o \leq (2-5) \times 10^{-2}$ cm.

Our investigations[3] of InGaAsP/GaAs QW DH lasers show that the whole value of I_{th} in short cavity lasers is determined only by the rate of radiative recombination in the active region as it is in lasers with longer cavities, i.e. increase of I_{th} in this case cannot be attributed to the rise of carrier leakage or the rate of nonradiative recombination.

In order to explain this and some other Features of the threshold behavior of QW DH lasers the calculations of the dependences of gain and radiative lifetimes on current densities have been performed for devices with different d_a. The calculations predict that the dependence of gain on current density has to become sublinear at the level of output losses typical for lasers with cavity length $L \leq L_o$. It should be noted that the influence of the complex structure of the A^3B^5 valence band and longitudinal hole masses modifications has been taken into account in a version of these calculations.

INTRODUCTION

This talk deals with the influence of gain saturation on the threshold behavior of single quantum well (SQW) separate-confinement (SC) heterostructure lasers.

Most of the studies of SC SQW laser threshold characteristics have

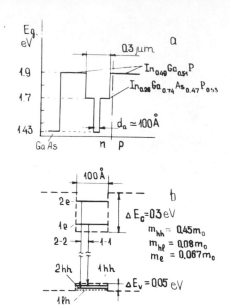

Fig. 1. a. Chemical composition and band gaps of heterostructure
layers. b. Energy diagram of the two-dimensional active region.

been carried out so far on AlGaAs/GaAs samples grown by MBE or MOCVD
techniques.

The lowest values of threshold current densities (J_{th}) for MBE and
MOCVD grown AlGaAs/GaAs SC SQW lasers were obtained by Tsang[1] and
Baldy et al.[2]. The dependences of J_{th} on the reciprocal cavity length
studied in Refs. (1,2) were interpreted by the authors as linear* just as
for the conventional DH lasers with thick active region.

When investigating SC InGaAsP/InP ($\lambda = 1.3$ μm) lasers prepared by a
modified variant of liquid phase epitaxy (LPE)[3-5] we observed a sharp
superlinear rise of thresholds with cavity length decrease[4,6]. It was
assumed then that the rise of J_{th} could be attributed to the enhancement
of Auger-recombination. However, we later found the same effect for
AlGaAs/GaAs SC SQW lasers grown by a similar LPE technique[7].

In 1986 Zory et al.[8] reported data especially devoted to this
problem. The superlinear increase of J_{th} with decrease of L was declared
by those authors to be a typical feature of threshold behavior of
AlGaAs/GaAs SC SQW lasers. Nevertheless, we believe that Zory's results,
as well as the data of our previous work[7], cannot be considered as
reflecting an intrinsic property of SC SQW lasers because the values of
J_{th} obtained there, even for the lasers with the longest cavities, are
three times higher than those reported by Tsang and Baldy.

It is evident that there should not be any nonradiative current

---

*It should be noted that after being replotted to a larger scale, the
dependence of $J_{th} = f(1/L)$ for the Baldy low threshold lasers (Fig. 24b
from Ref. 2) reveals a considerable deviation from linearity.

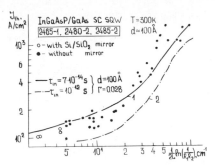

Fig. 2. Comparison of experimental data with calculated dependences of tje threshold current density on output losses α ($\alpha = 1/2L\ell n$ $(1/R_1 R_2)$). Curves 1 and 2 are calculated for $\tau_{in} = 7 \cdot 10^{-14}$ s and $\tau_{in} = 10^{-12}$ s, respectively.

leakages in the laser diodes chosen for the similar experiments.

One can see below that InGaAsP/GaAs ($\lambda = 0.86$ μm) SC SQW lasers studied in this work meet this fundamental condition. InGaAsP/GaAs heterostructures were grown by the above-mentioned variant of LPE in which growth of their thin GaAs-like active region ($d_a \approx 100$ Å) occurs when the substrate is being quickly moved under the melt placed in a special narrow slot[4,5,9]. The composition profile and band diagram of the InGaAsP/GaAs SC laser heterostructure are shown in Fig. 1a,b.

The photoluminescence (PL) studies of quantum size effects confirm that $\Delta E_c \geqslant 1/2\Delta E_g$ in these heterostructures[10]. PL studies have also revealed another very remarkable property of InGaAsP/GaAs DH´s: the internal quantum efficiency of radiative recombination for them remains extremely high, with the level close to 100% even for the well whose thickness is smaller than 100 Å [11,12]. A more detailed description of the results referring to undoped InGaAsP/GaAs DH can be found in a review paper published recently[12]. Some details concerning the technique of preparation of p-n junctions, ohmic contacts, and oxide stripe devices are given in Refs. 9 and 13.

Figure 2 demonstrates the dependence of J_{th} on the value of output losses for InGaAsP/GaAs SC SQW lasers ($\lambda = 0.86$ μm) at 300K. In order to obtain this dependence broad-area laser diodes with a wide range of cavity lengths were fabricated from several wafers with $d_a \approx 100$ Å. The dielectric (R $\approx$ 90%) Si/SiO$_2$ mirrors were deposited on the rear facets for several of these diodes.

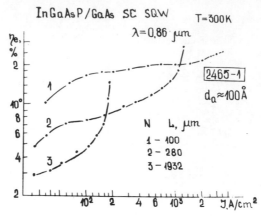

Fig. 3. External quantum efficiency for one mirror as a function of current density for three lasers with different cavity lengths.

When commenting on the data of Fig. 2 we should state primarily that injection lasers whose threshold current densities are not higher than 100 A/cm^2 at 300K were fabricated for the first time.

Further discussion will be devoted to another specific feature of these data — a sharp increase of J_{th} for the lasers with cavity lengths shorter than 0.5 mm. The data of Fig. 3 show that this superlinear increase of J_{th} does not result from enhancement of some kind of nonradiative current leakages. Indeed, one can see (curve 1) that the external quantum efficiency of spontaneous radiation (η_e) for the

nonlasing diode with the shortest cavity remains very high ($\eta_e \approx 2\%$) and nearly constant over all the range of current densities where generation is observed (curve 2, 3) for the lasers with longer cavities.

For QW laser diodes without current leakages the increase of current densities should lead to the rapid shift of quasi-levels and to appropriate changes in the shape of spontaneous spectra. As is seen from Fig. 4a the rise of current density will first result in spectrum broadening and then lead to the appearance of a second maximum in the higher energy region. Figure 4b shows that due to the reduction of cavity length and the rise of J_{th} the position of the generation peak also jumps to the same region of higher energies. The arrow in Fig. 4b refers to the laser with "intermediate" L for which generation is observed in both low- and high-energy regions.

In order to interpret quantitatively the experimental results described above, two models of the SQW heterostructure have been used. In the first model its active region has been considered as a potential well of finite depth (Fig. 1b) containing several parabolic subbands. Mixing of the valence bands has been taken into account when an infinite potential well has been considered within the scope of the second model[14]. The momentum selection rule is supposed to be fulfilled for these transitions. In order to take into account the broadening of the initial and final states of transitions due to intraband collisions the relaxation time (τ_{in}) has been introduced in the same way as it was previously done for conventinal DH lasers[15]. The absence of

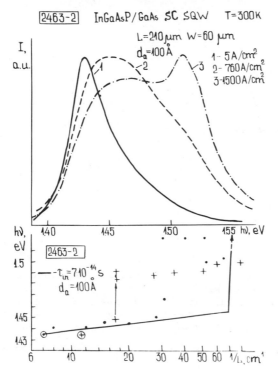

Fig. 4. a. Spontaneous radiation spectra for an oxide-stripe laser with a short cavity. Stripe width w = 60 μm. b. The position of the generation peak as a function 1/L. Points and crosses refer to sets of lasers obtained by cleaving two long cavity samples marked by circles. The arrow relates to a laser diode which has generation peaks in both energy regions. Calculated curve obtained with τ_{in} = $7 \cdot 10^{-14}$ s.

nonradiative leakages in the laser diodes under study makes it possible to state that their forward current is due entirely to the radiative transitions between electron and hole subbands in their active regions.

We shall consider primarily the results of calculations for the first model[16]. Figure 5a demonstrates the gain spectra for TE-polarization calculated for J = 0.8 kA/cm^2 with τ_{in} = ∞ (curve 1) and τ_{in} = $7 \cdot 10^{-4}$s (curve 2). The comparison with Fig. 5a shows that gain spectra have a structure similar to that typical of the spontaneous radiation spectra recorded at high ($\approx$ 1 kA/cm^2) current density. The result of calculation predicts that at J = 1 kA/cm^2 the electron quasi-level lies above the bottom of the second subband, and the gain peak corresponding to the transitions 2-2 (Fig. 1a) will dominate at higher current density. The decrease of current density in the range of J < 1 kA/cm^2 should lead to the narrowing of gain spectrum and only to the weak reduction of the amplitude of the first (1-1) gain peak.

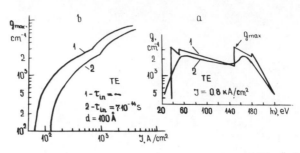

Fig. 5. a. Gain spectra and dependences g_{max} = f(J) calculated for τ_{in} = ∞ (curves 1) and τ_{in} = $7 \cdot 10^{-14}$ s (curves 2). Parameters of the well are shown in Fig. 1a.

The dependence of the value of the absolute maximum (g_{max}) of the gain spectrum on current density plotted in Fig. 5b reflects the above-mentioned features of the gain in the SQW lasers considered. As is seen from Fig. 5b, calculation predicts that the gain saturation for the 1-1 transition should be expected already in the range of J from

$J \gtrsim 200$–300 A/cm^2, while a contribution to g_{max} from transition 2-2

becomes considerable for $J \gg 1$ kA/cm^2.

Using the dependence g_{max} = f(J) it is easy to plot calculated

curves for J_{th} as a function of output losses (curves 1 and 2, Fig. 2). When calculating thse curves it was assumed that for separate confinement lasers all other kinds of losses are negligible and threshold conditions could be written in the simple form $\Gamma \cdot g = d$, where α and Γ are, respectively, the output losses and the optical confinement factor, which is equal to the fraction of the laser mode propagating within the active region. Using for our SC SQW laser with d_a = 100 Å the value of Γ calculated in Ref. (16) (Γ = 0.03) one can evaluate that in the operation of a laser with L_3 = 0.3 mm the gain in its active region should be not smaller than $\sim 10^3$ cm^{-1}. It is clearly seen in Fig. 5b that this value

of g corresponds to the region where the dependence g_{max} = f(J) is

saturated.

The saturation of gain is the origin of the superlinear rise of threshold current with decrease of L in the range L < 0.5mm (Fig. 2).

Curve 2 of Fig. 2 calculated for the value of τ_{in} = 7.10^{14} s gives the

best fit to the experimental dependence of threshold on output losses.

It should be noted that a calculation with the same τ_{in} leads to the

correct value for carrier mobilities in an electron-hole plasma[15].

A similar calculation makes it possible to plot the spectral position of g_{max} as a function of 1/L (Fig. 4b). The theoretically predicted value of 1/L at which the spectral position of the generation peak should be switched to the region of higher energy 2-2-transitions turns out to be larger than that experimentally observed. There are two

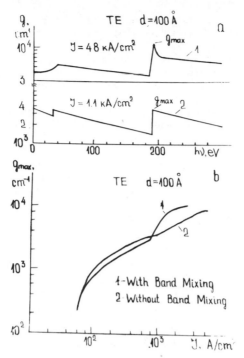

Fig. 6. a. Gain spectra for infinite well. Curves 1 and 2 – results of calculation for models with band mixing and for a simple parabolic subband model. b. The dependences g_{max} = f(J) calculated for the same models as curves 1 and 2 in Fig. 6a.

reasons which can explain this discrepancy: 1) in a real laser diode the total losses are always higher than the output losses; 2) as will be shown below, band mixing effects should lead to the relative enhancement of the 2-2 gain peak and, therefore, to a decrease in the value of 1/L corresponding to the switching of the generation peak.

Figure 6a demonstrates two gain spectra calculated for an infinite well with parabolic and mixed valence subbands. The difference between these spectra becomes considerable only in the spectral region of 2-2 transitions. Our calculation of band-mixing effects shows that as a result of the interaction of 2_{hh} and 1_{lh} subbands, the region of states with negative mass should originate near the bottom of the 2_{hh}-subbands. It is evident that due to this effect narrowing and enhancement of the 2-2 gain peak should occur.

The same effect is responsible for a divergence at high current density in the dependences g_{max} = f(J) (Fig. 5b) calculated for those two models which refer to the spectra in Fig. 6a. In the range of lower current densities where saturation of gain for the 1-1 transition takes place, the dependences compared practically coincide. Therefore, the band-mixing effects should not influence considerably the features of the SQW laser threshold behavior discussed above.

Summing up, fabrication and studies of SC SQW InGaAs/GaAs lasers make it possible to establish the following:

1. The increase of output losses leads to a superlinear rise of threshold and a step-like jump of generation peak to the higher energy region.

2. The lowest threshold density ($\approx$ 100 A/cm^2, 300K) can be obtained for the long cavity samples with small values of output losses.

3. The superlinear rise of thresholds with increase of ouput losses does not result from the enhancement of any kind of nonradiative current leakages.

4. It is the saturation of gain for transitions between the first subbands and jump of gain maximum to the second subband transitions that accounts for the threshold behavior features considered above.

In our opinion these features should be intrinsic for all types of $A^{III}B^{V}$ SQW lasers with well thickness smaller than 100 Å.

The authors wish to thank their co-authors whose works were referred to in the present paper.

REFERENCES

1. W. T. Tsang, Appl. Phys. Lett. <u>40</u>, 217 (1982).
2. M. Baldy, S. D. Hersee, and P. Assenat, Rev. Tech. THOMSON-CSF <u>15</u>, 5 (1983).
3. E. A. Rezek, N. Holonyak, V. A. Vojak et al., Appl. Phys. Lett. <u>31</u>, 288. (1977); E. A. Rezek, Jr., V. A. Vojak, R. Chin, N. Holonyak, and E. A. Samman, J. Electron Mater. <u>10</u>, 255 (1981).
4. D. Z. Garbuzov, I. N. Arsent´ev, V. P. Chalyi, A. V. Chudinov, V. P. Evtikhiev, and V. B. Khalfin, Fizika i Tekhnika Poluprovodnikov, <u>18</u>, 2041 (1984); [Sov. Phys. -Semicond. <u>18</u>, 1272 (1984).]
5. Zh. I. Alferov, D. Z. Garbuzov, I. N. Arsent´ev, B. I. Ber, L. S. Vavilova, V. V. Krasovskii, and A. V. Chudinov, Fisika i Tekhnika Popuprovodnikov <u>19</u>, 1108 (1985); [Sov. Phys. -Semicond. <u>19</u>, 679 (1985)].
6. V. P. Evtikhiev, D. Z. Garbuzov, Z. N. Sokolova, I. S. Tarasov, V. B. Khalfin, V. P. Chalyi, and A. V. Chudinov, Fisika i Tekhnika Popuprovodnikov <u>19</u>, 1420 (1985); [Sov. Phys. -Semicond. <u>19</u>, 8o73 (1985)].
7. Zh. I. Alferov, D. Z. Garbuzov, V. V. Krasovskii, S. A. Nikishin, D. V. Sinyavskii, and A. V. Tikunov Pis´ma v Zhurn. Tekhn. Fiz. <u>11</u>, 1409 (1985).
8. P. S. Zory, A. R. Reisinger et al., Electron Lett. <u>22</u> 475 (1986).
9. Zh. I. Alferov, I. N. Arsent´ev, L. S. Vavilova, D. Z. Garbuzov, and V. V. Krasovskii, Fizika i Tekhnika Poluprovodnikov <u>18</u>, 1655 (1984); [Sov. Phys. -Semicond. <u>18</u>, 1035 (1984)]; Zh. I. Alferov, I. N. Arsent´ev, L. S. Vavilova, D. Z. Garbuzov, A. V. Tikunov, and E. V. Tulashvili, Pis´ma v Zhurn. Tekhn. Fiz. <u>11</u>, 205 (1985).
10. I. N. Arsent´ev, N. Yu. Antoshkis, D. Z. Garbuzov, V. V. Krasovskii, A. V. Komissarov, and V. B. Khalfin, Fizika i Tekhnika Poluprovodnikov <u>21</u>, 178 (1987).
11. N. Yu. Antoshkis, I. N. Arsent´ev, D. Z. Garbuzov, V. P. Evtikhiev, V. V. Krasovskii, A. V. Chudinov, and A. E. Svetlokuzov, Fizika i Tekhnika Poluprovodnikov <u>20</u>, 708 (1986).
12. Zh. I. Alferov and D. Z. Garbuzov, 18th International Conference on the Physics of Semiconductors, Stockholm, Sweden (1986), p. 136.
13. Zh. I. Alferov, N. Yu. Antoshkis, I. N. Arsent´ev, D. Z. Garbuzov, V. V. Krasovskii, A. V. Tikunov, and V. B. Khalfin, Fizika i Tekhnika Poluprovodnikov <u>21</u>, 162 (1987).

14. M. I. Dyakonov and A. V. Khaetskii. Zhur. Eksp. i Teor. Fiz. 82, 1584 (1982).
15. M. Yamada and H. Ishiguro, Jap. J. Appl. Phys. 20, 1279 (1981); M. Yamada and Y. Suematsu, J. Appl. Phys. 52, 2653 (1981).
16. D. Z. Garbuzov, A. V. Tikunov, and V. B. Khalfin, Fizika i Tekhnika Poluprovodnikov 21, 1085 (1987).

SINGLE-FREQUENCY DIODE LASER WITH A HIGH TEMPERATURE STABILITY OF THE EMISSION LINE

Zh. I. Alferov, S. A. Gurevich, E. L. Portnoy, and F. N. Timofeev

A. F. Ioffe Physico-Technical Institute, Academy of Sciences
of the USSR, 194021 Leningrad, Politekhnicheskaja 26, USSR

ABSTRACT

Distributed Bragg reflector (DBR) injection lasers are considered as a promising light source for fiber optical communication systems and integrated optics. Here we report on the fabrication and investigation of a monolithic-hybrid DBR laser – heterostructure laser with DBR on sputtered dielectric waveguide.

In the monolithically-integrated hybrid (MIH) DBR diode laser, the five-layer Ga(Al)As heterostructure waveguide of the gain region was monolithically butt-joined on a common GaAs substrate with a highly-transparent corrugated dielectric-film waveguide consisting of sputtered SiO_2, Ta_2O_5, and evaporated (corrugated) As_2S_3 layers. The laser operated with the first-order grating under the pulsed current pumping at 300K. The efficient resonant mode conversion (70 percent in power) has been obtained at the interface between the heterostructure and dielectric waveguides. The fundamental transverse and single longitudinal mode output emission was obtained up to 160 mW (I_{th} = 120 mA) with external differential quantum efficiency $\eta \simeq 32$ percent.

The advantages of the dielectric-film waveguide DBR are demonstrated. The use of such DBR results in a high degree of sidemode suppression and stability of the average spectral position of the emission line under the temperature variation, the corresponding spectral shift being $\simeq 0.01$Å/K.

I. INTRODUCTION

Distributed feedback (DFB) and distributed Bragg reflector (DBR) heterostructure diode lasers containing corrugated optical waveguides are superior to conventional types of semiconductor lasers in spectral purity and modal stability. A number of DFB and DBR diode laser structures have been suggested and investigated. Recently many efforts have been made to achieve low threshold current, high external differential quantum efficiency, and adequate side-mode suppression in such single-frequency lasers. In practice, the temperature stability of the emission line is an important laser parameter. In this paper we show that a considerable improvement of heterostructure laser characteristics, especially of the temperature stability of emission line, can be obtained by the use of the

113

so-called monolithically integrated hybrid (MIH) DBR heterostructure lasers[1]. In these lasers the light amplification takes place in the semiconductor part of the heterostructure, that is in the heterostructure waveguide, while the distributed reflection is offered by a corrugated dielectric-film waveguide. This waveguide deposited on a common laser substrate serves as a very effective and highly stable DBR.

II. FABRICATION

The MIH DBR heterostructure diode laser is schematically shown in Fig. 1. One laser mirror is a DBR while the other (right in Fig. 1) is a conventional cleaved mirror. In the amplification area, a multilayer AlGaAs–GaAs heterostructure is used. In this structure a thin p-GaAs active layer is placed in the center of a symmetrical heterostructure waveguide (HW), containing n-$Al_{0.15}Ga_{0.85}As$ layers. This waveguide provides preferential excitation of the fundamental transverse mode. The DBR was formed on a GaAs heterostructure substrate by successive deposition of an SiO_2 (buffer), Ta_2O_5 (waveguide) film and of additional thin

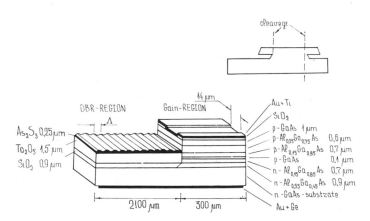

Fig. 1. Schematic diagram of a MIH DBR laser.

As_2S_3 layer. The SiO_2 and Ta_2O_5 films were deposited by reactive d.c. magnetron sputtering, and As_2S_3 by thermal evaporation. The corrugation was etched in the photosensitive As_2S_3 layer, providing a periodic perturbation of the whole SiO_2–Ta_2O_5–As_2S_3 dielectric waveguide (DW).

The mesa-structure of the gain area was formed using the micro-cleaving technique[2] illustrated by the inset in Fig. 1. During the process, the upper layers of the heterostructure were first removed stopping at the surface of the $Al_{0.5}Ga_{0.5}As$ sublayer. Then, using a selective etchant the sublayer was removed. As a result of additional treatment in this etchant the cantilevers were formed. The cantilevers were cleaved away to form the perfect and vertical side wall of the heterostructure mesa which makes it possible to obtain the efficient butt-coupling of HW and DW. To achieve efficient matching of HW and DW, the thickness of DW layers should be appropriately chosen and kept uniform up to the mesa edge. This important requirement was met by a certain choice of sputtering conditions.

III. MIH DBR LASER PERFORMANCE

For MIH DBR laser to operate in a single frequency regime a single transverse mode should be excited in DW. In our experiments the thicknesses of both HW and DW were chosen to be comparatively large, about 1.5 μm, to minimize the effect of tolerances on waveguide matching. Thus, both waveguides were multimode. As we have mentioned above, the fundamental transverse mode was preferentially excited in the HW due to the active layer placed in the center of the structure.

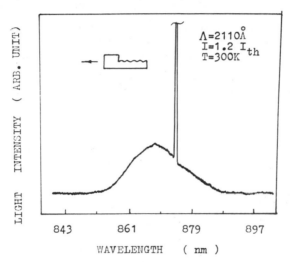

Fig. 2. Output emission spectrum of a MIH DBR laser operating with the first-order grating at pumping level 1.2 above threshold.

The analysis of HW and DW mode matching conditions shows[3] that the excitation of the fundamental mode of the DW by the fundamental mode of the HW, which takes place at the boundary, is not efficient due to the small overlap integral of these two modes' field patterns. In contrast, the resonant excitation of the second mode of DW can be obtained by the fundamental mode of HW. This kind of mode conversion was realized in the experiment by means of a careful control of DW layer thicknesses during its deposition. The measured power conversion coefficient was about 0.7, which is close to the theoretical limit.

The optical loss measured in the SiO_2-Ta_2O_5-As_2S_3 waveguides was about 1 dB/cm. Also, the film deposition processes used to fabricate the DW ensured a high degree of layer thickness uniformity. The use of such highly regular and transparent waveguides in the DBR region, alongside with the resonant mode conversion and first order distributed Bragg reflection, permits one to minimize the optical losses in the MIH DBR laser resonator and achieve highly efficient single-frequency generation. The first-order grating period in the DBR region was 2110 Å.

The emission spectrum of the MIH DBR laser operating in such conditions at room temperature is shown in Fig. 2. This single-frequency emission spectrum is characteristic of a MIH DBR laser. In this spectrum, the excitation of side-modes, which are the longitudinal modes of a gain region cavity, is not observed even on the amplitude scale of the spontaneous emission line. This result is evidence of the high reflectivity of the Bragg mirror. The Bragg mirror reflection coefficient R_B was obtained from the expression:

$$R_B = R_c \exp\left[\left(2\alpha_i h + \ln \frac{1}{R_c R_s}\right)\left(1 - \frac{\eta_1}{\eta_2}\right)\right] , \qquad (1)$$

where R_c and R_s are the power reflection coefficients at the waveguide discontinuity and at the main output cleaved mirror, respectively, α_i is the internal loss in the gain region, L is the gain region cavity length, η_1 and η_2 are the external differential quantum efficiencies of the laser operating which the corrugated DW DBR and regular DW, respectively. Substituting in (1) the values of all these parameters measured in the experiment one can find $R_B \approx 0.52$. Because of the obtained value of $R_B > R_s \approx 0.33$, the threshold current of MIH DBR laser was considerably decreased with respect to that of a conventional Fabry-Perot laser with cleaved mirrors.

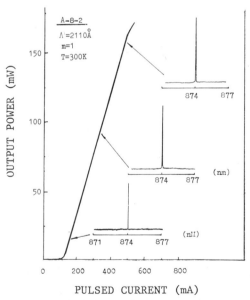

Fig. 3. Power-current characteristic and output spectra of the MIH DBR laser.

Figure 3 shows the light-output vs. current characteristic of the MIH DBR laser. This characteristic is linear up to the maximum output power of about 160 mW. From this characteristic the quantum efficiency was measured to be $\eta_1 = 0.32$.

Probably the most interesting feature of the MIH DBR laser is the temperature variation of its emission spectra. Figure 4a shows the example of the spectrum of spontaneous emission transmitted through the corrugated DW (in this particular laser sample the HW and DW were deliberately mismatched). The two dips which are the Bragg reflection stop bands of the corrugated waveguide are clearly observed. It is obvious that the laser emission line should be observed within the limits of the Bragg stop-band. The temperature variation of the spectral position of the Bragg stop-band centers measured in the 203-313 K range is shown in Fig. 4b. The average rates of the temperature shifts have been determined from the slope of the least-square fitted straight line drawn for each stop-band. These rates were smaller than 1.6×10^{-2} Å/K. This same quantity characterizes the upper limit of the laser emission line temperature shift rate. The extremely high temperature stability of

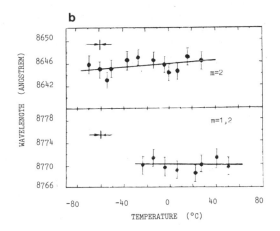

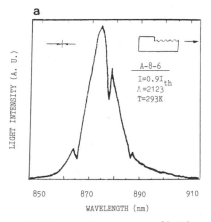

Fig. 4(a) Spectrum of the spontaneous radiation trasmitted through the corrugated dielectric waveguide (recorded below threshold); (b) the dependence of spectral position of Bragg stop-bands centers on temperature.

MIH DBR lasers, two orders of magnitude better than in ordinary semiconductor diode lasers, is due to a weak dependence of the corrugated DW
parameters on temperature. The reason for this is the fact that in a DW
the temperature variation of the layer refractive index is just compensated by the temperature variation of the corrugated period caused by the
thermal expansion of the substrate.

REFERENCES

1. S. A. Gurevich, S. I. Nesterov, E. L. Portnoy, V. I. Skopina, and F.
   N. Timofeev, Sov. Tech. Phys. Lett. 9, 4, 196 (1983).
2. S. A. Gurevich, E. L. Portnoy, N. V. Pronina, and V. I. Skopina, Sov.
   Tech. Phys. Lett. 8, 2, 83 (1984).
3. S. A. Gurevich, S. Yu. Karpov, S. I. Nesterov, E. L. Portnoy V. I.
   Skopina, and F. N. Timofeev, Sov. Tech. Phys. Lett. 11, 5, 218
   (1985).

RAMAN STUDIES OF PHONONS IN GaAs/AlGaAs SUPERLATTICES

M. V. Klein*, T. A. Gant, D. Levi*, and Shu-Lin Zhang[†]

Department of Physics and Materials Research Laboratory
University of Illinois at Urbana-Champaign (UIUC), 1110 W. Green
Urbana IL 61801 USA

T. Henderson, J. Klem, and H Morkoç

Department of Electrical Engineeering and Coordinated Science Lab.
(UIUC), Urbana IL 61801

ABSTRACT

The superlattice most studied has been that of GaAs/AlGaAs. Longitudinal phonon modes propagating parallel to the (001) growth direction have very simple properties that are shown both by calculations using simple models and by experiments. The acoustic modes are only weakly affected by the superlattice layering. Their dispersion curves are obtained by folding of those of the virtual crystal, and the phonon wavefunctions are nearly those of a single plane wave. For the nonresonant case, Raman activity is produced by the superlattice effect on the electronic structure, especially on the photoelastic constants. In resonance, complexity of the electronic structure is revealed. The optical modes are well-described as confined in the case of a square wave concentration profile. The effects of changing this profile by anealing on both acoustic and optical modes are studied. In the former case, only the intensities of the folded phonons change, in aproximate agreement with the photoelastic model. In the optical case, we have successfully explained the observed frequencies of the confined modes by an effective-mass-like theory for phonons.

INTRODUCTION

Techniques such as molecular beam epitaxy have stimulated much research on semiconductor heterostructures. Of the semiconductor superlattices, those from GaAs-AlGaAs have been the most studied. Major attention has been focussed on the electronic effects, many of which can be understood using effective mass theory. In this theory the band extrema act like effective potentials for the motion of the carriers with their appropriate effective masses. When the AlGaAs barriers are sufficiently thick, the GaAs wells are effectively decoupled for low energy electrons and holes, and one speaks of carrier "confinement" in the GaAs layers.

Our concern here is the effect of superlattice layering on phonons for the simple case where the propagation direction is normal to the interfaces. Raman scattering has many advantages for measuring phonon modes in superlattices[1,2] Phonons in GaAs-AlAs superlattices were first studied by Merz, Barker, and Gossard.[3] Brillouin zone folding of acoustic phonons was first reported by Narayanamurti *et al.* .[4] Colvard *et al.* [5] provide the first unambiguous Raman observation of phonon folding in semiconductor superlattices. The observation of downward shifts of LO phonon frequencies in superlattices was also first reported in Ref. 3. Superlattice effects on optical modes were first clearly identified using the notion of a phonon quantum well by Jusserand *et al.*,[6] then by Colvard *et al.*,[7] and Sood *et al.*. [8] The earliest resonant Raman scattering in a semiconductor superlattice was reported by P. Manuel *et al.* .[9] Then Colvard *et al.*,[10] Zucker *et al.*.,[11] and Sood *et al.* [8] did resonance Raman studies of semiconductor superlattices. Yip and Chang[12] calculated

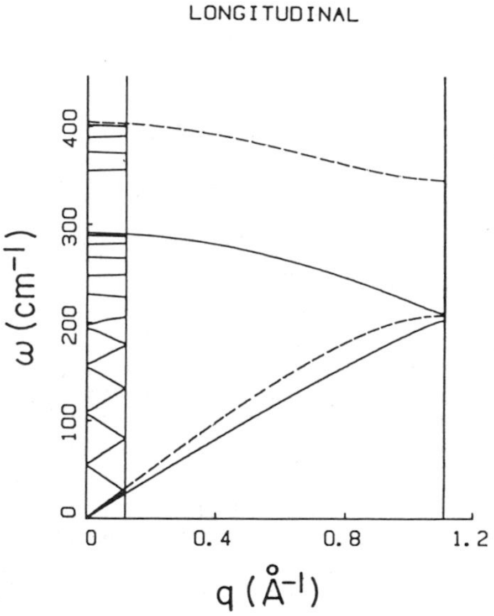

LONGITUDINAL

Fig. 1. Linear-chain model calculation of phonon dispersion curves for longitudinal modes.[7] Large zone: bulk GaAs (solid lines) and AlAs (dashed lines). Small zone: (5,4) superlattice.

LO mode frequencies of $(GaAs)_n (AlAs)_m$ superlattices for small numbers of layers n and m using a bond charge model and taking Coulomb interactions into account. With their dispersion curves for the bulk materials, one can estimate frequencies of phonons confined to GaAs or AlAs. A continuum theory for LO phonons in quantum wells and superlattices has been developed by Babiker.[13]

FOLDED ACOUSTIC PHONONS

Acoustic phonons become Raman active due to the folding of the dispersion curves into the smaller Brillioun zone formed through the additional periodicity of the superlattice layering. Calculated dispersion curves for a linear chain model[7] are shown in Fig. 1 for longitudinal modes propagating perpendicular to the interfaces in a superlattice with a repeat unit consisting of 5 layers of GaAs and 4 layers of AlAs. In the region below 200 cm^{-1} one sees that the dispersion curves for the acoustic modes of GaAs and AlAs are very similar. The resulting "folded acoustic phonon" dispersion curves for the superlattice show only very small minigaps at minizone center and boundary (at q=π/d).

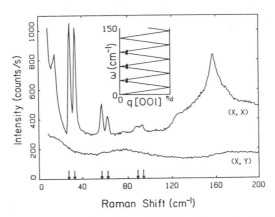

Fig. 2.   Raman spectra with 514 nm laser light from a nominal 42 Å GaAs and 8 Å Al$_{0.3}$Ga$_{0.7}$As sample.[7] Inset is Rytov model[14] calculation with experimental q values for a superlattice period of 52.2 Å determined by x-ray diffraction.

An experimental Raman spectrum with three folded acoustic phonon doublets is shown in Fig. 2.[7] The doublets are centered at 30, 60, and 90 cm^{-1}. The arrows mark the calculated position of the peaks. The inset gives dispersion curves calculated from the Rytov layered elastic continuum model.[14]

A good zero-order model for the wave function gives a single plane wave at wave-vector

$$q' = q \pm 2\pi m/d, \qquad\qquad (1)$$

where q is the wavevector transfer, m is an integer, and d is the superlattice period. The folded phonon frequencies seen in the Raman experiment are well-approximated by the relation

$$\omega = v \, |q'| \qquad\qquad (2a)$$

where v is the virtual-crystal sound velocity. Equations (1) and (2a) give doublets with a separation of

$$\Delta\omega = 2v \, q \qquad\qquad (2b)$$

121

We have established the usefulness of a photoelastic model for the Raman intensity of the doublets.[7] In this model the phonon wave functions are single plane waves, and light couples to the phonons via their strain through a local photo-elastic coefficient P(z). This is expected to hold away from resonance, where local optical properties may be defined. For a general periodic P(z) with Fourier coefficients P_m one can show that the mth-order folded acoustic doublet has intensity

$$I_m \propto k_B T \, |P_m|^2 \; , \tag{3}$$

where T is the temperature and k_B is Boltzmann's constant. If the photoelastic coefficient were a linear function of aluminum concentration, we would expect the Raman intensities of the folded phonon doublets to be proportional to the square of the Fourier components of the composition profile.

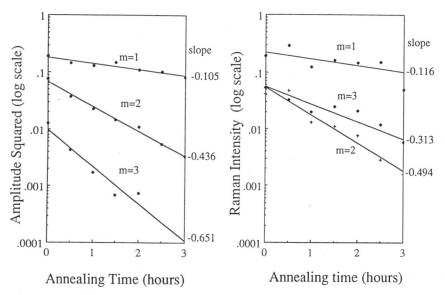

Fig. 3. Comparison of Fourier components from x-ray diffraction with intensities of folded phonons in Raman spectra excited at 514 nm. Lines are least-squares fits to the data. Slopes are shown to right of graphs. (a) Intensities of Fourier components from x ray diffraction plotted versus annealing time. (b) Intensities of folded phonon doublets in the Raman spectra plotted versus annealing time. Data are matched at m=1, t=0.

ANNEALING STUDY OF FOLDED ACOUSTIC PHONONS

Annealing of a superlattice rounds-off the concentration profile x(z) in a manner that can be followed by superlattice Bragg diffraction of x-rays.[15,16] We performed an annealing study at 850 °C on a sample that was initially a binary superlattice with $d_1 = 55$ Å thick layers of GaAs and $d_2 = 18$ Å thick layers of AlAs. Annealing was done in 30 minute increments.

Logarithmic plots of the the squares of the Fourier components of the superlattice profile and the folded phonon intensities vs. annealing time are shown in Fig.3. If the photoelastic theory were correct, the phonon intensities would decrease at the same rate as the squares of the corresponding Fourier components. The first two folded phonon doublet intensities match the Fourier components quite well in both slope and relative intensity. The third order folded phonon doublet is not only much larger in magnitude, but also decreases at less than half the rate of the third Fourier component. Since the thickness of the AlAs layer is somewhat less than one-third of the period, the Fourier coefficients x_3 [that of the composition profile x(z)] and P_3 will be quite sensitive to respective changes in composition and photoelastic coefficient in the AlAs layer. We expect the photoelastic coefficient P(z) at 5145Å to vary more strongly than linearly with composition x for x≈0.7, where the band gap of $Al_xGa_{1-x}As$ (2.42 eV) corresponds to the energy of the 5145Å laser photons. Thus in the Al-rich regions of the superlattice, we would not expect P(z) to faithfully follow x(z), and the third-order Fourier coefficient, which probes this region will be enhanced.

When the laser photon energy is in resonance with various electronic transitions across the band gap of the superlattice, we find very complicated changes in the spectra of the folded phonons. We find evidence of interference between HH2-CB2 and LH1-CB2 transitions due to valence band mixing. Calculations of this effect, which must also take excitonic effects into account, are underway.[17] We conclude that whereas the phonons are simple in thin superlattices, their electronic structure is not simple, and this will complicate the calculation of resonant Raman properties.

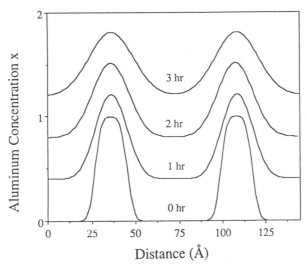

Fig. 4.   Composition profile of superlattice from fit to x-ray data for four annealing times. A value of 1.0 corresponds to 100% AlAs. Successive curves are shifted upwards by 0.40 for clarity.

ANNEALING EFFECT ON CONFINED LO PHONONS

The LO phonon frequencies in GaAs/AlGaAs superlattices are well-represented by a confinement model.[6,7,8] This model says essentially that there are m half wavelengths of the mode localized in an individual layer. The phonon frequency is given from the bulk dispersion curve for the LO phonon $\omega_{LO}(q)$ for the material of the layer [of thickness d_1] by use of the equation

$$\omega = \omega_{LO}(m\pi/d_1) , \qquad (4)$$

where the integer m is the confinement number of the mode. The validity of this picture rests on the fact that the dispersion curves of the LO branches of GaAs and AlAs are non-overlapping [Fig.1], leading to strongly-damped evanescent waves in the "wrong" material.

Through annealing we are able to systematically alter the interface width and monitor this width using x-ray diffraction. This is similar to an earlier study in which the interface was changed by changing the growth conditions.[18] In Ref. 18 calculations were made using a concentration-dependent linear chain model. Here we use the phonon analog of effective mass theory. To accurately model the confined phonon frequencies we must take into account the shift in frequency of the GaAs-like phonons in the alloy material present at the boundaries of the phonon quantum well. In the bulk AlGaAs alloy, the frequency of this mode shiftsdownwards upon increasing the Al concentration. We adopt a simple linear expression for its x-dependence[19,20]

$$[\omega_{LO}(x)]^2 = \omega_o^2(1-ax) \ , \tag{5}$$

where $\omega_o = 296$ cm^{-1} is the frequency of the LO phonon in bulk GaAs, x is the Al concentration, and a=0.20 is a linear fit to data[19]. We assume that in the uniform alloy the dispersion of the LO phonon is given by

$$\omega^2(x,q) = [\omega_{LO}(x)]^2 + (1-x)T(q) \ , \tag{6}$$

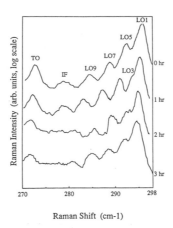

Fig. 5. Logarithmic plots of Raman spectra from optical phonons excited at 27K with 514 nm light showing spectra for four annealing times. Confinement order is indicated on plot. IF and TO indicate interface and transverse optical phonons, respectively.

where T(q) is the negative term that gives the dispersion for the GaAs LO phonon. The (1-x) factor allows for the fact that for low Ga concentration the entire GaAs-like LO branch tends towards a GaAs gap mode in AlAs. When x= x(z), Eq. (6) leads to an effective-mass-like equation for the phonon amplitude u(z):

$$\omega^2 u(z) = \{[\omega_{LO}[x(z)]^2\}u(z) + [1-x(z)]T(-id/dz)u(z) \tag{7}$$

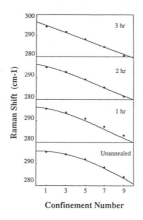

Confinement Number

Fig. 6. Experiment vs. theory for confined LO phonon dispersion. Points are experimental values from Raman scattering. Lines are an interpolation between corresponding points from effective mass calculation. Dispersion changes from parabolic curve of square well for the unannealed sample to the linear curve characteristic of a parabolic well for the 3 hour annealed sample.

To solve Eq. (7) we have used an error function concentration profiles[18] derived from the fit to the x-ray data shown in Fig. 4. ω^2_{LO} is modelled by Eq.(5), and $T(q) = -w^2q^2$ with $w = 3.00 \times 10^5$ cm/s. This value is based on recent neutron scattering measurements in bulk GaAs[21]. Figure. 5 shows the Raman spectra of the optical phonons in the "allowed" backsacttering geometry as a function of annealing time. The peaks labeled LO1,LO5,LO7, and LO9 in the unannealed sample all shift towards smaller wavenumbers as the sample is annealed. A shoulder labeled LO3 appears on the first order peak in the 1 hour annealed spectrum. As the annealing time increases this peak shifts and separates more from the LO1 peak. We interpret this annealing behavior as an indication of the confinement of the longitudinal optical phonons. To confirm our assignments we have performed a calculation of the frequencies of the confined optic modes as described above. The composition profiles obtained from x-ray diffraction shown in Fig. 4 were used to determine the shape of the phonon quantum well. The experimental results are presented on a logarithmic intensity scale in Fig. 5. Figure 6 compares the results of the calculation with the frequencies obtained experimentally. The squares are the experimental values while the lines are an interpolation between calculated points. The match between the two is exceptionally good. Note that there are no fitting parameters involved.

CONCLUSIONS

Through the combined measurement of Raman and x-ray diffraction spectra on successively annealed samples we have been able to test the accuracy of theories of phonons and Raman scattering in superlattices. It has been shown that a simple effective mass theory accurately predicts the energies of the confined optic phonons while the photoelastic theory of scattering by folded acoustic phonons is only strictly valid when the laser photons are far from resonance with any electronic transitions. The properties of propagation for acoustic phonons and confinement in the layer for optical phonons in GaAs-AlAs superlattices have been clearly demonstrated by the annealing study.

ACKNOWLEDGMENTS

We thank C. Colvard, R. Merlin, and A. C. Gossard for earlier collaborations, J. McMillan for technical assistance, and Prof. Y.C. Chang for helpful discussions. This research was supported by the NSF under DMR 83-16981, 85-06674, and 86-12860, by JSEP, and by AFOSR.

* Also at Coordinated Science Laboratory, UIUC.
†Permanent addresss: Department of Physics, Peking University, Beijing, P.R.China.

REFERENCES

1. J. M. Worlock, in *Proc. 2nd Int. Conf. Phon Phys.*, J. Kollár, N. Króo, N. Menyhárd, and T. Siklós, Eds. (World Scientific, Singapore, 1985) p. 506.
2. M.V. Klein, IEEE J. Quant. Elect. **QE-22**, 1780 (1986).
3. J.L. Merz, A.S. Barker, Jr., and A.C. Gossard, Appl. Phys. Lett., **31**,117 (1977); A.S. Barker, Jr.,J.L. Merz, and A.C.Gossard, Phys. Rev. **B17**, 3181 (1978).
4. V. Narayanamurtri, H.L. Störmer, M.A. Chin, A.C. Gossard and W. Wiegmann, Phys. Rev. Lett., **43**, 2012 (1979).
5. C. Colvard, R. Merlin, M.V. Klein, and A.C. Gossard, Phys. Rev. Lett., **45**, 298 (1980).
6. B. Jusserand, D. Paquet, and A. Regreny, Phys. Rev. B**30**, 6245 (1984).
7. C. Colvard, T.A. Gant, M.V. Klein, R. Merlin, R. Fisher, H. Morkoç, and A.C. Gossard, Phys. Rev B**31**, 2080 (1985).
8. A.K. Sood, J. Menéndez, M. Cardona, and K. Ploog, Phys. Rev. Lett., **54**, 2111 (1985).
9. P. Manuel, G.A. Sai-Halaz, L.L. Chang, C.-A.Chang, and L. Esaki, Phys. Rev. Lett. **37**, 1701 (1976).
10. C. Colvard, R. Merlin, M.V. Klein, and A.C. Gossard, J. Phys. **42**, C6, 631 (1981).
11. J.E. Zucker, A. Pinczuk, D.S. Chemla, A.C. Gossard, and W. Wiegmann, Phys. Rev. Lett., **51**, 1283 (1983).
12. S.K. Yip and Y.C. Chang, Phys. Rev. B**30**, 7037 (1984).
13. M. Babiker, J. Phys. C **19**, 683 (1986); M. Babiker J. Phys. C **19**, L339 (1986).
14. S. M. Rytov, Kust. Zh. **2**, 71 (1956) [Soviet Physics-Acoustics **2**, 67, (1956)].
15. L. L. Chang and A. Koma, Appl. Phys. Lett. **29**, 138 (1976).
16. R.M. Fleming, D.B. McWhan, A.C. Gossard, W. Wiegmann, and R.A. Logan, J. Appl. Phys. **51**, 357 (1980).
17. Y. C. Chang, private communication.
18. B. Jusserand, F. Alexandre, D. Paquet, and G. LeRoux, Appl. Phys. Lett., **47**, 301 (1985).
19. B. Jusserand and J. Sapriel, Phys. Rev. B**24**, 7194 (1981).
20. R. Tsu, H. Kawamura, L. Esaki, *Proc. 11th Int. Conf. on Pysics of Semiconducors*, p. 1135 (1972)
21. D. Neumann and J. E Clemans, private communication.

OPTICAL INTERACTIONS AT ROUGH SURFACES

[1]Alexei A. Maradudin, [2]A. R. McGurn, [3]R. S. Dummer, [3]Zu Han Gu, [4]A. Wirgin, and [5]W. Zierau

[1]Department of Physics, University of California, Irvine, CA 92717, USA. [2]Department of Physics, Western Michigan University, Kalamazoo, MI 49008, USA. [3]Surface Optics, Inc., P. O. Box 261602, San Diego, CA 92126, USA. [4]Laboratoire de Mécanique Théorique, Université Pierre et Marie Curie, 4 Place Jussieu, 75230 Paris Cedex 05, FRANCE. [5]Institut für Theoretische Physik II, Westfälisch-Wilhelms Universität, Wilhelm-Klemm-Strasse 10, 44 Münster, Federal Republic of Germany

ABSTRACT

A discussion is given of optical interactions at three types of rough planar surfaces: (a) a deterministic, nonperiodic surface; (b) a deterministic, periodic surface; and (c) a randomly rough surface. The scattering of p-polarized light from an isolated ridge or groove on an otherwise planar metal surface is studied, and the efficiency of exciting surface polaritons in this fashion is estimated. A strong enhancement of the electric field within the grooves of a perfectly conducting lamellar grating illuminated by s-polarized light is demonstrated. Then the Goos-Hänchen effect for a p-polarized, bounded (Gaussian) light beam incident from the vacuum side onto a random metal grating is investigated. The resulting lateral displacement of the reflected beam is negative, and is larger than when the metal surface is flat. Finally, the diffuse component of the light scattered from a two-dimensional, randomly rough metal surface is calculated. This diffuse component is shown to exhibit a maximum in the anti-specular direction (opposition effect) that can be associated with the localization of surface polaritons by the random roughness of the surface. Experimental results showing this opposition effect are presented.

I. INTRODUCTION

All surfaces, even the most carefully prepared ones, possess some degree of surface roughness. It is therefore at least of interest, and often of importance, to know how that roughness affects the optical properties of a surface. In addition, optical interactions at rough surfaces can sometimes be significantly enhanced over those at planar surfaces, and can give rise to new effects that have no counterparts at smooth surfaces.

To illustrate these points, in this lecture we describe the

scattering and diffraction of light from three types of rough surfaces, viz. (a) a deterministic, nonperiodic surface; (b) a deterministic, periodic surface; and (c) a randomly rough surface.

II. SCATTERING OF LIGHT FROM A RIDGE OR GROOVE ON A PLANAR METAL SURFACE[1]

We begin by studying the scattering of p-polarized light from an isolated ridge or groove on an otherwise planar metal surface, defined by the equation $x_3 = \zeta(x_1)$, when the plane of incidence is perpendicular to the ridge or groove. The metal is characterized by an isotropic dielectric constant $\varepsilon(\omega) = \varepsilon_1(\omega) + i\varepsilon_2(\omega)$ $(\varepsilon_2(\omega) > 0)$. The scattering of light from such a structure is believed to play a role in the formation of the periodic surface corrugations that are observed after repeated illumination of solid surfaces by intense laser pulses[2]. It has also been suggested as a way of exciting surface polaritons[3].

The single, nonzero component of the magnetic field has the form $H_2(x_1 x_3 | \omega) \exp(-i\omega t)$. In the region $x_3 > \zeta(x_1)_{max}$ the magnetic field

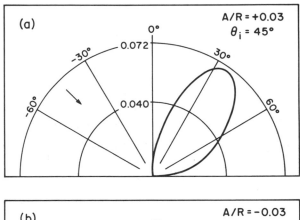

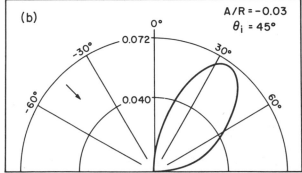

Fig. 1.   A polar plot of $L_1 I(q|k)_{diff}$ as a function of the scattering angle θ_s for the scattering of p-polarized light from a Gaussian groove ridge or a silver surface. The angle of incidence is 45°. The wavelength of the incident light is 4579Å, for which $\varepsilon(\omega) = -7.5 + i\ 0.24$. The value of R is 2500Å. (a) A/R = 0.03; (b) A/R = -0.03.

amplitude $H_2(x_1 x_3|\omega)$ is the sum of an incident field and a scattered field:

$$H_2^>(x_1 x_3|\omega) = e^{ikx_1} \left[e^{-i\alpha_o(k\omega)x_3} + \frac{\varepsilon(\omega)\alpha_o(k\omega)-\alpha(k\omega)}{\varepsilon(\omega)\alpha_o(k\omega)+\alpha(k\omega)} e^{i\alpha_o(k\omega)x_3} \right] -$$

$$- 2i \int \frac{dq}{2\pi} B(q|k)e^{iqx_1+i\alpha_o(q\omega)x_3}, \tag{2.1}$$

where $\alpha_o(q\omega) = (((\omega+i0)^2/c^2)-q^2)^{1/2}$ and $\alpha(q\omega) = ((\omega+i0)^2/c^2)-q^2)^{1/2}$. The efficiency of the diffuse scattering is then given by

$$I(q|k)_{diff} = \frac{4}{L_1} \frac{\alpha_o(q\omega)}{\alpha_o(k\omega)} |B(q|k)|^2, \quad |q| < \frac{\omega}{c}, \tag{2.2}$$

where L_1 is the length of the surface $x_3 = 0$ in the x_1-direction. The diffuse scattering amplitude $B(q|k)$ satisfies an integral equation that is obtained by combining Eq. (2.1) with Eqs. (1) and (3) of Ref. 1. We have solved this equation numerically, assuming a Gaussian form for $\zeta(x_1)$, viz. $\zeta(x_1) = A\exp(-x_1^2/R^2)$. Typical results are shown in Fig. 1. The angle of incidence is $45°$; the wavelength of the incident light is $\lambda = 4579$Å; the metal is silver with $\varepsilon(\omega) = -7.5 + i0.24$[4]; and the value of R is 2500Å. We see from this figure that the scattering is predominantly in the forward direction, and that for the values of A/R chosen there is little difference between scattering from a ridge or groove.

In the region $|q| > (\omega/c)$ the function $B(q|k)$ has the form (see Fig. 2)

$$B(q|k) = C(q|k)\left[\frac{1}{q-k_{sp}^{(1)}(\omega)-ik_{sp}^{(2)}(\omega)} - \frac{1}{q+k_{sp}^{(1)}(\omega)+ik_{sp}^{(2)}(\omega)} \right] \tag{2.5}$$

in the vicinity of $q = \pm k_{sp}(\omega)$, where $k_{sp}(\omega) = k_{sp}^{(1)}(\omega)+ik_{sp}^{(2)}(\omega)$ is the complex wave vector of a surface polariton of frequency ω at a planar vacuum/metal interface. As $\varepsilon_2(\omega) \to 0$ the efficiencies of excitation of surface polaritons propagating in the $\pm x_1$-directions can be shown to be

$$e_{\pm} = \frac{1}{L_1} \frac{1}{2} |\varepsilon(\omega)|^{1/2} \left(1 - \frac{1}{\varepsilon^2(\omega)}\right) \frac{|C(\pm k_{sp}(\omega)|k)|^2}{\alpha_o(k\omega)}. \tag{2.6}$$

An estimate of $C(\pm k_{sp}(\omega)|k)$ from the plots in Fig. 2 ((A/R) $= -0.03$) yields the following values for $e_{\pm}$:

$$e_+ = \frac{1}{L_1} (17.3 \text{ Å}), \quad e_- = \frac{1}{L_1} (2.73 \times 10^{-4} \text{ Å}). \tag{2.7}$$

It should be possible to extend the present calculation to obtain the scattering efficiency and the efficiency of excitation of surface

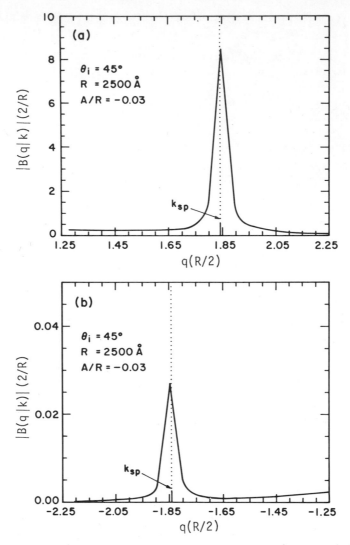

Fig. 2. Structure in the scattering amplitude $B(q|k)$ for $|q|$ $>(\omega/c)$ that is associated with the excitation of surface polaritons.

polaritons for a surface profile that consists of a finite number N of identical, equally spaced, grooves or ridges.

III.   DIFFRACTION OF S-POLARIZED LIGHT FROM A DEEP LAMELLAR GRATING

Much recent research on surface-enhanced effects[5]-[7] has focused on the role of rough surfaces, and gratings in particular, for coupling radiant energy into surface electromagnetic waves.

Until recently it has been thought that all resonant, near-field enhancement processes at bare classical gratings are possible only when the incident electromagnetic radiation is p-polarized (TM), with the plane of incidence perpendicular to the grooves of the grating. The first theoretical indication that the situation might be otherwise was provided by Hessel and Oliner[8], who predicted on the basis of a periodic surface impedance boundary condition that gratings are also capable of producing cavity-type resonances when illuminated by s-polarized light whose plane of incidence is perpendicular to the grooves of the grating. They predicted that such resonances occur only for gratings whose groove depth h satisfies $m\lambda_g/2 > h > (2m-1)\lambda_g/4$ (m = 1, 2, 3...), but did not explain the origin of this criterion nor did they define λ_g.

Subsequently Andrewartha et al.[9] showed that perfectly conducting lamellar gratings can give rise to s-resonances. However, the imaginary part of a wavelength resonant pole is generally so large that the corresponding s-resonance goes unnoticed in a far field observable such as the specular reflectance.

More recently it has been shown that s-resonances manifest themselves in the immediate vicinity of a grating where they give rise to large field enhancements when the groove depth satisfies the Hessel-Oliner criterion[10]-[12].

To see this we consider a scattering configuration in which the incident wave strikes a perfectly conducting lamellar grating (Fig. 3) normally, the grating period is a = 0.38 μm, the width of each groove is w = 0.35 μm, and its depth is h = 1 μm. The wavelength λ spans the visible range of the optical spectrum. The grating throws all of the incident energy into the specular (n = 0) spectral order, so that it is impossible to find any trace of s-resonances in a far-field intensity.

In Fig. 4 is plotted the electric field enhancement in the central groove as a function of groove depth at several fixed _relative_ observation ordinates. The wavelength of the incident light is very close to the wavelength at which the enhancements are greatest.

The condition of existence of a cavity resonance is found to be similar to that for a guided wave polariton[13], viz. $\tan(\gamma_1(\omega)h) < 0$, where $\gamma_1(\omega) = \left((\omega+i0)^2/c^2 - (\pi/w)^2\right)^{1/2}$. This is identical to the Hessel and Oliner criterion if λ_g is identified with $2\pi/\gamma_1(\omega)$.

In other recent work Glass et al.[14] found that the field enhancement associated with the grating-induced excitation of the plane surface polariton of a silver grating by incident light of p-polarization first

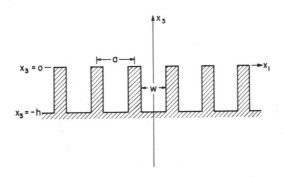

Fig. 3.   The lamellar grating studied in the present work.

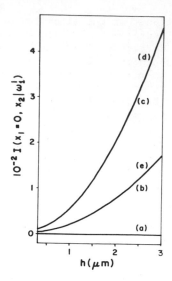

Fig. 4. Electric field enhancement versus groove depth at the wavelength $\lambda = \lambda_1$ and at fixed relative observation ordinates on the line $x_1 = 0$ within the central groove of a perfectly conducting grating exposed to a normally incident s-polarized plane wave. d = 0.38 μm, w = 0.35 μm. (a), (b), (c), (d), (e) apply to x_3 = 0-, -0.2h, -0.4h, -0.6h, -0.8h, respectively. [Ref.12].

increases with increasing groove depth h and then past some critical radiation damping. In contrast, we see from Fig. 4 that the field enhancement at the resonance wavelength of the incident light increases value of the latter begins to decrease with further increase in h due to monotonically with increasing h, with some tendency toward saturation noted in the results of Ref. 12, but with no evidence of a decrease for h greater than some critical value. This behavior is consistent with the finding of Andrewartha et al.[12] that the imaginary part of the complex zero of the dispersion relation that gives the wavelengths at which s-cavity resonances occur tends to zero as h → ∞.

A possible way of observing the s-resonance is to measure the enhancement in the intensity of light Raman scattered from molecules adsorbed within the grooves of a grating.

IV. THE GOOS-HÄNCHEN EFFECT FOR A RANDOM GRATING

The Goos-Hänchen effect[15] is the lateral displacement of the reflected beam, from the position predicted by geometrical optics considerations, when a bounded (e.g. Gaussian) light beam is incident from an optically denser dielectric medium onto its interface with a rarer dielectric medium. We shall consider in this section the incidence of a Gaussian light beam from vacuum onto a rough metal surface, and will emphasize the effects of the roughness-induced virtual excitation of surface polaritons on the lateral beam shift. It has been shown that in certain other experimental geometries in which surface polaritons can be resonantly excited, significant beam displacement arises from this excitation[16]. The present consideration of the coupling to polariton modes through roughness may be useful as a means of extracting information concerning the parameters that characterize the surface roughness.

Using the formalism developed in McGurn et al.[17] we can write the Fresnel coefficient for light reflected by a random metal grating described by the surface profile function $\zeta(x_1)$ as

$$R(k\omega) = \frac{\varepsilon(\omega)\alpha_o(k\omega) - [\alpha(k\omega) + m(k\omega)]}{\varepsilon(\omega)\alpha_o(k\omega) + [\alpha(k\omega) + m(k\omega)]}, \tag{4.1}$$

where

$$m(k\omega) = \frac{(\varepsilon(\omega)-1)^2}{\varepsilon^2(\omega)} \delta^2 \int \frac{dp}{2\pi} \frac{g(|k-p|)}{\varepsilon(\omega)\alpha_o(p\omega) + \alpha(p\omega)} [\varepsilon(\omega)kp - \alpha(k\omega)\alpha(p\omega)]^2 \tag{4.2}$$

and $g(|Q|) = \sqrt{\pi}\, a \exp(-a^2 Q^2/4)$. The function $m(k\omega)$, which represents the lowest order correction to the flat surface Fresnel coefficient due to surface roughness, has been obtained with the use of the assumption that $\zeta(x_1)$ is a Gaussianly distributed random variable with $\langle\zeta(x_1)\rangle = 0$ and $\langle\zeta(x_1)\zeta(0)\rangle = \delta^2 \exp(-x_1^2/a^2)$, where $\langle...\rangle$ indicates an average over the ensemble of realizations of $\zeta(x_1)$. Given the Fresnel coefficient in Eq. (4.1) and the form of the incident beam at the mean surface of the grating, one can calculate the reflected beam at the mean surface.

In Fig. 5 we plot $\Delta = D\cos\theta_i$ versus the angle of incidence θ_i, where D is the lateral displacement of the maximum in the intensity of the reflected beam at the mean surface from the position of the maximum intensity of the incident beam at the mean surface. This Δ represents the lateral shift in the plane perpendicular to the direction of the reflected beam. For these plots the intensity of the magnetic field of the incident beam at the mean surface is given by $|H(x_1,0)|_{inc.}^2 = \{\exp[-2\cos^2\theta_i(x_1^2/w^2)]\}/(\pi w^2)$, where $w(\omega/c) \gtrsim 75$. Two plots for $a = 1000$ Å, $\delta = 100$ Å, and $\lambda = 5000$ Å are shown, one corresponding to $\varepsilon = -20 + i$ and one to $\varepsilon = -50 + i$. The shift is seen be negative, to be enhanced by the surface roughness, to increase with increasing angle of incidence, and to be larger the more negative is the real part of the dielectric constant.

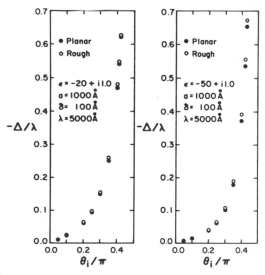

Fig. 5. Plot of $-\Delta/\lambda$ versus θ_i/π for $\varepsilon(\omega)$, a, δ and λ as noted on the figures for a flat and a rough surface [Ref. 18].

V. LOCALIZATION PHENOMENA IN LIGHT SCATTERING FROM A RANDOMLY ROUGH METAL SURFACE

In the present section we study the elastic scattering of light from a randomly rough vacuum/metal interface, and show that such scattering leads to an interesting backscattering effect that has recently been observed experimentally.

The efficiency of the diffuse (i.e. nonspecular) scattering of a light wave whose polarization is $\beta (= \parallel, \perp$ for p- and s-polarization, respectively) and the projection of whose wavevector on the mean surface $x_3 = 0$ is $\vec{k}_\parallel$, into a wave whose polarization is α and the projection of whose wave vector on the mean surface is $\vec{q}_\parallel$, can be shown to be [19,20]

$$I_{\alpha\beta}(\vec{k}_\parallel | \vec{q}_\parallel)_{diff} = \frac{4\alpha_0(q_\parallel \omega)\alpha_0(k_\parallel \omega)}{S} <|G_{\alpha\beta}(\vec{q}_\parallel | \vec{k}_\parallel)|^2>_{diff} . \qquad (5.1)$$

Here $G_{\alpha\beta}(\vec{q}_\parallel | \vec{k}_\parallel)$ is a surface polariton Green's function, $\alpha_0(k_\parallel \omega) = (((\omega+i0)^2/\omega^2)-k_\parallel^2)^{1/2}$, the angular brackets $<...>$ denote an average over the ensemble of realizations of the surface profile function, $<|G_{\alpha\beta}(\vec{q}_\parallel | \vec{k}_\parallel)|^2>_{diff}$ is the contribution to $<|G_{\alpha\beta}(\vec{q}_\parallel | \vec{k}_\parallel)|^2>$ that contains no factor of $(2\pi)^2 \delta(\vec{q}_\parallel - \vec{k}_\parallel)$, and S is the area of the mean surface $x_3 = 0$.

The two-surface polariton Green's function in Eq. (5.1) for polaritons traveling along a rough surface is of a form similar to the one encountered in the calculation of the conductivity of electrons moving among static, randomly distributed impurities. In fact, the same diagrammatic techniques used in the electron problem apply to our optics problem for surface polariton scattering from a random interface. The ladder diagram contributions to the two- particle Green's functions which yield the Drude conductivity in the electron problem are found in the

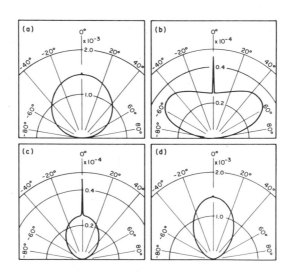

Fig. 6. The differential reflection coefficient versus the scattering angle θ in the degrees for $\delta = 29$Å, $a = 2000$Å, and light of wavelength $\lambda = 4579$Å incident normally on a silver surface with $\varepsilon(\omega) = -7.5+i0.24$. The polarization changes on scattering are (a) p to p; (b) s to p; (c) p to s; (d) s to s [Ref. 20].

optics problem to give rise to a general background of diffusely
scattered light for an arbitrary scattering angle measured from the[21]
normal to the mean surface. The set of maximally crossed diagrams[21]
which are responsible for localization effects in the electron problem
are found to cause a peak in the anti-specular (backscattering) direction
in the angular distribution of the intensity of the diffusely scattered
light in the optics problem. These backscattering effects are due to the
localization of surface polarizations by the rough interface.

Two limiting behaviors are encountered in the optics problem. If the
mean free path for polariton propagation due to dielectric losses is
shorter than that due to surface scattering we have weak localization.
In this limit a simple sum of maximally crossed diagrams suffices for the
determination of $I_{\alpha\beta}(\vec{q}_{\parallel}|\vec{k}_{\parallel})_{diff}$. In the opposite limit a fully self-
consistent calculation modeled after that of Vollhardt and Wolfle[21]
must be carried out. We shall treat only the weak localization limit
here. A complete discussion of the fully self-consistent theory is given
in Refs. 20 and 22.

In Fig. 6 we present theoretical results for the differential
reflection coefficient plotted versus the scattering angle θ measured
from the normal to the mean surface for light with λ = 4579 Å and
$\varepsilon(\omega)$ = − 7.5 + 0.24i, corresponding to a silver surface. The randomly
rough surface, using the same notation as in Section 4, has a =
1200 Å and δ = 29 Å, yielding results in the weak localization limit.
The light is incident normal to the mean surface and gives a backscattering peak of width $\widetilde{=}$ 1°.

The opposition effect in scattering of light from a randomly rough
metal surface, has been observed experimentally recently[23]. In Fig. 7
is plotted the differential reflection coefficient for the scattering of
p-polarized light from a rough copper surface. The angle of incidence is
20°. The value of the differential reflection coefficient at 1° away

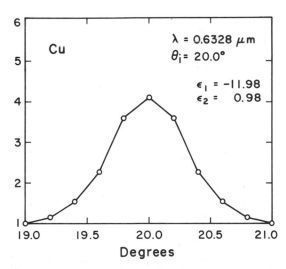

Fig. 7. The differential reflection coefficient versus scattering
angle θ in degrees for p-polarized light incident on a rough copper
surface. λ = 0.6328 μm and θ_i = 20°. [Ref. 23].

from the retroreflection direction, on either side of it, has been set equal to unity. Thus, there is a significant peak in the retroreflection (backscattering) direction. The values of the experimental parameters (wavelength of light, dielectric constant, roughness parameters) are such that the weak localization limit is realized. However, since the theory of Ref. 20 applies only to the case of normal incidence, a quantitative comparison between theory and experiment is not possible at this time.

In the future, detailed experimental studies of the opposition effect, together with theoretical analysis, should provide an additional method for the characterization of surface roughness on metal surfaces.

ACKNOWLEDGEMENTS

This research was supported in part by NSF Grant No. DMR 85-17634, and the Army Research Office Grant DAAG29-85R-0025.

REFERENCES

1. A preliminary account of this work was presented by W. Zierau and A. A. Maradudin, in Surface Waves in Plasmas and Solids, ed. S. Vukovic (World Scientific, Singapore, 1986), p. 687.
2. F. Keilmann and Y. H. Bai, Appl. Phys. A29, 9 (1982).
3. W. P. Chen and E. Burstein, in Electromagnetic Surface Modes ed. A. D. Boardman (John Wiley and Sons, New York, 1982) p. 550.
4. P. B. Johnson and R. W. Christy, Phys. Rev. B6, 4370 (1972).
5. See, for example, H. Metiu, in Surface Enhanced Raman Scattering, R. K. Chang and T. E. Furtak eds. (Plenum, New York, 1982), p. 1.
6. C. K. Chen, A. R. B. de Castro, and Y. R. Shen, Phys. Rev. Lett. 46, 145 (1981).
7. J. G. Endriz and W. E. Spicer, Phys. Rev. B4, 4159 (1971).
8. A. Hessel and A. A. Oliner, Appl. Opt. 4, 1275 (1965).
9. J. R. Andrewartha, J. R. Fox, and I. J. Wilson, Opt. Acta 26, 69 (1979).
10. A. Wirgin and A. A. Maradudin, Phys. Rev. B31, 5573 (1985).
11. A. A. Maradudin and A. Wirgin, Surf. Sci. 162, 980 (1985).
12. A. Wirgin and A. A. Maradudin, Prog. Surf. Sci. 22, 1 (1986).
13. If the structure consists of vacuum in the region $x_3 < 0$, a dielectric medium characterized by a dielectric constant $\varepsilon_d > 0$ in the region $0 < x_3 < h$, and an infinitely conducting substrate in the region $x_3 > h$, the dispersion relation is $\alpha(k\omega) + \alpha_{}(k\omega)\tan(\alpha(k\omega)h) = 0$, where $\alpha_o(k\omega) = (k^2-(\omega^2/c^2))^{1/2}$, $\alpha(k\omega) = (\varepsilon_d(\omega^2/c^2)-k^2)^{1/2}$.
14. N. E. Glass, M. Weber, and D. L. Mills, Phys. Rev. B29, 6548 (1984).
15. F. Goos and H. Hänchen, Ann. Physik 1, 333 (1947).
16. P. Mazur and B. Djafari-Rouhani, Phys. Rev. B30, 6759 (1984).
17. A. R. McGurn, A. A. Maradudin, and V. Celli, Phys. Rev. B31, 4866 (1985).
18. A. A. Maradudin and A. R. McGurn (unpublished work).
19. G. Brown, V. Celli, M. Haller, A. A. Maradudin, and A. Marvin, Phys. Rev. B31, 4993 (1985).
20. A. R. McGurn and A. A. Maradudin, J. Opt. Soc. Am. B (to appear).
21. D. Vollhardt and P. Wölfle , Phys. Rev. B22, 4666 (1980).
22. V. Celli, A. A. Maradudin, A. M. Marvin, and A. R. McGurn, J. Opt. Soc. Am. A2, 2225 (1985).
23. R. S. Dummer and Zu-Han Gu (unpublished work).

NONLINEAR DISTRIBUTED COUPLING INTO GUIDED WAVES

R. M. Fortenberry,* G. Assanto,† R. Moshrefzadeh,††
C. T. Seaton, and G. I. Stegeman

Optical Sciences Center
University of Arizona
Tucson, Arizona 85721

INTRODUCTION

Thin-film waveguides are a very attractive material geometry for performing nonlinear optics experiments for a number of reasons. They provide diffractionless propagation in one or more dimensions. Furthermore, efficient interactions are possible at low powers becasue of beam confinement to dimensions of the order of the wavelength of light. The application of second-order nonlinear interactions to optical waveguides is a well-developed field.[1,2] In contrast to this, interest and progress in third-order nonlinear guided wave phenomena is much more recent. Demonstrated to date have been degenerate four wave mixing in CS_2-covered[3] and semiconductor-doped glass[4] waveguides, coherent anti-Stokes Raman scattering with monolayer sensitivity,[5] nonlinear waveguiding phenomena,[6,7] and a variety of phenomena related to distributed coupling into nonlinear waveguides.[8-17] One of the disadvantages of distributed coupling via prisms or gratings is decreased coupling efficiency[11,12,16] at high laser powers, making efficient nonlinear interactions more difficult to achieve. On the other hand, the distributed coupling does lead to interesting phenomena including optical limiting,[12] optical bistability,[13,14] switching,[10,12-15,17] and measurement of waveguide nonlinearities.[8-10,12,16] In this paper we summarize our experiments on nonlinear distributed couplers in various time regimes.

SUMMARY OF THEORY

The physics of a nonlinear distributed-input coupler, such as a prism or a diffraction grating, is quite simple. For example, consider light incident inside a prism onto the prism base at an angle θ, such that the parallel component of the incident light wavevector is $n_p k_0 \sin\theta$, where n_p is the refractive index of the prism and $k_0 = \omega/c$ (see Figure 1). Since the transverse (z) dimension over which the incident and guided wave fields interact is of the order of the wavelength, only wavevector conservation parallel to the surfaces is required. Therefore efficient generation of the guided wave is obtained only when the guided-wave wavevector $\beta \cong n_p k_0 \sin\theta$. If the waveguide contains media whose refractive index depends on the local intensity, ($n = n_0 + n_2 I$

Permanent addresses:
*ICI Electronics Group, P. O. Box 11, The Heath, Runcorn, Cheshire, WA7 4QE, U.K.
†Dipartimento di Ingegneria Elettrica, Vialle Delle Scienze, Universita Degli Studi, 90128 Palermo, Sicilia, Italy
††3M Center, MS 201-2N-19, St. Paul, MN USA 55144

where I is the local intensity), then the guided-wave wavevector depends on the guided-wave power, $\beta \rightarrow \beta(P_{gw})$. Therefore, as the guided wave power grows under the coupler, wavevector matching parallel to the surface is lost, leading to a cumulative mismatch in phase between the growing guided wave and the incident field. This results in a decrease in coupling efficiency, an angular shift in the optimum coupling angle, and distortion in the pulse envelope in the case of pulsed excitation.

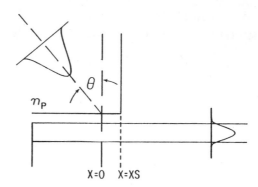

Figure 1. The coupling geometry. A beam, Gaussian along the x-axis, is incident at an angle θ onto the prism base, whose 90° corner is located at xs. A guided wave is generated and propagates for x>xs in the free waveguide.

The mathematical details of the theory of nonlinear distributed coupling can be found elsewhere.[18,19] Writing the field incident as

$$E(r,t) = \frac{1}{2}\hat{e}_{in}a_{in}(x,t)\,\exp[j(\omega t - n_p k_0[\sin\theta x - \cos\theta z])] + c.c. \tag{1}$$

and the guided wave field as

$$E(r,t) = \frac{1}{2}\hat{e}_{gw}a_{gw}(x,t)\,\exp[j(\omega t - \beta(x,t)x)] + c.c. \,, \tag{2}$$

leads to the coupled wave equation[18,19]

$$\frac{d}{dx}a_{gw}(x,t) = \Gamma\,a_{in}(x,t)\,f(z)\,\exp[j\phi(x,t)] - \left[\frac{\alpha}{2} + \frac{1}{\ell}\right]a_{gw}(x,t) \,. \tag{3}$$

Typically the incident light has a field distribution along the base of the prism which is Gaussian, $a_{in}(x) \propto \exp[-x^2/w_0^2]$. The time dependence is either Gaussian with full width at half intensity Δt for pulsed excitation, or a constant for the cw case. For the guided wave field: α is the guided wave attenuation (intensity) coefficient; ℓ is the distance over which a guided wave field amplitude falls to 1/e of its initial value because of reradiation back into the prism alone; $a_{gw}(x,t)$ is the guided wave field amplitude normalized so that $|a_{gw}(x,t)|^2$ is the guided wave power in watts/meter²; and $f(z)$ is the transverse field distribution.

The nonlinear evolution of the guided wave is governed by the dynamics of $\phi(x,t)$. Neglecting diffusion effects, and including nonlinearities with relaxation times (τ) both faster $(\tau_f \gg \Delta t)$ and slower $(\Delta t \gg \tau_s)$ than the pulse width (or power ramping rate), the evolution of the guided wave phase $\phi(x,t)$ is given for the general case by[19]

$$\frac{\partial}{\partial x}\phi(x,t) = \beta_0 - n_p k_0 \sin\theta + k_0\left[A_f|a_{gw}(x,t)|^2 + A_s\left(\frac{1}{\tau_s}\right)\int_{-\infty}^{t}dt'|a_{gw}(x,t')|^2\right] \,. \tag{4}$$

138

$$A = \frac{\int_{-\infty}^{\infty} dz \; n_2(z) |f(z)|^4}{\int_{-\infty}^{\infty} dz \; |f(z)|^4} \; ,$$

(5)

where $n_2(z)$ is the appropriate coefficient for the fast (usually Kerr law) or slow (termed "integrating") nonlinearity. From Eq. (4) it is useful to define for integrating nonlinearities

$$n_{2eff} \cong n_2 \; \frac{\Delta t}{\tau_s} \; ,$$

(6)

which shows clearly that the effective nonlinearity for pulses of width Δt is reduced by the ratio of the pulse width to the relaxation time.

These equations can be interpreted as follows. For the instantaneously responding nonlinearity, it is possible to define a nonlinear guided-wave effective index $N(x,t) = N_0 + A_f |a_{gw}(x,t)|^2$, in analogy with an intensity-dependent refractive index $n = n_0 + n_2 I$. Therefore the in-coupled guided wave power reflects instantaneously the power of the incident field and the resulting wavevector mismatch leads to reduced coupling efficiency. For the integrating nonlinearity, the change in the effective index accumulates over the duration of the pulse. Therefore, the effect of the nonlinearity becomes progressively more pronounced, resulting in pulse distortion, as well as reduced coupling efficiency. Thermally induced changes in refractive index subsequent to guided wave absorption are the most common type of integrating nonlinearities.

EXPERIMENTS

To date we have performed nonlinear distributed coupling experiments on five different material systems: liquid-crystal cladding layers,[11] ZnO[12] and ZnS[13] thin-film waveguides, and ion-exchanged semiconductor-doped glass waveguides.[17] In most cases, thermal effects were the dominant nonlinearity encountered, especially at high laser powers. We separate the discussion into experiments on a) integrating nonlinearities in which the pulse width Δt is shorter than the nonlinearity "turn-off" time τ_s, and b) Kerr-like situations in which the optical power is varied slowly relative to the relaxation time τ_f.

Integrating Nonlinearities

Extensive experiments have been performed on thin-film ZnO waveguides. They were fabricated by magnetron sputtering, had propagation losses of the order of dB/cm, were epitaxial with the c axis normal to the substrate surface, and exhibited bulk ZnO refractive indices to ±0.001. Strontium titanate prisms and holographically formed, ion-milled gratings were used for input and output coupling.

Large changes in the coupling efficiency with incident pulse energy were found for both nanosecond (15 ns) and picosecond (25 ps) pulses of 530-nm wavelength (see Figure 2). (The coupling efficiency was first optimized at low powers.) These results are in good agreement with theory, based on either a Kerr law or an integrating nonlinearity, and yield a nonlinearity (effective for the integrating case) with absolute value 2×10^{-16} m²/W. Nanosecond pulse experiments at 570 nm and 750 nm yielded similar results, showing that the nonlinearity is only weakly wavelength dependent. The peak powers obtained in the picosecond experiments imply that two photon effects should be contributing in the ZnO[20] films, and indeed the difference between the nanosecond and picosecond results is indicative of interference between two photon and absorptive nonlinear mechanisms.

139

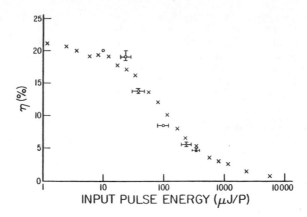

Figure 2. Measured coupling efficiency versus pulse energy at λ = 532 nm for prism coupling into the ZnO waveguides. The x identifies 10 ns pulses and the o the 25 ps pulses.

The time response of the nonlinearity in the coupling region was probed by simultaneously in–coupling a 632–nm He–Ne laser beam overlapped in the coupling region with the high-power beam. The recovery time for the coupling of the He–Ne beam was measured to be $\cong 1$ μs, identifying the nonlinearity as integrating with a relaxation time $\tau_s = \cong 1 \mu$s. Model calculations for thermal diffusion in the ZnO films predicted a τ of 0.6 μs, implying that the mechanism is thermal.

The variation in the coupling efficiency with incidence angle at low and high powers is shown in Figure 3. Both a broadening of the response curve and a shift in the peak coupling angle are obtained. The peak shift gives $n_{2\text{eff}} = +2\times10^{-16}$ m^2/W, which is in agreement with the coupling efficiency measurements. This yields $n_2 = 2\times10^{-14}$ m^2/W for the thermal nonlinearity.

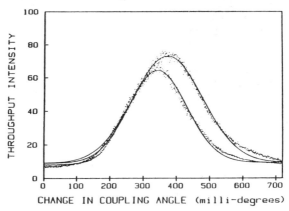

Figure 3. In-coupling efficiency versus coupling angle at low and high powers. The vertical scale of the higher energy pulse (broader one) was adjusted to match the rising edge of the lower energy curve.

Integrating nonlinearities are predicted to produce asymmetric pulse distortion by means of Eq. (3). For a coupler optimized at low powers, the evolution of the temporal envelope of the in–coupled guided wave pulse with increasing energy is shown in Figure 4. Switching up and down within the pulse profile is observed, with switching times which decrease with increasing pulse energy. At high energies, sub-nanosecond switching within the 15–ns incident pulse envelope was measured with a streak camera. At the highest energies used, three maxima were observed in the pulse envelope. Such

140

maxima are indicative of a travelling–wave interaction between the incident field and the guided wave field, with cumulative phase differences between them of multiples of π. The theoretically predicted pulse profiles are also shown in Figure 4, and the agreement with experiment is excellent. Good agreement with theory was also obtained at fixed input energy and variable incidence angle. Although this switching can be fast (sub–nanosecond), the material does not relax for the order of a microsecond. For this time period, a spatial temperature and hence refractive index distribution exists in the coupler region. Therefore, reproducible switching characteristics will not be obtained for pulses whose separation is less than the characteristic relaxation time: thus this phenomenon is not useful for switching high–speed serial data streams.

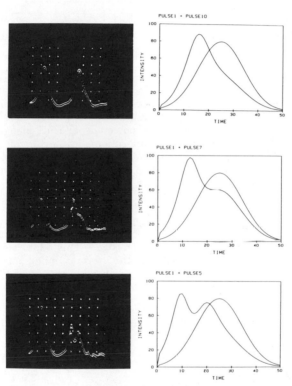

Figure 4.   Comparison of experimental and theoretical temporal–pulse profiles at three different energy levels. The first experimental trace in each pair corresponds to the input pulse. Horizontal scale is 10 ns/division.

Grating couplers were also used to verify that the nonlinear coupling process is common to distributed couplers. The coupling grating had a periodicity of 0.34 μm and was ion–milled into the ZnO film to a groove depth of 0.02 μm. The shape of the coupling efficiency versus pulse energy and pulse distortion were qualitatively similar to that obtained with the prism couplers.

Similar phenomena have been observed in coupling to ion–exchanged semiconductor–doped glass waveguides,[21] also with excellent agreement between experiment and theory, for pulse envelopes of hundreds of nanoseconds duration with mode–locked pulses.[17] For single 70–ps pulses,[22] an interference between the very fast electronic and slow thermal nonlinearities was observed in the coupling efficiency experiments (see Figure 5). Furthermore, large changes in the pulse profile were observed when the incident beam was translated relative to the 90° edge of the coupling prism, all of which is in agreement with theory.

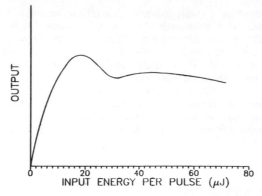

Figure 5.   The output versus input energy for prism coupling 100-ps pulses into a semiconductor-doped glass waveguide.

Kerr–Like Nonlinearities

Our most extensive experiments[13] have been performed on ZnS_0 film waveguides which were fabricated by ion-assisted deposition for the first 1000 Å and by normal thermal evaporation for the balance.   The power-dependent change in the refractive index has been identified as thermal (via absorption), leading to an intensity-dependent absorption associated with a temperature-induced shift of the band gap toward the excitation wavelength.   The angular variation in the coupling efficiency was performed with precise angular positioning and computer-controlled scanning.

The variation in coupling efficiency to the TM_0 mode for increasing incident power (20 mW → 2.6 W) under steady-state conditions is shown in Figure 6.   The optimum coupling angle shifts to higher angles and a progressively stronger asymmetry with increasing power occurs on the high angle edge.   This edge steepens into a vertical switching characteristic for angular shifts in the peak coupling angle greater than the angular width of the low-power coupling peak.   These results are in excellent agreement[13] with the theory outlined here.

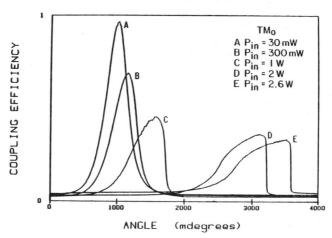

Figure 6.   The measured angular variation in coupling efficiency with increasing incident power for a ZnS waveguide.

As the angle for the high-power case is tuned towards the optimum coupling angle, progressively more of the guided wave is coupled in and hence absorbed, raising the film temperature.   This in turn increases the effective guided wave index and moves the peak coupling angle (for that temperature) to higher angles.   Ultimately the incidence angle corresponds to the optimum coupling angle. A further increase in incidence angle produces run-away switch-down because the in-coupled power is reduced, which results in a temperature drop and hence a reduction in optimum coupling angle, which in turn moves the coupling angle further from resonance, resulting in yet less in-coupled power. The net result is switch-down to the low-coupling efficiency associated with the low-power case.

Further experiments at large powers indicated angular bistability as the angle was tuned from above and below the low-power optimum coupling angle. This can be explained either in terms of increasing absorption bistability, or in terms of longitudinal thermal feedback in the coupling region. As expected, power bistability was observed for an incidence angle detuned by 5 linewidths from the low-power case (see Figure 7). No bistability was observed unless the incidence angle was sufficiently detuned to positive angles to intercept a switching characteristic in the angular scans.

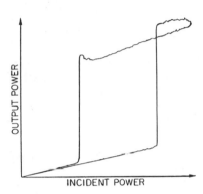

Figure 7. The output versus input energy for both increasing and decreasing power at a detuning of ≅5 linewidths from the optimum low-power coupling angle for a ZnS waveguide.

The reduced coupling efficiency with incident power is an impediment to using optical waveguides for nonlinear optics interactions in nonlinear spectroscopy.[5]   As discussed before, the problem is that the effective index N varies with position in the coupler region.   However, if for grating coupling the grating wavevector κ (and therefore periodicity) can be varied with distance so that

$$k_0 \sin\theta + \kappa(x) = \beta_0 + A_f |a_{gw}(x)|^2 \, , \qquad (7)$$

the synchronous coupling condition can be maintained throughout the coupling region and the coupling efficiency remains optimized.   The required form for $\kappa(x)$ is complicated, but it can be approximated by a linear chirp,[18] that is a linear variation in grating periodicity with distance.

This concept was tested with spun-on polystyrene waveguides which exhibit a useful thermal nonlinearity of $n_2 = -10^{-12}$ m^2/W.   The procedures outlined in Ref. 23, in conjunction with Reactive-Ion-Beam-Etching, were used to fabricate chirped gratings (chirp rate of ≅4×10^{-3} over a 2-mm grating) in a substrate, prior to overcoating with the polystyrene guiding film.   The coupling efficiency for the chirped grating, illuminated with a 2-mm-diameter laser beam at λ = 0.515 μm, was measured as a function of incident power, and the results are shown in Figure 8.   In good agreement with theory, the coupling efficiency has a definite optimum which occurs at a well-defined input power.   Therefore using chirped gratings it is possible to couple high power beams efficiently into nonlinear waveguides.

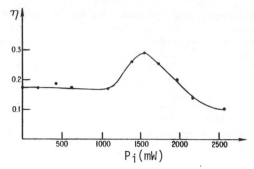

Figure 8. Grating coupling efficiency versus input power for a chirped grating. The nonlinear waveguide was a polystyrene film.

SUMMARY

Prism or grating coupling into integrated optics waveguides is complicated by the presence of waveguide media that exhibit an intensity-dependent refractive index with relaxation ("turn-off") time longer and/or shorter than the rate at which the optical power changes. In all cases, the coupling efficiency decreases with increasing power (for fast nonlinearities) or increasing energy (slow nonlinearities). However, the coupling efficiency can be re-optimized at high powers by using appropriate chirped gratings.

Asymmetric pulse distortion is produced on coupling into waveguides which exhibit nonlinearities with "turn-off" times much longer than the pulse widths. Fast switching and multiple maxima within the pulse envelope are observed because of cumulative phase mismatches in the travelling wave interaction between the incident and guided wave fields.

Under certain conditions, bistability can be obtained in the coupling process.

ACKNOWLEDGMENTS

This research was supported by the Joint Services Optics Program of ARO and AFOSR, the Air Force Office of Scientific Research (AFOSR-84-0277), the NSF/Industry Optical Circuitry Cooperative, and the National Science Foundation (ECS-8304749).

REFERENCES

1. W. Sohler and H. Suche, Integrated Optics III, *Proc. Soc. Photo-Opt. Instr. Eng.* 408:163 (1983).
2. G. I. Stegeman and C. T. Seaton, *Appl. Phys. Rev.* 58:R57 (1985).
3. C. Karaguleff, G. I. Stegeman, R. Zanoni, and C. T. Seaton, *Appl. Phys. Lett.* 7:621 (1985).
4. A. Gabel, K. W. Delong, C. T. Seaton, and G. I. Stegeman, Efficient degenerate four-wave mixing in an ion-exchanged semiconductor-doped glass waveguide, *Appl. Phys. Lett.*, submitted.
5. W. M. Hetherington III, Z. Z. Ho, E. W. Koenig, R. M. Fortenberry, and G. I. Stegeman, *Chem. Phys. Lett.* 128:150 (1986).
6. H. Vach, C. T. Seaton, G. I. Stegeman, and I. C. Khoo, *Opt. Lett.* 9:238 (1984).
7. I. Bennion, M. J. Goodwin, and W. J. Stewart, *Electron. Lett.* 21:41 (1985).
8. Y. J. Chen and G. M. Carter, *Appl. Phys. Lett.* 41:307 (1982).
9. G. M. Carter, Y. J. Chen, and S. K. Tripathy, *Appl. Phys. Lett.* 43:891 (1983).

10. Y. J. Chen, G. M. Carter, G. J. Sonek, and J. M. Ballantyne, *Appl. Phys. Lett.* 48:272 (1986).
11. J. D. Valera, C. T. Seaton, G. I. Stegeman, R. L. Shoemaker, Xu Mai, and C. Liao, *Appl. Phys. Lett.* 45:1013 (1984).
12. R. M. Fortenberry, R. Moshrefzadeh, G. Assanto, Xu Mai, E. M. Wright, C. T. Seaton, and G. I. Stegeman, *Appl. Phys. Lett.* 49:6987 (1986).
13. G. Assanto, B. Svensson, D. Kuchibhatla, U. J. Gibson, C. T. Seaton, and G. I. Stegeman, *Opt. Lett.* 11:644 (1986).
14. W. Lukosz, P. Pirani and V. Briguet, *in:* "Optical Bistability III," H. M. Gibbs, P. Mandel, N. Peyghambarian, and S. D. Smith, eds., Springer-Verlag, Berlin (1986).
15. F. Pardo, A. Koster, H. Chelli, N. Paraire and S. Laval, *in:* "Optical Bistability III," H. M. Gibbs, P. Mandel, N. Peyghambarian, and S. D. Smith, eds., Springer-Verlag, Berlin (1986). F. Pardo, H. Chelli, A. Koster, N. Paraire and S. Laval, *IEEE J. Quant. Electron.*, QE-23:545 (1987).
16. S. Patela, H. Jerominiek, C. Delisle, and R. Tremblay, *J. Appl. Phys.* 60:1591 (1986).
17. G. Assanto, A. Gabel, C. T. Seaton, G. I. Stegeman, C. N. Ironside, and T. J. Cullen, *Electron. Lett.* 23:484 (1987).
18. C. Liao and G. I. Stegeman, *Appl. Phys. Lett.* 44:164 (1984). C. Liao, G. I. Stegeman, C. T. Seaton, R. L. Shoemaker, J. D. Valera, and H. G. Winful, *J. Opt. Soc. Am.* A2:590 (1985)
19. G. Assanto, R. M. Fortenberry, R. Moshrefzadeh, C. T. Seaton, and G. I. Stegeman, *J. Opt. Soc. Am. B*, submitted.
20. E. W. van Stryland, H. Vanherzeele, M. A. Woodall, M. J. Soileau, A. L. Smirl, S. Guha, and T. F. Boggess, *Opt. Eng.* 24:613 (1985).
21. T. J. Cullen, C. N. Ironside, C. T. Seaton, and G. I. Stegeman, *Appl. Phys. Lett.* 49:1403 (1986).
22. G. Assanto, C. T. Seaton, and G.I. Stegeman, unpublished.
23. Xu Mai, R. Moshrefzadeh, U. J. Gibson, G. I. Stegeman, and C. T. Seaton, *Appl. Opt.* 24:3155 (1985).
24. R. Moshrefzadfeh, B. Svensson, Xu Mai, C. T. Seaton, and G. I. Stegeman, "Chirped gratings for efficient coupling into nonlinear waveguides," *Appl. Phys. Lett.*, in press.

BISTABILITY ON A SURFACE IN NONLINEAR DIFFRACTION

V. M. Agranovich, A. I. Voronko, and T. A. Leskova

Institute of Spectroscopy, USSR Academy of Sciences, Troitsk,
Moscow r-n, 142092, USSR

ABSTRACT

We discuss the optical phenomena that arise due to the diffraction
of a surface polariton (SP) by the impedance step created when a thin
film of thickness d (d $\ll$ λ, where λ is the vacuum wavelength of the SP)
is applied on a metal surface. The frequency of the excitations in the
film is assumed to be in the frequency range of the SP. The theory of SP
diffraction by an impedance step has been developed in[1] for the case
when the optical nonlinearity of the film may be neglected. On the basis
of this linear theory we have succeeded in studying analytically the
nonlinear diffraction of SPs by the impedance step produced by applying
an optically nonlinear film with a specific form of the nonlinearity on
an optically linear substrate (metallic substrate). Namely, in the
resonance region $\omega \approx \omega_o$, where $\varepsilon_1(\omega_o) = 0$, the dielectric constant of the
film is assumed to have the form $\varepsilon^{NL} = a[\omega - \omega_o(Y)]$, where $\omega_o(Y) = \omega_o - Y$,
$Y = \alpha \omega_o N$, N is the concentration of SPs and α is the nonlinearity
constant, and N = $U(\omega)/\hbar\omega$, where $U(\omega)$ is the density of the electromagnetic field energy carried by the SP in the film. In the frequency
region under discussion ($\omega \approx \omega_o$) the existence of surface waves of
several kinds is possible, so the expression assumed for ε^{NL} corresponds
to the disregard of wave interference. To estimate the role of this
effect the problem of nonlinear diffraction has been solved numerically
for a nonlinearity of the form $Y = \beta|\vec{E}|^2$, where $\vec{E}$ is the total electric
field in the film. The analytical study, as well as the numerical calculations, shows a bistable dependence of the intensity of the transmitted
(or diffracted) radiation on the intensity of the incident SP.

1. LINEAR DIFFRACTION OF SURFACE POLARITONS

In the presence of a transition layer or a thin film on a metal
surface the spectrum of surface polaritons can change drastically. The
effect of a thin film is most prominent when the frequencies of electric-
dipole-active excitations in the film fall in the frequency range in
which surface polaritons on the metal surface can exist. As was shown

in[1] such a resonance leads to the appearance of a gap in the surface
polariton spectrum whose width is $\Delta \sim \sqrt{d/\lambda}$, where d is the thickness of
the film, λ is the vacuum wavelength of the surface polariton, and d
$\ll \lambda$. This splitting was observed in both the infrared region[2] and in
the spectral region of electronic transitions[3,4]. Furthermore,
additional surface waves appear in the same spectral region[5]. Their
appearance however, is not connected with the inclusion of spatial
dispersion in the substrate or film, but follows from the fact that there
are terms in the surface polariton dispersion law that are linear in k.
These terms are of the order of $|k|d$, where k is the wave vector of the
surface polariton. The inclusion of spatial dispersion in the film only
increases the number of additional surface waves[6].

It is known that in bulk crystal optics with spatial dispersion, the
inclusion of additional waves results in appreciable modifications of the
Fresnel formulae that determine the amplitudes of the radiation reflected
from and transmitted through the crystal boundary. It turns out that
analogous effects also occur in the reflection and refraction of surface
waves at the boundary between surfaces. However, an important feature of
surface polariton refraction is the possibility of the conversion of
surface polaritons into bulk radiation, and this makes the mathematical
treatment of the problem much more complicated.

A theory of the diffraction of surface polaritons by the impedance
step created by applying a thin film on a metal surface was developed
in[6-9]. The thin film was assumed to be described by a dielectric
constant containing a zero at the frequency ω_0. An analytic solution for
the diffraction problem was obtained by the Wiener-Hopf technique, and
the energy transformation coefficients for surface polaritons impinging
on the step either from the region of the clean metal surface or from the
region of the metal surface coated with the film were calculated, as well
as the energy transformation coefficients for a bulk electromagnetic wave
incident onto the step from the vacuum above the surface.

Figure 1 illustrates schematically the process of wave transformation when a surface polariton with a wave vector k (ω) is incident onto
the edge of the film from the region of the clean metal surface. The
dispersion curve for surface polaritons on the clean metal surface is
shown in Fig. 1b, while the dispersion curves for surface polaritons on
the metal surface coated with the film in which at the frequency ω_0 there
is a dipole-active excitation polarized perpendicularly to the plane of
the film are shown in Fig. 1c. In this case a gap and an additional
surface wave appear in the surface polariton spectrum. Therefore, for
frequencies below the gap two transmitted surface waves with wave
vectors $k_1(\omega)$ and $k_2(\omega)$ may propagate in the region x > 0. When the
frequency of the incident surface polariton is far from the resonance
frequency most of the incident energy is carried away by the transmitted
ordinary surface polariton. As the frequency approaches the gap most of
the incident energy is radiated into the vacuum (Fig. 2). For the
reverse process, viz. the diffraction of bulk plane waves by the
impedance step, the anomaly in the coefficient of specular reflection
from the region of the impedance step is of special interest. In Fig. 3
the reflection coefficient is plotted as a function of the frequency for
different angles of incidence. It is possible by decreasing the angle of
incidence to bring up the dip in the frequency dependence of the
reflection coefficient to the natural width which is determined by
damping.

We have already pointed out that in the presence of spatial
dispersion in the film not two but four surface waves can exist in a
certain frequency region (the dispersion curves for surface polaritons

148

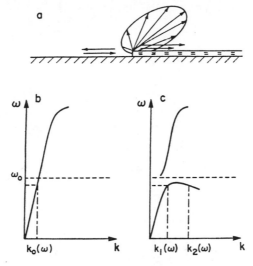

Fig. 1. Scheme of wave transformations by an impedance step (a). The dispersion curves for surface polaritons on a clean metal surface (b) and on a metal surface coated with a film (c).

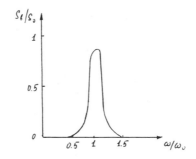

Fig. 2. Frequency dependence of the intensity of the radiation.

are shown in Fig. 4). In addition, spatial dispersion leads to the appearance of edge modes, i.e. quasilocal, one-dimensional oscillations of the excitonic polarization near the edge of the film. The excitation of edge modes results in an additional resonance structure in the frequency dependence of the energy transformation coefficients for surface polaritons. For example, the transmission coefficients for the case when the edge mode is absent (curve I) and for the case when there

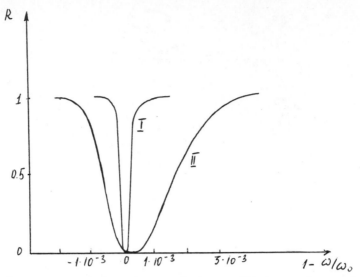

Fig. 3. Frequency dependence of the coefficient of specular reflection I – an angle of incidence of 3º, II – an angle of incidence of 10º.

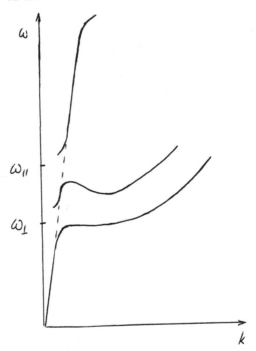

Fig. 4. The dispersion curve for surface polaritons in the presence of spatial dispersion in the film.

is an edge mode at the frequency ω_n (curve II) are shown in Fig. 5.

A film, whose length in the direction of propagation of surface polaritons is finite, deposited on a metal surface is an interferometer for surface polaritons. In this case, the intensities of the transmitted and reflected surface polaritons oscillate as functions of either the length of the film or the frequency of the incident surface polariton. There can be two types of interference effects. The first one is due to

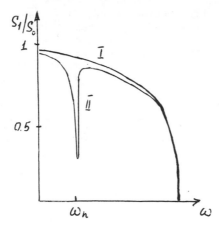

Fig. 5. Frequency dependence of the transmission coefficient. I – in the absence of an edge mode, II – in the presence of an edge mode.

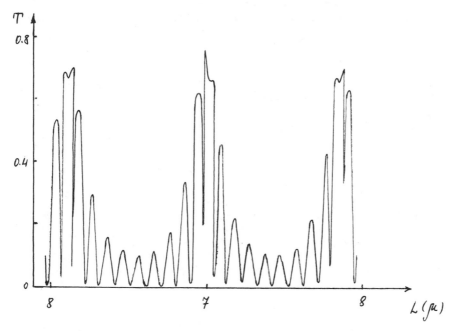

Fig. 6. The transmission coefficient for surface polaritons as a function of the interferometer length L.

the multiple reflection of surface waves from the boundaries of the interferometer. The second is due to the fact that bulk radiation appearing at the front edge of the film can excite a transmitted surface wave at the back edge. This rediffraction leads to the appearance of oscillations of an anomalously large period. These oscillations were observed in [10]. The presence of the additional surface wave leads to the appearance of resonance structure in the interference pattern (Fig. 6).

2. NONLINEAR DIFFRACTION OF SURFACE POLARITONS

Up to this point we have not taken into account optical nonlinearity, i.e. the dependence of the dielectric constants of the media on the radiation intensity. In this section we will discuss some features of the nonlinear diffraction of surface polaritons by the impedance step created by applying a thin nonlinear film on a linear metallic substrate. In this case and for a specific form of the nonlinearity the diffraction problem can be studied analytically. A general form of the nonlinearity requires the use of numerical methods. The results obtained in this case will be presented below.

In the vicinity of the resonance $\omega \approx \omega_o$ ($\varepsilon_1(\omega_o) = 0$)) the dielectric constant of the film can be written in the form

$$\varepsilon_1(\omega) = a(\omega - \omega_o + Y), \tag{1}$$

where the nonlinear shift of the frequency is

$$Y = \alpha \omega_o (N_1(\omega) + N_2(\omega)), \tag{2}$$

and N_1, N_2 are the concentrations of the surface polaritons ($i = 1, 2$) in the film. Here and in what follows we consider the frequency region below the gap in the surface polariton spectrum, $\omega < \bar{\omega}$, where $\bar{\omega}$ is the frequency of the lower edge of the gap. In (2) α is the nonlinearity constant and

$$N_i(\omega) = \frac{1}{\hbar\omega} U_i(\omega), \tag{3}$$

where $U_i(\omega)$ is the density of the energy of the electromagnetic field of the surface polariton in the film[11].

As the dielectric constant of the film depends on Y, all the fields in the film are functions of Y. Consequently, the definition (2) is, actually, an equation for Y

$$Y = F(Y). \tag{4}$$

At a fixed frequency ω, $\omega < \bar{\omega}$, Y does not depend on the coordinate x, so that the dielectric constant $\varepsilon_1(\omega,Y)$ also does not depend on x. Hence, in order to calculate the function F(Y) we can use the results of the theory of the linear diffraction of surface polaritons[7,8] at fixed values of Y. Taking into account all that has been said we can write for the function F(Y) the following expression

$$F(Y) = 4\alpha S_o \frac{a\kappa_o(\kappa_1 + \kappa_2)^2}{h\varepsilon_1^2(\omega,Y)k_o(k_1 + k_2)} \left(\frac{k_o^3 k_2}{(k_1 + k_o)(k_2 - k_o)^2} + \right.$$

$$\left. + \frac{k_1^3 \kappa_1^2 (k_2 + k_o)^3}{k_o(k_2 + k_1)^2(\kappa_1 + \kappa_o)^2(\kappa_2 + \kappa_o)^2} \right), \tag{5}$$

where S_o is the energy flux in the incident surface polariton, k_i ($i = 0,1,2$) are the wave vectors of the incident and transmitted surface waves, respectively, and $\kappa_i = (k_i^2 + \frac{\omega^2}{c^2})^{1/2}$. Note that the wave vectors k_i ($i = 1, 2$) are the solutions of the dispersion relation for fixed values of Y.

The function $F(Y)$ given by (5) increases rapidly as Y approaches the critical value $Y_c = \bar{\omega} - \omega$. Consequently, there can be a different number of solutions of Eq. (4) depending on the parameter S_o. (For illustration, the right- and left-hand sides of Eq. (4) are plotted in Fig. 7.) It is straightforward to determine the solutions of Eq. (4) graphically. When this is done we can determine the function $Y(S_o)$ and, subsequently, the energy fluxes in the diffracted surface and bulk waves as functions of the incident energy S_o.

Note that the results obtained up to this point are based on the assumption that the dielectric constant of the film does not depend on x because $\omega \leqslant \bar{\omega}$. If $\omega \leqslant \bar{\omega}$, but $Y > Y_c = \bar{\omega} - \omega$, then the dielectric constant depends on x. Indeed, in this case due to the nonlinear shift of the resonance frequency $\omega_o - Y$, the frequency ω falls in the region of the gap in the surface polariton spectrum. So for a surface polariton of frequency ω a barrier appears at the impedance step, and the width of

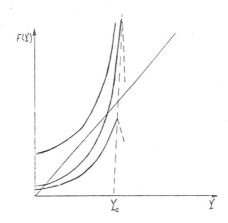

Fig. 7. Illustration of the graphical solution of Eq. (4).

this barrier equals approximately the penetration depth of the radiation. Far away from the edge of the film the surface polaritons can be excited either due to penetration through the barrier or due to rediffraction of the surface polaritons. It follows from what has been said that the results of the theory of linear diffraction cannot be used to calculate the function $F(Y)$ in this region of Y. However, we can approximate $F(Y)$ by means of the following qualitative arguments. Let $Y = Y_1 > Y_c$, but differ only slightly from Y_c, so that $Y_1 - Y_c \ll \Delta$, where Δ is the width of the gap. In this case the penetration depth is large so the amplitudes of the fields at the impedance step should be the same as those for $Y = Y_2 < Y_c$, $Y_c - Y_2 \ll \Delta$. Consequently, the function $F(Y)$ in the vicinity of Y_c can be regarded as a symmetric function (see Fig. 7). Of course, all these arguments can be verified by means of numerical calculations. Assuming the symmetry of the function $F(Y)$ we obtain a third root of Eq. (4).

In Fig. 8 the energy flux in the transmitted ordinary surface polariton (i=1) is shown as a function of the incident energy S_o. A bistable behavior occurs also for all the energy fluxes in the diffracted

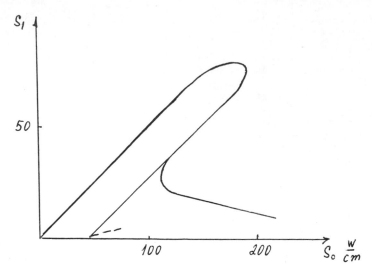

Fig. 8. The energy flux in the transmitted surface polariton as a
function of the incident energy. I according to an analytical
theory, II from numerical calculations.

waves. The portion of the curve corresponding to values of $Y > Y_c$, is
denoted by the dotted line. The results presented in Fig. 7 and Fig. 8
have been obtained for the system consisting of a CuCl film of thickness
$d = 40$ Å on an Al surface. A value of the nonlinearity constant $\alpha = 1.4$
10^{-20} cm^3 was assumed[12]. We would like to point out that the nonlinear
contribution to the dielectric constant of the film does not exceed the
value $\Delta\varepsilon_1 \simeq 10^{-3}$. Finally, we note that the critical values of
either α or S_0 for which the phenomena under consideration can be
observed are determined by the inequality $F(Y_c) > Y_c$.

All of the results presented have been obtained under the assumption
of a specific form of the nonlinearity. In this case the interference of
the fields in the film does not contribute to the dielectric constant.
In order to take into account effects of this interference we obtained a
numerical solution for the problem of nonlinear diffraction for the case
when the nonlinear frequency shift is given by

$$Y = \beta|E(r,t)|^2, \tag{6}$$

where $E(r,t)$, the electric field strength in the film, equals the sum of
the fields of the surface and bulk waves. The numerical solution was
obtained by means of the mode method[12]. The results obtained are shown
in Fig. 8 (curve II). It can easily be seen that the results obtained
from the numerical calculations coincide qualitatively with those
obtained on the basis of the simple model of nonlinearity. Hence, one
can conclude that the interference effects lead to a weak redistribution
of the intensities of the diffracted waves.

Note that the effects described above occur due to the fact that the
energy density increases rapidly as the frequency approaches the gap in
the surface polariton spectrum. Such an effect occurs also for bulk
polaritons in dielectrics in the region of excitonic resonances with a
negative effective mass. The situation is quite analogous to that
described above, and bistability of the reflection and transmission
coefficients for the bulk radiation is found to be possible.

REFERENCES

1. V. M. Agranovich and A. G. Mal´shukov, Optics Commun. <u>11</u>, 169 (1974).
2. V. A. Yakovlev, V. G. Nazin, and G. N. Zhizhin, Optics Commun. <u>15</u>, 293 (1975).
3. T. Lopez-Rios, F. Abeles, and G. Vuye, J. de Phys. <u>39</u>, 645 (1978).
4. I. Pockrand, A. Brillante, and D. Mobius , J. Chem. Phys. <u>77</u>, 6289 (1982).
5. V. M. Agranovich, Zh. Eksp. Teor. Fiz. <u>77</u>, 1124 (1979).
6. V. M. Agranovich, V. E. Kravtsov, and T. A. Leskova, Zh. Eksp. Teor. Fiz. <u>84</u>, 103 (1983).
7. V. M. Agranovich, V. E. Kravtsov, and T. A. Leskova, Zh. Eksp. Teor. Fiz. <u>81</u>, 1828 (1981).
8. V. M. Agranovich, V. E. Kravtsov, and T. A. Leskova, Solid State Commun. <u>40</u>, 687 (1981).
9. T. A. Leskova and N. I. Gapotchenko, Solid State Commun. <u>53</u>, 351 (1985).
10. Z. Schlesinger and A. J. Sievers, Appl. Phys. Lett. <u>36</u>, 409 (1980).
11. V. M. Agranovich and V. L. Ginzburg, <u>Crystal Optics With Spatial Dispersion and Excitons</u>. (Springer-Verlag, Berlin, 1984) .
12. D. Sarid, N. Peyghambarian, and H. M. Gibbs, Phys. Rev. B<u>28</u>, 1184 (1983).
13. A. I. Voronko, L. G. Klimova, and G. N. Shkerdin, Dokl. Akad. Nauk SSSR <u>287</u>, 1362 (1986).

PICOSECOND NONLINEAR OPTICAL SPECTROSCOPY OF SEMICONDUCTOR

SURFACE STRUCTURE TRANSFORMATIONS

S.A. Akhmanov, S.V. Govorkov, N.I. Koroteev, and I.L. Shumay

R.V. Khokhlov Institute of Optics
Moscow State University, Moscow

ABSTRACT

The efficiency of nonlinear optical spectroscopy is demonstrated for studying the modification of semiconductor surface layers under the influence of charged particle beams, laser pulses, etc. The method is shown to provide an opportunity for real-time-study of the above mentioned processes on a picosecond time-scale.

The complex tensor nature of the nonlinear response is especially sensitive to structural changes and lattice inhomogeneities. This is a key-point in the nonlinear diagnostics of surface modification.

Silicon surface ion-beam randomization and laser amorphization are studied by quadrupole optical second harmonic generation in reflection. The appearance of a strong dipole contribution to the nonlinear response is discovered in a medium with inversion symmetry under the influence of mechanical stress.

The Si crystal lattice dynamics excited by picosecond light pulses is studied by CARS in reflection.

The perspectives of the technique for studying thin films and layered structures including superconducting films is discussed.

1. INTRODUCTION

The demonstration of pulsed laser annealing of semiconductors[1] stimulated wide interest in the problem of material surface-laser interaction. The most interesting problems are to determine the details of the relaxation of energy absorbed by a near surface layer in electron and phonon systems of a crystal and the consequent chain of phase transitions. From the technological point of view it is important to find optimal regimes of laser annealing, to study the dynamics of laser-induced phase transitions, to control the quality of laser-annealed samples, and to find means of optical diagnostics of impurity concentration distribution, etc. The development of non-disturbing optical methods of technological processes control for use in modern semiconductor electronics is of prime importance. These methods have unique temporal resolution which can provide an opportunity to study various fast processes on a subpicosecond time-scale in real time and also have sufficient spatial and spectral resolution. Unfortunately, the usual linear optical methods of diagnostics are practically insensitive to the lattice structure transformation in the case of most commonly used crystals with inversion symmetry. Nonlinear optical methods based on measurement of nonlinear optical susceptibilities might be useful in this case. The symmetry of third and higher order nonlinear susceptibilities reflect crystal symmetry. Thus, the absolute value and relations between various nonvanishing tensor components reveal crystal structural perfection and (or) crystal symmetry distortion under external influence. In so doing nonlinear optical diagnostics turns out to be a powerful tool for study of semiconductor surface modification during ion implantation, laser amorphization, pulsed annealing, film deposition, etc. Being an optical method it has also high temporal, spatial and spectral resolution.

We report here on experimental investigation of the Si surface layer randomization as a result of ion implantation and the creation of strong internal stress in Si surface layer after oxide or silicide film deposition by second optical harmonic generation in reflection from the sample. We report also on picosecond Si lattice excitation by picosecond laser pulses studied by coherent anti-stokes Raman spectroscopy (CARS) in reflection. Experimental results demonstrate the high efficiency and sensitivity of the nonlinear laser spectroscopy of semiconductor surface layers and their structure modification.

2. History

The solution of the problem of nonlinear reflection from the surface of a nonlinear medium was obtained by Bloembergen and Pershan[2] in 1962. The first experiments carried out by Bloembergen et al.[3] did not show any anisotropy of the second harmonic generation in reflection (SHG) from Si surface. However, with more advanced techniques the same experiment[4] did show the antisotropy of the SHG signal. This result was later confirmed in Tom, et al.[5] Shen et al[6] studied molecules adsorbed at the interface. We used SHG in reflection to study the dynamics of pulsed laser annealing of ion-implanted GaAs.[7] Shank et al.[8] studied the dynamics of Si melting under strong laser irradiation on a sub-picosecond time-scale by SHG in reflection. Malvezzi et al.[9] studied the dynamics of GaAs melting by monitoring the dependence of SHG signal on laser intensity. We have shown[10,11] that SHG and the sum-frequency generation in reflection could be used for diagnostics of: non-centrosymmetric semiconductor crystal (GaAs) surface randomization during ion implantation, and also the quality of laser-annealed crystals. Heinz et al.[12] used SHG to study Si surface reconstruction in ultra-high vacuum. The application of third harmonic generation in reflection for the study of surface randomization in ion-implanted Si crystals was shown in.[13,14] Besides information on semiconductor surface layer structures, nonlinear optical diagnostics (for example CARS in reflection) can provide data on energy relaxation during short pulse laser irradiation.[15]

Crystals with inversion symmetry (Si) have electric dipole forbidden second order nonlinearity. However, they can reveal one if the material is undergoing stress caused by film deposition on its surface or by impurity ion intrusion into the crystal during pulsed-laser annealing. These processes cause a dramatic increase of SHG intensity.[16]

The effects mentioned above will be discussed in this report.

3. <u>Si Surface Randomization and Amorphization Studied by SHG in Reflection</u>

Crystals without inversion symmetry (GaAs) have electric dipole allowed second order nonlinearity which is responsible for effective SHG generation in reflection.[7,11] In this case surface randomization as a result of ion implantation introduces partial isotropy into its surface layer symmetry thus leading to the decrease of the effective second order susceptibility $\chi^{(2)D}_{ijk}$ and to the decrease of SHG intensity.[10,11] For crystals with inversion symmetry (Si, Ge) the second order susceptibility is electric dipole forbidden. However, Tom et al.[5] have shown that Si has appreciable quadrupole susceptibility $\chi^{(2)Q}_{ijkl}$ which has anisotropy leading to pronounced dependence of SHG signal intensity on relative orientation of crystal axes and incident light plane of polarization. Thus the SHG intensity would demonstrate modulation when the crystal is rotated about its normal described by

$$I(\psi) \sim |\zeta\cos3\psi + A|^2 \qquad (1)$$

Here ζ is the $\chi^{(2)Q}$ tensor anisotropy, A is an isotropic contribution independent of crystal rotation and ψ is the angle of rotation.

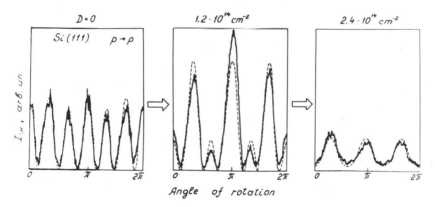

Fig. 1. Orientational dependence of SH signal

a. from pure Si(111)

b. from Si(111) implanted with P^+ (E = 80 kev, D = 1.2 x 10^{14} cm^{-2})

c. from Si(111) implanted with P^+ (E = 80 kev, D = 2.4 x 10^{14} cm^{-2})

According to (1) the SHG intensity is expected to have 6 lobes per period 2π if a P-polarized SHG component is registered with P-polarized incident light, and this is really the case (see Fig. 1a). The data were obtained using actively mode-locked Q-switch Nd:YAG laser as a probe source (λ = 1,06 μm). The laser produce picosecond pulse (τ ~ 150 ps) trains with a repetition rate of about 1 kHz and with an energy 1 mJ per train. The probe pulses were focussed on the sample so that it would not be damaged.

The orientational dependence changed dramatically when the crystal was ion-implanted. Fig. 1b shows that under P^+ implantation (E = 80 keV) with a dose 1.2 x 10^{14} cm^{-2} even peaks reduce in intensity with respect to the odd ones. With implantation-dose 2.4 x 10^{14} cm^{-2} only three peaks are seen (Fig. 1c). The transformation of the SHG orientational dependence is a result of Si surface layer randomization leading to an increase of the isotropic contribution (A) relative value in (1). Thus, a relative decrease of anisotropy in SH signal enables one to introduce a quantitative value describing surface randomization which can be compared with a model of surface randomization suggested by Gibbons.[18] In this model each ion causes atoms to be knocked-out off its sites and creates a large number of radiation defects of various types. The latter are localized in a tunnel shaped volume along the ion trace in the crystal having a cross-section δ_i. However, in order to achieve complete amorphization of a material in a tunnel the energy deposition per unit volume should exceed a definite value. In the case of light ions this value is achieved when several tunnels are superimposed. With this concept in mind Gibbons suggested a formula for the relative area of the amorphous material depending on the dose of ion implantation

$$S = 1 - \sum_{k=0}^{m} \frac{(\delta_i D)^k}{k!} \exp(-D\delta_i) \tag{2}$$

where (m + 1) is a number of superimposed tunnels necessary to achieve complete amorphization. This value of S can be related to the value S' which describes a relative value of the isotropic contribution to SH signal

$$S' = 1 - \frac{|\zeta A|}{|\zeta_0 A_0|} \tag{3}$$

where ζ_0 and A_0 are anisotropic and isotropic components of the SH signal in pure non-implanted silicon.

The dependence of S' on the dose of ion implantation D, calculated from our experimental data[17] is shown in Fig. 2. It is compared with a computer simulation of the dependence S(D) using (2). The m and δ_i were varied to achieve the best fit to the experimental points. Change of δ_i cause the curve S(D) to shift along D-axis as a whole while variation of m determines the initial tilt of the curve. The values obtained from a numerical fit are m + 1 = 3 and δ_i = $60A^2$. Thus, we can estimate the critical energy of amorphization.

$$E_\alpha \sim (m+1)E/\delta_i \ell$$

where ℓ is the mean path of an ion in a crystal. With E = 80 kev and ℓ ≈ 1000 $\overset{o}{A}$ we obtain E_α ≈ 4 x 10^{21} kev/cm^3 which is in good agreement with other data.

Surface melting under a strong UV short laser pulse illumination can lead to surface amorphization as a result of extremely rapid solidification following laser-induced melting. Fig. 3 shows the orientational dependence of SHG from amorphous film on Si surface. We attribute the transformation of orientational dependence to an amorphous material contribution and possibly to a contribution of bulk Si influenced by thin amorphous film on its surface.

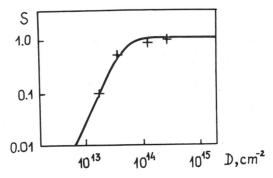

Fig. 2   Si relative suface randomization.
          Points-experiment, curve-model calculation.[18]

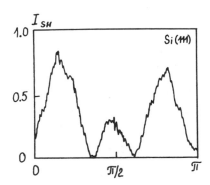

Fig. 3.   Orientational dependence of SH from laser amorphized Si.

4.   Anomalous Increase of SH Intensity from Si with Surface Deposited
      Films and from Ion-Implanted Laser-Annealed Si:The Electric Dipole
      Contribution

      Deformation of a crystal lattice caused by mechanical stress in the
near   surface   layers   of   multi-layered   epitaxial   structures   and
semiconductor-dielectric or semiconductor-metal systems due to various
lattice parameters or inter-atomic distance in different materials at the
interface is a difficult problem in semiconductor technology. This stress
could be controlled by optical means.

      We discovered that SiO_2 film-formation during thermal oxidation of Si
and the resulting appearance of strong mechanical compression (~ 10 kbar)
reveals itself in a dramatic increase (by a factor of about 20) of SH
intensity and a pronounced transformation of its orientational dependence.

      The Si (111) samples with a thickness of about 0,3 mm were studied.
Those were oxidized in a dry O_2 atmosphere at a temperature of ~ 1100°C.
The oxide film had a thickness ~ 500 A°.   The experimental orientational
dependence of the SH signal is shown in Fig. 4.   In this case the P-
polarized SH component was registered with a P-polarized incident beam.
Fig. 4a shows the orientational dependence of the SH signal from the Si
surface with removed oxide layer (however, it had a natural oxide film on
it), and Fig. 4b show orientational dependence from the thermally oxidized
Si.

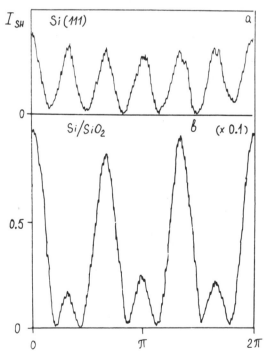

Fig. 4.   Orientational dependence of SH

         a. from pure Si(111)

         b. from Si(111) with thermal oxide film (thickness 500 A°).

We attribute the increase of SH intensity and its orientation dependence transformation to lattice deformation in a near surface layer which leads to violation of its symmetry, resulting in the appearance of strong electric dipole second order nonlinearity forbidden in the bulk material. Theoretical analysis of the induced dipole tensor $\chi^{(2)D}{}_{def}$ shows[19] that the resultant orientational dependence of the SH signal is similar to (1) but in this case an isotropic contribution is expected to be comparable to the anisotropic one ($|\zeta'| \approx |A'|$). Experiment shows that it is really the case.

Inhomogeneous mechanical stress in the near-surface layer is seen also in spontaneous Raman spectra. The observed spectral broadening (≈ 5 cm^{-1}) enables one to estimate the peak value of mechanical stress (~ 10 kbar) since according to Anastassakis et al.[20] the pressure-induced Raman frequency shift is equal to $0,47$ cm^{-1} kbar. We observed similar effects from Si samples with silicide film deposited on its surface. SH intensity increased in this case by more than 2 orders of magnitude.[16]

Probably the most striking effect due to dipole contribution to SHG was observed with ion-implanted laser-annealed Si.[16]

Fig. 5 shows the increase of SH intensity with respect to SH signal from pure Si in ion-implanted laser-annealed Si depending on the dose and type of implanted ions. With As$^+$ implantation (E = 40 kev, D = 1.8 x 10^{16} cm^{-2}) and subsequent pulsed ruby laser annealing SH intensity increased by nearly 3 orders of magnitude, when second harmonic from the fundamental Nd:YAG laser output (λ = 1.06 μm) was registered (process $\omega \rightarrow 2\omega$). However, no increase in SH intensity was observed when SH signal at λ = 0.26 was recorded (process $2\omega \rightarrow 4\omega$). The SH signal from ion-implanted Si without pulsed laser annealing did not show any increase either. Finally, there were no anomalous increase in SH intensity from ion-implanted thermally annealed samples. The orientational dependence shown in Fig. 6 has only three strong peaks per period compared with 6 from pure Si.

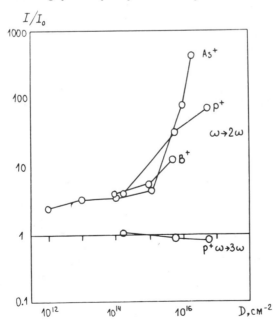

Fig. 5. Relative increase of SH intensity from ion-implanted pulsed-laser annealed Si.

The observed peculiarities can be explained if one assumes the appearance of an electric dipole allowed second order nonlinearity caused by inhomogeneous lattice deformation due to impurity ions. The theoretical analysis[19] shows that this inhomogeneous deformation must be proportional to the gradient of concentration of impurity ions having size and force constants different from that of the host material. During nanosecond pulsed-laser annealing doped Si samples can have impurity concentrations exceeding equilibrium.[1] Besides, due to extremely short (~ 10^{-8}s) melt duration the depth distribution of the impurity does not change appreciably. The maximum value of gradient of impurity concentration is achieved under the sample surface (Z ~ 1000 $\overset{\circ}{A}$) as a result of it, while the concentration gradient is nearly zero at the interface. Thus, the strongest deformation of Si lattice causing the appearance of dipole contribution to $\chi^{(2)}$ (ω) is expected at some depth inside the bulk. This explains why SH is strong in the case of the $\omega \rightarrow 2\omega$ process and does not show any increase in the case of $2\omega \rightarrow 4\omega$. The reason is strong absorption of pulses so that very this surface layer (d ~ 100 $\overset{\circ}{A}$) is probed by $2\omega \rightarrow 4\omega$ process while in the case of $\omega \rightarrow 2\omega$ process the crystal thickness probed is an order of magnitude larger. Taking into account that the impurity concentration gradient is nearly zero at the interface we conclude that there should be no dipole contribution to $\chi^{(2)}$ in the case of UV SH generation.

In the case of thermally annealed samples impurity ions can drift relatively large distances during the process (~ 30 min) thus leading to lower concentration gradient values and small dipole contribution to $\chi^{(2)}$.

Note that the SH intensity increase is not due to the decrease of fundamental bandgap in ion-implanted Si which could cause a resonant increase of $\chi^{(2)}Q$. Fig. 5 shows that there is no increase of third harmonic intensity ($\omega \rightarrow 3\omega$) though it must be expected if the resonant increase of nonlinear susceptibility really happened.

The orientational dependence of the SH signal (Fig. 6) is in agreement with the predicted structure of the dipole tensor induced by inhomogeneous uniaxial stress along the surface normal.[19] The wavelength dependence of the SH signal also shows that silicon-impurity complexes do not play a considerable role in SH increase since in this case the dipole contribution is expected to have the maximum value at the interface where the impurity concentration is the largest. At the same time coherent and incoherent effects due to cluster formation need further investigation.[21]

5. The Dynamics of Si Lattice on a Picosecond Time-Scale:Picosecond CARS in Reflection

Picosecond CARS study of Si under strong picosecond laser excitation is an exciting demonstration of possibilities provided by nonlinear optical spectroscopy in the study of lattice dynamics.[15] Computer simulation of the observed Raman spectra transformation enabled us to get data on temperature, electron hole plasma concentration and amplitude of mechanical stress generated in Si via the deformation potential mechanism.

Two RH6G dye lasers synchronously pumped by a pulse train of the second harmonic of the passively modelocked Nd:YAG laser output were used as coherent sources with frequencies ω_1 and ω_2 in the traditional CARS arrangement. One of the lasers (ω_2) was tuned in the experiment to get anti-Stokes spectrum at frequency $\omega_a = 2\omega_1 - \omega_2$ when the difference frequency $\omega_1 - \omega_2$ was scanned in the vicinity of the ≈ 520 cm^{-1} Raman resonance in Si. The wavelength of the second laser was fixed at the maximum of RH6G emission and its output power was sufficiently strong to cause melting of the sample surface. The computer was used to accumulate photon counts at each point of the CARS spectrum, to monitor laser energy and to tune the dye lasers and spectrometer.

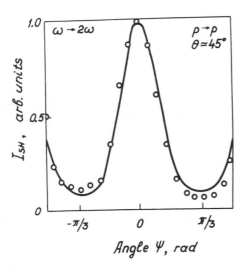

Fig. 6. Orientational dependence of SH signal from ion-implanted pulsed-laser annealed Si.

The Si optical phonon spectra were recorded at various excitation laser pulse fluences E from E = $(0.30 \pm 0.05)E_0$ to E = E_0, where $E_0 = 0.2$ Jc/cm^2 - the laser illuminated spot on the sample surface had the diameter of about 700 μm. At room temperature without unusual excitation the Si optical phonon spectrum is known to be rather narrow (3 cm^{-1}) strong Raman line with frequency of 520 cm^{-1}. As one can see from Fig. 7 increase of the excitation laser pulse fluence induces a dramatic change of both the line shape and width as well as decrease of the Raman line intensity compared with nonresonant electronic background. The Si lattice heating up is known to cause phonon mode "softening" and line broadening and shift to the lower frequency side.[22] On the contrary in our experiment the Raman spectrum remains centered at 520 cm^{-1}. We believe that his feature is a result of mechanical stress generated in a sample in the presence of the dense electron hole plasma leading to Raman frequency shift toward higher frequency.[20]

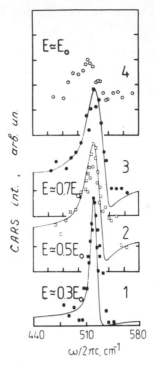

Fig. 7.   Experimental (dots) and computer simulated (curves) CARS spectra
of laser excited Si for various laser fluences E(E_O - threshold
for melting).

The inhomogeneous nature of laser excitation is known to cause
considerable distortion of the spectral information obtained from Raman
spectra.[23]   Thus, in order to get a quantitative estimate of temperature,
concentration of photo-excited carriers n and amplitude of mechanical
compression P in Si we used computer simulation of the CARS spectra.   In
so doing we took into account the temperature dependence of optical
parameters of Si, dependence of Raman line position and width on T and P
and also spatial and temporal distribution of laser intensity and
corresponding distribution of n.   Thus the CARS spectra, both experimental
and computer simulated, were essentially integrated spectra.

In order to get data on temporal and spatial distribution of T, P and
n we carried out numerical solution of the following set of basic
equations:

$$\frac{\partial T}{\partial t} = \frac{\hbar\omega - E_g}{\hbar\omega} \frac{1-R}{d_o \rho C_p} I(t,z)$$

$$\frac{\partial n}{\partial t} = -\gamma n^3 + \frac{1-R}{d_o \hbar\omega} I(t,z)$$

$$\frac{\partial^2 U}{\partial t^2} - C_o^2 \frac{\partial^2 U}{\partial Z^2} = \frac{\theta_o}{\rho} \frac{\partial n}{\partial Z}$$

Here U is the atom displacement, E_g is the fundamental band gap, Z is the
distance along the surface normal, ρ is the density, R is the
reflectivity, γ- Auger recombination rate, θ_o- deformation potential, d_o
depth of probing, C_p is specific heat, $I(t,z) = I_o(t)e^{-z/d_o}$ laser intensity
with Gaussian profile in time and in transverse plane.

An example of the calculated dependences n(t) and T(t) with
E = ∫ I(t)dt = 0.5 E_O is shown in Fig. 8.

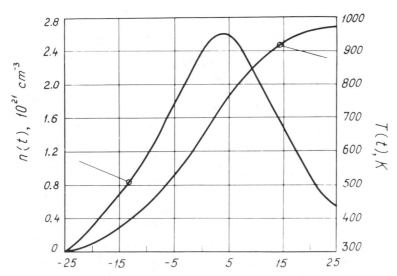

Fig. 8. Calculated dependences of free carrier concentration (n) and surface temperature (T) for laser excited Si with E = 0.5 E_O.

The deformation potential value θ_O and the depth of probing were varied in the calculation in order to get the best fit to the experimental spectra. The values obtained are θ_O = 15 ev, d_O = 1000 A°.

Thus, the experimental results show that the phonon mode is heated up during a laser pulse. Computer simulation enabled us to deduce data on n, T, P, characteristic of strongly excited Si. Application of subpicosecond laser pulses for CARS experiments might enable one to probe the dynamics of excitation in phonon branches.

6.   Conclusion

Nonlinear optical diagnostics of semiconductor surface compared with other methods of surfaces diagnostics has unique temporal resolution (~ 10^{-14}s), enables one to change the depth of probing changing the probe pulse wavelength, has high spectral resolution, is able to probe interfaces between two dense media, does not need high vacuum, and is relatively simple.

Many important problems could be studied using nonlinear optical diagnostics of semiconductors. Among them are processes of electron-photon and phonon-phonon relaxation under strong laser excitation, mechanisms of nonlinear relaxation, lattice dynamics on a femtosecond time scale (for example, using CARS), nonlinear response of thin films (including superconducting films). A number of problems could be solved using combined methods, for example, combining nonlinear pico- and femtosecond spectroscopy with electron and X-ray spectroscopy.

Contribution from V.N. Zadkov and G.I. Petrov are gratefully acknowledged. We are thankful also to I.B. Khaibullin, Y. Fattakhov and M.F. Galijautdinov for preparation of samples.

References

1. I.B. Khaibullin, E.I. Shtyrkov, M.M. Zaripov et al., Rad. Effects 36:225 (1978).
2. N. Bloembergen, P. Pershan. Phys. Rev. 128:606 (1962).
3. N. Bloembergen, R. Chang, S. Jha, C. Lee, Phys. Rev. 174:813 (1968).
4. D. Guidotti, T.A. Driscoll, H.J. Gerritsen, Sol. State Commun. 46:337 (1983).
5. H. Tom, T. Heinz, Y.R. Shen, Phys. Rev. Lett. 51:1983 (1983).
6. C. Chen, T. Heinz, D. Ricard, Y.R. Shen, Phys. Rev. Lett. 46:1010 (1981).
7. S.A. Akhmanov, M.F. Galijautdinov, N.I. Koroteev, G.A. Paitian, I.B. Khaibullin, E.I. Shtyrkov, I.L. Shumay, Sov. Kvant. Electron 10:1077 (1983); Opt. Commun. 43:202 (1983).
8. C.V. Shank, R. Yen, C. Hirlimann, Phys. Rev. Lett. 51:900 (1983).
9. A. Malvezzi, J. Liu, N. Bloembergen, Appl. Phys. Lett. 45:1019 (1984).
10. S.A. Akhmanov, M.F. Galiautdinov, N.I. Koroteev, G.I. Paitian et al., Sov. J. Tech. Phys. Lett. 10:1118 (1984).
11. S.A. Akhmanov, M.F. Galiautdinov, N.I. Koroteev, G.A. Paitian, I.B. Khaibullin, E.I. Shtyrkov, I.L. Shumay, J. Opt. Soc. Amer. B 2:283 (1985); Sov. Phys. Izvestia 46:506 (1985).
12. T. Heinz, M. Loy, W. Thompson, Phys. Rev. Lett. 54:63 (1985).
13. D.J. Moss, H.M. van Driel, J.E. Sipe, Appl. Phys. Lett. 48:1150 (1986).
14. C.C. Wang, J. Bomback, W.T. Donlon, C.R. Huo, J.V. James, Phys. Rev. Lett. 57:1647 (1986).
15. S.V. Govorkov, V.N. Zadkov, N.I. Koroteev, I.L. Shumay, Sov. JETP Lett. 44:98 (1986); Sol. State Commun. 62:331 (1987).
16. S.V. Govorkov, N.I. Koroteev, G.I. Petrov, I.L. Shumay, Report at VII International School on Coherent Optics, Tbilisi, 1987.
17. S.V. Govorkov, N.I. Koroteev, I.L. Shumay, Sov. Phys. Izvestia 50:683 (1986).
18. J.F. Gibbons, Proc. IEEE, 60:1062 (1972).
19. V.I. Yemel'anov, N.I. Koroteev, V.V. Yakovlev, Optica Spectr. (in Russian), 62:1188 (1987).
20. E. Anastassakis, A. Pincuk et al., Sol. State Commun. 8:133 (1970).
21. T. Heinz, M. Loy, Rep. at VII Int. School on Coherent Optics, Tbilisi, 1987.
22. M. Balkanski, R.F. Wallis, E. Haro, Phys. Rev. B 28:1928 (1983).
23. G. Wartman, M. Kemmler, D. von der Linde, Phys. Rev. B 30:4850 (1984).

RAMAN SCATTERING OF LIGHT FROM COHERENTLY EXCITED SURFACE POLARITONS

N.I. Lipatov, Yu. N. Polivanov, and R. Sh. Sayakhov

Institute of General Physics of the Academy
of Sciences of the USSR, Moscow

There is currently active interest in the study of the surface polaritons (SP) which propagate along the interface between two dielectric media.[1] In this paper we present preliminary results of the first experimental study of the Raman scattering (RS) of a probe beam from surface phonon polaritons coherently excited by infrared laser radiation. This process can be viewed either in term of Stokes and anti-Stokes RS from hot SP or as optical mixing (generation of sum and difference frequencies) of bulk and surface polaritons via the second-order nonlinear susceptibility. The main advantages of this technique over spontaneous RS are determined by a considerable and selective enhancement of the scattered light intensity. In this way, in particular, it is possible to observe the RS from SP in bulk samples. (Experiments on spontaneous RS from SP were performed so far only on thin samples).[1]

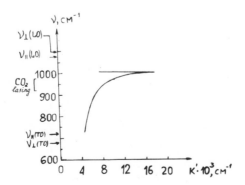

Fig. 1. Calculated dispersion curve of surfaces polaritons in BeO for the case c perpendicular to the crystal surface. The tunable range of the generation of CO_2 laser is shown on the left.

169

In our experiments the SP were excited in a BeO single crystal by the frequency tunable repetitively-pulsed TEA CO_2 laser (with a repetition rate about 10 Hz) using an attenuated total reflection method (Otto's method).[7] BeO was chosen since it is non-centrosymmetric (it has a second-order nonlinearity) and the emission spectrum of a CO_2 laser overlaps the SP spectrum of this crystal (see Fig. 1).

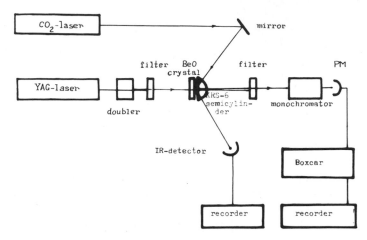

Fig. 2. Experimental setup.

The dispersion of the surface polaritons at the interface of the uniaxial BeO crystal and vacuum with orientation of the optic axis (c axis) perpendicular to the crystal surface are given by

$$\kappa = \kappa' + i\kappa'' = 2\pi\nu \left[\frac{\varepsilon_{\shortparallel}(\nu)\varepsilon_{\perp}(\nu) - \varepsilon_{\shortparallel}(\nu)}{\varepsilon_{\shortparallel}(\nu)\varepsilon_{\perp}(\nu) - 1} \right]^{1/2}$$

In the following calculations we assume that the dispersion laws for $\varepsilon_{\shortparallel}(\nu)$ and $\varepsilon_{\perp}(\nu)$ are given by

$$\varepsilon_{\shortparallel,\perp}(\nu) = \varepsilon_{\shortparallel,\perp}(\infty) \frac{\nu^2_{\shortparallel,\perp}(LO) - \nu^2 - i\nu\Gamma_{\shortparallel,\perp}(LO)}{\nu^2_{\shortparallel,\perp}(TO) - \nu^2 - i\nu\Gamma_{\shortparallel,\perp}(TO)}$$

because in this crystal Γ (LO) is not equal to Γ (TO). The calculated dispersion curve of the surface polaritons is shown in Fig. 1. The parameters used in the calculation were taken from our earlier measurements: $\nu_{\shortparallel}$ (TO) = 678 cm^{-1}, $\Gamma_{\shortparallel}$ (TO) = 2.3 cm^{-1}, $\nu_{\shortparallel}$ (LO) = 1078 cm^{-1}, $\Gamma_{\shortparallel}$ (LO) = 16 cm^{-1}, $\nu_{\perp}$ (TO) = 723 cm^{-1}, $\Gamma_{\perp}$ (TO) = 1.2 cm^{-1}, $\nu_{\perp}$ (LO) = 1097 cm^{-1}, $\Gamma_{\perp}$ (LO) = 11.5 cm^{-1}, and $\varepsilon_{\shortparallel}$ (∞) = 2.99, $\varepsilon_{\perp}(\infty)$ = 2.95.

The experimental setup for observation of the RS from coherently excited SP is shown on Fig. 2. The second harmonic of a repetitively-pulsed Q-switched YAG laser, synchronised with the CO_2 laser was the probe of the hot SP with generation of a difference (Stokes) frequency. The power of the probe beam was about 10 KW. The RS light was examined in the near-forward scattering configuration with normal incidence of a probe beam on the crystal surface (see the inset on Fig. 3).

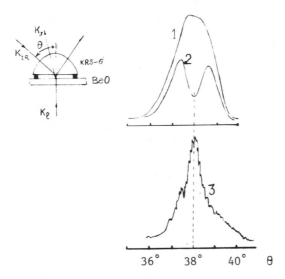

Fig. 3. The inset shows the scattering configuration, where $\vec{k}_1$, $\vec{k}_S$ and $\vec{k}_{IR}$ are respectively the probe beam, scattered light and infrared beam. 1 - the intensity of the internally reflected p- polarized IR radiation versus θ at ν_{IR} = 975 cm^{-1} without surface polariton excitation. Angular width is determined by an angular aperture of the IR detector. 2 - the loss of the internally reflected intensity near θ = 38° is due to excitation of the surface polaritons at $\nu_{sp} = \nu_{IR}$ = 946 cm^{-1}. 3 - the intensity of the RS from coherently excited SP versus θ at ν_{sp} = 946 cm^{-1}.

The SP were excited using KPS-6 semicylinder with gap about 3 μm between the sample and the semicylinder. Optically flat BeO crystal was cut into a parallelepiped measuring 1 x 1 x 0.3 cm^3 with its c axis normal to the large faces. Excitation of the SP was controlled by decreasing the internally reflected p-polarized infrared intensity (see Fig. 3). We have found that the RS of a probe beam was observed only if the SP were excited. In our experiments the scattered signal was about 10^3 - 10^4 photons per single laser pulse. This value is considerably higher compared to spontaneous RS intensity.

In conclusion, we have observed RS from SP in a bulk sample via enhanced RS intensity under coherent excitation of SP. The effect of RS from hot SP may be a useful spectroscopic technique for the study of different phenomena in which the SP are involved. In particular, the SP can be excited directly by coupling of SP with light via roughness of the surface (see e.g.).[1] Then, the spectra of RS from the hot SP will contain information on the roughness structure of the surface, the change of the dispersion relation of SP on rough surfaces, the damping produced by emission of light due to the decay of the nonradiative SP into photons via roughness, etc. It would be possible to obtain useful information characterizing the nature of ripple formation on laser-treated surfaces of solids (see e.g.[8],[9] and references therein).

References

1. Surface polaritons:Electromagnetic waves of surfaces and interfaces. Eds. V. M. Agranovich, D. L. Mills - In "Modern Problems in Condensed Matter Sciences," 1, North-Holland Publishing Company, 1982.
2. W. L. Faust, C. H. Henry, Phys. Rev. Lett. 17:1265 (1966).
3. D. M. Hwang, S. A. Solin., Solid State Commun. 13:983 (1973).
4. W. D. Wagner, R. Claus, Phys. Status Sol.(a) 64:647 (1981).
5. F. Bogani, M. Colocci, M. Neri, R. Querzoli, Nuovo Cimento 4D:453 (1984).
6. L. E. Zubkova, A. A. Mekhnatyuk, Yu. N. Polivanov, K. A. Prokhorov, R. Sh. Sayakhov, Piz'ma Zh. Eksp. Teor. Fiz. 45:47 (1987).
7. A. Otto, Z. Phys. 216:398 (1968).
8. F. Kleinman, Y. H. Bai, Appl. Phys. A29:9 (1982).
9. A. M. Prokhorov, V. A. Sitchugov, A. V. Tischenko, A. A. Khakimov, Piz'ma Zh. Techn. Fiz. 8:961 (1982).

RESONANT RAMAN SCATTERING AS A SPECTROSCOPIC PROBE OF

THE VALENCE ELECTRON EXCITATIONS OF ADSORBATES ON METALS[+]

A. Brotman[*], E. Burstein and J.D. Jiang

Physics Department and Laboratory for Research on
    the Structure of Matter
University of Pennsylvania
Philadelphia, PA 19104

INTRODUCTION

The lack of specific information about the electronic structure and valence electron excitations of molecules adsorbed on metal surfaces has been a major barrier to efforts to elucidate the mechanisms that play a role in their optical properties (eg, Raman scattering, second harmonic generation, etc). The optical properties of molecules adsorbed on metals are most appropriately viewed as the properties of adsorbate-substrate (A-S) complexes whose "intramolecular" and "intermolecular" valence electron excitations are coupled to the single particle and collective electron excitations of the underlying metal[1]. Moreover, the A-S complexes that are formed correspond to donor-acceptor complexes[2] in which the atoms of the metal substrate may serve either as the donor, or the acceptor. As a consequence, the A-S complexes can exhibit intermolecular (adsorbate-substrate) charge-transfer excitations that involve the transfer of an electron from the donor to the acceptor, or the back transfer of an electron from a negatively charged acceptor to the positively charged donor in cases where there is a transfer of charge from the donor to the acceptor in the ground state.

Our group has been using measurements of the Raman scattering excitation-profile (RS-EP) spectra of molecules adsorbed on "smooth" metal surfaces, ie, in the absence of enhancement of the EM fields by "surface roughness", to obtain information about the valence electron excitations of the A-S complexes that are formed, and about the underlying Raman scattering (RS) mechanisms[3,4]. The measurements of the excitation-profile spectra for RS by the vibration modes of a molecule (eg, the dependence of the intensities of the vibration mode Raman peaks on the excitation frequency) is, in fact, a form of modulation spectroscopy in which the modes of vibration provide the modulation fields[5,6]. The resonances in the RS-EP spectra for the vibration modes of different symmetry and different spatial configuration yield information not only about the energies and widths of the valence electron excitations, but also about the underlying Raman scattering mechanisms and the vibronic character of the electron states that are involved in the RS processes.

---

[*]Present address: AT&T Technology Systems, P.O. Box 241, Reading, PA 19603

THEORETICAL CONSIDERATIONS

There are several mechanisms that contribute to the Raman scattering by an A-S complex[1]. These include mechanisms that involve the modulation of the electric susceptibility of the underlying metal substrate by the vibration modes of the A-S complex, as well as mechanisms that involve the electronic excitations of the complex. The mechanisms that involve the modulation of the electric susceptibility of the metal substrate exhibit a dependence on the excitation frequency which is determined by the dielectric response of the substrate. In the absence of structure in the frequency dependence of the electric susceptibility of the substrate (eg, at frequencies away from the onset of interband transitions) the contributions to the RS by the A-S complex from these mechanisms will exhibit only a slow variation with frequency. On the other hand, the contributions to the RS from the mechanisms that involve the electronic excitations will exhibit "resonances" at the frequencies of the valence electron excitations of the A-S complex.

The RS by the A-S complex, via the electronic excitations of the complex, is a two-step process involving a virtual transition from a vibronic level in the ground state to a vibronic level in the excited state, accompanied by the anihilation of a photon of the incident radiation, followed by a transition from the vibronic level in the excited state to a vibronic level in the ground state (which differs by one or more vibrational quanta from the initial vibronic level), accompanied by the emission of a photon of the scattered radiation (Fig.1). The contribution

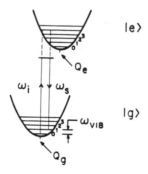

Fig. 1 Energy versus configuration coordinate diagram for the ground and excited vibronic states of a molecule, showing the two-step virtual electronic excitations that are involved in Raman scattering.

to the transition susceptibility for the RS by the j vibration mode $X_{A-S,j}$ has the same form for intramolecular and intermolecular electronic excitations of the A-S complex, and is given by[7,8]:

$$X_{AS,j} \propto \sum_{e,v_e} \frac{\langle g, v_g | r_i | e, v_e \rangle \langle e, v_e | r_j | g v'_g \rangle}{E_{g,v_g - e,v_e} - \hbar \omega_i + {}_i \Gamma_{g v_g - e v_e}}$$

where $|g,v_g\rangle$ and $|e,v_e\rangle$ are the ground-state and excited-state vibronic wavefunctions of the A-S Complex whose symmetry is, in general, different from that of the free adsorbate; $\langle g,v_g | r | e,v_e \rangle$ is the optical matrix element in the dipole approximation for the vibronic transitions between the ground and excited valence electron states; $E_{g,v_g} - e,v_e$ is the energy of the vibronic transition; w_i is the frequency of the incident radiation; $\Gamma_{g,v_g} - e,v_e$ is the Lorentzian width of the vibronic transition.

On expressing the vibronic wavefunctions in the adiabatic

approximation as products of an electronic wavefunction and a vibrational wavefunction, and expanding the electronic wavefunctions in terms of electron–vibration mode coupling matrix elements, one obtains to zero order in the normal coordinates, the following expression for the contribution transition electric susceptibility, which corresponds to the Franck–Condon mechanism:

$$(X_{AS,k})_{FC} \propto \sum_{e_0,v_e} \frac{\langle g_0|r_i|e_0\rangle\langle e_0|r_j|g_0\rangle}{E_{g,v_g-e,v_e} - \hbar\omega_i + i\Gamma_{gv_g-ev_e}} \langle v_g|v_e\rangle\langle v_e|v_g'\rangle.$$

$X_{(A-S,j)FC}$ involves the dipole matrix elements of the electronic excitations for fixed nuclei and Franck–Condon factors (eg, vibrational overlap integrals), and exhibits resonances at hw_i = E_{g,v_s-e,v_e}. The magnitudes of the Franck–Condon factors depend on on $<e|(dH/dQ_j)_0|e>$ which determines the change in the equilibrium configuration coordinate of the jth vibration mode in the excited state relative to that in the ground state, and which is non-zero only for totally symmetric vibration modes.

The expression for the transition susceptibility to first order in the normal coordinates, which corresponds to the contribution from the Herzberg–Teller mechanism, involves factors representing the vibronic coupling of electronic states, eg, $<s|(dH/dQ_j)_0|t>/(E_t - E_s)$, which involve the coupling matrix elements and the energy separation of the coupled electronc states. $(X_{A-S,j})_{HT}$ can be non-zero for both non-totally symmetric and symmetric vibration modes. Since the Herzberg–Teller mechanism is a higher order mechanism, it is weaker than the Franck–Condon mechanism when both are applicable.

RAMAN SCATTERING BY MOLECULES ADSORBED ON SMOOTH METALS

Measurements of the RS by "transparent" molecules (eg, benzene) adsorbed on a smooth single crystal Ag surface in UHV have been carried out at a single excitation frequency by Campion et al[9,10] using incident and scattered planes that are orthogonal and p-polarized incident and scattered radiation, and making use of an optical multichannel analyser and other techniques for increasing detectivity. Their data show that the symmetry of the adsorbed benzene molecules which lie flat on the substrate is lowered and, moreover, that the Raman tensor of the totally symmetric breathing mode has xz, yz and zz components which are absent for the "free" molecules. However, their data do not provide any information about the specific valence electron excitations that contribute to these Raman tensor components.

We describe here the results of mesurements of the RS–EP spectra of free-base phthalocyanine (H_2Pc) and crystal violet chloride (CV–Cl), adsorbed on smooth noble metal substrates in UHV, dye molecules that are resonant in the visible where we have tunable dye lasers. Our data on the polarization dependence of the resonance peaks in the RS–EP spectra provide information about the valence electron excitations that are involved in the RS by the adsorbed molecules.

Free-Base Phthalocyanine

The phthalocyanines are rigid planar molecules. The $\pi - \pi^*$ singlet excitations of metal phthalocyanines, which have D_4 symmetry, occur in the visible (Q band) and in the UV (Soret band) and are doubly degenerate. In the case of free-base phthalocyanine (H_2Pc) which has D_2 symmetry (Fig. 2), the x,y degeneracy is split by the configuration of the two central H atomes, and the Q_x and Q_y bands are separated by 750 cm^{-1} [10].

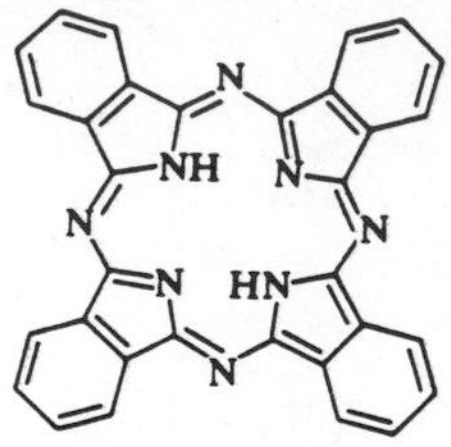

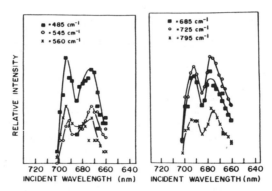

Fig. 2 Schematic structure of free-base phthalocyanine showing the bonding of the two central H atoms.

Although the optical matrix elements are quite large, the RS by a monolayer of H_2Pc molecules adsorbed on a smooth Ag surface is quite weak. Measurements of the RS-EP were accordingly carried out for H_2Pc molecules adsorbed on a Ag island film in UHV. The RS-EP spectra for the dominant totally symmetric and non-totally symmetric vibraton modes of the adsorbed H_2Pc molecules, whose frequencies are the same within 1 cm^{-1} of those of the free H_2Pc molecules, all exhibit resonance peaks at 693 $\pm$ 5 nm and 674 $\pm$ 5 nm which have a FWHM of ~300 cm^{-1} (Fig. 3). (Since these resonances are very much narrower than the width (FWHM ~6000 cm^{-1}) of the localized plasma resonances of the Ag island films, we believe that there is little distortion of the shape or position of the resonance peaks).

Fig. 3 Raman scattering excitation-profile spectra for various vibrational modes of free-base phthalocyanine adsorbed on a Ag island film in UHV at a frequency in the vicinity of the singlet Q-band electronic excitations.

By working at frequencies in the vicinity of the resonances in the RS-EP and using a computer interfaced to our Raman spectrometer and long data acquisition time, we were able to measure the polarization dependence of the RS for the vibration modes of H_2Pc molecules adsorbed on a smooth Ag surface with the largest RS crossection. On the basis of the excitation-profile data for various angles of incidence and scattering and for different combinations of s- and p-polarization for the incident and scattered light, we are led to the conclusion that the xz, yz and zz component of the contributions to the RS tensors are very much smaller in magnitude than the xx and yy components. For example, the ratio of the RS intensities for p-polarized to s-polarized is, within experimental error, the same as the ratio of the intensities derived from the relevant Fresnel factors for the component of the electric field parallel to the surface. We conclude, from the absence of the zz components of the Raman tensor, that the adsorbed molecules lie flat on the surface, and that, as in the case of the free molecules, the transition moments for the $\pi - \pi^*$ resonances lie in the plane of the molecules.

The fact that the vibration mode frequencies of H_2Pc are not affected by chemisorption on the silver surface, and the absence of any appreciable xz, yz and zz components of the Raman tensors, lead us to the conclusion that the singlet electronic excitations of the adsorbed H_2Pc molecules are similar in character to those of the free molecules. We therefore identify the RS–EP resonances at 692 nm and 674 nm with the singlet Q_x and Q_y bands of the adsorbed molecules. The energies of the Q_x and Q_y electronic excitations (661 nm and 630 nm) are, however, red-shifted relative to the Q_x and Q_y electronic excitations of the isolated molecules, and the widths of the excitations (250 cm^{-1}) are appreciably broadened relative to the widths of the excitations (1 cm^{-1}) of the isolated molecule. Moreover, the separation of the Q_x and Q_y resonances of the adsorbed H_2Pc molecules (400 cm^{-1}) is smaller by a factor of two than the separation (750 cm^{-1}) of the excitations of the isolated molecules. We believe that the decrease in splitting is due to the bonding of the H_2Pc to the substrate via the lone pair electrons of the central nitrogen atoms, which weakens the nitrogen-hydrogen bond and, thereby, decreases the x-y asymmetry of the adsorbed molecules.

From the fact that the RS–EP spectra exhibit the same resonances for the non-totally and totally symmetric vibration modes, and that there is little deformation of the H_2Pc molecules in the excited state, as indicated by measurements of fluorescence excitation spectra for the isolated molecules in a supersonic jet[12], we attribute the contribution from the Q_x and Q_y resonances to the RS to a Herzberg–Teller mechanism involving the vibronic coupling of the nearly degenerate Q_x and Q_y states with one another and with higher excited states.

Crystal Violet Chloride

Crystal violet chloride (CV–Cl) molecules dissociate in aqueous solution into carbonium ions (CV^+) and Cl^- ions. The first singlet $\pi - \pi^*$ electronic excitation of the CV^+ ions, which have a propeller-like (D_3 symmetry) structure (Fig. 4), occurs in the visible at 590 nm. Theoretical

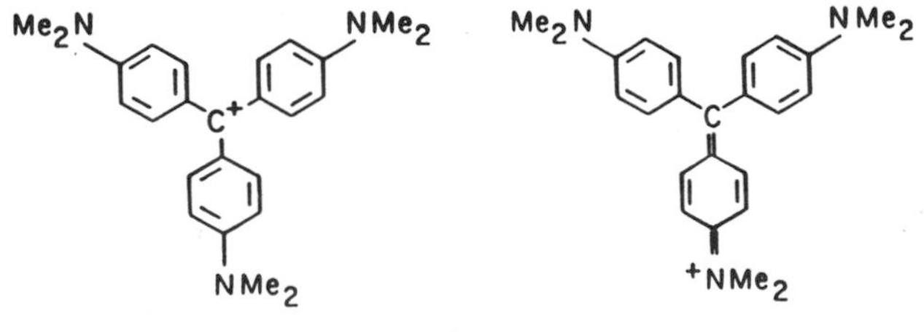

Fig. 4   Schematic structure of the carbonium (CV^+) ion, which is a linear combination of the structure with the positive charge located at the central carbon atom, and the three equivalent structures with the positive charge located at different nitrogen atoms.

calculations of the electronic structure of the free CV^+ ion[13,14] indicate that the positive charge in the ground state is located predominantly at the central-carbon atom. The singlet $\pi - \pi^*$ electronic excitation, whose transition dipole moment lies in the molecular (eg, x-y) plane, leads to an excited state in which i) the positive charge is located predominantly on the nitrogen atoms and ii) the central carbon–benzene bonds are appreciably deformed. Polarization studies of the resonance RS by CV[+] ions in solution show that the Raman tensor for the totally symmetric central-carbon breathing (ccb) mode has only xx and yy components[15].

Unlike H_2Pc, the RS by CV–Cl adsorbed on a smooth metal surface is readily observable, and it was possible to obtain polarization-resolved RS-EP spectra for the adsorbed CV–Cl molecules. Two RS configurations were used in the measurements of the excitation-profile spectra: a "parallel-plane" configuration in which the incident and scattered planes are parallel, and a "perpendicular-plane" configuration in which they are perpendicular to one another. The non-zero Raman tensor components which contribute to RS in these configurations are shown in Table I.

TABLE I. ACTIVE RAMAN TENSOR COMPONENTS

PARALLEL–PLANE CONFIGURATION		PERPENDICULAR PLANE CONFIGURATION	
incident plane (yz)		incident plane (xz)	
scattered plane (yz)		scattered plane (yz)	
polarization	tensors	polarization	tensor
$s_i // s_s$	xx	$s_i \perp s_s$	xy
$s_i // p_s$	xy, xz	$s_i \perp p_s$	xx, xz
$p_i // s_s$	yx, zx	$p_i \perp s_s$	yy, zy
$p_i // p_s$	yy, yz, zz	$p_i \perp p_s$	xy, zy, zz

The excitation-profile spectrum for the totally symmetric ccb mode of the adsorbed molecules (208 cm^{-1}), that is obtained in the parallel plane configurations for $s_i s_s$, exhibits a pronounced resonance peak at 606 nm (Fig. 5). This feature is not present in the excitation-profile spectra for the non-totally symmetric central-carbon bending mode (338 cm^{-1}).

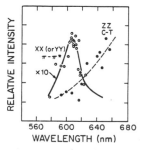

Fig. 5 Excitation profile spectra for the RS by the ccb mode of CV[+] adsorbed on a smooth Au surface. The solid line with the resonance peak at 606 nm is the spectrum for the xx component of the Raman tensor obtained in a parallel-plane configuration. The dashed line is the spectrum for the zz component obtained in a perpendicular-plane configuration. The relative intensity of the two spectra was normalized by taking into acount differences in the Fresnel factors and in the collection efficiencies of the two scattering configurations.

We conclude from this that the resonance Raman scattering by the ccb mode is dominated by the Franck-Condon mechanism. The frequencies of the peaks in the Raman spectrum of the adsorbed molecules are not appreciably shifted from those observed in the corresponding Raman spectrum of the CV^+ ions in solution[15]. This, together with the fact that the frequency of the resonance peak in the excitation-profile spectrum (606 cm^{-1}) is not appreciably different from the frequency of the first singlet $\pi - \pi^*$ electronic excitation of the CV^+ ions in aqueous solution (590 cm^{-1}), indicates that the molecules are adsorbed on the Au surface as CV^+ ions.

Measurements of the polarization dependence of the resonance Raman scattering by the ccb mode in the vicinity of the $\pi - \pi^*$ resonance indicated that the RS intensity for the $p_i // p_s$ configuration was appreciably stronger than that for the $s_i // s_s$ configuration, indicating thast the Raman tensor of the ccb mode may have sizeable zz and xz (or yz) components. In order to get further information about the relative magnitudes of the different components of the Raman tensor, we carried out RS intensity measurements in the perpendicular-plane configuration. In this configuration, a sizeable Raman scattering by the ccb mode was observed only for $p_i \perp p_s$ indicating that the Raman tensor of the ccb mode of the adsorbed molecules has a sizeable zz component, but that the xz and yz components are either very weak or zero.

To ascertain the origin of the zz component of the Raman tensor of the ccb mode, a measurement of the excitation-profile spectrum for the ccb mode was carried for the $p_i \perp p_s$ configuration. The resulting excitation-profile spectrum does not exhibit a $\pi - \pi^*$ resonance feature but, rather, it exhibits a steadily increasing scattering intensity with wavelength (Fig 5). Moreover, the RS intensity at the freqency (606 nm) of the $\pi - \pi^*$ resonance is appreciably stronger than the intensity of the $\pi - \pi^*$ resonance peak in the $s_i // s_s$ excitation-profile spectrum.

The fact that we were able to obtain an RS-EP spectrum for the ccb mode using an $s_i // s_s$ (eg, xx or yy) polarization configuration (in spite of the small Fresnel factors for the components of the EM field parallel to the surface of the metal substrate), is attributed, on the one hand, to the large oscillator strength and moderate width of the singlet $\pi - \pi^*$ electronic excitation and, on the other hand, to the large vibronic interaction of the ccb vibration mode with the π and π^* states of the CV^+ ion. The fact that the singlet $\pi - \pi^*$ resonance peak is only observed in the xx (or yy) component excitation-profile spectrum implies that the plane of the molecule is parallel to the surface.

The CV^+ ions are bonded to the Au surface by Coulomb and van der Waals interactions. Since the frequencies of the vibration modes in the ground state are not appreciably different from those in the free molecules, we surmise that the bonding does not have any appreciable covalent character. The CV^+-Au^- adsorbate-substrate complex that is formed is, in effect a donor-acceptor type complex in which there is a sizeable charge-transfer from the donor (CV) to the acceptor (Au) in the ground state. On this basis, we attribute the sizeable zz component of the Raman matrix element of the ccb vibration mode to an "intermolecular" charge-transfer electronic excitation of the CV^+-Au^- adsorbate-substrate complex involving a back transfer of an electron from Au^- to CV^+. Such an intermolecular charge-transfer excitation may be expected to have a transition dipole which is directed along z and, therefore, to contribute to the zz component of the Raman tensor of the ccb mode. Both the intermolecular charge-transfer excitation and the $\pi - \pi^*$ excitation, which involves an intramolecular charge-transfer, result in a large deformation of the configuration coordinate of the ccb mode.

RS measurements were also carried out for CV-Cl deposited onto smooth Cu and Ag surfaces. The polarization dependence of the RS by CV^+ adsorbed on a smooth Cu surface in UHV is similar to that for CV^+ adsorbed on Au. On the other hand, we were not able to observe RS by CV^+ adsorbed on a smooth Ag surface in UHV. We believe that this is due to the preferential adsorption of Cl^- ions by the Ag surface, and to the consequent low CV^+ coverage on the Ag surface.

CONCLUDING REMARKS

The use of RS-EP measurements as a spectroscopic probe of molecules adsorbed on metal surfaces is limited only by the observability of the resonance RS of the A-S complexes that are formed. It is therefore limited to A-S complexes with electronic transitions that have large dipole matrix elements, narrow widths and strong electron-vibration mode interactions. In general, small molecules will tend to have electronic excitations with smaller dipole matrix elements, but larger electron-vibration mode interactions. The widths of the electronic excitations depend on the strength of the interaction with the underlying metal substrate, as well as on the nature of the excitation, ie, whether it is an intramolecular excitation, or an adsorbate-substrate intermolecular interaction. In general, molecules which are chemisorbed will have larger excitation widths than molecules which are physisorbed. Since the Fresnel factors for the electric field components parallel to the surface are much smaller than those for the electric field component perpendicular to the surface, the electronic excitations of an A-S complex whose dipole matrix elements have sizeable z components, and whose Raman tensors have a sizeable zz component, are more apt to have observable RS intensities. We note, however, that it should be possible, by adsorbing molecules on very thin metal substrate (ie, thickness much less than the optical skin depth so that the reflectance is small) to avoid the reduction in the components of the electric field parallel to the surface.

We acknowledge valuable discussions with Dr. R. Messmer and Dr. H. Kobayashi and with Professor H. Yamada of Kwansei Gakuin University, Japan.

REFERENCES

[+] Research supported in part by ONR and by the NSF-MRL Program at Penn under Grant No. DMR-821678.

1. S. Brotman, E. Burstein and J.D. Jiang, Surf. Sci. 158: 1 (1985).
2. J.B. Birks, "Photophysics of Aromatic Molecules" (Wiley-Interscience, New York 1970).
3. A. Brotman and E. Burstein, Physica Scripta 32: 385 (1985).
4. J.D. Jiang, E. Burstein and H. Kobayashi, Phys. Rev. Lett. 57: 1793 (1986).
5. M. Cardona, Surf. Sci. 37: 100 (1973).
6. W. Siebrand and M. Zgierski in "Excited States" Vol 4, ed. by E.C. Lin (Academic Press, New York 1979) p. 1.
7. A. Albrecht, J. Chem. Phys. 34: 1476 (1961).
8. L.D. Ziegler and A. Albrecht, J. Chem. Phys. 70: 2634 and 2644 (1979).
9. A. Campion, V.M. Grizzle, R.D. Mullins and J.K. Brown, J. Phys. (Paris) Colloq. 44: C10-341 (1983).
10. V.M. Hallmark and A. Campion, Chem. Phys. Lett. 110: 561 (1984).
11. B.D. Berezin, "Coordination Compounds of Porphyrins and Phthalocyanines", (John Wiley & Sons, New York 1981).

12. P.S.H. Fitch, L. Wharton and D.H. Levy, J. Chem. Phys. 69: 3424 (1978).
13. F.C. Adams and W.T. Simpson, J. Mol. Spect. 8: 305 (1959).
14. W. Salaneck and J.L. Bredas (private communication).
15. L. Angeloni, G. Smulevich and P.M. Marzocchi, J. Raman Spect. 8: 305 (1979).

LASER-STIMULATED IONIZATION AND DESORPTION OF MOLECULES

IN AN ELECTRIC FIELD

S. R. Egorov, V. S. Letokhov, and E. V. Moskovets

Institute of Spectroscopy, USSR Academy of Sciences
142092 Troitsk, Moscow Region, USSR

The field ionization and field desorption of atoms in electric fields of the order of 10^8 V/cm were first observed by Muller.[1,2] Based on these phenomena, the field-on microscopy technique was developed allowing single atoms of a metal surface to be observed with a resolution of around 2 Å.[3]

An ion microscope, by virtue of its simplicity, was more than once considered a likely instrument for studying the structure of biological molecules. One principal difficulty in realizing ion microscopy of biomolecules is associated with their rapid field destruction and removal from the field-emitter tip in fields with an intensity of the order of 10^8 V/cm (the best image voltage for tungsten being 4.5×10^8 V/cm in helium and 2.2×10^8 V/cm in argon.[3]

Attempts at realizing biomolecular ion microscopy experimentally were based on different approaches. In the works reported in,[4-6] they walled up the molecules placed on the tip surface with a metal layer and then, as this layer underwent field evaporation, observed in the ion image specific features with a characteristic size of the order of that of the molecules under analysis. Another approach based on the registration of the early stages of ion image formation at the minimal field intensities at the tip[7] proved a complete failure. The images obtained practically could not be reproduced and had nothing in common with the known structure of the molecules studied. The approach developed by Panitz[8] turned out to be most successful. The molecules to be investigated were placed on a cooled field-emitter tip and covered with a layer of benzene molecules. The images of the molecular contours were observed as the benzene layer was gradually removed through controlled field desorption. In this way, reproducible images of ferritin clusters were obtained with a resolution of about 3 nm at a magnification of 2×10^5. To obtain an image by the Panitz method, conditions must be found for some range of field intensities at the field-emitter tip in which the molecules under study suffer practically no field destruction and desorption, whereas the absorbed molecular layer undergoes a sufficiently intense field desorption.

An alternative approach to the ion microscopy of biomolecules was suggested in 1975 by Letokhov.[9,10] In the papers cited, he considered the possibility of applying laser radiation in a field-electron emission or a field-ion microscope to effect selective detachment of an electron or ion from macromolecules. It was suggested that the selective photoionization of certain chromophore groups should be carried out by a multiple-step scheme using several laser pulses differing in frequency. At the same time, consideration was also given to another laser modification of the field-ion projector allowing the electric field strength necessary for field ionization of the imaging gas atoms to be substantially reduced. It was suggested for the purpose that the imaging gas atoms should be raised with laser radiation to quantum states close to the ionization limit. The work by Panitz on the ion microscopy of biomolecules has recently stimulated the advent of a new laser ion projector version.[11] It is suggested that the cooled field-emitter tip with adsorbed macromolecules of interest should be coated with a thin (a few monolayers thick) layer of vacuum-deposited physisorbed molecules or atoms. The deposited particles should have a small size (≤ 5 Å), a low adsorption energy, and a low ionization potential. Certain chromophore groups of the macromolecule can be ionized with laser radiation. A two-dimensional image of the spatial arrangement of these groups will be formed on the screen of the field-ion microscope as a result of the field desorption of the adsorbate layer transparent for the radiation, stimulated by the tunnel transfer of positive charges onto the adsorbate under the action of a pulsed electric field. At the present time, this variant of macromolecular laser field-ion microscopy seems most promising. One may hope to obtain with it spectral information about adsorbed molecules with a spatial resolution much better than the radiation wavelength used.

To find out whether it is possible to realize the laser field-ion microscope idea, it is necessary to investigate the process of field desorption from adsorbed molecular layers and determine the efficiency of various nonthermal mechanisms of molecular ionization and desorption in an electric field under the action of laser radiation.

Even in the early experiments on field desorption it was found that the field desorption rate could be increased by the action of light. In most cases, the stimulation of the desorption process was explained as resulting from the heating of the surface by the radiation.[11,12] That the quantum stimulation of the process is possible was indicated by the results of the work reported in,[13] in which the flux of molecular ethylene ions from a silver surface was observed to increase under the effect of a nonresonant laser radiation, and in proportion to the photon flux in the laser pulse. Antonov and Letokhov et al.[14,15] were the first to investigate the possibility of stimulating field desorption by way of the resonant photoexcitation and ionization of the adsorbate. They managed to observe experimentally the two-photon stimulation, by resonant radiation from a KrF excimer laser, of field desorption of molecular tetracene ions from molecular layers deposited on the cutting edge of a razor blade. The one-photon stimulation of field desorption of molecular anthracene layers adsorbed on a tungsten tip was reported in.[16] A photostimulation signal linear in the laser fluence was observed with the tip being irradiated with excimer and dye laser pulses at characteristic fluences up to 1 mJ/cm². The efficiency of the observed process was reported to drop as the radiation wavelength used was substantially increased.

In the text below, we summarize the results of our experimental and theoretial studies of the field desorption of molecular layers and analyze the mechanisms of laser stimulation of this process.

The experimental setup is a field-ion microscope equipped with quartz windows to let laser radiation in and out. The electric fields 10^7-10^8 V/cm in strength necessary to observe field desorption were produced on the apex of a tungsten tip with a radius from a few hundred to a few thousand Ångstrom units. The tips used in our experiments were prepared by the standard electrochemical polishing technique, and cleaned of surface impurities prior to experiment through controlled field evaporation of the tip material at field strengths around 6 V/Å. The surface morphology was monitored with an atomic-scale resolution by helium field-ion imaging at a pressure of 10^{-5} Torr. The absolute field strength at the tip surface and the tip radius were found accurate to within 15-20% from the best image voltage (BIV). Measured to the same degree of accuracy, the best image voltages for tungsten are $E_{BIV\ He}$ = 4.5 V/Å and $E_{BIV\ Ar}$ = 2.2 V/Å.[17] The tip was placed in a high-vacuum chamber evacuated to a pressure of 10^{-8} Torr. The adsorbate was deposited on the tip either from a heated microcell set at a distance of 0.5 cm from the tip or by means of a retractable capillary metering device similar to the one described in.[18]

When studying field desorption, we measured the ion current as a function of the linearly increasing tip voltage and the total number of ions desorbed during the pulse as a function of the pulsed electric field amplitude. At field strengths below 1 V/Å used to observed field desorption, the field-desorption signal was easy to separate from the background noise due to field ionization in the gas phase.

In studying photostimulated processes, use was made of pulsed excimer and dye lasers. The ions produced were registered by means of two microchannel plates (MCP) in a chevron arrangement and a phosphor screen. The distance between the apex of the tip and the front edge of MCP was $1 = 6$ cm. The time of flight of the ions, $t = 1(2Ue/m)^{-1/2}$, varied over the range 0.5-2 μs, depending on the tip voltage U, and considerably exceeded the characteristic laser pulse width of 20 ns. This made it possible to carry out the time-of-flight mass-separation of the ions. The accuracy of ion mass determination was ± 10%.

The ramped field-desorption technique (measuring ion currents as a function of linearly increasing tip voltages) proposed by Panitz[18] is outwardly similar to thermal desorption spectroscopy (see, for example,).[19] With the latter technique, the temperature of the surface with adsorbed molecules under analysis is raised in a linear fashion, and the flux of molecules desorbing from the surface is registered. The position and shape of the thermal-desorption peaks are governed by the adsorption characteristics of the molecules (their adsorption energy and its dependence on coverage). But we have demonstrated[20] that the position and shape of the field-desorption spectrum are determined by the electronic parameters of the adsorbed molecular layer. The field absorption of a physisorbed molecular layer takes place according to the following scheme: field tunnel ionization in the layer → migration of positive charges to the surface → desorption of ions over the potential barrier reduced by the field. In many cases of practical interest, the ion desorption rate at a given field value materially exceeds the field ionization rate, the ion current being limited by this latter rate. Thus, the method of field desorption of adsorbed layers allows one to investigate tunneling processes in condensed media in electric fields up to 10^7-10^8 V/cm. An important advantage of this technique as compared with other methods for studying tunneling phenomena in solids[21] is the possibility of visual monitoring of the spatial distribution of the desorbing ions and establishing the electric field strength range within the limits of which no avalanche ionization occurs in a layer of a given thickness.

Figure 1A shows a ramped field-desorption spectrum of benzene adsorbed on a cooled tungsten field-emitter tip at a temperature around 80 K. The bottom curve was obtained by ramping the tip voltage a second time without any additional deposition of benzene. As can be seen, there

is practically no ion current in the same field strength range during the
second ramp. The numerical modeling of the field desorption process
allows the field strength dependence of the ion current to be described
qualitatively.[20] In the general case, to describe the shape of the peak,
allowance must be made for thermal diffusion of the molecules over the
tip surface. The role of thermal diffusion is reduced where the surface
temperatures are low and the molecules are desorbed by a pulsed electric
field. Figure 1B presents the relationship between the integral ion
signal and the pulsed electric field amplitude. This curve was taken with
the coverage (the adsorbed layer thickness) being held constant by making
the interval between successive pulses long enough to exceed the time it
took the coverage to get completely restored on account of thermal
diffusion (see Fig. 2). The flattening out of the integral ion signal at a
pulsed electric field amplitude in excess of 5 V/nm corresponds to
complete desorption of the molecules from the tip surface area being
observed, $S = 2\pi R^2 (1 - \cos \theta)$, where $2\theta = 36^O$ is the viewing angle at the
microchannel plate. The maximum ion current in pulsed field desorption is
proportional to the adsorbed layer thickness, and so can be used to
monitor the coverage.

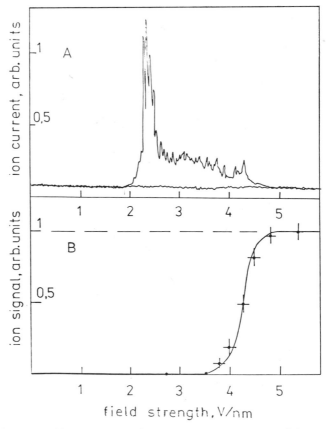

Fig. 1. (A) Ramped field-desorption spectrum and (B) pulsed-field
desorption of benzene layer from tungsten tip at 80 K. Ramp rate
0.2 V/(nm.s); electric field pulse is approximately sine in shape
with $\tau = 0.5$ µs (FWHM).

Figure 2 presents the results of studies of the thermal desorption
of benzene molecules over the tip surface at T = 80 K. Applied to the
tip were two successive pulses 6 V/nm in strength with a controlled delay

186

time between them. As can be seen, complete restoration of the adsorbed layer through thermal diffusion takes around a minute. This makes it possible to carry out measurements with one and the same original coverage without additional deposition.

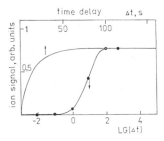

Fig. 2. Two-pulse field desorption of benzene layer from tungsten tip at 80 K. Pulse amplitude 6 V/nm; pulse width 0.5 μs.

Since the field desorption rate is limited by the field ionization rate, field desorption can be stimulated by laser radiation. Figure 3 illustrates schematically the possible elementary processes of ionization of an adsorbed layer in an electric field involving the absorption of a photon: (1) photoionization; (2) photogeneration of electron-hole pairs in the emitter material, followed by the injection of holes into the adsorbed layer (tunneling of electrons onto the vacant level in the emitter metal); (3) resonance absorption of photons in the layer, followed by the tunnel ionization of excitons; and (4) tunneling accompanied by absorption of photons (Franz-Keldysh effect).

A detailed experimental study of laser-stimulated field desorption was carried out with monomolecular anthracene layers adsorbed on tungsten at room temperature.

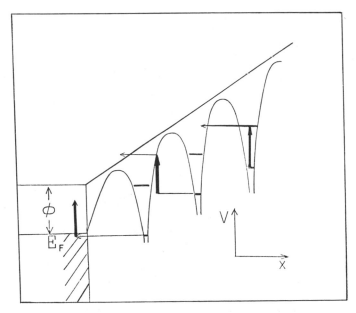

Fig. 3. Mechanism of one-photon laser stimulation of ionization in a molecular layer adsorbed on a metal surface in the presence of a strong electric field (see text).

Figure 4 presents the relationship between the laser stimulation efficiency γ and the adsorbed layer thickness. The layer thickness was monitored by measuring the integral ion signal, with the anthracene molecules being completely desorbed by the pulsed electric field. The fact that the efficiency is linear in coverage indicates that the laser stimulation mechanism is related to the ionization of the adsorbed molecules.

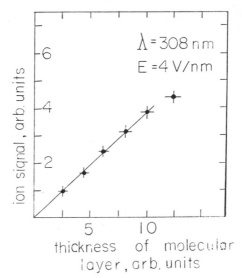

Fig. 4. Laser-stimulated field desorption signal (total number of ions per pulse) as a function of the absorbed anthracene layer thickness.

The efficiency γ for various laser wavelengths was measured relative to the stimulation efficiency at the XeCl excimer laser wavelength λ = 308 nm ($\hbar\omega$ = 4 eV); $\gamma = (N/N_{308})(F_{308}/F) = N/F\gamma_{308}$. Here, N stands for the total number of molecular anthracene ions produced during the laser pulse and F is the laser fluence.

An analysis of the spectrum of the efficiency γ in several intervals 25-30 nm in width shows that there are no distinct resonant bands typical of the absorption spectra of crystalline anthracene. Figure 5 shows the laser stimulation efficiency spectrum in the wavelength range 390-415 nm (PBBO-dye laser). For comparison, the figure also shows the absorption band spectrum of cyrstalline anthracene.[22] It can be seen that the efficiency of laser stimulation of field desorption in this region varies but little, while the absorption spectrum exhibits a clearly observed resonant band. Note that this spectral region corresponds to the long-wavelength absorption boundary of crystalline anthracene. Thus, Fig. 5 indicates that laser stimulation occurs even below the zero-field long-wavelength absorption boundary of crystalline anthracene.

In the study of the efficiency γ, it was found to rise monotonically with decreasing laser wavelength over a wide region of the spectrum. Figure 6 shows the efficiency γ as a function of the energy of photons generated by various excimer and dye lasers. The energy scale shows the lower singlet state energy, the conduction band boundary, and the ionization potential of cyrstalline anthracene[22] and also the difference between the ionization potential and the work function of tungsten.

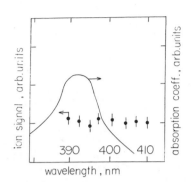

Fig. 5. Efficiency of laser-stimulated field desorption as a function of
laser wavelength (solid circles), and the absorption band of
crystalline anthracene (solid curve).

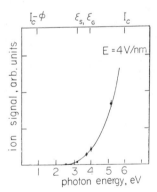

Fig. 6. Efficiency of laser-stimulated field desorption as a function of
photon energy. Solid curve – results of fitting experimental data
to the expression for the spectral dependence of the Franz-Keldysh
effect.

Let us summarize the main characteristic features of the laser-stimulation desorption process observed:

(1) Linear dependence of desorption on the laser fluence, indicative of a single-photon character of the process;[6]

(2) Linear dependence of desorption on the thickness of the adsorbed layer, indicative of the fact that stimulation is associated with ionization in the bulk of the adsorbed layer;

(3) Occurrence of laser stimulation below the zero-field long-wavelength absorption boundary of crystalline anthracene (and moreover, below the conduction band boundary);

(4) Absence of distinct resonant bands and monotonic rise of stimulation efficiency with increasing photon energy.

The above experimental facts can be most adequately described qualitatively in terms of the Franz-Keldysh effect. Analyzed below is the expression for the spectral dependence of the Franz-Keldysh effect and the best results of fitting the experimental data on the laser field-desorption stimulation efficiency spectrum to this expression.

The Franz-Keldysh effect (shift of the intrinsic light absorption boundary in a semiconductor or a dielectric toward the long-wavelength side of the spectrum under the effect of an external electric field) was theoretically predicted to exist by Franz[23] and Keldysh[24] in 1958. Experimental results were obtained for many inorganic semiconductor materials, primarily for Si, Ge, and GaAs in fields of 10^5–10^6 V/cm. The respective shifts were found to be in the range 10^{-2}–10^{-1} eV (see ref[25] and references therein.) In the fields 10^7–10^8 V/cm in strength that can be realized in photostimulated field-desorption experiments, the respective shifts must be considerably greater. Indeed, the expression for the spectral dependence of the Franz-Keldysh effect contains an exponent of the form

$$\alpha(\omega) \sim \exp\{-[4(2\mu)^{1/2}(\varepsilon_0 - \hbar\omega)^{3/2}]/(3Ee\hbar)\}$$

Here, $\mu^{-1} = m_e^{-1} + m_h^{-1}$ is the reduced mass of the electron and hole, E the electric field strength in the dielectric layer, and ε_0 the distance between the bands linked by the tunneling accompanied by the absorption of a photon with an energy of $\hbar\omega$

$$\alpha(\omega) \sim \exp\{-[(\varepsilon_0 - \hbar\omega)/(\Delta\varepsilon)]^{3/2}\}$$

$$\Delta\varepsilon = [9e^2E^2\hbar^2/(32\mu)]^{1/3} = 1.29\ E^{2/3}\mu^{-1/3}$$

In the latter expression, $\Delta\varepsilon$ is expressed in eV, the field strength E in 10^8 V/cm = 1 V/Å, and μ in terms of the free electron mass. The best results of fitting the expression for $\alpha(\omega)$ to the experimental data on $\Upsilon(\omega)$ yield the following values: $\varepsilon_0 = (5.9\pm0.5)$ eV, $\Delta\varepsilon = (1.1\pm0.1)$ eV. The resulting curve is presented in Fig. 6. The values of ε_0 coincide to good accuracy with the ionization potential of crystalline anthracene and not with the gap $\varepsilon_g = 4$ eV. This result seems fairly natural. Indeed, even in crystalline anthracene, the excited electronic state bands, the conduction bands included, are not very wide – around 0.1 eV; the respective free electron and hole masses are 5–20 times the free electron mass for electrons at the bottom of the conduction band and for holes at the top boundary of the valence band (see ref.[22], pp. 268-269). The thin anthracene layer absorbed on tungsten is apparently amorphous enough. The energy levels of individual molecules are chaotically distributed over a width of the order of 0.1 eV and the electrons and holes are therefore localized. In these conditions, photon-assited tunneling can proceed effectively to ionized states. The ionization energy of the amorphous anthracene layer approximates that of crystalline anthracene accurately enough.

The lack of information on the strength of the internal field in the anthracene layer at electric fields of 10^7-10^8 V/cm in vacuum and on the reduced mass μ of the electrons and holes makes it impossible to calculate $\Delta\epsilon$. The principal uncertainty in our estimations stems from the lack of data on the dielectric constant k. Putting for estimation purpose $\mu = 1$ and $k = 1$, we get $\Delta\epsilon = 0.7$ eV. Thus, the experimentally obtained values of $\Delta\epsilon$ seem quite reasonable. The method of field desorption of adspecies from the surface of a controlled-thickness layer, used in conjunction with the ion energy analysis technique, makes it possible to measure the dielectric constant of adsorbed layers in electric fields of the order of 10^8 V/cm. With the value of k known, one can find μ from the photostimulated field desorption spectrum.

References

1. E. W. Muller, Z. Phys. 131:136 (1951).
2. E. W. Muller, Phys. Rev. 102:618 (1956); Advances in Electron. and Electron Phys. 13:83 (1960).
3. E. W. Muller, T. T. Tsong, Field Ion Microscopy (Elsevier, New York 1969).
4. T. Gurney, F. Hutchinson Jr., R. Young, J. Chem. Phys. 42:3939 (1965).
5. E. W. Muller, K. Rendulic, Science 156:961 (1967).
6. W. R. Graham, F. Hutchison Jr., D. Reed, J. Appl. Phys. 44:5155 (1973).
7. E. S. Machlin, A. Freilich, D. Agrawal, V. Burton, C. Briant; J. Microsc. 104:127 (1975).
8. J. Panitz, J. Microsc. 125:3 (1982).
9. V. S. Letokhov, Phys. Lett. 51A:231 (1975).
10. V. S. Letokhov, Kvantovaya Elektronika (Russian) 2:930 (1975).
11. W. Drachsel, S. Nishigaki, J. H. Block, Int. J. Mass Spectrom. Ion Phys. 32:333.
12. G. L. Kellog, T. T. Tsong, J. Appl. Phys. 51:1184 (1980).
13. S. Nishigaki, W. Drachsel, J. H. Block, Surf. Sci. 87:389 (1979).
14. V. S. Antonov, V. S. Letokhov, Multiphoton Processes, Springer Series on Atoms and Plasmas, Vol. 2 (Springer-Verlag, Berlin, New York, tokyo 1984).
15. V. S. Antonov, V. S. Letokhov, E. V. Moskovets, Izv. Akad. Nauk SSSR, ser. fiz. (Russian) 50:690 (1968).
16. S. E. Egorov, V. S. Letokhov, E. V. Moskovets, Appl. Phys. A (in press).
17. T. Sakurai, E. W. Muller, Phys. Rev. Lett. 30:532 (1973).
18. J. Panitz, J. Vac. Sci. Technol. 16:868 (1979).
19. P. A. Redhead, Vacuum 12:203 (1962).
20. S. E. Egorov, E. V. Moskovets, Surf. Sci. (to be published).
21. E. Burstein, S. Sundquist, Eds., Tunneling Phenomena in Solids (Plenum Press, New York 1969).
22. M. Pope, C. E. Swenberg, Electronic Processes in Orgahic Crystals Claredon Press, Oxford, (1982).
23. W. Franz, Z. Naturforsch 13A:484 (1958).
24. L. V. Keldysh, Zh. Eksp. Teor. Fiz. (Russian) 34:1138 (1958).
25. V. A. Tyagay, O. V. Snitko, Elektrootrazhenie sveta v polyprovodnikakh (Electrical Reflection of Light in Semiconductors) (Russian) (Naukova Dumka, Kiev, 1980), p. 47.

NONLINEAR OPTICAL PHENOMENA IN SINGLE MICRON-SIZE DROPLETS

Richard K. Chang

Yale University
Section of Applied Physics and Center for Laser Diagnostics
New Haven, Connecticut 06520, USA

INTRODUCTION

In spite of the short interaction lengths in micron-size spheres, nonlinear optical effects can be readily observed. The curved liquid-air interface modifies the internal electromagnetic distributions of the incident radiation (assumed to be a plane wave) and of the internally generated nonlinear optical radiation described through the nonlinear source polarization (P^{NLS}).

A plane wave (with intensity I_0 and $\lambda = 0.532$ μm) incident on a transparent droplet (with radius $a = 35$ μm) is concentrated at three locations:[1] (1) outside the shadow face with intensity in excess of $10^3 I_0$; (2) inside the shadow face with intensity in excess of $10^2 I_0$; and (3) inside the illuminated face with intensity less than $10^2 I_0$. The internal intensity distribution along the principal diameter is shown in Fig. 1. Nonlinear optical "pumping" is localized in the two high internal intensity regions which are confined near the droplet interface. Both the external and internal enhanced intensity (near the shadow face) can cause laser-induced breakdown (LIB) in the surrounding gas and/or in the liquid. Hence, it is no surprise that the LIB threshold of a gas is greatly lowered by the presence of droplets.

The internally generated nonlinear radiation via P^{NLS} is also affected by the spherical liquid-air interface. The internally "trapped" radiation is confined near the liquid-air interface. For particular wavelengths which satisfy the morphology dependent resonances (MDR's), the droplet can be envisioned as an optical cavity.[2] The wavelength separation between MDR's of adjacent mode numbers with the same mode order is nearly constant, analogous to a Fabry-Perot interferometer. The density of MDR's within 10 cm^{-1}, when different mode numbers and orders are considered, has been calculated.[3] The Q-factor of the droplet cavity can be calculated from the poles of the internal or external field coefficients for elastic scattering[2,3] and can be in excess of 10^4. Such high Q-factors can modify the Einstein A and B coefficients of an emitting atom with two energy levels,[4] analogous to the Rydberg atom placed in a metal cavity.[5] Manifestation of MDR's in linear spectroscopy has been reported for the optical levitation,[6] elastic scattering,[7] fluorescence,[8] energy transfer between molecules,[9] and Raman scattering.[10]

We will briefly review the nonlinear optical phenomena in single micron-size liquid droplets upon irradiation by a high intensity laser with output in the visible spectrum. The lower intensity levels, which leave the droplet shape intact, and the higher intensity levels, which cause explosive vaporization will be discussed.

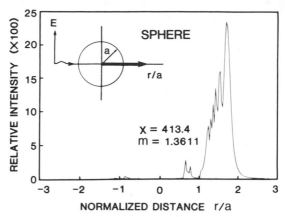

Fig. 1. Relative intensity on a line through the center of a dielectric sphere for a = 35μm with vertical incident polarization, x = 413.4 and m = 1.3611. r/a = ±1 defines the boundary of the sphere. Incident intensity is unity. The maximum intensities at the internal and external near-field peaks are 279 and 2356, respectively.

NONLINEAR OPTICAL EFFECTS

(a) At Lower Intensity

The report of lasing from ethanol and water droplets containing Rhodamine dye[11] was soon followed by the report of stimulated Raman scattering (SRS) from water and ethanol droplets.[12] Both the lasing and SRS thresholds were surprisingly low and emission occurred only at those wavelengths corresponding to MDR's which provided the optical feedback. The multiorder SRS from CCl_4 indicated that not only can the internal fields at the incident wavelength (enhanced and localized) serve as the pump but also the internal fields of the preceding SRS order (intense and uniformly distributed near the interface) can serve as the pump for the next first-order SRS.[13] The intense fields at both the incident and Raman-shifted wavelengths can cause phase-modulation broadening in CS_2 via its large intensity-dependent index of refraction coefficient.[14]

Nonlinear optical processes which require phase-matching have also been observed with micron-size droplets. Coherent anti-Stokes Raman scattering (CARS) and coherent Raman gain (i.e., additional gain provided by another intense beam at a different wavelength) have been observed.[15] These results from droplets indicate that the external angle phase-matching curve is broad, asymmetrical, and displaced toward the collinear angle when compared with the phase-matching curve of the same liquid in an optical cell.[2,15] The broad, asymmetrical, and displaced angular-tuning curve is the combined manifestation of the spread in the k-vectors within the droplet due to the curved interface, of the short interaction region due to the localization of the pump and Stokes fields, and of the overlap of the pump and Stokes high intensity regions which favor the collinear geometry.[2,15] No MDR's were observed in the CARS spectra, indicating that the standing waves of a sphere (the vector spherical harmonics needed to describe the MDR's) are not compatible with the plane waves (normally used to consider wave-vector matching in an extended medium).

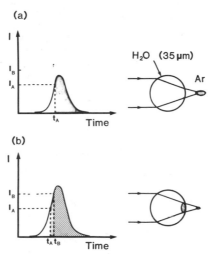

Fig. 2. Schematic of the LIB process for a droplet with higher LIB threshold than the surrounding gas. (a) At lower I_0, LIB is initiated outside the shadow face when the intensity reaches I_A, and the remaining portion of the pulse (shaded area) sustains the external plasma. (b) At higher I_0, LIB is also initiated inside the shadow face when the intensity reaches I_B. The pulse energy between t_A and t_B (shaded area) sustains the external plasma, and the remaining portion of the pulse energy with $t > t_B$ (diagonal lines area) sustains the internal plasma. The internal plasma blocks the remaining portion of the laser pulse from reaching the gas region outside the shadow face. For a droplet with lower LIB threshold than the surrounding gas, LIB occurs only within the droplet.

(b) At Higher Intensity

When I_0 is increased to a specific range so that the enhanced intensity inside and/or outside the shadow face causes LIB in the liquid and/or air, plasma generation via multiphoton ionization results and is followed by cascade multiplication. Whether the LIB occurs inside or outside the shadow face is dependent on the enhanced intensity as well as on the optical breakdown parameters of the liquid and surrounding gas. During the rising portion of the laser pulse, although LIB occurs first in the gas (see Fig. 2), once the LIB threshold has been reached within the shadow face, the internal plasma density can rapidly reach 10^{18} cm^{-3}, corresponding to a plasma frequency in the visible wavelength range. The transparent droplet at lower intensity now becomes absorbing and blocks the remaining portion of the laser pulse from reaching the region external to the shadow face (see Fig. 2).[16] The external plasma, which was initiated before the internal plasma, can no longer be sustained by the remaining portion of the laser pulse. The major portion of the remaining pulse will sustain the internal plasma to cause shock waves, which can expel plasma and neutral material from the droplet shadow faces.

Spatially resolved spectroscopy has been used to determine the location of the initiation of LIB. The experimental arrangement consists of imaging the droplets and expelled plasma plumes onto the vertical slit of the spectrograph. A vidicon camera is placed at the exit plane of the spectrograph (see. Fig. 3). The emission spectra from various locations along the laser beam direction can then be determined in one laser shot. The spectral evidence that LIB has occurred is the presence of a plasma continuum (extending to the UV) and of discrete emission lines from neutral and atomic species comprising the hot dense plasma.[16]

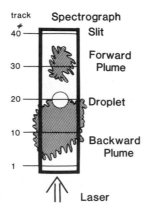

Fig. 3. Schematic of the exploding droplet, which is rotated by 90° and imaged onto the entrance slit of the spectrograph. The vidicon tracks are numbered.

CONCLUSION

Micron-size droplets provide an interesting medium for studying nonlinear optical interactions. The combination of the enhanced intensity and the optical feedback provided by the spherical interface leads to a series of nonlinear optical effects such as lasing, SRS, CARS, coherent Raman gain, and phase-modulation. The internally and externally enhanced intensity leads to the lowering of the LIB threshold. Spatially resolved spectroscopy provides detailed information on the dielectric breakdown process as well as the location of LIB, whether it occurred inside and/or outside the shadow face of the transparent droplet.

ACKNOWLEDGMENTS

We gratefully acknowledge the partial support of this research by the U.S. Air Force Office of Scientific Research (Contract No. F49620-85-K-0002), the U.S. Army Research Office (Contract No. DAAG29-85-K-0063), and the U.S. Army Research Office DoD-University Research Instrumentation Program (Grant No. DAAL03-86-G-0104).

REFERENCES

1. D. S. Benincasa, P. W. Barber, J.-Z. Zhang, W.-F. Hsieh, and R. K. Chang, Spatial Distribution of the Internal and Near-Field Intensities of Large Cylindrical and Spherical Scatterers, _Appl. Opt._ 26:1348 (1987).
2. S.-X. Qian, J. B. Snow, and R. K. Chang, Nonlinear Optical Processes in Micron-Size Droplets, _in_: "Laser Spectroscopy VII," T. W. Hansch and Y. R. Shen, eds., Springer-Verlag, Berlin (1985).
3. S. C. Hill and R. E. Benner, Morphology-dependent Resonances Associated with Stimulated Processes in Microspheres, _J. Opt. Soc. Am. B_ 3:1509 (1986).
4. S. C. Ching, H. M. Lai, and K. Young, Dielectric Microspheres as Optical Cavities: Einstein A, B Coefficients and Level Shifts, _J. Opt. Soc. Am. B._, in press.
5. W. Jhe, A. Anderson, E. A. Hinds, D. Meschede, L. Moi, and S. Haroche, Suppression of Spontaneous Decay at Optical Frequencies: Test of Vacuum-field Anisotropy in Confined Space, _Phys. Rev. Lett._ 58:666 (1987).
6. A. Ashkin, Applications of Laser Radiation Pressure, _Science_ 210:1081 (1980).
7. J. F. Owen, P. W. Barber, B. J. Messinger, and R. K. Chang, Determination of Optical Fiber Diameter from Resonances in the Elastic Scattering Spectrum, _Opt. Lett._ 6:272 (1981); A. Ashkin and J.M. Dziedzic, Observation of Optical Resonances of Dielectric Spheres by Light Scattering, _Appl. Opt._ 20:1803 (1981).
8. R. E. Benner, P. W. Barber, J. F. Owen, and R. K. Chang, Observation of Structure Resonances in the Fluorescence Spectra from Microspheres, _Phys. Rev. Lett._ 44:475 (1980); J. F. Owen, P. W. Barber, P. B. Dorain, and R. K. Chang, Enhancement of Fluorescence Induced by Microstructure Resonances of a Dielectric Fiber, _Phys. Rev. Lett._ 47:1075 (1981).
9. L. M. Folan, S. Arnold, and S. D. Druger, Enhanced Energy Transfer within a Microparticle, _Chem. Phys. Lett._ 118:322 (1985).
10. J. F. Owen, R. K. Chang, and P. W. Barber, Morphology-Dependent Resonances in Raman Scattering, Fluorescence Emission, and Elastic Scattering from Microparticles, _Aerosol Sci. Technol._ 1:293 (1982); R. Thurn and W. Kiefer, Structural Resonances Observed in the Raman Spectra of Optically Levitated Liquid Droplets, _Appl. Opt._ 24:1515 (1985); T. R. Lettieri and R. E. Preston, Observation of Sharp Resonances in the Spontaneous Raman Spectrum of a Single Optically Levitated Microdroplet, _Opt. Commun._ 54:349 (1985).
11. H.-M. Tzeng, K. F. Wall, M. B. Long, and R. K. Chang, Laser Emission from Individual Droplets at Wavelengths Corresponding to Morphology-Dependent Resonances, _Opt. Lett._ 9:499 (1984); S. X. Qian, J. B. Snow, H.-M. Tzeng, and R. K. Chang, Lasing Droplets: Highlighting the Liquid-Air Interface by Laser Emission, _Science_ 231:486 (1986); H.-B. Lin, A. L. Huston, B. L. Justus, and A. J. Campillo, Some Characteristics of a Drop-

let Whispering-Gallery-Mode Laser, <u>Opt. Lett.</u> 11:614 (1986).

12. J. B. Snow, S.-X. Qian, and R. K. Chang, Stimulated Raman Scattering from Individual Water and Ethanol Droplets at Morphology-Dependent Resonances, <u>Opt. Lett.</u> 10:37 (1985).

13. S.-X. Qian and R. K. Chang, Multiorder Stokes Emission from Micrometer-Size Droplets, <u>Phys. Rev. Lett.</u> 56:926 (1986).

14. S.-X. Qian and R. K. Chang, Phase-Modulation-Broadened Line Shapes from Micrometer-Size CS_2 Droplets, <u>Opt. Lett.</u> 11:371 (1986).

15. S.-X. Qian, J. B. Snow, and R. K. Chang, Coherent Raman Mixing and Coherent Anti-Stokes Raman Scattering from Individual Micrometer-Size Droplets, <u>Opt. Lett.</u> 10:499 (1985).

16. J. H. Eickmans, W.-F. Hsieh, and R. K. Chang, Laser-Induced Explosion of H_2O Droplets: Spatially Resolved Spectra, <u>Opt. Lett.</u> 12:22 (1987).

OPTICAL SPECTROSCOPY OF SIZE EFFECTS IN SEMICONDUCTOR MICROCRYSTALS

A. I. Ekimov[1] and Al. L. Efros[2]

[1]A. I. Vavilov State Optical Institute, 199164, Leningrad
USSR, [2]A. F. Ioffe Physico-Technical Institute, 194021
Leningrad, USSR

ABSTRACT

Size effects in semiconductor systems with reduced dimensions have
attracted considerable attention within the last few years. Besides the
well known quantum-well and quantum-wire structures[1,2], semiconductor
microcrystals prepared in aqueous and gaseous media were investigated[3,4]. The purpose of this paper is to investigate the quantum
confinement effects in semiconductor microcrystals grown in an optically
transparent matrix of oxide glass. The preparation technique developed
makes it possible to vary the size of microcrystals in a controlled
manner from a few tens to thousands of angstroms[5].

I. GROWTH OF MICROCRYSTALS

Microcrystals were grown in a multicomponent oxide glass in which
the semiconductor phase was dissolved during the high temperature
melting. The nucleation and growth of semiconductor particles were
performed in the course of a heat treatment of the glass samples due to
the phase decomposition by diffusion of the supersaturated solid
solution. Figure 1 shows experimental dependences of the average radii
of CdS and CuBr microcrystals as a function of the duration of heat
treatment for different temperatures – (a,b) and as a function of the
reciprocal temperature at fixed time duration – (c). Mean values of the
radius of the microcrystals for each sample were determined by the method
of small angle x-ray scattering in the approximation of spherical
particles. These experimental dependences are in good agreement with the
theory developed by Lifshitz and Slezov for the recondensation stage of
the process of diffusive phase decomposition of a supersaturated
solution[6]. This stage begins when the degree of supersaturation of the
solution becomes sufficiently small and large particles grow at the
expense of the dissolution of small ones. The kinetics of the growth is
described by the following expression[6]:

$$\bar{a} = \left(\frac{4}{9} \sigma \, DC\tau\right)^{1/3}, \tag{1}$$

where $\bar{a}$ is the mean radius of the particles; σ is proportional to the
interfacial surface tension, the diffusion coefficient, D, and the
equilibrium concentration of the solution, C, depend exponentially on

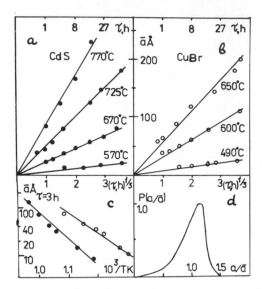

Fig. 1. Mean microcrystal radius vs heat treatment duration τ at different temperatures: (a) CdS, (b) CuBr; (c) the mean microcrystal radius vs reciprocal temperature of heat treatment (t = 3h); (d) size distribution function $P(a/\bar{a})$.

temperature. So, the experimental data do in fact reveal the recondensation nature of the growth of semiconductor particles in our case. It was shown by Lifshitz and Slezov that a steady-state size distribution is formed in the course of the recondensation growth, and an analytical expression for it was obtained. The distribution function is shown in Fig. 1d.

Thus, the technique of diffused-controlled growth of semiconductor microcrystals in the glassy matrix developed here makes it possible to vary the size of particles merely by choosing the temperature and the duration of heat treatment. The steady-state size distribution of microcrystals is rather narrow and may be taken into account in calculations, since an analytical expression for it is known.

II. OPTICAL SPECTRA OF SEMICONDUCTOR PARTICLES

Since oxide glass is transparent in the optical spectral range, it is possible to employ optical spectroscopy methods to study such heterophase systems[7]. Figure 2 shows the absorption spectra measured on the glassy samples containing CdSe, CdS, CuBr and CuCl particles. As can be seen, at sufficiently large dimensions of microcrystals, the absorption spectra reveal the typical excitonic structure of near-band-gap optical transitions. This structure is related to the spin-orbit splitting of the valence band in cubic materials and to the spin-orbit and crystal field splitting in hexagonal materials. These spectra, as well as luminescence and Raman spectra, demonstrate that the semiconductor particles grown in a glassy matrix have the crystalline structure of bulk materials and sufficiently high spectroscopic quality.

III. QUANTUM CONFINEMENT EFFECTS

As was shown earlier[8], a variety of qualitatively different confinement effects may occur that depend on the ratio of the microcrystal radius a and the exciton radius r_{ex}. So, in analyzing the experimental data it is necessary to consider three different cases:

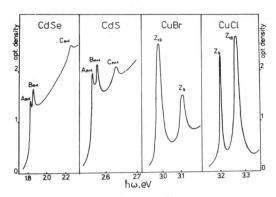

Fig. 2. Low temperature (T = 4.2K) absorption spectra of glasses with microcrystals: CdSe ($\overline{a}$ = 380Å), CdS ($\overline{a}$ = 320Å), CuBr ($\overline{a}$ = 240Å), and CuCl ($\overline{a}$ = 310Å).

r_{ex} < a, r_{ex} ~ a and r_{ex} > a. Because the values of the exciton radius are different for the enumerated semiconductor materials, each of the cases may be studied for a certain material.

1. <u>Size quantization of excitons</u>. The exciton radius in CuCl crystals is r_{ex} = 8Å. So, for this material the relationship r_{ex} < a holds for all microcrystal sizes investigated. Figure 3a shows the size

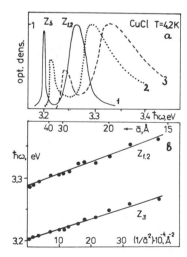

Fig. 3. (a): absorption spectra of glasses with CuCl microcrystals (a = 310Å (1), 29Å (2), 20Å (3)). (b): exciton lines spectral position vs $1/\overline{a}^2$ in CuCl.

dependence of absorption spectra of CuCl microcrystals. As can be seen, the absorption is of excitonic nature down to the smallest sizes. The decrease of the radius of the microcrystals leads to a high energy shift for both of the excitonic lines. The value of the shift is about 50 meV. This effect is due to the size quantization of an exciton as a whole[9].

Figure 3b represents a spectral position of absorption lines as a function of the reciprocal mean square radius of the microcrystals. It is seen that both dependences are linear for values for microcrystal radius down to about $\bar{a} \sim 15$Å.

For excitons originating from the upper doubly-degenerate valence subband Γ_7(Z_3 - line) the slope of the curve is determined by the exciton translational mass M_S:

$$\hbar\omega_{Z_3} = E_g - E_{ex} + 0.67 \frac{\hbar^2}{2M_S \bar{a}^2} \pi^2 , \qquad (2)$$

where E_g is a band gap and E_{ex} is the exciton binding energy. The size dependence of the spectral position of excitons originating from the lowest four-fold degenerate valence subband Γ_8 ($Z_{1,2}$ - line) is given by:

$$\hbar\omega_{Z_{12}} = E_g - E_{ex} + \Delta + 0.67 \frac{\hbar^2}{2M_r \bar{a}^2} \left[\phi(M_\ell/M_h)\right]^2 , \qquad (3)$$

where M_h and M_ℓ are the "heavy" and "light" exciton masses, Δ is the value of spin-orbit splitting, and $\phi(M_\ell/M_h)$ is a solution of the transcendental equation given in Ref. (10). The numerical factor in Eqs. (2)-(3) results from the averaging of the Lifshitz size distribution function.

Comparison between experimental and theoretical results enables us to determine the masses of excitons for all three exciton subbands[10]. The value M_S = (1.9±0.2)m_0 is in good agreement with available data. As far as we know, the value of the "heavy" exciton mass, M_h = 2.6±0.2)m_0, and that of the "light" exciton mass, M_ℓ = (1.5±0.2)m_0, have been determined for the first time.

2. Effect of localization of hole by the electron-hole interaction. It is convenient to study the intermediate situation, corresponding to the case when the microcrystal size is of the order of the exciton size ($a \sim r_{ex}$), by taking CuBr crystals, where the exciton radius is $r_{ex} \approx 18$Å. Figure 4a shows the size dependence of the absorption spectra of CuBr microcrystals. The decrease of microcrystal size leads to a high energy shift of the absorption lines which amounts to ~ 150 meV. The excitonic structure of the absorption spectra practically disappears for microcrystals with radius close to the exciton radius.

Figure 4b shows the size dependence of the spectral position of the absorption lines. As can be seen, within the range of microcrystal sizes much larger than the exciton radius, there is a slight shift, which is due to the quantization of an exciton as a whole. But upon decreasing $\bar{a}$ to as low as 50Å, a drastic decrease of the slope of the curve occurs.

Such behavior of the spectral position of exciton lines as a function of microcrystal size was described under the assumption of localization of the hole in the center of a microcrystal due to its interaction with an electron[11]. This effect was predicted theoretically for the case when the hole mass, M_h, is much larger than the electron mass M_e, ($M_h \gg M_e$). In this case the exciton may be treated as

202

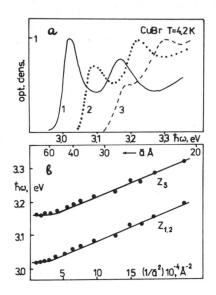

Fig. 4. (a): absorption spectra of glasses with CuBr microcrystals
(a = 240Å (1), 36Å (2), 23Å (3)). (b): exciton lines' spectral
position vs $1/a^2$ in CuBr.

a donor atom, situated at the center of a microcrystal. The rapid shift
of absorption lines is due to the size dependence of the energy of the
ground state of such a donor-like exciton. The calculations of this
dependence were made with the use of a variational technique. The best
fit, shown in Fig. 4b by solid line, was obtained for values of the
dielectric constant $\kappa = 8.0$ and $M_e = 0.25\ m_o$, which are in good agreement
with data available for this material.

3. Quantization of the energy spectrum of holes in the adiabatic
potential of the electron. The last case, when the microcrystal radius
is smaller than the exciton radius, will be studied for CdS crystals,
where the exciton radius $r_{ex} = 30$Å. Figure 5a shows the size dependence
of the absorption spectra of CdS microcrystals. It is seen that the
decrease of microcrystal size leads to a large high energy shift of the
absorption band edge by about ~ 1 eV. Oscillations in the absorption
spectra are due to the transitions to quantum sublevels of the conduction
band[12]. A size dependence of the absorption edge and spectral position
of the oscillations may be described in the first approximation by the
following expression[13]

$$\hbar\omega_{\ell,r} = E_g + 0.71 \frac{\hbar^2}{2M_e a^{-2}} \phi_{\ell,r}^2 \ , \tag{4}$$

where $\phi_{\ell,r}$ are the roots of the Bessel function

$(\phi_{0,1} \approx 3.14;\ \phi_{1,1} \approx 4.49;\ \phi_{2,1} \approx 5.76)$.

The slope of the curves in Fig. 5b are in good agreement with the
value of the electron effective mass, $M_e = 0.2m_o$, which is well-known for
CdS. The deviations of the experimental points from the theoretical

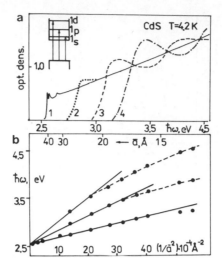

Fig. 5. (a): absorption spectra of glasses with CdS microcrystals (a = 330Å (1), 23Å (2), 15Å (3), 12Å (4)) (b): interband absorption spectra and second derivative spectra in CdS microcrystals.

curves may be due to a number of reasons; among them the non-parabolicity of the electronic band far away from the Γ point is the most significant.

Next we shall discuss the role of the electron-hole Coulomb interaction. In the range of microcrystal sizes a < r_{ex} the energy of motion of an electron in a quantum well is considerably higher than the energy of the Coulomb interaction of the electron with a hole. When $M_h \gg M_e$, the Coulomb potential which affects the hole may be assumed to be averaged over the fast motion of the electron (adiabatic approximation). If the electron is situated in the lower size-quantization level, this potential will have a minimum at the center of the microcrystal, and near the bottom may be written in the form[14]

$$V(r_h) = -\frac{2.42e^2}{\kappa a} + \frac{\pi^2}{3}\frac{e^2}{\kappa a}\frac{r_h^2}{a^2}, \qquad (5)$$

where κ is the dielectric constant of the crystal, the numerical factor 2.42 results from the electron charge density distribution, and where r_h is the distance of the hole from the center of the microcrystal. This potential has a form of the potential of a three-dimensional harmonic oscillator.

Thus, in this case a hole moves in a potential formed by an electron rather than in the potential well itself. The energy spectrum of the hole moving in this potential is a system of equidistant levels[15]. The wave functions of the hole corresponding to these states differ markedly from the wave functions which describe their motion in the absence of Coulomb interaction. As a result, the selection rules change, and optical transitions from each level to the lowest electron size quantization level become possible. This fine structure was observed in the second derivative spectra of CdS microcrystals with sizes ranging from 25 to 15Å. (See Fig. 6).

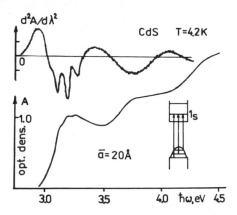

Fig. 6. Interband absorption spectra and second derivative spectra in CdS microcrystals.

IV. SURFACE PHENOMENA IN SEMICONDUCTOR MICROCRYSTALS

The small size of microcrystals under investigation makes surface phenomena important in the formation of their optical spectra. Now, the intrinsic and impurity luminescence spectra of microcrystals will be discussed.

1. Surface exciton mode. A doublet structure of the resonant exciton luminescence of CuCl microcrystals was observed. Figure 7a shows the size dependence of the structure. For $\bar{a} < 60$Å, we have observed a singlet whose spectral position is determined by exciton quantization and depends linearly on $1/\bar{a}^2$ according to Eq. (2) (see Fig. 7b). For $\bar{a} \sim (70-80)$Å a second line appears. Its spectral position does not depend on microcrystal size (see Fig. 7b), and its intensity increases and comes to dominate, while the mean radius increases.

The observed structure was interpreted within the framework of polariton size quantization theory[16]. It was shown that the appearance of the second line is due to the sharp decrease of the radiative lifetime of a polariton when its size quantization energy levels coincide with the surface exciton energy level given by

$$\hbar\omega_s = \hbar\omega_T + \frac{\hbar\omega_{LT}}{1+2\kappa_m/\kappa_\infty} , \tag{6}$$

where $\hbar\omega_T$ is the position of the bottom of the exciton subband, $\hbar\omega_{LT}$ is the longitudinal-transverse splitting of the exciton ($\hbar\omega_{LT} = 5.7$ meV for CuCl), κ_∞ and κ_m are the dielectric constants of the semiconductor and the matrix respectively.

2. Surface-influenced impurity luminescence. Low temperature luminescence spectra of CdS microcrystals were studied (see Fig. 8 inset). A strong dependence of spectral position and relative intensity of an exciton, band-impurity (D-V), and inter-impurity (D-A) luminescence transitions on the size of microcrystals was observed. The size dependence of donor-acceptor band spectral position is shown in Fig. 8.

205

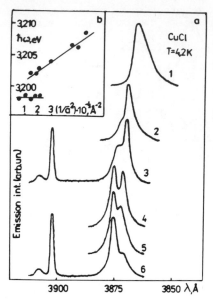

Fig. 7. (a): Resonance exciton luminescence spectra in CuCl microcrystals ($\bar{a}$ = 45Å (1), 56Å (2), 70Å (3), 76Å (4), 95Å (5), 132Å (6)). (b): fine structure maxima vs $1/a^2$.

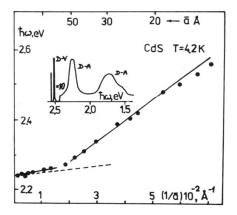

Fig. 8. Donor-acceptor transition line spectral position vs $1/\bar{a}$.

There are two regions on this curve. The first of them corresponds to large microcrystal sizes ($a > a_B$ where a_B is the Bohr radius of the donor) and could be ascribed to donor acceptor distance confinement by microcrystal dimensions[17]. As a result, the energy of the transition should exceed the value

$$\hbar\omega_{D-A} = E_i - E_A + \frac{e^2}{2\kappa a}, \qquad (7)$$

where E_i is the donor energy measured from the top of the valence band, and E_A is the acceptor binding energy. Equation (7) is plotted in Fig. 8 by a dashed line.

The second region corresponds to small microcrystal sizes ($\overline{a} < a_B$), where the donor energy, E_i, is dependent on the microcrystal size a and the distance (r_D) between donor and the center of microcrystal

$$E_i = E_g + \frac{\hbar^2 \pi^2}{2M_e a^2} - \frac{e^2}{\kappa a} f\left(\frac{r_D}{a}\right) , \qquad (8)$$

where $f(0) = 2.42$ and $f(1) = 1$.

This leads to a large additional shift and broadening of D-A lines within the region of sizes $a < a_B$.

V. CONCLUSION

Our results show that all the features of the quantum confinement effects in semiconductor microcrystals grown in a glassy matrix can be well-described within the framework of the effective mass approximation. They further show that this approximation is valid for particles with sizes as small as 25Å.

REFERENCES

1. K. Ploog and C. N. Döhler, Adv. Phys. 32, 285 (1983).
2. M. Laviron and P. Averbuch, et al. J. Phys. Lett. 44, L-1021 (1983).
3. L. E. Brus, J. Chem. Phys. 79, 5566 (1983).
4. P. Martin, Adv. Phys. 34, 216 (1985).
5. V. V. Golubkov, A. I. Ekimov, A. A. Onushchenko, and V. A. Tsekhomskii, Fiz. Khim. stekla 7, 397 (1981).
6. I. M. Lifshitz and V. V. Slezov, Zh. Eksp. Teor. Fiz. 35, 479 (1958).
7. A. I. Ekimov, A. A. Onushchenko, and V. A. Tsekhomskii, Fiz. Khim. stekla 6, 511 (1980).
8. A. I. Ekimov, Al. L. Efros, and A. A. Onushchenko, Sol. St. Comm. 56, 921 (1985).
9. A. I. Ekimov and A. A. Onushchenko, Pis´ma ZhETF 34, 363 (1981).
10. A. I. Ekimov, A. A., Onushchenko, A. G. Plukhin, A. L. Efros, ZhETF, 88, 1490 (1985).
11. A. I. Ekimov, A. A. Onushchenko, S. K. Shumilov, Al. L. Efros, Pis´ma ZhETF, 13, 281 (1987).
12. A. I. Ekimov, A. A. Onushchenko, In: Proc. All-Union Conf. Phys. Semiconductor (ELM, Baku, 1982) p. 176.
13. A. I. Ekimov, and A. A. Onushchenko, Pis´ma ZhETF, 40, 337 (1984).
14. A. I. Ekimov, A. A. Onushchenko, Al. L. Efros, Pis´ma ZhETF, 43, 292 (1986).
15. Al. L. Efros and A. L. Efros, Fis. Tekn. Polupr., 16, 1209 (1982).
16. A. I. Ekimov, A. A. Onushchenko, M. E. Raikh, and Al. L. Efros, ZhETF, 90, 1795 (1986).
17. N. Crestnow, L. E. Brus, T. D. Harris, and R. J. Hull, Phys. Chem., 90, 3393 (1985).

COOPERATIVE BEHAVIOR OF CONFIGURATIONAL DIPOLE DEFECTS

IN PLASTIC DEFORMED SEMICONDUCTOR CRYSTALS

Yu. A. Ossipyan, V. D. Negriy and N. A. Bul'enkov

Academy of Sciences of the USSR, Institute of Solid State Physics,
Chernogolovka, Moscow District, 142432, USSR

INTRODUCTION

The development of the technique of mechanically loading crystals, placed in a low-temperature optical cryostat, with simultaneous recording of time-dependent, spectral and polarization characteristics of photoemission of a low-temperature deformed crystal enabled the observation of a number of new effects.[1,2] An essential specific feature of our technique is that all these optical characteristics may be observed by light emission from a fixed local crystal region and correlated with its structural singularities. The application of this space-resolved photoluminescence method made it possible to observe emission bands arising in a rather narrow spectral region in deformed CdS crystals.[3,4]

Below are given the main results summing up the characteristics of the emission observed.

1. The photoluminescence spectra of low-temperature plastically deformed crystals exhibit the formation of a new group of lines, missing in purposely undeformed crystals (see Fig. 1). These lines (in the region λ = 505-510 nm) manifest themselves only in the spectra of crystals with deformation-induced dislocations (with basal or prismatic glide.) We termed this spectrum as dislocation photoemission spectrum.[5]

2. Subsequent experiments suggest the conclusion that the emission sources are not so much the dislocations themselves as the traces left behind by the dislocations when they are gliding in the glide planes. Utilizing this emission, we have developed a technique for the observation of the low-temperature motion of not only the whole bands, but also, of isolated dislocations (see the videomovie). The movie[*] showed the cases of fast and slow motion of individual dislocations and their clusters as a function of the optical excitation level, the temperature (within 4.2 - 77 K) and the magnitude of the external stresses as well as the sample crystallographic orientation and the type of gliding dislocations.

---

[*]Editor's note: The film was displayed at the Symposium.

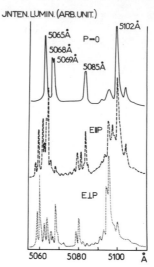

Fig. 1.   Dislocation photoemission spectrum CdS crystal at different
               elastic stresses: P = 0, P = 100 MPa.

3. Further understanding of the emission mechanism was facilitated by
the investigation of the polarization of emitted light. In the unpolarized
light under the conditions of a low constant excitation level (w ≤ 100
W/cm²) and any deformation geometry, all dislocation traces look like
continuously light emitting bands.

     In polarized light the traces of basal dislocations are divided into
domains with two different polarizations of the emitted light.   The
domains appear in polarized light as separate, independently scintillating
segments.   Their number and size can both increase and decrease
spontaneously.   The   nonstationary   behaviour   of   the   domains   is
significantly dependent on the temperature and the optical excitation
level.   At 4,2 K and a low level of excitation, when the scintillations
were quite rare, it was established by direct observation that the
extinction of domains in one of the polarizations indicated corresponds
uniquely with their flare-up in another polarization, and vice versa (Fig.
2).   The projections of the electric polarization vector of the basal
dislocation traces on (1100) form angles of ∓ 60° with the a-axis, the
projections on a (1120) -- ∓ 45°.

     Traces   of   prismatic   dislocations   are   unpolarized.   However,
polarization arises if the crystal is subjected to a small external
uniaxial elastic deformation.

4. Along with a uniaxial plastic deformation, we studied the traces of
dislocations, induced by indentation of a diamond pyramid into the CdS
crystal – the well known case of a dislocation rosette. The indentation
by a diamond needle was performed on a (0001) face at 300 K (Fig. 3.). A
significant difference from a uniaxial plastic deformation is that even in
the absence of an external loading each trace of prismatic dislocation in
the rosette has its strictly definite polarization.   Fig. 4 demonstrates
schematically the formation of a dislocation rosette, the direction of
typical traces and their emission polarization orientation are indicated.
The traces of <1120> glide bands and the orientation of the polarization
vectors projections correspond to the hexagonal symmetry of a (0001)
face.

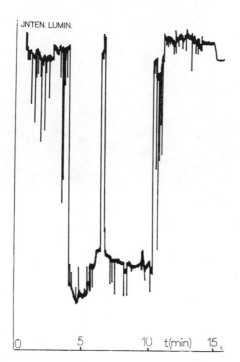

Fig. 2. Fragment of the time dependence of the intensity of
polarized radiation from one
of the domains of the dislocation trace at 4,2 K.

Fig. 3.   Photograph of the dislocation rosette light emission in the range
λ = 505-510 nm.

Such are the experimental facts. We propose the following model for
their explanation.

When moving, both prismatic and basal dislocations generate in the
glide plane closely spaced specific dipole defects, which under optical
and electronic excitation act as light emission sources (505-510 nm). The
emission of each source is polarized. Due to the mutual interaction the
system of closely spaced defects may become orientationally ordered.
This implies that the emission from all the defects has the same
polarization. Under external effects this system can be reoriented and,
as a consequence, it changes the polarization direction. Like in any
ordered system, the separation into domains (with respect to the light
polarization) and the motion of the domain walls can take place.

Basal dislocations leave behind the orientation-ordered system of
defects the light emission of which is always polarized. On the contrary
prismatic dislocations generate a system of dipole defects which may have
several possible orientations. Therefore in the absence of an external
elastic field this system is orientation-degenerated. Its light emission
is unpolarized. The degeneracy is eliminated by applying an external
elastic field. The system becomes preferentially oriented and its light
emission becomes polarized. For the same reason (because of the internal
elastic stresses) the dislocation-rosette emission is always polarized.

The above stated model concerning the characteristics of the defect
induced by the low temperature dislocation motion directly follows from
the treatment of the experimental results. It sets aside the principal
questions, namely, what is the microscopic nature of these light emitting
dipole defects? How are they formed and what is their atomic structure?

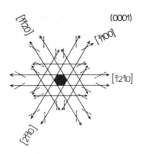

Fig. 4. Schematic presentation of the six-branch rosette formation on a
(0001) plane and the polarization of the dislocation traces emission.

The attempts to answer all these questions have necessitated the
creation of a microscopic model of a light-emitting dipole centre based on
the idea of a new type of a configurational defect in the diamond-type
structure. We termed this defect: bimodulus.

The atomic structure of the bimodulus for the diamond lattice was
discussed earlier.[6,7] One can construct in the diamond-type structure the
configurationl defects caused only by a small shift with respect to the
matrix in the direction of one of the three double axes. The formed
bimodulus has D_2 symmetry and in the direction of the remaining two of
the three double axes conserves a coherent bond with the matrix. To
attain the bimodulus-matrix coherent conjugation in the "shift axis"
direction as well, two pairs of five and seven-element cycles in the
antisymmetric position must be introduced each on the right and the left
of the bimodulus.

Now we consider the structure of the analogous bimodulus in the
hexagonal lattice of wurtzite. It can also be constructed from two (left
and right) disparate moduli with breaking of two chemical bonds. The
evolution of the perfect matrix to a bimodulus with two dangling bonds
for the hexagonal wurtzite lattice is shown in Fig. 5. This
reconstruction is attained by an antiparallel displacement of two atom
pairs by a distance of 2/5 $\bar{b}$ (F' and M' rightwards, R' and n' leftwards.)
The reconstruction back to the perfect matrix is attained by a recurrent
displacement of the same atoms with "closing" of a chemical bond. Like in
the case of the cubic diamond structure, to attain coherency between the
modulus and matrix in the wurtzite lattice two pairs of five- and seven-
element cycles in the antisymmetric position must be introduced on the
right and the left of the bimodulus. This is necessary for conjugation of
the incoherent cycles "twist-chair" (t-c) of the bimodulus and the normal
cycles "bath" (b) of the matrix. These five- and seven-element and seven-
element cycles have the atomic structure corresponding to an edge
dislocation core structure.[8] So, each bimodulus contains two atoms with
dangling bonds. This atom pair may be represented as a dipole, which is
the light emitting source. Upon the dipole formation the shift occurred,
for example, in the direction <1120>, so the presence of two equivalent
dipole orientations in bimoduli is possible for each of the three
directions <1120>.

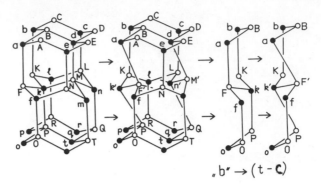

$$_{\text{\textbf{n}}}b^{\prime\prime} \rightarrow (t - \mathbf{c})$$

Fig. 5. Formation of the bimodulus with dangling bonds in the wurtzite lattice.

Now, we have to answer the principal question. How can bimoduli be formed in a sufficiently great number in the course of plastic deformation?

Below we propose a scheme for a microscopic mechanism of the bimoduli chain formation following a gliding edge prismatic dislocation (Fig. 6). During the dislocation shift by two translations in the [1120] direction (Fig. 6a) two five- and seven-element cycles in the antisymmetric position are formed between the starting and final position of the dislocation. The dislocation shift by one more translation in the same direction may lead to the fact that rather great elastic stresses, which arise in the vicinity of the dislocation core, can be relieved by breaking of one bond with the formation of a bimodulus (Fig. 6b). This breaking must naturally be favored by both the presence of thermal fluctuations and by the encounter of the gliding dislocation with an acceptor impurity.

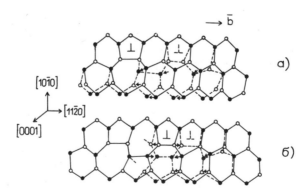

Fig. 6. Schematic presentation of the bimodulus formation at dislocation shift in the [1120] direction: a) by two lattice constants, b) by three lattice constants.

Further rightward motion of the dislocation (Fig. 7a) results in the formation of a row of bimoduli with five- and seven-element cycles at its sides. This row of bimoduli dipoles can, in particular, break away from the gliding dislocation (Fig. 7b) and exist individually as a chain of a finite length.

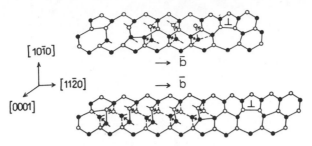

Fig. 7. Schematic presentation of a bimoduli chain (a) and its break away
from the gliding dislocation (b).

Inasmuch as during the motion in the glide plane the dislocation can
form in some points of its length individual bimoduli with different
dipole orientation, then at further motion this dislocation will "print"
fragments of bimoduli chains with different dipole orientation from the
atoms with dangling bonds (domains.) A similar scheme may also be
proposed for the case of the basal dislocation motion. It is quite
naturally to presume that under the action of a directed stress the
bimoduli chain will be "domainized" also due to straightening of bimoduli
traces and elongating of the portions with identical orientation of the
direction of dipole polarization.

As for basal dislocations apparently there is strong interaction
between bimoduli, which gives rise to the appearance of a domain
structure of dislocation traces.

So, our model the origin and multiplication of bimoduli at low
temperature plastic deformation, containing dipoles with two atoms with
dangling bonds, may explain consistently and unambiguously all the
experimental facts observed and described above. In particular, a
sufficiently low energy of the bimoduli chain explains the disappearance
of this type of defect under a comparatively small increase of the
temperature and optical excitation level.

In conclusion we point out that the bimoduli model of dipole defects
in diamond-type semiconductors can have rather general character and if
subsequently confirmed, it can find a wide application for explaining the
structure-dependent properties of semiconductors, including electrical,
optical and magnetic ones. It can also amplify our ideas about the
mechanism of plastic deformation of semiconductors and about the
interaction of their structural defects with impurities.

References

1. V. D. Negriy and Yu. A. Ossipyan, Pis'ma Zh. Eksp. Teor. Fiz. 35:484
      (1982); JETP Lett. 35:598.
2. V. C. Negriy and Yu. A. Ossipyan, Fizika Tverd. Tela 24:344 (1982).
3. V. D. Negriy and Yu. A. Ossipyan, Fizika Tverd. Tela 20:744 (1978).
4. L. N. Golovko, V. D. Negriy and Yu. A. Ossipyan, Fizika Tverd. Tela
      20:1717 (1986).
5. V. D. Negriy and Yu. A. Ossipyan, Phys. Stat. Sol. (a) 55:583 (1979).
6. N. A. Bul'enkov, Dokl. Acad. Nauk 284:1392 (1985).
7. N. A. Bul'enkov, Dokl. Acad. Nauk 290:605 (1986).
8. Yu. A. Ossipyan and I. S. Smirnova, J. Phys. Chem. Solids 32:1521
      (1971).

THE SPECTRUM OF LIGHT PROPAGATION TIMES THROUGH A DISORDERED MEDIUM

AND MESOSCOPIC FLUCTUATIONS

Altshuler B.L.[*], Kravtsov V.E., and Lerner I.V.

Institute of Spectroscopy, USSR Academy of Sciences, Troitsk
Moscow obl., 142092, USSR
[*]Leningrad B.P. Konstantinov Institute of Nuclear Physis
USSR Academy of Sciences, Leningrad obl., Gatchina

INTRODUCTION

Considerable progress has been made in the last decade toward understanding quantum transport in disordered media (see for reviews Ref. 1,2). The phenomenon of Anderson localization[3] arises from quantum interference under multiple electron scattering by impurities. Therefore, localization may be regarded as a result of the creation of standing waves by a set of random scatterers which acts as a random interferometer (Fig. 1). As the localization arises only from the wave

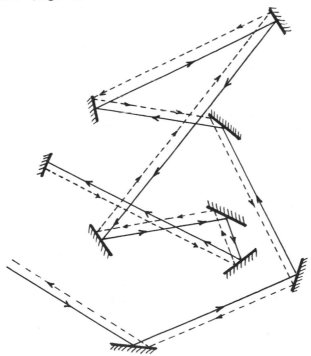

Fig. 1. Anderson localization as a standing wave creation in a random interferometer.

nature of the electrons, it should reveal itself for any wave propagation in the presence of random scatterers. The object here of particular interest is light propagation through disordered media. However, interference effects under multiple light scattering will be essential only if the mean free path for elastic scattering ℓ is much smaller than that for inelastic scattering (e.g. for absorption) ℓ_{inel}. Unfortunately, under experimental conditions these lengths obey, as a rule, the opposite inequality: $\ell \gg \ell_{inel}$. Nevertheless, there exist specially prepared highly inhomogeneous media where, in addition, ℓ_{inel} is long as compared to ℓ. In studies of light propagation through such media, one meets with some advantages in searching for the Anderson localization effects as against the conventional case of electron localization. For one thing, photon-photon interaction is negligible and hence "pure" localization effects may be studied; while for electrons these effects should be distinguished from those of Coulomb interaction. For another, there is no need for subkelvin temperatures which have been required for revealing the wave nature of electrons. In addition, one can change the photon wavelength and intensity continuously which is quite impossible for electrons in a metal. Thus, the experiments on photon localization may provide valuable information about the Anderson transition. For all these reasons, the problem of photon localization has attracted considerable interest during the last years.[4,5]

The main experimental efforts have been focused on studying both backscattering enhancement of light scattered from highly inhomogeneous media and reproducible asperiodic oscillations in the angular dependence of the scattered light intensity[5]; these oscillations being analogous to the mesoscopic fluctuations of disordered metal conductivity.[6] Another possible way of investigating photon localization effects has been recently proposed in experiments on short light pulse propagation through media with strong scattering and weak absorption.[7] A diffusion coefficient D for propagation of light may be determined from the time-dependence of output intensity. When the parameters of the system are close to the Ioffe-Regel localization criterion $\lambda/2\pi\ell \sim 1$ localization effects should result in D decreasing with increase of a propagation length L which should influence the shape of the output signal. A similar way of investigating the electron localization effects has been proposed[8] in theoretical study of the current relaxation in disordered metals. Localization effects have been shown, however, to result in a more complicated picture than that the only change is of the diffusion coefficient changing with scale. In particular, long-time tails of the relaxation current turn out to decay considerably slower than exponentially.[8]

In this paper we will theoretically consider the long-time-tails problem for a short input pulse of light propagating with negligible absorption through disordered media. This problem, similarly to that of the current relaxation, proves to be rather complicated so that one fails to treat it within the framework of the conventional[1,2] localization theory.

Let us begin with a statement of the problem. The present considerations deal with the pulse of light propagating through an inhomogeneous sample of size L, the source of short pulses being placed inside the sample. Interest will be concentrated on the time-dependence of the output signal. The initial problem is governed by the usual wave equation

$$(\omega^2 \varepsilon(r)/c^2 + \nabla^2)E(r,\omega) = 0, \tag{1}$$

where $\varepsilon(r)$ is the randomly inhomogeneous permitivity, $E(r,\omega)$ is the amplitude of the electromagnetic wave at the frequency ω. Dealing only with the intensity I of multiply scattered light, one applies as usual the diffusion approximation:

$$(\frac{\partial}{\partial t} - D_0 \nabla^2)\, \rho\, (r,t) = 0,$$ (2)

where $I(r,t) = -D_0 \Delta \rho\, (r,t)$, $\rho\, (r,t)$ is the photon density at point (r,t). On solving this equation with boundary conditions $\rho\, (t) = 0$ and an initial condition $\rho\, (r,0) \sim \delta\, (r)$ one obtains (omitting irrelevant details) the time dependence of the output intensity as follows:

$$I(t) \sim t^{-3/2} \exp(- \frac{t_D}{t} - \frac{t}{t_D})$$ (3)

with $t_D = L^2/D_0$ being the time of diffusion propagation through the sample. (A more accurate expression for $I(t)$ with effects of the sample geometry taken into account has been given in Ref. 7).

On deriving the diffusion equation (2) from the initial wave equation (1), one has neglected both the interaction of diffusion modes and higher order time and spatial derivatives. The inclusion of diffusion modes interaction which is required for obtaining localization effects is most rigorously made by the field theoretical approach in the framework of the nonlinear σ model. This model has been proposed for the localization problem in Ref. 9 and was developed in a set of papers[10] including those concerning light propagation through disordered media.[4] Such considerations enabled us to confirm the one-parameter scaling ansatz[1,2] which formed the basis of localization theory. Within this concept, all the change of the expression (3) due to localization effects reduces to a substitution of the diffusion coefficient $D(L)$ depending on scale L for the bare one $D_0 = lc/3$.

However, the solution to a set of problems in localization theory requires exceeding the limits of one-parameter scaling. One point is that higher order time and spatial derivatives which have been neglected in the conventional scaling description are relevant to all these problems.[8,11,12] For the problem under consideration we will be interested in higher order time derivatives. Their importance is due to the fact that a coefficient C_n attached to the n-th derivative undergoes an extremely rapid increase with n:

$$C_n \sim \exp(2un^2),$$ (4)

where

$$u = \ln\frac{D_0}{D(L)}.$$ (5)

This increase has been found to be responsible for the instability of the one-parameter scaling theory of localization in Ref. 11; a similar increase is inherent in some other problems of localization theory.[8,12,13] The derivation of the expression (4) has required a rather complicated renormalization group treatment of the nonlinear σ- model complemented with an infinite number of additional terms, all these terms having been derived from the initial wave equation (1) (or equivalent Schroedinger equation for electrons) on averaging over all the realizations of disorder. Naturally, we will not adduce here such a derivation (the analogous one may be found in Ref. 12). Instead, we will describe how the law (4) for the time derivative coefficients changes the time-dependent shape of the output signal.

With the higher order derivative terms added, the diffusion equation for the photon density (2) is substituted by the following one:

$$\frac{\partial \rho}{\partial t} + \sum_{n=1}^{\infty} C_n \tau^n \frac{\partial^{n+1} \rho}{\partial t^{n+1}} = D \Delta\rho, \tag{6}$$

where $\tau = l/c$ is the time of mean free path, $D = D(L)$ is the diffusion coefficient reduced due to localization effects. Note, that if a time interval t considerably exceeds τ, then the n-th term is small compared to the first one in the parameter $(\tau/t)^{n-1}$. Therefore, only the very fast increase (4) of the coefficients C_n which is due to the interaction of diffusion modes (i.e. to the interference effects) forces one to keep the higher order terms. On solving equation (6) with the initial and boundary conditions mentioned above one obtains the photon density evolution as follows:

$$\rho(r,t) = \frac{1}{L^3} \int_{-\infty}^{+\infty} \frac{d\omega}{2\pi} \sum_{k} \frac{f(\omega)}{DK^2 - i\omega f(\omega)} e^{i(kr - \omega t)}. \tag{7}$$

Here the summation is performed over momenta running from $k = \pi/L$ to $k \sim \pi/l$ (the momentum $k = 0$ is excluded by the boundary conditions), the function of $f(\omega)$ is given by the following series:

$$f(\omega) = 1 + \sum_{n=1}^{\infty} C_n (-i\omega\tau)^n. \tag{8}$$

In the region $t \gg t_D = L^2/D$, the integral over frequencies is contributed to mainly by the frequency region $\omega \ll t_D^{-1}$. At such frequencies one may neglect the term $-i\omega f(\omega)$ in the denominator in Eq. (7). Then the time dependence of $I(r,t) = -D \Delta\rho(r,t)$ is factorized from the spatial one and given as follows

$$I(r,t) = I(r) \sum_{n} \int_{-\infty}^{+\infty} \frac{d\omega}{2\pi} e^{-i\omega t} (-i\omega\tau)^n C_n. \tag{9}$$

Here $I(r) \sim r^{-2}$. Before carrying out the summation, it is useful to compare these asymptotic series to those for power law or exponential time decay of $I(t)$. If $I(t)$ were proportional to $t^{-\alpha}$, then for $n \geq \alpha - 1$ the coefficients C_n would diverge. If $I(t)$ were proportional to $\exp(-t/t_0)$ then the coefficients C_n should be proportional to $n! \sim \exp(n \log n)$. With C_n being finite but increasing more rapidly than n , the output signal $I(t)$ should decay slower than exponentially but faster than from a power law. Indeed, the summation of the series (9) is performed with the help of a trick similar to the Borel summation by making use of the identity

$$e^{2un^2} = \frac{1}{(8\pi u)^{1/2}} \int_{0}^{\infty} x^{n-1} \exp\left[-\frac{1}{8u} \ln^2 x\right] dx, \tag{10}$$

which is substituted for C_n into Eq. (9). Then a change of order of summation and integration yields the following logarithmically normal asymptotics:

$$I(t) \sim \frac{1}{\sqrt{t\tau}} \exp\left[-\frac{1}{8u} \ln^2 t\ \tau\right]. \tag{11}$$

We should emphasize this law is only asymptotic, due to contributions from high order terms in the series (8). Naturally, it remains also the exponential contribution (3) to the output signal, which is governed by the polar part of the integral (7) with $f(\omega) = 1$. A complete time dependence of the output is given then by a sum of the expressions (3) and (11). In the very far asymptotic regime the logarithmically normal tail (11) will always prevail. But it will also prevail even for $t \geq t_D$, if the parameter u (5) is sufficiently large. On approaching the Anderson transition, u increases markedly which results in changing the asymptotics from the exponential (3) to the logarithmically normal one (11) for t not too large compared to t_D. The possibility of approaching the conditions for the Anderson localization of photons in various systems has been recently discussed by a number of authors.[4,14] Far from the transition u is rather small thus making the asymptotics (11) unobservable. In the very same region, $D = D(L)$ differs from D_0 only in a small parameter thus making the deviation of the exponential decay (3) from purely diffusive one unobservable also.

The nonexponential decay of the output signal points to the fact that pulse propagation through disordered media is characterized by the spectrum of various propagation times t_{pr} instead of a single diffusion time $t_D = L^2/D$. This spectrum is governed by the distribution function $P(t_{pr})$ which has a comparatively sharp maximum at $t_{pr} \sim t_D$ and a logarithmically normal tail at large t_{pr}. On using this distribution one may represent the sum of the expressions (3) and (11) as follows:

$$I(t) \sim \int_0^\infty e^{-t/t_{pr}}\ P(t_{pr})\ dt_{pr} \tag{12}$$

A similar distribution has previously been obtained for the relaxation times in disordered conductors[8] as well as for mesoscopic fluctuations of conductance.[12]

The logarithmically normal tail of the distribution function $P(t_{pr})$ indicates sufficiently large propagation times t_{pr} existing with a probability which is not exponentially small. It means physically the presence of the specific regions in the system where interference effects forces the time of photon random walks to exceed considerably a mean diffusion time through such regions. These regions will be identified as "interference traps".

How do interference traps reveal themselves in light propagating through a particular sample? The question arises naturally because all the above results apply to the intensity averaged over the all realizations of disorder. It is well known that the dispersion of various static quantities from realization to realization may be essential;[6] for a particular sample it reveals itself as "mesoscopic fluctuations" on changing, e.g., magnetic field.[6] To answer the question one should calculate such a dispersion for time-dependent intensities I(t). The calculations which are similar to those made for electric current relaxation[8] show that at $t > t_D$ the dispersion from realization to realization exceeds the mean value. For a particular sample one can change realization by changing a point of incidence or varying angle or frequency of the incident light. The light pulse is characterized, however, not by a single frequency but by a whole frequency interval $\delta\omega$

which is inversely proportional to a pulse duration δt. Then the
dispersion should reveal itself as aperiodic reproducible oscillation of
the output signal with typical period of δt (Fig. 2) which are directly
analogous to the mesoscopic fluctuations. It seems rather natural as the
output signal consists of multiple echoes of the input so that its long-
time-tail is sensitive to the duration δt. If the averaging is made over a
time interval Δt >> δt (such an averaging may be made by an inertial
detector), then the amplitude of oscillation is reduced by a factor of
$\sqrt{\overline{\Delta t}}$ δt. These reproducible oscillations may be distinguished from usual
noise by accumulation of output signals. Note in conclusion that the time
dependence of the output signal averaging over sufficiently large time
intervals Δt would be close to exponential (3) if the localization effects
are weak but it would be close to logarithmically normal asymptotic (11)
on approaching the Anderson transition.

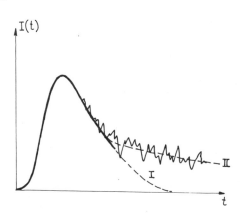

Fig. 2.
Time dependence of output intensity
(for clarity not to scale):

I - exponential decay

II - logarithmically normal decay
obtained after averaging
over periodic oscillations.

References

1. Altshuler B.L., Aronov A.G., Khmelnitskii D.E., Larkin A.I. in: Quantum
   Theory of Solids, edited by I.M. Lifshits, Moscow, Mir Publishers,
   1982.
2. Lee P.A., Ramakrishnan T.V., Rev. Mod. Phys. 57:287 (1985).
3. Anderson P.W., Phys. Rev. 109:1492 (1958).
4. John S., Phys. Rev. Lett. 53:2169 (1984); Phys. Rev. B 31:304 (1985).
5. Kuga Y., Ishimaru A., J. Opt. Soc. Am.A 1:831 (1984); van Albada M.P.,
   Lagendijk A., Phys. Rev. Lett. 55:2692 (1985); Wolf P.E., Maret G.,
   Phys. Rev. Lett. 55:2696 (1985); Fajans J., Wurtele J., Bekefi G.,
   Knowles P.S., Xu K., Phys. Rev. Lett. 57:575 (1986); Kaveh M.
   Rosenbluh M., Edrei I., Freund I., Phys. Rev. Lett. 57:2049 (1986);
   Akkermans E. Wolf P.E., Maynard R., Phys. Rev. Lett. 56:1471
   (1986).
6. Altshuler B.L., Sov. Phys. JETP Lett. 41:649 (1985); Lee P.A., Stone
   A.D., Phys. Rev. Lett. 55:1622 (1985); Webb R.A., Washburn S.,
   Umbach C.P., Laibowitz R.B., Phys. Rev. Lett. 54:2696 (1985).
7. Watson G., Fleury P., McCall S., Phys. Rev. Lett. 58:945 (1987).
8. Altshuler B.L., Kravtsov V.E., Lerner I.V., Pis'ma Zh. Exp. Teor. Fiz.
   45:160 (1987).
9. Wegner F., Z. Phys. B 35:207 (1979).
10. Efetov K.B., Larkin A.I., Khmelnitskii P.E., Sov. Phys. JETP 52:568
    (1980); Shäfer L., Wegner F., Z. Phys. B 38:113 (1980); Hikami S.,
    Phys. Rev. B 24:2671 (1981).
11. Kravtsov V.E., Lerner I.V., Sov. Phys. JETP 61:758 (1985).
12. Altshuler B.L., Kravtsov V.E., Lerner I.V., Zh. Exp. Teor. Fiz. 91:2276
    (1986).
13. Wegner F., Z. Phys. B 36:209 (1980).
14. Arya K., Su Z.B., Birman J.L., Phys. Rev. Lett. 57:2725 (1986); John
    S., Phys. Rev. Lett. 58, 2486 (1987); Agranovich V.M., Kravtsov
    V.E., Lerner I.V., Phys. Lett. 1987 (in press).

PHOTON LOCALIZATION IN TIME AND SPACE

P. A. Fleury, G. H. Watson, S. L. McCall, and K. B. Lyons

AT&T Bell Laboratories
Murray Hill, New Jersey 07974

INTRODUCTION

Wave propagation in strongly scattering random media has been a subject of continuing interest to physicists for decades. Anderson's predictions[1] that electron wave function behavior in the presence of strong disorder can lead to particle localization have led to a variety of experiments and theories aimed at elucidating coherent effects in disordered systems. Although many phenomena (coherent backscattering, mobility edge, universal conductance fluctuations) have been at least qualitatively understood by examining electronic transport,[2] detailed comparisons with theory are still lacking partly because theories have not treated the strong effects of electron-electron and electron-phonon interactions.[3] Indeed the nature of the transition from extended to localized electronic states in three dimensions still involves some open questions. Even if the transition is continuous, the value of the critical exponent remains very much in dispute. In principle, photons offer many advantages for improved quantitative study of localization phenomena: First, their self-interaction is quite negligible; second, their energies ($h\nu$) and momenta ($h\vec{k}$) can be easily measured, controlled and varied; third, detailed studies of subtle wave interference phenomena over several decades in intensity are easy; fourth, the relevant spatial scale on which randomness is important ($\sim\lambda$) is subject to control and characterization.

The perturbative approach to light propagation through optically disordered media applies only when the amplitude of refractive index variation is small (Born approximation). Strong multiple scattering from large amplitude variations has been considered without regard for coherence effects until relatively recently. Only within the last 3 years has the coherent backscattering responsible for the strongly peaked ($\sim 2x$) enhancement in the reflection of an electromagnetic wave from a random medium been observed.[4-6] The familiar "laser speckle" observed in both reflection and transmission from a random optical medium is the analogue to the universal conductance fluctuations only recently observed in electron transport experiments.[7]

The similarities of these phenomena suggest that much can be learned about random media and localization phenomena by studies of photon transport. General criteria for localization of waves (particles) of wavelength λ include:[2] a long mean free path for inelastic (energy changing) processes $\ell_i \gg \lambda$; and a short mean free path for elastic (energy conserving) processes $\ell_e \lesssim \lambda$. Because meeting the condition $\ell_e < \lambda \ll \ell_i$ has proven impossible thus far for optical materials, no observations of strong photon localization have yet been made. Enhanced coherent backscattering provides a qualitative example of "weak localization" but suffers some limitations. It does not probe the random medium beyond a relatively few elastic mean free path lengths, and provides no means for quantitative determination of inelastic absorption effects. Some recent experiments[8] have attempted to explore localization by examining the thickness dependence of optical transmission

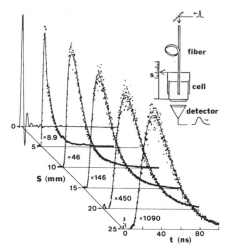

Fig. 1. Transmitted pulse shape evolution of 50 psec pulses injected via optical fiber [inset] into a 9.6% (by volume) aqueous suspension of latex spheres, for various distances s from the fiber tip to the exit window. Vertical scales amplified by amounts shown. Solid curves are best fits to diffusion model.

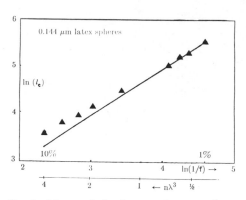

Fig. 2. Measured elastic mean free path ℓ_e as a function of filling fraction, for an aqueous suspension of latex spheres, illustrating departure from Mie theory as the number of scatterers per cubic wavelength exceeds unity.

through a slab of disordered dielectric. Unfortunately these experiments have lacked the sensitivity or dynamic range required to separate out absorption effects as well. Finally, some observations of laser speckle have been made but the theoretical analysis required to relate speckle patterns to photon transport parameters has not appeared heretofore.

In this paper we describe a novel time-domain experimental method[9] to observe diffusive photon transport in a strongly scattering random medium. It probes the medium on length scales from a few to a few million ℓ_e's, provides a quantitative determination of absorption effects, and permits a sensitive search for onset of strong localization. We also present preliminary observations of the frequency dependence of laser speckle of a system which has been well characterized by the time domain method. A preliminary theory for speckle spectra in the weak limit has been formulated and its major predictions may be tested by comparing the speckle and the time domain experimental results. The goal of both approaches is to provide methodology and insight for the slight departures from simple diffusive propagation expected as ℓ_e is decreased towards $\lambda/2\pi$.

TIME DOMAIN EXPERIMENTS

The samples were contained in a 5 cm diameter, 5 cm long cylinder whose internal surface was gold plated to inhibit chemical deterioration of the sample. In the time domain experiment light pulses of .532 μm vacuum wavelength and 50 ps in duration from a mode-locked YAG laser were injected into the interior of the cylindrical scattering medium[10] through an optical fiber with its end a distance s from the exit window of the cell. This arrangement avoids the large reduction in signal transmission which results from the high reflectivity ($R=1-\ell/L$) which is characteristic of an interface between a homogeneous medium (like air) and a strong multiple scattering medium of thickness L with mean free path ℓ. The light emitted at the exit window was detected by a biplanar photodiode (rise time 0.25 ns). In the absence of scattering, the propagation time from fiber to boundary would be $t_p = s/c_n$. For a random sample which scatters light elastically with a momentum-exchange mean-free-path length ℓ_e, the propagation time is stretched to approximately $t_d=(s/\ell_e)t_p$. Figure 1 shows the temporal character of the pulses detected as the distance s is varied. Clearly a fraction of the photons detected have traveled more than 18 m since delays exceeding 80 nsec are observed, corresponding to nearly 10^6 scattering events ($\ell_e \sim 21$ μm, see below).

More quantitatively, for a dilute suspension of scatterers in a container whose dimensions are large compared to ℓ_e the diffusion equation applies,

$$\frac{\partial \Phi}{\partial t} = D\nabla^2\Phi - \gamma\Phi + N\delta(x)\delta(y)\delta(z-s)\delta(t) \tag{1}$$

where Φ is the density of photons at a point x,y,z at time t. The term $-\gamma\Phi$ describes the loss due to absorptive processes; $D = \ell_e\,c_n/3$ is the photon's diffusion coefficient; and the delta function term represents the short input pulse of N photons injected at t=0 on the cylinder axis. The momentum exchange mean-free-path length is $\ell_e = 1/n\sigma_m$, where n is the density of scatterers and

$$\sigma_m = <\int \frac{d\sigma}{d\Omega}(1-\cos\theta)d\Omega> \ . \tag{2}$$

In Eq. (2) σ is the scattering cross section of an individual particle and the average is taken over all particles. The appropriate boundary conditions are Φ=0 for t<0, for z=0 or L, and for $x^2 + y^2 = R^2$, and where L and R are the length and radius of the sample cylinder. For t>0 the solution is

$$\Phi(z,r,t) = NG_z(z,t)G_r(r,t)e^{-\gamma t} \tag{3}$$

where $r = \sqrt{x^2+y^2}$. The functions G_z and G_r are given by

$$G_z = \frac{1}{\sqrt{4\pi Dt}} \sum_{m=-\infty}^{\infty} \left\{ \exp\left[\frac{-(z-2mL-s)^2}{4Dt}\right] - \exp\left[\frac{-(z+2mL+s)^2}{4Dt}\right] \right\} \tag{4}$$

$$G_r = \sum_i \frac{1}{\pi R^2} \frac{1}{[J_1(x_i)]^2} J_0(x_i r/R) \exp\left[\frac{-Dx_i^2 t}{R^2}\right] \tag{5}$$

where J_0, J_1 are cylindrical Bessel functions, and x_i, i = 1,2..., are the roots of $J_0(x) = 0$. The light flux escaping the window at z = 0 is $F(r,t) = D\,\partial\Phi/\partial z$. To obtain the detected signal I(t,s), F is multiplied by a function A(r) describing our detector aperture and sensitivity and the product is integrated over the window aperture $0 < r \leq R$.

As reported earlier[9] we have examined a variety of optically disordered systems with this technique. Fig. 1 shows the exit pulse evolution with increasing s compared to the theoretically expected shape (solid curves) for a 9.6% aqueous suspension of 0.198 μm latex spheres. The fitted value of $\ell_e = 23 \pm 1$ μm is in excellent agreement with the value of $\ell_e = 21$ μm calculated from Mie theory. The value of γ obtained is 0.036 ns^{-1}, corresponding to an inelastic absorption length of about 6 m. Experiments on related systems verify quantitatively that the rising edge of the pulse is more sensitive to D, while the tail is sensitive to γ. We find excellent agreement between our time domain observations and the predictions of simple theory obtained for all systems in which the particle size and volume fraction gives an average density of one scatterer or less per cubic wavelength — i.e. particle-particle correlation contributions are negligible. We explored the effects of departure from this condition by successively diluting a well characterized 10% suspension of 0.144 μm latex spheres. The results are shown in Fig. 2, where the effects of $n\lambda^3 > 1$ result in a clear *increase* in ℓ_e due to correlation effects.

Although the monodispersity of the latex suspensions facilitates comparison of experimental results with theoretical predictions, optically disordered materials with higher refractive index (which are thus more likely to exhibit strong localization) are available albeit with much broader distributions of scatterer sizes. It is possible to estimate the effect of such polydispersity as follows. For small particles ($d\ll\lambda$), the single particle scattering cross section σ goes as d^6, while the number density n goes as d^{-3} for a constant filling fraction f. Thus, the total cross section is $\ell_e^{-1}=\sigma n\sim <d^6>f/<d^3>$, where $<>$ denotes averaging over the particle distribution. Thus, for monodisperse particles the mean free path ℓ_o goes as d_o^{-3}. For example, using a 5% (by volume) sample of a-SiO$_2$ particles in air (Degussa Aerosil OX-50; surface area 50 m^2/g), we calculate ℓ_o=340 μm using the average particle size d_o=0.055 μm, assuming them to be monodisperse, and ignoring particle correlation effects. Using the manufacturer's provided particle size distribution for which $<d^6>^{1/6} = 0.070$ μm and $<d^3>^{1/3} = 0.061$ μm, we obtain $<\ell> = 110$ μm. Our observed value, however, was $\ell_e = 49$ μm, more than a factor of two smaller. Were the particles to be uniformly distributed in space, $n\lambda^3$ would substantially exceed unity but this would have no effect on

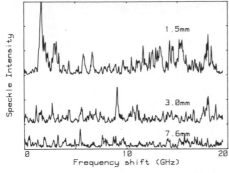

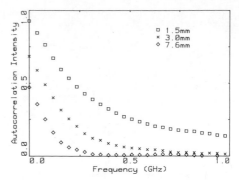

Fig. 3. Laser speckle spectra I($\nu+\Delta\nu$) obtained from polydisperse aerosil (fumed silica) powder in air, showing variation with path distance s. The pinhole diameter is 50 μm. Traces are offset vertically for clarity.

Fig. 4. Intensity autocorrelation functions $g_2(\Delta\nu)$ as defined in Eq. (6) from the sample used in Fig. 3. Several hundred spectra were averaged in each plot; the ensemble members were formed by sampling different regions of the speckle pattern using an 80 μm translation of the 25 μm pinhole between frequency sweeps.

ℓ_e in the absence of particle correlation. We conclude, then, that the mean free path observed in the a-SiO_2: air system is anomalously low by considerably more than a factor of two. This can be explained by inhomogeneous clumping of the dry a-SiO_2 particles on spatial scales smaller than an optical wavelength, with the result that locally (on a scale of many wavelengths) the mean free path is substantially shorter than average. In such regions "pockets of localization" may exist which stretch out the time delay and give the overall effect of a smaller ℓ_e.

These experiments verify the sensitivity of the time domain method to determination of D and γ, and even to quite small departures from simple diffusive transport and point the way toward material systems which might exhibit stronger localization. Directions for additional research include: fabrication of even stronger elastically scattering random media to reduce ℓ_e below $\lambda/2\pi$; utilization of resonantly enhanced or nonlinear scattering processes; study of fractally random media, possibly even while aggregation is proceeding.

FREQUENCY DOMAIN EXPERIMENTS

Thus far the effects of coherence and phase information on the transmitted light have not been treated. The familiar phenomenon of laser speckle[10] is a manifestation of these effects, and in principle laser speckle spectroscopy (intensity autocorrelation function measurements) can be used to extract the same information as the time domain method described above. In this case the intensity transmitted through a small aperture in the far field is recorded as the incident frequency is swept. In such experiments extreme care is needed to ensure that illumination conditions do not change as the frequency is swept. In order to ensure such stability in the present case, a single-mode fiber was used to deliver the incident light inside the sample. The sample cell is the same as above, but the sample is a dry powder rather than a suspension. Typical laser speckle spectra are shown in Fig. 3, where the fluctuations in intensity received through a small aperture (50 μm diam. pinhole) placed 1.7 m from the cell are displayed as the highly monochromatic (linewidth $\sim$ MHz) cw laser wavelength is scanned over 20 GHz. Qualitatively one expects that the intensity will fluctuate from bright to dark on a scale $P \cdot \delta k \simeq 2\pi$, where $\delta k = 2\pi\delta\nu/c$ and P is the total average photon path length traversed inside the medium, $P \simeq s^2/\ell_e$. Thus, in the absence of absorption, one might expect $\delta\nu \propto D/s^2$.

A more quantitative approach to the speckle problem lies in measurement of the autocorrelation function $g_2(\Delta\nu) \equiv \langle I(\nu)I(\nu+\Delta\nu)\rangle/I^2 - 1$, which may be constructed from scans such as those displayed in Fig. 3. In order to determine $g_2(0)$ correctly, free of a quantum shot noise peak, we recorded each trace as N evenly spaced point *pairs* $I_j(\nu_i)$ where j=1,2; i=1,2, $\cdots$,N; and $\nu_i \equiv i \cdot d\nu$. The function $g_2(\Delta\nu)$ is then defined as

$$g_2(m \cdot d\nu) = \sum_{i=0}^{N-m} I_1(\nu_i)I_2(\nu_{i+m})/[(N-m)<I>^2] - 1 \qquad (6)$$

where the average intensity $<I>$ is obtained by averaging over all the individual speckle patterns, such as the ones shown in Fig. 3. Samples of $g_2(\Delta\nu)$ are shown in Fig. 4, for three representative values of s. To first order g_2 may be characterized by its half width, ν_c, defined by $g_2(\nu_c)/g_2(0)=0.5$. Shapiro[11] has shown that $\nu_c=\alpha D/s^2$, where $\alpha=(\ln 2)^2/4\pi=0.038$ for the case of a point source and point detector, both inside an infinite lossless random medium. However, the realistic experimental situation is more complicated from a theoretical standpoint, due to boundary conditions and size effects.

Preliminary experimental results are shown in Fig. 4 for various values of s. Although they show a qualitative decrease in ν_c as s is increased, the expected s^{-2} dependence is not observed. A full theoretical treatment shows that absorption can modify this dependence substantially. A full discussion is beyond the scope of this paper and will be presented elsewhere.[12] It must suffice here to say only that analysis of $g_2(\Delta\nu)$ requires accurate account of several experimentally controllable variables, including aperture size, source position, and finite sample thickness, before meaningful values of the physically interesting parameters γ and D may be extracted. Only if all of these dependences are quantitatively understood and accounted for, can laser speckle spectroscopy be used to measure the photon transport in a disordered medium. However, the connections which this work and Ref. 12 have established between the time domain and frequency domain (speckle) techniques should open to study a wider range of systems.

CONCLUSION

In conclusion we have demonstrated that time domain transmission measurements give quantitative values for photon diffusion and absorption coefficients in weakly localizing random media. Effects of particle density correlations, particle polydispersity, and departure from the dilute limit have been demonstrated. Furthermore the relation of laser speckle autocorrelation measurements to the optical properties of disordered optical media has been elucidated both theoretically and experimentally. These results should permit more quantitative understanding of localization phenomena as optical media with smaller ℓ_e are fabricated.

REFERENCES

1. P. W. Anderson, Phys. Rev. *109*, 1492 (1958).
2. See for example, P. A. Lee and T. V. Ramakrishnan, Rev. Mod. Phys. *57*, 287 (1985).
3. D. Chowdhury, Comments on Solid St. Phys. *12*, 69 (1986).
4. Y. Kuga and A. Ishimaru, J. Opt. Soc. Am. A *1*, 831 (1984).
5. M. P. van Albada and A. Lagendijk, Phys. Rev. Lett. *55*, 2692 (1985).
6. P. E. Wolf and G. Maret, Phys. Rev. Lett. *55*, 2696 (1985).
7. R. A. Webb et al. Phys. Rev. B*30*, 4048 (1984).
8. A. Z. Genack, Phys. Rev. Lett. *58*, 2043 (1987).
9. G. H. Watson, P. A. Fleury and S. L. McCall, Phys. Rev. Lett. *58*, 945 (1987).
10. S. Etemad, R. Thompson and M. J. Andrejco, Phys. Rev. Lett. *57*, 575 (1986).
11. B. Shapiro, Phys. Rev. Lett. *57*, 2168 (1986).
12. S. L. McCall, G. H. Watson, P. A. Fleury and K. B. Lyons, to be published.

CLASSICAL DIFFUSIVE PHOTON TRANSPORT IN A SLAB

Melvin Lax[a, b], V. Nayaranamurti[b, c], and R.C. Fulton[b]

a. Physics Department, City College of New York, New York 10031

b. AT&T-Bell Laboratories, Murray Hill, New Jersey 07974-2070

c. Now at Sandia National Laboratory, Albuquerque, NM 87185

The problem of localization of light in a random atmosphere requires, for comparison, a knowledge of the classical transport of light by an iso-tropically scattering medium in a bounded medium such as a slab. [See V. Sobolev, *Light Scattering in Planetary Atmospheres*, Pergamon (1975)] We consider the time-dependent Milne problem generalized to a slab with absorbing boundary conditions. We can adapt results obtained for phonon transport in GaAs at low temperatures to the corresponding optical problem. Transport of a pulse across the slab is evaluated by Monte Carlo methods, as well as by analytic approximations. Although the behavior is diffusive, when the detector is moved transversely, the mean time of arrival is found to be linear in the distance from point source to point detector!

1. INTRODUCTION

Recent efforts on electron localization in random systems[1] have stimulated proposals that photon localization can also be achieved[2,3]. Preliminary measurements exhibiting weak localization have been made[4,5] in essentially classical systems. In order to be sure one has observed local-ization, it is necessary to know what photon transport is to be expected when localization is absent.

Because almost no analytic solutions exist to most transport prob-lems, one often resorts to Monte Carlo calculations to obtain insight. We have recently performed a series of Monte Carlo calculations[6] on transport in a more complicated system – phonons at low temperatures in GaAs. These calculations permitted the possibility of down-conversion i.e. break-up of a phonon into two lower frequency phonons. In order to detect, and remove, possible errors in the Monte Carlo code we needed a special case to check, in which theoretical estimates were available.

The check calculations performed were on a system in which down-conversion was prevented and isotropic, elastic scattering is all that is permitted. To our chagrin, even for this simple, special case no analytic solution was available. However, by extending the concept of *extrapolation length* that arose out of an analysis of the Milne problem[7,8] in astrophysics (light scattering in a semi-infinite atmosphere) and was heavily exploited in neu-tron transport theory[9,10] we found an approximate solution to the transport of particles in a slab with absorbing boundaries.

Some rather startling Monte Carlo results were found, and confirmed by our semi-analytic procedures. We shall report these results here because they are equally applicable to photon transport.

2. DESCRIPTION OF THE PROBLEM

We consider a slab whose walls are perfectly absorbing to photons hitting them from inside. Photons hitting the left wall are absorbed. Those hitting the right wall are "detected" and absorbed. In between the walls, isotropic scattering takes place. A photon pulse (a delta function in time) enters the slab at a point from the left. We would like to obtain the shape and normalization of the emerging pulse at any point on the right hand surface. The shape of this pulse is a solution of the notoriously difficult first passage time problem![11] The time integral over the pulse and over some area of the right hand surface indicates the total number of photons that would be counted by a detector covering that area. If the area covers the entire right hand surface, the integral determines the total number of photons that survive the passage across the slab. If the results are normalized to one input photon traveling toward the right (at a random angle) we obtain the *fraction, F,* of photons that survive the passage across the slab.

A second objective is the dependence of the flux on the transverse radial distance R or the actual distance $r = \sqrt{R^2 + w^2}$ from the point of origin, where w is the slab width.

A third objective is the dependence of the peak or the mean time of arrival, $\bar{t}$, on the radial distance r from source to detector. The time $\bar{t}$, in random process literature, is known as the mean first passage time.

3. SURPRISING RESULTS

A. The Fraction of Surviving Photons

The fraction of photons that survive a passage across the slab is clearly a function, $F(w/\lambda)$, of the ratio of the slab thickness w to the mean free path λ. Moreover, $F(0) = 1$ and $F(\infty) = 0$. But how does the decrease take place? The most natural choice obtained by analogy from other areas of physics would be $F(w/\lambda) = \exp(-w/\lambda)$. For $w/\lambda \gg 1$, our modified diffusion approximation yields:

$$F(w/\lambda) = \frac{.8126}{1 + .7034(w/\lambda)} \tag{3.1}$$

See Table I for a comparison with Monte Carlo results.

TABLE I. Fraction F of surviving photons as a function of w/λ

ω	σ	λ	w	w/λ	F	$\dfrac{.8126/F - 1}{w/\lambda}$
1.5	1	6.182×10^{-4}	.02	32.36	.034	.708
1.5	1/3	1.854×10^{-3}	.02	10.79	.0945	.704
1.5	0.1	6.182×10^{-3}	.02	3.236	.249	.699
0.7	1.0	3.833×10^{-2}	.2	5.218	.1747	.701

ω and σ refer to phonon calculations, and are used here only to correlate information in Table I with that in Table II.

TABLE II. Fraction of photons detected at different distances $r = \sqrt{R^2 + w^2}$

Case 1: $\omega = 1.5$ $\sigma = 1$ Input # = 1468170 Detected # = 50000

r	Output #	Fraction of Input	Fraction of Detected
0.02	48398	3.30×10^{-2}	.968
0.04	1522	1.04×10^{-3}	.0304
0.06	78	5.31×10^{-5}	.00156
0.08	2	1.36×10^{-6}	.00004

Case 2: $\omega = 1.5$ $\sigma = 1/3$ Input # = 528942 Detected # = 50000

r	Output #	Fraction of Input	Fraction of Detected
0.02	47561	8.99×10^{-2}	.951
0.04	2268	4.32×10^{-3}	.0457
0.06	143	2.70×10^{-4}	.00286

Case 3: $\omega = 1.5$ $\sigma = 0.1$ Input # = 200791 Detected # = 50000

r	Output #	Fraction of Input	Fraction of Detected
0.02	44728	2.23×10^{-1}	.895
0.04	4618	2.30×10^{-2}	.0924
0.06	573	2.85×10^{-3}	.0115
0.08	70	3.49×10^{-4}	.0014
0.10	22	1.10×10^{-4}	.00044

Case 4: $\omega = 0.7$ $\sigma = 1$ Input # = 572361 Detected # = 100000

r	Output #	Fraction of Input	Fraction of Detected
0.20	92905	1.62×10^{-1}	.929
0.40	6441	1.13×10^{-2}	.0644
0.60	4512*	9.51×10^{-4}	.00544
0.80	410*	8.64×10^{-5}	.00049
1.00	35*	7.38×10^{-6}	.00004

*Based on 4,743,386 input photons

B. The Transverse Dependence of Detected Photons

For purposes of comparison between Monte Carlo and an approximate theory, we used five detectors in the form of rings from $R_0 = 0$ to R_5. These were chosen to produce a linear increase in the radial distance r_n with n:

$$r_n = nw \quad , \quad R_n = \sqrt{r_n^2 - w^2} = w\sqrt{n^2 - 1} \tag{3.2}$$

Table II indicates an approximately exponential falloff with R. This is to be expected in view of our diffusion approximation for the time integrated intensity:

$$I(R) = \frac{1}{\pi^2 w^2} \sum_{m=1}^{\infty} \theta(m) \sin\theta(m) K_0 \left(\frac{R\theta(m)}{w}\right) \Delta\theta ; \quad \theta(m) = \frac{\pi wm}{d}; \quad \Delta\theta = \frac{\pi w}{d} \tag{3.3}$$

where $d = w + 2z_0$ is the slab width w, extended by an extrapolation length[12] $z_0 = .71045\lambda$ on each side.

C. Mean Arrival Time $\bar{t}$ Versus Distance r

Since our approximate analytic solution is based on a diffusion approximation valid when w is several mean free paths, λ, or more, we would

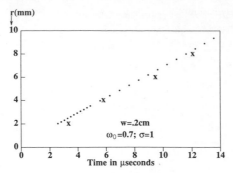

Fig.1 X's Monte Carlo, Dots Calculated.

expect the distance traveled in a time t to be order $\sqrt{t}$, or conversely that $t \propto r^2$. Instead, we find, to an excellent approximation, (see Fig. 1) that $\bar{t}$ is linear in r.

Our quasi-analytic expression for $\bar{t}$ is

$$\bar{t} = \frac{Rd}{2\pi D} \frac{\sum\limits_{m=1}^{\infty} \sin\theta(m) K_1(R\theta(m)/w)}{\sum\limits_{m=1}^{\infty} m\sin\theta(m) K_0(R\theta(m)/w)} \xrightarrow[r \to \infty]{} \frac{rd}{2\pi\upsilon\lambda/3} \tag{3.4}$$

where the diffusion constant D is given by $\upsilon\lambda/3$ and υ is the photon velocity. For $R \gg w$, K_0 and K_1 decrease exponentially as $\exp(-\pi Rm/d)$, so that only the $m = 1$ term is important, and $K_1 \approx K_0$.

What is not at all clear, is that Eq. (3.4) is approximately linear in r when $R \ll w$, and the Bessel functions are nearly singular. See Fig. 1.

D. Pulse Shape

The exact solution of the semi-infinite Milne problem demonstrates that far from the surface, the exact steady state solution can be represented as a solution of the diffusion equation. But the extrapolation of the diffusion solution vanishes outside the medium at a distance z_0. We take this z_0, which is known only in the steady state (or zero frequency) and treat it as if it were a fixed value, even for time dependent problems. Thus we seek a solution of the diffusion problem over the region $0 \le z \le d$ that vanishes at 0 and d, where the actual slab occupies the space from z_0 to $d - z_0$ or $w = d - 2z_0$.

The solution of the time dependent diffusion equation can be obtained, and the current density at the surface $z = d - z_0$ is evaluated using Fick's law:

$$J_z(R, t) = -D \partial n/\partial z \mid_{z = d - z_0} \tag{3.5}$$

The total detected current is given by

$$I(t) = 2\pi \int_0^{\infty} J_z(R, t) R dR \tag{3.6}$$

This total current is plotted against time and represents the upper solid curve in Fig. 2. The shape is approximately correct, but the total current is overestimated.

232

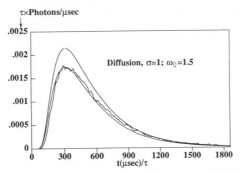

Fig. 2 Lower solid curve corrected from upper by Eq. (3.7).

The reason for the overestimate is that it is not valid to use the diffusion solution within a mean free path of the surface. Yet, the current in Eq. (3.5) is evaluated right at the surface. For the Milne solution of the transport problem, the exact photon flux is known. The ratio of exact flux to diffusion estimate is found to be[12]

$$\frac{J \, \text{exact}}{-D \, \partial n / \partial z} = .8126 \qquad (3.7)$$

It seems appropriate then, to apply this reduction ratio to the complete $I(t)$ pulse output. When this is done, we obtain the unexpected agreement between the Monte Carlo pulse shape and the calculated pulse shape shown in Fig. 2. We say unexpected, because it is not clear that it was permissible to use either z_0 or the reduction ratio from the time independent solution and apply it to the pulse problem. The factor .8126 appears in Eq. (3.1) for the same reason, but with better justification because the time integrated flux corresponds to a steady state solution.

4. QUALITATIVE EXPLANATIONS

A. The Fraction of Surviving Photons

When the total number of photons is counted, the result is independent of where on the left hand surface the source is located. Averaging over all input positions, we have a truly one-dimensional problem. In effect, a random walk is being performed with absorbing boundaries.

The problem is analogous to a gambling problem in which one starts with capital C, and hopes to win an amount W. The game ends when one loses all of one's capital, or one has the total sum $C + W$. If the game is fair, the probability of winning is

$$P = \frac{C}{C + W} = \frac{1}{1 + W/C} \qquad (4.1)$$

independent of the amount of the individual bets or their odds, provided that each individual bet is fair. In this context, an inverse linear function is the one that appears natural.

In the original Monte Carlo calculation, we only include a first step that moves away from the boundary. At this point, the photon is in a distance of the order λ. We would therefore expect a probability of order $(1 + aw/\lambda)^{-1}$ where a is a numerical coefficient of order unity.

B. Transverse Dependence of Detected Photons

If the slab thickness were small compared to the mean free path, travel over an interval dR has the probability dR/λ of a collision, with a probability of near unity of hitting a wall. Thus the probability of traveling dR and not hitting the wall is $1 - dR/\lambda$. When this probability is compounded there is a probability $\exp(-R/\lambda)$ of surviving out to R.

When the thickness of the wall is not small, the probability of escape is reduced by the factor $\lambda/(\lambda + w)$, roughly the probability that the photon is within one mean free path of the wall. Thus we get a survival probability of

$$\exp\{- (R/\lambda) [\lambda/(\lambda + w)]\} = \exp[- R/(w + \lambda)] \tag{4.2}$$

in rough agreement with the first term in Eq. (3.3).

C. Mean Arrival Time $\bar{t}$ vs Distance r.

Aside from numerical factors, the mean arrival time was found to be $\bar{t} \approx (r/\upsilon) \cdot (w/\lambda)$. If a single step takes roughly τ seconds, and $\upsilon\tau = \lambda$, the number of steps is then $n \approx \bar{t}/\tau = (r/\lambda) \cdot (w/\lambda)$ instead of the expected $(r/\lambda)^2$.

If the steps were fixed in length, (r/λ) would be the minimum number of steps. Because of the diffusive forward and backward motion roughly (r/λ) steps are needed for each of the minimum number of steps when no boundaries are present. Because of the boundaries, roughly (λ/w) of the photons are absorbed in each step. Thus the *surviving* photons only take about (w/λ) redundant steps to achieve each useful step.

This view is corroborated in our extrapolation length model. If the entire right hand surface is used as a detector the mean time of arrival is

$$\bar{t} = \frac{1}{2} \frac{w (w + 4z_0)}{\upsilon\lambda} \tag{4.3}$$

in good agreement with Eq. (3.4).

5. DIFFUSION THEORY EXPLANATION OF THE LINE SHAPE

We apply our modified diffusion theory by requiring the concentration to vanish, at both $z=0$ and $z=w+2z_0=d$, where the actual slab occupies the region from $z=z_0$ to $z=w+z_0$. The particle density solution to the diffusion equation at (x, y, z) with source at (x', y', z') can be written by the method of images in the form

$$n (R, z, t) = \frac{1}{8} \frac{S}{(\pi Dt)^{3/2}} \exp\left[- \frac{R^2}{4Dt}\right] \tag{5.1}$$

where $R^2 = x^2 + y^2$ and S represents the sum

$$S = \sum_{n=-\infty}^{\infty} \left\{\exp[-(2nd + z' - z)^2/4Dt] - \exp[-(2nd - z' - z)^2/4Dt]\right\} \tag{5.2}$$

where n runs over all integral values. Here $D=\upsilon\lambda/3$ is the diffusion constant.

The sum S can also be written as a sum over reciprocal space by using the Poisson sum formula.[13, 14] The result is

$$S = 2\frac{\sqrt{\pi Dt}}{d} \sum_{m=-\infty}^{\infty} \sin\frac{\pi m z'}{d} \sin\frac{\pi m z}{d} \exp\left(-\pi^2 Dt m^2/d^2\right) \tag{5.3}$$

We now place our source at $z' = z_0$, $x' = y' = 0$, and evaluate the z component of the current density, Eq. (3.5):

$$J_z (R, t) = \frac{\exp[-R^2/4Dt]}{4\pi^2 w^2 t} \sum_{m=1}^{\infty} \theta(m) \sin\theta(m) \exp[-Dt\theta(m)^2/w^2] \Delta\theta \tag{5.4}$$

The total current may be obtained by integrating Eq. (5.4) over the complete area of the right hand surface. The result is:

$$I(t) = \frac{D}{\pi w^2} \sum_{m=1}^{\infty} \theta(m) \sin\theta(m) \Delta\theta \exp[-Dt\theta(m)^2/w^2] \tag{5.5}$$

The zeroth and first moment of Eq. (5.5) with respect to time can be easily be obtained, and the series summed analytically. The result for $\bar{t}$ is that given in Eq. (4.3).

REFERENCES

1. See for example P. A. Lee and T. V. Ramakrishnan, Rev. Mod. Phys. 287, (1985)
2. K. Arya, Z. B. Su and J. L. Birman, Phys. Rev. Lett. 2725-2728, (1986)
3. Ibid 1559, (1985)
4. M. P. Van Albada and Ad Lagendijk, Phys. Rev. Lett. 2692, (1985)
5. D. E. Wolf and G. Mart, Phys. Rev. Lett. 2696, (1985)
6. M. Lax, V. Narayanamurti, R. C. Fulton and N. Holzwarth, Monte Carlo Calculations of Phonon Transport, Phonon Scattering in Condensed Matter (1986)
7. V. Sobolev, Light Scattering in Planetary Atmospheres, Pergamon, (1975)
8. S. Chandrasekhar, Radiative Transfer, Oxford, (1950)
9. B. Davison and J. B. Sykes, Neutron Transport Theory, Oxford, (1957)
10. See M. M. R. Williams, Mathematical Methods in Particle Transport Theory, Wiley, (1971)
11. See W. Feller, *Probability Theory and Its Applications*, John Wiley (1950)
12. See Williams (Ref. 10) p. 339 and M. M. R. Williams, Nucl. Sci. Engineering 260, (1964)
13. See E.C. Titchmarsh, Theory of the Fourier Integral, Oxford, (1937,1948)
14. M. Lax, Symmetry Principles in Solid State and Molecular Physics, Wiley (1974) p. 198.

ANDERSON LOCALIZATION OF CLASSICAL PHOTONS IN A DISORDERED DIELECTRIC

K. Arya, Z.B. Su[*], and Joseph L. Birman

Department of Physics, The City College of the City
University of New York, New York 10031
[*]Institute of Theoretical Physics, Academia Sinica
Beijing, China

ABSTRACT

The Localization of an electromagnetic wave in a medium of randomly distributed dielectric spheres is investigated. The t-matrix approach is used for the calculation of the scattering cross-section form one sphere; multiple scattering from different spheres is calculated by a diagrammatic Green's function method which gives the criterion for strong localization of an e.m. wave. We find that this criterion can be satisfied for metallic particles because of the large scattering cross-section from a single sphere due to the Mie resonance. Localization effects can be observed in transmission experiments.

I. INTRODUCTION

Recently, there has been a growing interest in studies of Anderson localization of classical waves, such as electromagnetic (em) waves in a disordered dielectric or elastic waves in solids with a random ionic potential.[1-8] The physical basis for this localization is essentially the same as in the case of an electron in a random impurity potential, i.e. the transport diffusion coefficient vanishes due to the coherent interference between waves scattered from random scatterers.[9] The vanishing diffusion coefficient has been observed in many experiments in the case of electrons where one sees a sharp decrease in the electron conductivity at the mobility edge. However, there is as yet no such sucessful experiment in the case of em waves in which the Anderson transition or the vanishing of the transmission coefficient has been reported.

Instead, in recent experiments, weak coherent interference effects have been directly observed in case of scattering of em waves from a concentrated solution of polystyrene particles.[7] These show an increase in the scattered intensity in the backward direction by a factor of about 2 compared to the diffuse scattered intensity. This agrees with the theory of weak localization that includes both ladder and maximal cross diagrams (the latter corresponds to interference effects of a scattered wave with another scattered wave propagating in reversed time order).[5,6] The angular width of the peak is also reported to be in agreement with the theory, i.e., width $\sim \lambda/2\pi l$ where λ and l, respectively, are the wave length and scattering length of the em wave. However, the small angular width of $\sim 1°$ (i.e., $l \gg \lambda$) in these experiments indicates that even for a 10% concentration of polystyrene solution, one is far from the Anderson transition that occurs for $l \approx \lambda$.

Furthermore, for the studies of strong localization, transmission experiments rather than scattering experiments are more suitable because only the former can give information about the diffusion of the em energy through the medium. Recently there have been a few experiments along these lines and a decrease in diffusion constant due to coherent interference has also been reported.[10,11] However, the Anderson transition has not yet been observed in case of an em wave. This is because the scattering cross section in these materials is rather small and the criterion for an Anderson transition is not satisfied. Studies of other dielectrics with larger scattering cross section is, therefore, necessary for strong localization.

In this paper we discuss a t-matrix approach for calculation of the scattering length from a random medium consisting of dielectric spheres. Earlier theoretical discussions are based on Maxwell-Garnett theory which is valid only for a weakly disordered dielectric. By using the t-matrix and diagrammatic Green's function method we also derive the criterion for Anderson localization of an em wave which is similar to that for an electron in a random potential. We show that this condition of localization for an em wave can be satisfied by a suspension of metallic particles, e.g., those of Ag, Au etc. in a medium. This is mainly because of the large scattering cross-section due to the Mie resonance present in case of a metal sphere with negative dielectric function.

II. AVERAGE ONE-PHOTON GREEN'S FUNCTION AND SCATTERING LENGTH

We consider a medium consisting of a random dispersion of spherical particles each of radius a. To include multiple scattering, we use a Green's function method, where the Dyadic Green's function $\overleftrightarrow{d}$ is the solution of Maxwell's equation

$$-\nabla \times \nabla \times \overleftrightarrow{d}(\mathbf{r}, \mathbf{r}', \Omega) + \frac{\Omega^2}{c^2}\varepsilon(\mathbf{r}, \Omega)\overleftrightarrow{d}(\mathbf{r}, \mathbf{r}', \Omega) = 4\pi\delta(\mathbf{r} - \mathbf{r}')\overleftrightarrow{I} , \qquad (1)$$

where

$$\varepsilon(\mathbf{r}, \Omega) = 1 + \sum_i \varepsilon(\Omega)\Theta(|\mathbf{r} - \mathbf{R}_i| - a) . \qquad (2)$$

R_i is the position of the ith sphere and is given by the random distribution of particles in space. $\varepsilon(\Omega)$ is the bulk dielectric constant of the particles which we assume to be frequency (Ω) dependent only. The spatial dependence of the medium dielectric function $\varepsilon(\mathbf{r},\Omega)$ thus comes only through the position of the metal particles. To obtain the average Green's function, we first convert Eq. (1) into an integral equation for $\overleftrightarrow{d}$

$$\overset{\leftrightarrow}{d}(\mathbf{r}, \mathbf{r}', \Omega) = \overset{\leftrightarrow}{d}_0(\mathbf{r}, \mathbf{r}', \Omega) - \frac{\Omega^2}{4\pi c^2} \int d^3 r'' \, \overset{\leftrightarrow}{d}_0(\mathbf{r}, \mathbf{r}'', \Omega) \Delta\varepsilon(\mathbf{r}'', \Omega) \cdot \overset{\leftrightarrow}{d}(\mathbf{r}'', \mathbf{r}', \Omega),$$

$$(3)$$

where $\overset{\leftrightarrow}{d}_0$ is the vacuum Green's function, i.e., the solution of Eq. (1) with $\varepsilon(\mathbf{r},\Omega) = 1$ and $\Delta\varepsilon(\mathbf{r},\Omega) = \varepsilon(\mathbf{r},\Omega) - 1$. We now take the average over all possible configurations due to the random position of the particles. As shown in Ref. (12), the average $\langle \overset{\leftrightarrow}{d}(\mathbf{r}, \mathbf{r}', \Omega) \rangle \equiv \overset{\leftrightarrow}{d}(\mathbf{r}-\mathbf{r}', \Omega)$ satisfies the Dyson equation:

$$\overset{\leftrightarrow}{d}(\mathbf{r}-\mathbf{r}', \Omega) = \overset{\leftrightarrow}{d}_0(\mathbf{r}-\mathbf{r}', \Omega)$$

$$+ \int d^3 r'' \, d^3 r''' \, \overset{\leftrightarrow}{d}_0(\mathbf{r}-\mathbf{r}'', \Omega) \cdot \overset{\leftrightarrow}{\Sigma}(\mathbf{r}''-\mathbf{r}''', \Omega) \cdot \overset{\leftrightarrow}{d}(\mathbf{r}'''-\mathbf{r}', \Omega).$$

$$(4)$$

Using Eq. (3), one can then write the self-energy Σ in the form of an expansion

$$\int d^3 r'' \, \overset{\leftrightarrow}{\Sigma}(\mathbf{r}-\mathbf{r}'', \Omega) \cdot \overset{\leftrightarrow}{d}(\mathbf{r}''-r', \Omega) = \frac{-\Omega^2}{4\pi c^2} \langle \Delta\varepsilon(\mathbf{r}, \Omega) \rangle \overset{\leftrightarrow}{d}_0(\mathbf{r}-r', \Omega)$$

$$+ \left[\frac{\Omega^2}{4\pi c^2} \right]^2 \int d^3 r'' \langle \Delta\varepsilon(\mathbf{r}, \Omega) \overset{\leftrightarrow}{d}_0(\mathbf{r}-\mathbf{r}'', \Omega) \Delta\varepsilon(\mathbf{r}'', \Omega) \rangle \cdot \overset{\leftrightarrow}{d}_0(\mathbf{r}''-r', \Omega) + \dots\dots,$$

$$(5)$$

where the first term on the right hand side essentially corresponds to some average dielectric constant of the randomly distributed particles. The second term corresponds to fluctuations and it is this term which is of importance to us. In fact, one can neglect the first term by redefining $\overset{\leftrightarrow}{d}_0$, with reference to a constant effective dielectric function (which can be estimated, for example, by using Maxwell-Garnett theory) instead of the vacuum value, and then considering only the second term.

To calculate the second term, we use a t-matrix approach. Following Ref. (13), one can rewrite the Fourier transform as

$$\overset{\leftrightarrow}{\Sigma}(\mathbf{k}, \Omega) = \frac{f}{v} \int \frac{d^3 k'}{(2\pi)^3} \, \overset{\leftrightarrow}{t}(\mathbf{k}, \mathbf{k}', \Omega) \cdot \overset{\leftrightarrow}{d}_0(\mathbf{k}', \Omega) \cdot \overset{\leftrightarrow}{t}(\mathbf{k}', \mathbf{k}, \Omega) g(|\mathbf{k}-\mathbf{k}'|),$$

$$(6)$$

where

$$g(\mathbf{R}) = \int \frac{d^3 k}{(2\pi)^3} \, e^{i\mathbf{k}\cdot\mathbf{R}} g(|\mathbf{k}|),$$

$$(7)$$

is the correlation function between the positions of the spheres and is assumed in our case to depend on the distance between the centers of the spheres, v is the volume of one sphere and f is the volume fraction of the dielectric material. $\overset{\leftrightarrow}{t}$ is the scattering matrix for the optical wave from a single sphere.

Furthermore, in the spirit of Ref. (14), we need only calculate the imaginary part of the self-energy and that only for $\Omega/c = k$. For this we need the scattering matrix $\overleftrightarrow{t}(\mathbf{k}, \mathbf{k}', \Omega)$ only for $k=k'=\Omega/c$. This can be calculated exactly from the exact Green's function for the em field in the presence of one dielectric sphere, which is well known since the eigenfunctions of Maxwell's equations are known from Mie theory. Thus we have[13,15]

$$\overleftrightarrow{t}(\mathbf{k}, \mathbf{k}', \Omega) = \frac{4\pi}{k} \sum_{l,m} \left[\hat{k} \times \mathbf{X}_{lm}(\Omega_k) \frac{c_{kl}^E}{1+ic_{kl}^E} k' \times \mathbf{X}_{lm}^*(\Omega_{k'}) + \mathbf{X}_{lm}(\Omega_k) \frac{c_{kl}^M}{1+ic_{kl}^M} \mathbf{X}_{lm}^*(\Omega_{k'}) \right],$$

$$(8)$$

where $\mathbf{X}_{lm}(\Omega) = \sqrt{1(1+1)}\, L\mathbf{Y}_{lm}(\Omega)$ is the vector spherical harmonic function. Coefficients c_{kl}^σ ($\sigma = E$ and M which correspond to electrical and magnetic modes in Mie theory) are obtained by matching the Maxwell boundary conditions at the surface of the dielectric sphere and are[15]

$$c_{kl}^E = \frac{-\varepsilon(\Omega) j_l(k_i a)\left[kaj_l(ka)\right]' + j_l(ka)\left[k_i aj_l(k_i a)\right]'}{\varepsilon(\Omega) j_l(k_i a)\left[ka\eta_l(ka)\right]' - \eta_l(ka)\left[k_i aj_l(k_i a)\right]'},$$

$$(9)$$

$$c_{kl}^M = \frac{j_l(k_i a)\left[kaj_l(ka)\right]' + j_l(ka)\left[k_i aj_l(k_i a)\right]'}{-j_l(k_i a)\left[ka\eta_l(ka)\right]' - \eta_l(ka)\left[k_i aj_l(k_i a)\right]'},$$

$$(10)$$

where $k_i = \sqrt{\varepsilon(\Omega)}\, k$. $j_l(x)$ and $\eta_l(x)$ are spherical Bessel functions and the prime on the brackets in Eqs. (9) and (10) denotes differentiation with respect to the argument of the Bessel function.

Inserting Eq. (8) in Eq. (6) and also using a partial wave expansion for the correlation function $g(|k-k'|)$, one can perform the angular integration in Eq. (6). After some simplification one finds for the imaginary part of the self energy

$$\text{Im}\,\overleftrightarrow{\Sigma}(\mathbf{k}, \Omega) = S(k, \Omega)(\overleftrightarrow{I} - \hat{k}\hat{k}),$$

$$(11)$$

where

$$S(k, \Omega) = \frac{3}{2k}\frac{f}{v} g_0(k, k) \sum_{\sigma, l} \left| \frac{c_{kl}^\sigma}{1+ic_{kl}} \right|^2.$$

$$(12)$$

In writing this we have retained only the $l=0$ term in the expansion of the correlation function $g(|\mathbf{k}-\mathbf{k}'|)$ which gives the largest contribution. The Fourier transform of the average Green's function $\overleftrightarrow{d}(\mathbf{r}, \Omega)$ can be written as

$$\overset{\leftrightarrow}{d}(\mathbf{k}, \Omega) = d(k, \Omega)(\overset{\leftrightarrow}{I} - \hat{k}\hat{k}) \; ; \qquad d(k, \Omega) \approx \frac{2\pi c^2}{\Omega(\Omega - \Omega_k + i\gamma)} \;, \tag{13}$$

where $\Omega = ck$ and $\gamma = \dfrac{2\pi c^2}{\Omega} S(\Omega/c, \Omega/c)$. The scattering length l is thus defined as

$$l = \frac{c}{2\gamma} = \frac{c}{\Omega}\left[\frac{9f}{2}\left(\frac{1}{ka}\right)^3 \sum_{\sigma, l}\left|\frac{c_{kl}^{\sigma}}{1 + ic_{kl}^{\sigma}}\right|^2\right]^{-1} . \tag{14}$$

III. ANDERSON LOCALIZATION

Based on the above approach and following the diagrammatic Green's function method we have earlier discussed the localization of an em wave in a disordered dielectric.[1] The em energy density still satisfies the diffusion equation but the renormalized diffusion coefficient $D(q,\omega)$ gets renormalized due to to the maximal cross-diagrams representing coherent interference. We have[1]

$$D(q, \omega) = D_0\left[1 - \frac{l_c^2}{l}\int_0^{1/l_c} dk \frac{k^2}{k^2 - i\omega/D(q, \omega)}\right] , \tag{15}$$

where $D_0 = cl/3$ is the bare diffusion constant and $l_c = (3/\pi)^{1/2}c/\Omega$. Eq. (15) gives the condition for localization, i.e., in the limit $\omega \to 0$, $D(q,\omega) \longrightarrow 0$ only if $l < l_c \approx \lambda/2\pi$. This agrees with the localization condition derived in case of electron with λ corresponding to Fermi wave length. The localization length $\xi = \lim(iD/\omega)^{1/2}$ is then given by

$$(1 - l^2/l_c^2) = (l/\xi)\tan^{-1}(\xi/l) , \tag{16}$$

which for $l < l_c$ diverges as $(l_c - l)^{-1}$.

IV. RESULTS

We now discuss the possibility of whether in realistic experiments the criterion $(l/l_c < 1)$ for localization can be satisfied for the optical wave. The scattering length l calculated from Eq. (14) for polystyrene particles ($f=0.1$, $a=500$ Å, $\varepsilon(\Omega)=2.53$, $\lambda=6330$ Å) is ~ 35000 Å which is almost 35 times l_c. It can be reduced somewhat by suitably chosing the particle radius and the optical wave-length. But it seems difficult to attain the localization condition in case of polystyrene due to the small scattering cross-section from a single sphere. To see how this scattering can be increased we write c_{kl}^{σ} for $ka < 1$ for which the dominant contribution comes from $l=1$ and $\sigma=E$. Thus from Eq. (9)

$$c_{kl}^{E} = \frac{2}{3}(ka)^{3}\frac{1-\varepsilon}{2+\varepsilon} ,$$

(17)

which can be increased by

a)   increasing the particle radius within the approximation ka < 1 for which Eq (17) is valid. For ka>1, the scattering cross-section reaches the maximum value given by geometrical optics.

b)   using a dielectric material with larger ε. There have been attempts, e.g., to use particles of TiO_2 with $\varepsilon(\Omega)$=7.3 and apprreciable decrease in l is claimed.

c)   The other possibility we propose is to use metal particles. For metals with negative $\varepsilon(\Omega)$, l can decrease by an order of magnitude due to the factor $(1-\varepsilon(\Omega))/(1+\varepsilon(\Omega))$ appearing in Eq. (17). It is well known from Mie theory that for a metal sphere (i.e., with a negative dielectric constant) the em field near its surface is quite large, especially when incident frequency is close to Mie resonance (Note c_{kl}^{E} has a pole for each l value which corresponds to the Mie resonance; for ka<1, Mie resonance is given by $\varepsilon(\Omega) = -(l+1)/l$ (see also Eq. (17)). This gives a large scattering cross-section from one sphere and hence l can become sufficently small even for small metal concentration. From Eq. (14), we have calculated l for a suspension of Ag particles, with radii a=300, 400, 500, 600, 700 Å and f=0.2 and these value are given in Table I. The dielctric function $\varepsilon(\Omega)$ of Ag is taken from Ref. (16). These parameters are experimentally accessible and we find that in a reasonable frequency range it is possible to satisfy the condition ($l\sim l_c$) for strong localization. Similarly this condition can be satisfied in other metals like Au, Cu, etc.

Table I

Ω (eV)	$\varepsilon(\Omega)$	a=300 Å	a=400 Å	a=500 Å	a=600 Å	a=700 Å
				l/l_c		
2.7	−7.5	10.74	3.57	1.55	1.06	1.15
2.8	−6.5	7.71	2.48	1.12	0.92	1.16
2.9	−6.0	5.98	1.87	0.91	0.89	1.24
3.0	−5.0	3.63	1.09	0.66	0.86	1.33
3.1	−4.5	2.41	0.73	0.59	0.92	1.47
3.2	−3.5	0.78	0.34	0.60	1.10	1.71
3.3	−3.0	0.25	0.35	0.77	1.30	1.81

One however, should note that an optical wave has to make a large number of multiple scatterings to build up strong localization. Therefore, the losses due to inelastic scattering in the dielectric should be minimum. Metals usually have large imaginary part of the dielectric function, however, for noble metals, in particular, for Ag these losses are very small in the optical frequency range.

Instead of using a steady state wave for the studies of Anderson localization which has been discussed in this paper, similar studies in the time domain can also be of interest. For example, by using an optical picosecond pulse, one should be able to observe directly the building up of localization due to multiple scattering both in reflection and transmission experiments. Similar experiments should also be possible with nanosecond pulse in the microwave region. These investigations are in progress and we will report this work in a separate paper.

ACKNOWLEDGEMENTS

This research was supported in part by FRAP CUNY-PSC and NASC.

REFERENCES

1.   K. Arya, Z.B. Su, and J.L. Birman, Phys. Rev. Lett. $\underline{57}$ 272(1986).
2.   S. John, Phys. Rev. Lett. $\underline{53}$, 2169 (1984), and Phys. Rev. B $\underline{31}$, 304(1985).
3.   P.W. Anderson, Philos. mag. B $\underline{52}$, 505(1985).
4.   K. Arya, Z.B. Su, and J.L. Birman, Phys. Rev. Lett. $\underline{54}$, 1559 (1985).
5.   E. Akkermans, P.E. Wolf, and R. Maynard, Phys. Rev. Lett. $\underline{56}$, 1471 (1986).
6.   M.J. Stephen and G. Cwilich, Phys. Rev. B $\underline{34}$, 7564 (1986).
7.   M.P. Van Albada and Ad Lagendijk, Phys. Rev. Lett. $\underline{55}$, 2692 (1985); P.E. Wolf and G. Mart, Phys. Rev. Lett. $\underline{55}$, 2696 (1985).
8.   S. John and M.J. Stephen, Phys. Rev. B $\underline{28}$, 6358 (1983).
9.   For example, see G. Bergmann, Physics Reports 107, $\underline{1}$ (1984).
10. G.H. Watson, Jr., P.A. Fleury, and S.L. McCall, Phys. Rev. Lett. Phys. Rev. Lett. $\underline{58}$, 945(1987).
11. A. Z. Genack, Phys. Rev. Lett. $\underline{58}$, 2043 (1987).
12. K. Arya and R. Zeyher, Phys. Rev. B $\underline{28}$, 4090 (1983).
13. V.A. Davis and L. Schwartz, Phys. Rev. B $\underline{31}$, 5155 (1985).
14. D. Vollhardt and P. Wölfle, Phys. Rev. B $\underline{22}$, 4666 (1980), and Phys. Rev. Lett. $\underline{48}$, 699 (1981).
15. K. Arya and R. Zeyher, in Light Scattering in Solids IV, edited by M. Cardona and G. Guntherodt (Springer-Verlag, Heidelberg, 1984).
16. P.B. Johnson and R.W. Chersty, Phys. Rev. B $\underline{6}$, 4370 (1972).

ELECTRON SINGLE-PARTICLE LIGHT SCATTERING IN InP CRYSTALS AND

IN MIXED CRYSTALS $Ga_xIn_{1-x}P$

B. H. Bairamov and I. P. Ipatova
A.F. Ioffe Physical Technical Institute
Leningrad 194021, USSR

In addition to traditional Raman scattering from lattice vibrations there has appeared a new field such as quasi-elastic light scattering from single particle electronic exitations in semiconductors. The energy conservation law for single particle scattering has the form

$$E(\vec{p}+\vec{q}) - E(\vec{p}) = \hbar\omega = \hbar(\omega^S - \omega^I). \tag{1}$$

Here $\vec{q}$ is the momentum transferred to the electron plasmas; $E(p)$ is the energy of the electron with momentum $\vec{p}$; ω^I, ω_S being frequencies of the incident and scattered light. The scattering spectrum forms a wide band extending about 100 cm^{-1} on either side of the laser line.[1,2]

In semiconductor plasmas light scattering experiments are done very often in the regime of degenerate plasmas with strong screening where

$$qr_S \ll 1, \tag{2}$$

r_S being the screening radius.

The non-screened mechanism of single particle light scattering results from valley-to-valley fluctuations of the electron density in many-valley semiconductors.[1] The scattering of this type has been observed by Chandrasekhar et al.[3] In direct gap semiconductors the unscreened single particle scattering arises either from spin density fluctuations[4] or from the energy density fluctuations[5] caused by the nonparabolicity of the electron energy spectrum. Both mechanisms were identified in GaAs, see e.g.[6]

An InP crystal is another direct gap material with smaller band gap than GaAs. The nonparabolicity of the electron energy spectrum appears to be more pronounced in InP. It creates favorable conditions for obtaining light scattering spectra from fluctuations of the electron energy. On the other hand, the spin orbit interaction in InP which is responsible for the light scattering from the spin density fluctuations is somewhat lower then in GaAs.

In the present investigation we studied both experimentally and theoretically the inelastic light scattering from free carriers in InP single crystal doped with shallow donors up to the concentrations 1.1 x 10^{19} cm^{-3} and from $Ga_xIn_{1-x}P$ mixed crystals with x = 0.05 - 0.06 and n = (3 ~ 4) x 10^{17} cm^{-3}.

In a degenerate free electron gas the electrons from a spherical layer with the depth $q = \dfrac{\hbar\omega}{v_F}$ near the Fermi surface contribute to the light scattering, v_F - being the electron Fermi velocity. Since the number of the electrons near the Fermi surface increases linearly with ω, the light cross section should increase linearly with ω. It follows from the conservation law (1) that the cross section for light-scattering vanishes sharply at $\omega = qv_F$.[1]

Single particle light scattering spectra in InP and $Ga_xIn_{1-x}P$ appear to be different in this behavior.

Our experiments have shown that at 300 K the spectra are given by well shaped Lorentzians. Similar Lorentzian profiles are known in the radiation spectra of gaseous atomic spectra obtained under the condition of frequent collisions of the radiating atom with atoms of the buffer gas.[7] The width of the Lorentzians was shown to be

$$\Gamma = q^2 D \tag{3}$$

where D is the diffusion coefficient of the atom. Lorentzian shaped light scattering spectra in degenerate semiconductor plasmas were predicted in[8] and observed experimentally in n-Si[9] and n-Ge.[10]

For the carrier densities used in our experiment the condition of frequent carrier-defect collisions was fulfilled

$$q\ell \ll 1, \tag{4}$$

ℓ being the mean free path of the carrier.

The Lorentzian profiles observed show a considerable narrowing with increase of dopant concentration[11] (see Fig. 1). Theoretical consideration has shown that this narrowing means collision controlled regime of carriers in InP and $Ga_xIn_{1-x}P$.

1. Experimental

Single crystals of InP were studied: these included the semi-insulating sample N 1 with resistivity $4,7 \times 10^7$ Ω cm, and samples NN 2-4 with carrier concentration in the range from $5,0 \times 10^{17}$ cm^{-3} to $9,44 \times 10^{18}$ cm^{-3}. We also studied samples of solid solutions $Ga_xIn_{1-x}P$ with x = 0,05 and x = 0,06 grown from a nonstoichiometric melt by the modified Bridgman method.[12]

The composition of the samples was determined by chemical analysis and by X-ray structure microanalysis. Homogenity was verified with an electron microscope. The radial distribution of the composition was found to be constant. This enable us to cut up an as-grown ingot into layers 200-300 μm thick, 0,5 cm^2 in area. The deviation of the composition from the nominal x value in each sample did not exceed 2%. Electrical measurements showed that the samples had n-type conductivity with n = (3 ~ 4) $\times 10^{17}$ cm^{-3}.

The spectra were excited in the transparency range of the crystal using the neodymium-doped yttrium aluminium garnet cw laser with the wavelength λ_I = 1,064 μm. The light scattered at right angles was analyzed with a double diffraction monochromator having resolution 1 cm^{-1} and using a cooled photomultiplier in the photon counting regime. The power was sufficiently low to exclude the heating of the samples in the focus of a beam. The heating of the sample was controlled for each material by the corresponding TO (Γ)-phonon line.

It is seen from Fig. 1 that there is no single particle scattering for the semi-insulating sample N 1. The light scattering spectrum shows the spectral resolution of the detection system in this case. It is well seen that for the samples NN 2-4 near the laser line there appears single particle quasi-elastic electronic scattering. The shape of each spectrum is Lorentzian. Along with the concentration increase there is a visible narrowing of the Lorentzian profiles.

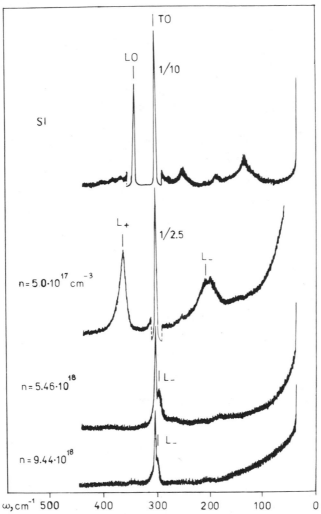

Fig. 1. Single - partial light scattering spectra of n-InP with n = 1.1 x 10^{18} cm^{-3} (1) and n-Ga$_x$In$_{1-x}$P with x = 0.06 and n= 3 x 10^{17} cm^{-3} (2). Points (3) are theoretical fits to the experimental data. N is the noise level.

2. Theoretical

The light cross section for quasi-elastic electronic scattering can be conveniently expressed in terms of the correlation function of the electron polarizability fluctuations[10]

$$\frac{d^2 \Sigma}{d\omega d\Omega} = \frac{1}{2\pi} \frac{(\omega^I)^4}{c^4} e_i^I e_s^I e_j^I e_s^-_n \int_{-\infty}^{+\infty} d\tau \int d^3 r e^{-i(\omega t - \vec{q}.\vec{r})}\times$$

$$\times \langle \delta\chi_{ij}(\vec{R} + \vec{r}, t + \tau)\delta\chi_{kn}(\vec{R},t)\rangle. \tag{5}$$

Here angular brackets mean the correlation function of electron susceptibilities taken in the Heisenberg representation.

When the condition of frequent collisions (4) holds, the electron gas reacts to the electromagnetic wave like a continuous medium. Therefore the correlation function can be found from the macroscopic hydrodynamic equations. The general approach for considering unscreened hydrodynamic fluctuations with the condition of strong screening (2) consists of determining the $\delta\chi_{ij}$ which are statistically independent of the charge density fluctuations δn.

The unscreened contribution to the symmetric second rank tensor χ_{ij} comes from nonparabolicity of the electron energy spectrum. The fluctuations of the corresponding electron susceptibility has the form

$$\delta\chi_{ij}(E) = - \frac{4\pi\delta n(E)}{m^*(E)(\omega^I)^2} \delta_{ij} \approx 4\pi \left[\frac{e^2}{m^*(0)(\omega^I)^2}\right] \delta n(E)B(e)\delta E\delta_{ij} \tag{6}$$

Here $\delta E = E - \langle E \rangle$ is the fluctuation of the electron energy; $m^*(E)$ is the energy-dependent electron effective mass; $\delta n(E)$ is the concentration fluctuation for electrons with the energy E. Under the condition of strong screening (2) the neutrality condition is the following:

$$\int dE\delta n(E) = 0 \tag{7}$$

But the statistically independent fluctuating quantity $\frac{\delta n(E)}{m(E)}$ does not vanish on the average

$$\int dE \frac{\delta n(E)}{m(E)} \neq 0. \tag{8}$$

According to Eq. (6) the ratio $\frac{\delta n(E)}{m(E)}$ is proportional to the electron energy fluctuation δE.

The relaxation of the energy fluctuations δE is caused by the thermal conductivity equation

$$\frac{d}{dt} \delta E + \text{div } \bar{q} = 0 \tag{9}$$

where the thermal current $\bar{q}$ equals

$$\bar{q} = - \kappa \text{ grad } T. \tag{10}$$

It has been shown[12] that due to Eq. (6) the correlation function from (5) obeys the same Eqs. (9) and (10) as the energy fluctuation δE.

The initial conditions for the correlation function depend only on the temperature

$$\langle \delta E^2 \rangle_{\substack{\omega=0 \\ \bar{q}\to 0}} \sim \left(\frac{T}{E_q}\right)^2 \tag{11}$$

248

Calculations show that the corresponding light scattering cross section has the form

$$\frac{d^2 \Sigma}{d\omega d\Omega} = V \left(\frac{e^2}{m_0 c^2}\right)^2 (\bar{e}^I \cdot \bar{e}^S) B^2_E < (\delta E)^2 >_{\substack{q \to 0 \\ \omega = 0}} \frac{q^2 \chi}{(\omega^2 + (q^2 \chi))^2} \qquad (12)$$

Here $\chi = \frac{\kappa}{C_V}$ is the thermal diffusivity of the electron gas. This scattering occurs with parallel polarization: $\bar{e}^I || \bar{e}^S$.

Eqs. (9) and (10) do not take into account the energy fluctuation relaxation through energy transfer to the lattice. This is true when $T << \hbar\omega_{opt}$, and therefore $\tau_{e-ph} >> \tau_{ee}$.

The second unscreened contribution to the electronic light scattering is known to come from the spin density fluctuation in the conduction band. The neutrality condition has the form

$$\delta n = \delta n_\uparrow + \delta n_\downarrow = 0 \qquad (13)$$

Here $\delta n_\uparrow$ and $\delta n_\downarrow$ are fluctuations of the electron density with spin "up" and spin "down", respectively.

The difference of spin subband population is statistically independent of the magnitude of the electron density fluctuation, so

$$\delta n_\uparrow - \delta n_\downarrow = T_r \sigma_z \sim e_i^I \chi_{ij} e_j^S, \qquad (14)$$

where σ_z is the Pauli matrix.

In the hydrodynamic limit (3) the fluctuation of the difference $(\delta n_\uparrow - \delta n_\downarrow)$ satisfies the continuity and diffusion equations

$$\frac{\partial}{\partial t} (\delta n_\uparrow - \delta n_\downarrow) + \text{div } \vec{j} = 0$$
$$\vec{j} = D \text{ grad } (\delta n_\uparrow - \delta n_\downarrow), \qquad (15)$$

where D is the electron diffusion coefficient.

The initial condition for Eq. (15) is the value of the temperature dependent correlation function

$$< (\delta n - \delta n_\uparrow)^2 >_{\substack{\omega = 0 \\ q \to 0}} = T \frac{\partial n}{\partial \zeta}, \qquad (16)$$

ζ being the chemical potential of electrons.

Using a Fourier transformation one can solve Eq. (15) and find the light differential cross section for light scattering as:

$$\frac{d^2 \Sigma}{d\omega d\Omega} = TV \left(\frac{e^2}{M_0 C^2}\right) (\bar{e}^I x \bar{e}^S) B_\sigma^2 (\omega^I) \frac{\partial n}{\partial \zeta} \frac{q^2 D}{\omega^2 + (q^2 D)^2} \qquad (17)$$

It follows from Eq. (17) that the scattering from the spin density fluctuations occurs with crossed polarizations $e^I \perp \bar{e}^S$. The spectrum has Lorentzian shape with width $\Gamma = q^2 D$. When the concentration of defects in the lattice increases, the diffusion coefficient D decreases and the Lorentzian contour becomes less broadened. The narrowing observed experimentally indicates directly the diffusional behaviour of carriers in crystals.

Lorentzian profiles for $Ga_xIn_{1-x}P$ mixed crystals with n = 3 x 10^{17} cm^{-3} and x = 0.06 and InP with n = 1.1 x 10^{18} cm^{-3} are shown in Fig. 2 together with the corresponding theoretical Lorentzians. Considerable narrowing of the $Ga_xIn_{1-x}P$ spectrum with respect to the InP spectrum is attributed to the additional scattering of carriers by fluctuations of the mixed crystal composition.

The values of the carrier mobilities calculated with the help of the Einstein relation and measured values of the diffusion coefficient are given in Table 1. These data are in reasonable agreement with the values obtained from electrical measurements.

Table 1

Material	Mobility (units: cm^2/V.s)		Concentration (cm^{-3})
	Light Scattering	Electrical Measurements	
n-InP	1700	2000	1.1 x 10^{18}
n-$Ga_xIn_{1-x}P$ x = 0.06	1470	1540	3 x 10^{17}

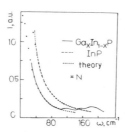

Fig. 2. Single - partical light scattering spectra semi-insulating (SI) and doped n-InP crystals. T = 300 K.

References

1. P. M. Platzman, Phys. Rev. 139A:379 (1965)
2. A. Mooradian, in: Light Scattering Spectra of Solids, ed. by G. B. Wright, Springer Verlag, p.285-295 (1969).
3. M. Chandrasekhar, M. Cardona, E. O. Kene, Phys. Rev. B 16:3579 (1977).
4. D. C. Hamilton, A. L. McWhorter, in: Light Scattering Spectra of Solids, ed. by G. B. Wright, Springer-Verlag, p.309 (1969).
5. P. A. Wolff, Phys. Rev. 171:436 (1968).
6. M. V. Klein, in: Light Scattering Spectra of Solids, ed. by G. B. Wright, Springer-Verlag, p.309 (1969).
7. R. H. Dicke, Phys. Rev. 89:472 (1953).
8. V. A. Voitenko, I. P. Ipatova, A. V. Subashiev, Sov. Phys.-Trans. of Academy of Sciences of the USSR 48:749 (1984); B. H. Bairamov, V. A. Voitenko, I. P. Ipatova, A. V. Subashiev, V. V. Toporov, E. Jahne, Sov. Phys. Solid State 28:720 (1986).
9. G. A. Contreras, A. K. Sood, M. Cardona, Phys. Rev. B 32:924,930 (1985).
10. N. Mestres, M. Cardona, Phys. Rev. Lett. 55:1132 (1985).
11. G. Voight, R. Raidht, H. Peibst, H. Menninger, L. Hildish, Phys. St. Sol. 2(a):36,173 (1976).
12. E. M. Lifshitz, L. P. Pitaevskii, Physical Kinetics, Pergamon Press, Oxford (1981).

DYNAMICAL LIGHT SCATTERING AT THE

NONEQUILIBRIUM CRYSTAL-MELT INTERFACE*

H.Z. Cummins, Henry Chou, G. Livescu[+], O. Mesquita[++],
M. Muschol and M.R. Srinivasan[+++]
Department of Physics
City College of the City University of New York
New York, NY   10031

The interface between a growing crystal and its melt has long been a system of both fundamental and practical interest. A large body of theoretical literature exists on the crystal growth process, originating with the earliest model of Wilson (1900) and Frenkel (1932). Experimental studies have, until recently, been largely limited to photographic observations and to kinetic studies, i.e. analysis of the functional relationship between growth velocity and the undercooling of the interface.

In 1978, Bilgram, Guttinger and Kanzig first applied the quasielastic laser light scattering technique known as photon correlation spectroscopy (PCS) to the crystal-melt interface and discovered a new and unexpected phenomenon[1]: a laser beam directed at the surface of an ice crystal growing along its c-axis into highly purified water exhibited no visible scattering when the crystal was stationary or growing slowly. However, when the growth speed surpassed a critical value of V_C ~ 1.5 μm/sec, strong scattering appeared, localized at the growing interface, and its intensity was found to increase monotonically with increasing growth speed.

The intensity autocorrelation function of the scattered light was found to be exponential: $C_I(\tau) = B[1+a \exp(-2\Gamma\tau)]$ with $\Gamma=Dq^2$. Here q is the scattering vector $q = 2k_0\sin(\theta/2)$ where k_0 is the wavevector of the incident light in the medium and θ is the scattering angle. The "diffusion constant" D was found to be 4×10^{-8} cm^2/sec.

This relaxational scattering was further investigated at the ice-water interface[2,3,4,5], in salol[6,7,8,9,10], and, recently, in cyclohexanol[11], and biphenyl[12].

The essential elements of the experimental apparatus are illustrated in figure 1. In our experiments with salol, which is optically highly anisotropic, changing the polarization of the incident laser beam which approached the interface from the solid side allowed us to have either partial transmission at the interface or total internal reflection within the crystal. Only for the former was the strong relaxational scattering observed which proved that it originates in the fluid boundary layer and <u>not</u> at the crystal surface[6].

The principal results of dynamic light scattering experiments on ice, salol, cyclohexanol and biphenyl crystal-melt interfaces performed at the E.T.H.-Zurich, City College of New York, University of California-Davis and the

Universidade Federal de Minas Gerais-Brazil can be summarized as follows:

(a) It occurs in the fluid boundary layer and coexists with translational dynamics on the surface.
(b) It is much stronger than scattering from either the crystal surface or the bulk fluid.
(c) It requires a minimum velocity and a delay time on the order of one hour for onset.
(d) The correlation function is approximately a single exponential and the decay rate scales approximately with the square of the scattering vector: $\Gamma = Dq^2$ (although there is often considerable scatter in Γ values.)
(e) It occurs on both atomically smooth and rough interfaces.
(f) D is typically 10^4 times smaller than the self-diffusion coefficient and 10^6 times smaller than the thermal diffusion coefficient, but often varies considerably for a given system.

Several explanations for this dynamic interfacial light scattering were proposed between 1978 and 1986. In their original paper, Bilgram, Guttinger and Kanzig proposed that the anomalous scattering is produced by dynamic surface corrugations (i.e. capillary waves) on the interface[1]. However, the observations that established property (a) disproved this suggestion and it was subsequently retracted[4]. Boni et al. have also suggested that a new state of matter might exist in the fluid boundary layer with properties that differ from the liquid and the solid with isothermal compressibility 700 times greater than that of water[4]. This suggestion of an enhanced thermal diffusion mode is, however, inconsistent with the observed dependence of intensity on scattering angle to be discussed below. Keizer, Mazur and Morita developed a theory based on diffusion of defects within the crystal surface[13], but this explanation is also ruled out by property (a).

In our first report of our salol experiments, we suggested that the advancing solidification front might induce partial orientation of the molecules in the fluid boundary layer, and that the anomalous scattering might therefore arise from orientation fluctuations similar to those observed in liquid crystals[6]. In order to test this hypothesis, we undertook a study of the plastic crystal cyclohexanol. Since there is no molecular orientation in plastic crystals, there is no reason to expect orientational ordering in the boundary layer. We observed anomalous scattering in cyclohexanol, however, similar to that observed in water and salol. Thus all four mechanisms that were proposed to explain relaxational light scattering at the nonequilibrium crystal-melt interface are in conflict with the experimental results.

It has been suggested that the scattering might arise from small-scale turbulence, or from the diffusive motion of small particles (e.g. precrystalline clusters) in the boundary layer. The diffusion mechanism seemed particularly attractive since it would explain the Dq^2 behaviour of $C_I(\tau)$, but it requires the presence of particles ~ 3,000A in diameter with a narrow size distribution, and no plausible mechanism for generating such particles had been recognized.

In 1986 we proposed a new explanation for this phenomenon based on the existence of a small residual concentration of dissolved gas in the melt[10]. Solute rejection during solidification can produce a high concentration of dissolved gas in the boundary layer ahead of the advancing solidification front. If the concentration should exceed the saturation concentration C_S, it would be possible for bubbles to nucleate in the fluid boundary layer. We were led to this idea by a study of the angular dependence of the scattering intensity $I(\theta)$. If the scattering is caused by a homogenous fluctuation phenomenon such as entropy fluctuations, $I(\theta)$ should follow the Ornstein-Zernike form

$$I(\theta) = I(0)/[1+q^2\zeta^2] \tag{1}$$

where ζ is the correlation length. However we found that $I(\theta)$ decreases with increasing θ much more rapidly than Eq. (1) predicts regardless of the size of ζ. On the other hand, $I(\theta)$ is in reasonable agreement with the Mie theory for scattering from spheres of radius $R \approx 3{,}000A$ which is the size obtained by combining the observed diffusion constants with the Stokes-Einstein equation[10].

The nucleation of gas bubbles in the liquid boundary layer during crystal growth is a phenomenon which has been known and investigated for over 10
0 years. As cited in a review of gas bubble nucleation by Wilcox and Kuo[14], it was studied in 1870 by Bunsen who first obtained bubble-free ice by boiling water in the growth tube and sealing it to exclude air. Growth of bubbles in the interface region has also been studied recently by Geguzin and Dzuba[15].

One apparent problem with the bubble model was the difficulty of explaining the uniformity of bubble sizes in the interface. If surface tension is neglected, a bubble of any radius R would be in equilibrium with the fluid if the concentration C were exactly the saturation concentration C_S. When surface tension is included, however, bubbles can only exist if the liquid is supersaturated, i.e. if $f = C/C_S > 1$. If the fluid pressure is p and the surface tension is σ, then the effective pressure inside the bubble will be $p' = p + 2\sigma/R$, and bubbles can exist in (unstable) equilibrium only if $f = p'/p = 1 + (2\sigma/Rp)$, or equivalently, if

$$R_0 = \frac{2\sigma}{[p(f-1)]} . \tag{2}$$

Bubbles with $R < R_0$ will collapse, while those with $R > R_0$ will grow[16]. This suggests that once
$f > 1$, a distribution of bubble sizes should be present in the interface.

The resolution of this apparent difficulty lies in another well-known phenomenon: gas bubbles formed in the fluid boundary layer can be trapped by the advancing crystallization front, leading to voids in the crystal. In practical crystal growth procedures, one tries to avoid such trapped bubble formation which impairs crystal quality[14].

In a comprehensive review by Chernov and Temkin, the theory and experimental facts concerning the formation of inclusions in crystals was presented[17]. They pointed out that any particles (including bubbles) in the melt can be repulsed by the advancing crystal front if the growth rate does not exceed a critical capture value V_C which decreases with increasing particle size. For particles with sizes of 10^{-4} to 10^{-2} cm, a typical value of V_C is ~ 10^{-5} to 10^{-3} cm/sec. It is therefore the growth velocity V_G which sets the upper limit on bubble size in the interface region. Our result that at a growth velocity of 0.4 µm/sec the average bubble radius is ~ 3,000A indicates that at this velocity, bubbles which significantly exceed 3,000A in radius will be captured by the advancing solidification front.

The bubble model explanation of relaxational interfacial light scattering has several attractive features. First, it provides a mechanism for generating many particles (bubbles) with a narrow size distribution whose diffusive motion can explain the correlation function of the observed dynamic light scattering. Second, it explains the delayed onset of anomalous scattering since it will take some time for the interface concentration to reach its terminal supersaturation value. The narrowness of the region in which scattering is observed is due to the strong concentration gradient. Third, it suggests why the observed D varies slightly from sample to sample and also depends on the temperature gradient, since the balance of transport and solute rejection would be expected to be sensitive to these factors. Fourth, it suggest that the steady-state bubble concentration should be an increasing function of growth velocity so that the scattered intensity should increase with V_G, as is observed. Fifth, it shows why no scattering is observed at very low growth rates since the nucleation rate of bubbles on the interface decreases exponentially as $V_G \to 0$[17]. Sixth, it explains why relaxational scattering is also observed during melting since bubbles trapped in the crystal during growth will be released when the crystal melts.

We have observed the following additional aspects of the phenomenon which support the above model. First, when salol samples which have been repeatedly zone refined are grown very rapidly ($V_G > 3 \mu m/sec$), gas evolves visibly at the interface and large bubbles can be seen moving upwards through the fluid. Second, the intensity of the scattering is not constant[9]. We monitored the intensity at a fixed scattering angle continuously and found that at $V_G = 0.43$ $\mu m/sec$ the intensity increased by about 300% during six hours. This effect is reproducible. We have also noted that the scattering does not begin uniformly in the boundary layer. It begins instead in a small isolated patch and then spreads out across the interface.

Although the observations described above provide substantial support for our bubble model, we recognize the clear need for additional theoretical analysis and experiments. We have assembled a new growth tube which includes a pressure gauge and is permanently attached to a vacuum station during the experiment. With this apparatus we can study the effect of changing the pressure and gas composition on the light scattering intensity and correlation functions. These experiments are currently in progress and will be described in a future publication.

REFERENCES
1. J.H. Bilgram, H. Guttinger and W. Kanzig, Phys. Rev. Letters 40, 1394 (1978).
2. H. Guttinger, J.H. Bilgram and W. Kanzig, J. Phys. Chem. Solids 40, 55 (1979).
3. J.H. Bilgram and P. Boni, in: Light Scattering in Liquids and Macromolecular Solutions edited by V. DeGiorgio, M. Corti and M. Giglio (Plenum, 1980).
4. P. Boni, J.H. Bilgram and W. Kanzig, Phys. Rev. A 28, 2953 (1983).
5. R.A. Brown, J. Keizer, U. Steiger and Y. Yeh, J. Phys. Chem. 87, 4135 (1983).
6. O.N. Mesquita, D.G. Neal, M. Copic and H.Z. Cummins, Phys. Rev. B 29, 2846 (1984).
7. O.N. Mesquita and H.Z. Cummins, Physico-Chemical Hydrodynamics 5, 389 (1984).
8. J.H. Bilgram, in: Nonlinear Phenomena at Phase Transitions and Instabilities edited by T. Riste (Plenum, 1982) p. 343.
9. U. Durig, J.H. Bilgram and W. Kanzig, Phys. Rev. A 30, 946 (1984).
10. H.Z. Cummins, G. Livescu, Henry Chou and M.R. Srinivasan, Solid State Communications 60, 857 (1986).
11. G. Livescu, M.R. Srinivasan, Henry Chou, H.Z. Cummins and O. Mesquita (submitted to Phys. Rev. B)
12. O. Mesquita - in preparation
13. J. Keizer, P. Mazur and T. Morita, Phys. Rev. A 32, 2944 (1985).
14. W.R. Wilcox and V.H.S. Kuo, J. Crystal Growth 19, 221 (1973).
15. Ya E. Geguzin and A.S. Dzuba, J. Crystal Growth 52, 337 (1981).
16. P.S. Epstein and M.S. Plesset, J. Chem. Phys. 18, 1505 (1950); H-C Chang and L-H Chen, Phys. Fluids 29, 3530 (1986).
17. A.A. Chernov and D.E. Temkin, Capture of Inclusions in Crystal Growth, in: Current Topics in Material Science, Vol.2 edited by E. Kaldis (North-Holland, 1977) p.4.

SELECTIVE LASER HEATING AND NONLINEAR LIGHT SCATTERING

IN A HOMOGENEOUS MEDIUM

M. V. Belyayev, A. P. Maiorov, V. A. Smirnov and V. P. Chebotayev

Institute of Thermophysics, Siberian Branch of the USSR
Academy of Sciences, 630090 Novosibirsk, USSR

INTRODUCTION

In the present paper we report on observations of nonlinear light scattering on spatial-temporal fluctuations of the medium permittivity that occur due to local heating of the medium around impurity molecules resonantly absorbing the incident radiation and transmitting the absorbed energy into the medium. In its physical content, the observed scattering differs qualitatively from both the nonlinear scattering on small particles[1] and the scattering due to the thermal gratings produced by interference of light beams.[2] Experimental results of observation of selective reaction on the surface of exposed photomaterials irradiated with a high-power IR laser are presented.

1. Let us consider a transparent liquid that contains a small amount of molecules absorbing the light with wavelength λ. Since with quenching collisions practically the whole absorbed energy is converted into heat, each absorbing molecule is a point source of heat. In the steady-state case,

$$T(r) = Q/4\pi\kappa r, \tag{1}$$

where Q is the amount of heat released per unit time, κ is the thermal conductivity, $r = 2.5 \sqrt[3]{P_1}$ is the minimum radius for which the formula is applicable, and P_1 is the liquid molecule concentration.

Medium heating in the nearest vicinity of the radiation-absorbing molecule can be considerable. For instance, for water molecules: radius $r = 7.8 \times 10^{-8}$ cm, absorption cross-section $\sigma = 1.5 \times 10^{-15}$ cm^2. With the intensity of laser radiation $I_0 = 1$ GW/cm^2, $P = \sigma I_0 = 15$ erg/s, $T(r) \approx 260$ K. Considerable local heating in the nearest vicinity of the absorbing molecule results in changing the refractive index of the medium in this region. Each absorbing molecule is the center of optical inhomogeneity that is proportional to the incident radiation intensity.

Let us estimate the effect. Assume that $1 = p^{-1/3}_{eff} \ll \lambda$, $p_{eff} \ll p$, 1 is an average distance between absorbing molecules, p_{eff} is the concentration of absorbing molecules, p is the overall concentration of molecules. Assume also that I_{p_1}, I_{p_2}, ... I_{p_p} are added incoherently. In this case one can use the known formula for Rayleigh scattering of linearly polarized light:[3]

$$I_S = I_0 \frac{\pi^2}{\lambda^4 r^2} N (\Delta\varepsilon)^2 V^2 \sin^2\phi. \tag{2}$$

Here I_0 and I_S are the intensities of incident and scattered light, N is the number of inhomogeneities in the observed volume, $(\Delta\varepsilon)^2$ is the mean-square deviation of permittivity in the inhomogeneity, V is the inhomogeneity volume, and ϕ is the angle between the vector of the incident radiation field and the direction of observation.

The quantity $B = N(\Delta\varepsilon)^2 V^2$ consists, in our case, of two parts: $B_0 = VA$ is an ordinary molecular scattering,

$$B_i = V_{Peff} (\int \Delta\varepsilon dV)^2 = V_{Peff} (\frac{d\varepsilon}{dT})^2 (\int \Delta T dV)^2 \tag{3}$$

is the scattering initiated by laser radiation.

Here

$$\int T dV = \tau \sigma I_0 \text{ for } \tau \ll 1^2/4a$$

$$= \frac{1^2 \sigma I_0}{2\kappa} \text{ for } \tau \gg 1^2/4a.$$

c is the specific heat, a is the thermal conductivity, ρ is the medium density, τ is the pulse duration of incident radiation, and σ is the cross-section of light absorption of given λ.

Finally, the formula allowing for the scattering on thermal inhomogeneties acquires the following form:

$$I_S(I_0) = I_0 \frac{\pi^2 V}{\lambda^4 r^2} A \sin^2\phi(1 + c(I_0)), \tag{4}$$

$$c(I_0) = 1/3 \frac{Peff}{A} (\frac{d\varepsilon}{dT})^2 (\frac{\tau\sigma I_0}{c\rho})^2 \text{ for } \tau \ll 1^2/4a \tag{5}$$

$$= \frac{Peff}{A} (\frac{d\varepsilon}{dT})^2 (\frac{1^2\sigma I_0}{2\kappa})^2 \text{ for } \tau \gg 1^2/4a. \tag{6}$$

The estimates for given conditions of our experiment will be presented below.

2. Experimentally the scattering was observed with the apparatus presented schematically in Fig. 1. We have obtained the dependences of the intensity of the scattered radiation I_S on that of the incident radiation I_0 for distilled water and for a weak solution of iodine ($p = 2.8 \times 10^{17}$ cm^{-3}). They are given in Fig. 2.

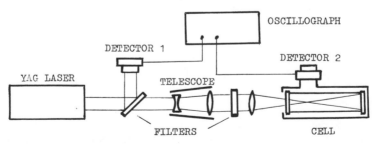

Fig. 1. Experimental installation. The YAG laser with Q-switching and radiation frequency doubling.

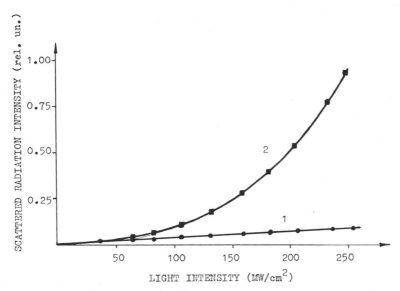

Fig. 2. Experimental dependence of the scattered radiation intensity on the intensity of incident laser radiation. 1 - distilled water, 2 - iodine solution (n = 2.8×10^{17} cm^{-3}).

The coefficient C is determined experimentally in equation (2) for $I_{max} \cdot C(I_{max}) \approx 15$. The absorption factor $\alpha = 0.2$ cm^{-1} and the absorption line halfwidth $\Delta\omega = 3.4 \times 10^{14}$ s^{-1} were measured for iodine concentration p = 2.8×10^{17} cm^{-3}.

We observed the dependence of α on I_0. No saturation was observed. With focusing the laser radiation by a cylindrical lens, the scattering intensity is independent of the angle between the cylinder axis and the observation direction. A conclusion can be made that the observed scattering is not induced, as follows from our model.

The above-described experiments were carried out in a solution passing through filters with 0.8 μm pores. With filters having 10 μm pores we observed, when studying the scattering, the following phenomenon. At the first moment after the laser has started operating, the scattering intensity is maximum, then for 0.2 - 0.3 s it decreases and becomes steady-state. The steady-state value of the scattering intensity corresponds to the scattering intensity in a well-filtered solution. The dependence of the gain of the scattering intensity at the first moment on the intensity of incident laser radiation is practically a linear one. The duration of establishing the steady-state value of the scattering intensity decreases with an increase in laser radiation intensity. A flash-up in the intensity was observed when the cell with the solution was moved perpendicularly to the laser beam.

These factors appear to be explained by an ordinary Rayleigh scattering on complexes of 1-10 μm in size occurring in a poorly filtered solution. Under laser irradiation these complexes are destroyed which results in establishing the steady-state conditions.

Let us estimate the value of the coefficient C which takes into account the temperature inhomogeneity for the conditions of our experiment. First of all, we shall estimate the value of p_{eff}. Let us assume that the relaxation time for an iodine molecule is $\tau_p = 10^{-12}$ s. Accordingly, $\Gamma = 10^{12}$ s^{-1}. The absorption factor is found from the formula

$$\alpha = \frac{8\pi^2 |d|^2 p_{eff}}{\lambda h \Gamma}. \tag{7}$$

where d is the dipole moment, p_{eff}, $= pg\Gamma/\Delta\omega$. Low-frequency oscillations typical of liquids[3] are allowed for by using a statistical weight g. According to,[4] for the electron $B \rightarrow X$ transition of J_2 $d \approx 10^{-18}$ cm g s. Consequently, from the formula for α $p_{eff} = 14 \times 10^{14}$ cm^{-3}, which corresponds to g = 0.17. Further for J_2 $1 \approx 2 \times 10^{-5}$ cm, $\dagger^2/4a$ = 65 ns, $\sigma = \alpha /p_{eff} = 1.5 \times 10^{15}$ cm^2; σI_0 = 3.75 erg/s.

Since $\tau < 1^2/4a$ the expression to estimate C is of the form

$$C(I_0) = (1/3) \frac{p_{eff}}{A} \left(\frac{d\varepsilon}{dT}\right)^2 \left(\frac{\tau \sigma I_0}{c \rho}\right)^2.$$

Substituting $A = 10^{-24}$ cm^{-1} (see ref. 5, page 121) and $(d\varepsilon/dT)^2 = 5.23 \times 10^{-8}$ K^{-2} we obtain $C(I_{max}) = 7.8$ as compared to $C(I_{max}) \approx 15$ measured experimentally. With allowance for the estimation character the agreement should be considered to be good.

With I_{max} = 250 MW/cm^2, ΔT in the center of the inhomogeneity is 65 K, with an average water heating in the focus of 0.4 K.

3. The selective absorption accompanied by considerable heating of the nearest neighborhood of the absorbing molecule allows selective initiation of thermal chemical reactions in substances. Ref. 6 describes the influence of local temperature fluctuations on the chemical reactions under radiation absorption. The thermal chemical reaction probably occurs on the surface of photomaterial under irradiation with a high-power IR laser.

Photomaterials (film, paper) are widely used for visualization of the IR radiation from a pulsed Nd laser. It was observed that the laser beam produces a print only on the exposed photomaterial. Nonexposed photomaterial reveals no trace even with the radiation flow of 10^{10} W/cm^2.

Experiments with different photomaterials showed that the transparency of these materials for IR radiation decreases after exposure. This decrease is negligible for normally exposed photomaterials and grows with the increase of exposure time, as shown in Fig. 3.

The dependence of the observed intensity of IR radiation on the exposure time of photomaterial is given in Fig. 4. With a small exposure time the sensitivity of photomaterial to IR radiation is low. It can be attributed to a slight absorption of IR radiation in the photosensitive layer of the material, as is shown in the plot.

The time-dependent process of visualization for laser IR radiation was studied experimentally. The experiment is schematically presented in Fig. 5. The exposed photomaterial was placed in a cassette between two light filters KS-19 passing IR radiation. A weak laser radiation (pulse duration of 300 ns) was directed at an angle to the cassette. The radiation passed through the photomaterial, was collected by a lens, and focused to a photoreceiver I-LDF-2. The rapid-record oscillograph S7-19 recorded the signal from the photoreceiver. The oscillograph was started by the signal from the photoreceiver II which recorded the high-power radiation from laser II visualized on the photomaterial. The optical paths of the radiation of laser I and II from the cassette with photomaterial to the appropriate photoreceivers were chosen to be equal.

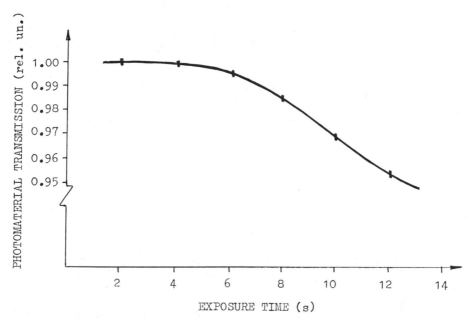

Fig. 3. Experimental dependence of the photomaterial transmission for IR radiation on the exposure time.

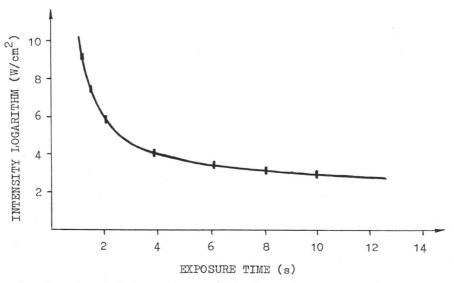

Fig. 4. Experimental dependence of the intensity of the visualized laser radiation (λ = 1.06 μm) on the photomaterial exposure time.

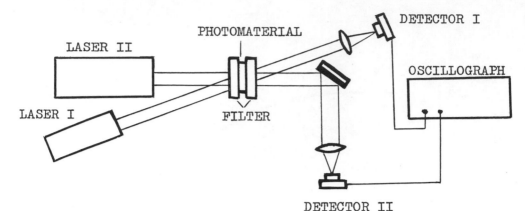

Fig. 5. Schematic representation of the experimental installation for investigation of the duration of the visualization process of IR radiation.

The photomaterial in the region of high-power radiation from laser II become dark and its transmission decreased for the radiation from laser I. The pulse duration of laser II was 0.7 ns.

The experimental results have shown that the visualization starts practically at the moment of radiation action on the photomaterial and lasts for a short period (about 2 ns). Such a picture was observed for different photomaterials and it remained the same with variation of the exposure time of photomaterials.

The visualization mechanism of IR radiation on the exposed photomaterials can be as follows. During the photomaterial exposure chemical transformations of silver salts occur in the photosensitive layer under the light effect, which gives new compounds, the centers of IR radiation absorption. Under irradiation of the exposed material with laser IR radiation its energy is absorbed partially. The energy converts into heat and thereby accelerates the chemical reactions near the absorption center, which results in visualization of the radiation on the photomaterial. The obtained prints of radiation on a nonsensitized photomaterial allow a conclusion that only a gelatin layer with silver salts participate in the process of IR radiation visualization.

The dependence of the photomaterial sensitivity to IR radiation on the exposure duration can be used to visualize a latent photographic image. The experiment was performed with photoplates FT-41 in a high-power Nd-glass laser. The negative image was projected onto a photoplate via a photomultiplier. Then the exposed photoplate was placed into a cassette with the light filter KS-19 transmitting the IR radiation and cutting off the visible spectrum. The cassette was exposed to a laser beam. After radiation a distinct image appears on the photoplate. Fig. 6 shows the picture of the photoplate with the visualized image. The image was obtained with one laser pulse.

Fig. 6. Picture of the photoplate FT-41 with the visualized image. The parameters of the visualized laser radiation are as follows: pulse duration 10^{-9} s, intensity 2×10^9 W/cm^2, beam aperture 60 mm.

References

1. Yu. K. Danileiko, A. A. Manenkov, V. S. Nechitailo, V. Ya. Khaimov-Mal'kov, ZhETF 60:1245 (1971).
2. S. A. Akhmanov, N. I. Koroteyev, Methods of Nonlinear Optics in Scattered Light Spectroscopy, (Nauka, Moscow, 1981).
3. M. F. Vuks, Light Scattering in Gases, Liquids, and Solutions, (Nauka, Leningrad, 1977).
4. B. Ya. Zeldovich, I. I. Sobelman, Usp. Fiz. Nauk 101:3 (1970).
5. J. Tellinghuisen, J. Chem. Phys. 58:2821 (1973).
6. V. N. Sazonov, ZhETF 82:1092 (1982).

DYNAMIC CENTRAL MODES AND PHOTOREFRACTIVE EFFECTS AT T_I AND T_c IN BARIUM SODIUM NIOBATE[*]

W. F. Oliver and J. F. Scott

Dept. of Physics
Univ. Colorado
Boulder, CO 80309-0390

Scott Lee and Stuart Lindsay

Dept. of Physics
Arizona State Univ.
Tempe, AZ 85287

ABSTRACT

A dynamic central mode near T_I=573K in $Ba_2NaNb_5O_{15}$ is studied via a nine-pass interferometer. It has width of order 20GHz that is nearly T-independent in the IC phase and intensity which increases as $T \rightarrow T_I^-$. It couples strongly with the longitudinal acoustic (LA) phonons. An unrelated central mode is observed near T_c=836K. It is broader (~60GHz) and couples to neither TA or LA phonons. Within ±15K of T_c an unexpected phase-matched two-wave mixing is observed, attributed to a charged defect modulation along the polar axis; this process has a time delay of 6 sec at input powers of order 100mW of blue light.

I. INTRODUCTION

Barium sodium niobate ($Ba_2NaNb_5O_{15}$) exhibits[1] two truly incommensurate (IC) phases - a lower one from ~20K to 105K, and an upper one from ~543K to ~573K - plus a defect-stabilized nearly commensurate phase from 105-543K. In this paper we present some laser spectroscopic studies near T_I=573K and T_c=836K.

----

* Work supported in part by NSF grant DMR86-06666.

Section II below describes dynamic central mode scattering studies near T_I. Interest exists in the temperature and wave vector dependences of width, intensity, and coupling with the LA phonons.

Section III describes analogous studies on an unrelated central mode near the Curie temperature T_C. Barium sodium niobate is an unusual incommensurate crystal in that its incommensurate phases are simultaneously ferroelectric (FE); hence T_C lies above T_I. In most incommensurate insulators the IC phase is a phase of relatively narrow temperature width lying just above the FE phase.

Section IV presents some preliminary data on photorefractive effects in the nearly commensurate phase (ambient temperatures) and at high temperatures very near T_C.

II. PHASON SCATTERING NEAR $T_I \overset{\sim}{=} 573K$

As T_I^- is approached from below, a dynamic central mode appears in the Brillouin spectra of $Ba_2NaNb_5O_{15}$. The spectra shown in Figs. 1,2 were obtained with the nine-pass Fabry-Perot interferometer[2] at ASU. The intensity of the central mode grows as T_I^- is approached. The width, however, must be extracted by careful analysis of the data, due to strong coupling with the LA phonon at ~40GHz. The coupling produces an asymmetric broadening of the LA phonon response. In Figs. 1 and 2 the smooth curve is the calculated spectra from a standard coupled mode analysis[3] which treats the LA phonon as a simply damped harmonic oscillator. The fitting parameters for several temperatures and scattering angles are given in Table I. Qualitatively, the effect is to add intensity on the low-frequency side of the LA phonon and subtract it from the high-frequency side.

The parameters obtained from these fits are fairly unambiguous and involve seven fitting parameters: a background level, uncoupled frequencies W_a and $1/\tau_c$ for the LA phonon and central mode, uncoupled scattering amplitudes P_a and P_c, and LA width Γ_a and coupling term $W_a\Gamma_{ac}$ (taken as real).

It is appropriate to comment upon the temperatures involved here. Barium sodium niobate has a nearly commensurate structure at room temperature, thought to be defect-stabilized, with critical wave vector q_0 within 0.01 of commensurate. At $T_{lock}=543K$ it becomes truly incommensurate and at T=573K there is a minimum in $C_{22}(T)$, usually taken

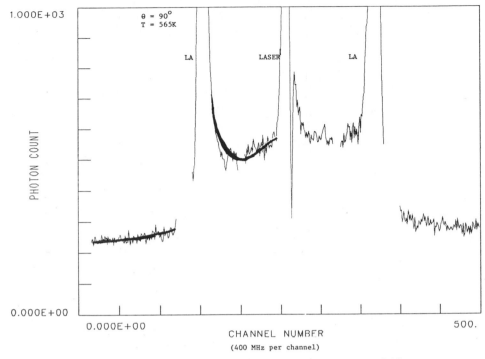

Fig. 1. Central mode scattering spectra at T=565K. Solid curve
is the fit to the equation and parameters given in Table
I. 90° scattering.

as T_I, the normal-incommensurate transition. However, some evidence for
residual IC structure persists until $T \approx T_I$ + 9K, at which point $C_{11} = C_{22}$
and the structure is tetragonal and fully commensurate. T_c is much
higher, at ~836K in our specimen. The data we present in Table I are all
taken on heating (there are hysteresis effects) and correspond to two just
below T_{lock}, one in the middle of the IC phase, and one 2.5K above T_I but
below the point where $C_{11} = C_{22}$.

More extensive data will be presented elsewhere, but the data in Table
I are sufficient to show the following: the central mode width varies
only slightly within the IC phase from T_{lock}=543K to T_I=573K $T \rightarrow T_I^-$ (a
finite width is observed exactly at T_I, which may imply a gap in the
phason dispersion curve or be due to phason-phonon interactions); the
phason does not borrow all its intensity from the LA phonon (a fit to the
spectra _can_ be made assuming zero intrinsic phason scattering intensity,[4]
but it is slightly worse at each of our measured temperatures); there is a
q-dependence to the phason width τ^{-1}(q), and it is very nearly as q^2,
suggesting the existence of a rather small gap; however, the overall
phason cross-section P_c^2 increases with scattering angle in very good

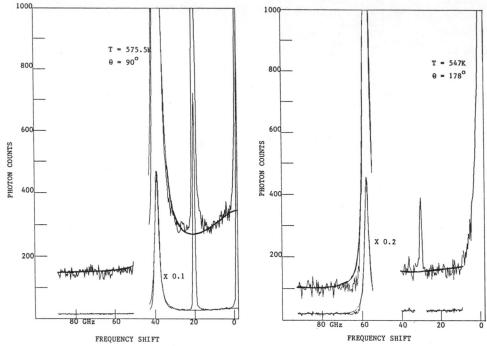

Fig. 2a. As in Fig. 1 at T=546K.   Fig. 2b. T=546K; $\Theta = 178^{\circ}$.

agreement with the theoretical prediction[5] of Levanyuk that

$$I_{phason}(q) = Aq^2/(E_g + \hbar^2 q^2).\qquad (1.)$$

This equation predicts q-independent intensities for systems with gapless phasons and $I \sim q^2$ for systems with large gaps. Our data show $I(178^{\circ})/I(90^{\circ}) \approx 2$ in agreement with the large gap prediction of Eq.(1).

It is useful to compare this dynamic central mode width with the value predicted from the diffusion constant $D \sim 1.3 \pm 0.3 cm^2/s$ indirectly inferred earlier from the changes in sound velocity and Brillouin linewidths <u>below</u> $T_I = 573K$. Our measured widths agree numerically with this earlier prediction[6] (there is an independent one[7] based upon soft mode linewidths and relaxation effects in elastic coefficients just <u>above</u> T_I but the physical mechanisms may be unrelated). Both values are at least compatible with our theoretical model prediction[9] that kink diffusion is faster than thermal diffusion of entropy fluctuations very near T_I.

266

Table I. Dynamic Central Model Parameters near T_I=573K in $Ba_2NaNb_5O_{15}$

$$S(w,T) = \frac{-kT}{\pi\hbar w} \sum_{ij=1}^{2} P_i P_j \, \text{Im}\left\{G_{ij}(w,T)\right\}$$

$$\text{with } G_{ij}^{-1} = \begin{bmatrix} w_a^2 - w^2 + iw\Gamma_a & -w_a\Gamma_{ac} \\ -w_a\Gamma_{ac} & w_a^2(1+iw\tau_c) \end{bmatrix}$$

	T = 527.7K $\theta = 90°$	T = 538.6K $\theta = 90°$	T = 546.0K $\theta = 90°$	T = 546.0K $\theta = 178°$	T = 565K $\theta = 90°$	T = 575.5K $\theta = 90°$
P_a^2	2.1 units	2.9	3.2	1.5	2.3	2.9
W_a	41.7 GHz*	41.8	41.6	60.4	42.0	41.8
P_c^2	.20 units	.36	.44	.81	.41	.49
P_c^2/P_a^2	.096	.125	.134	.54	.177	.170
τ_c^{-1} (GHz)	8.9	14.4	18.3	44.4	19.1	20
$-\Gamma_{ac}$ (GHz)	+3.22	5.63	6.03	6.6	6.11	7.44
Coupling Term $-\sqrt{w_a\Gamma_{ac}}$ (GHz)	11.6	14.4	15.8	20.0	16.0	17.6

* Note that the value of w_a (uncoupled) is nearly independent of temperature, showing that the LA renormalization at T_I is entirely due to relaxation. Γ_a =1.77 GHz is also independent of T.

III. FERROELECTRIC CENTRAL MODE SCATTERING NEAR T_c=836K

Fig. 3 shows dynamic central mode scattering in the FE commensurate phase between 573 and 856K. The width is generally greater than that of the "phason" scattering in Figs. 1,2. Note also that there is no coupling with either the LA or TA phonons. [In addition to this central mode there is a much broader one that arises from the relaxing self energy of the lowest-frequency TO phonon, as can be shown from detailed fitting (not shown) of the asymmetric TO phonon lineshape.] Our widths agree only semi-quantitatively with earlier studies of this FE central mode, obtained with a grating spectrometer with a claimed resolution of 12 GHz.[8]

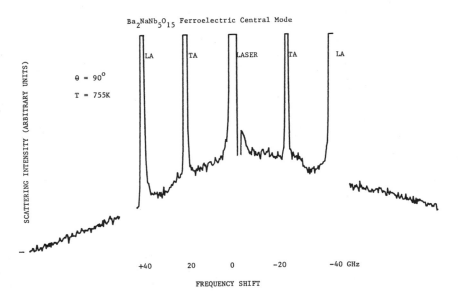

Fig. 3. Central mode scattering in the commensurate ferroelectric phase at 755K.

IV. PHOTOREFRACTIVE EFFECTS

A. In the Nearly-Commensurate Phase

We have found[10] that four-wave mixing in $Ba_2NaNb_5O_{15}$ at 293K converts up to 25% of an input signal of an Ar laser at ~100mW into a second similar beam at a crossing angle of ~$30°$. The sign of the mobile carriers responsible for this photorefractive effect is inferred to be the same as that of the electrooptic constants r_{13}, r_{23}, r_{33}, if the model of Kuchtarev[11] is used.

We have attempted to measure these effects as the sample is heated through the transition temperatures 543 and 573K to see if the incommensurate modulation affects the diffusivity of the charged defects responsbile for photorefraction. Unfortunately, convection currents precluded quantitative results. We would like to see if only electrons (or holes) are mobile in this material (as is assumed[11,12] in the donor-acceptor models; see the paper by Hellwarth in this volume), or if charged ions are dragged along by the mobile antiphase boundaries (APBs) in the IC phase, which decorate the standing wave produced by the crossed Ar laser beams.

B. Near T_c

Although our results to date are inconclusive, we do observe an unexpected high-temperature photorefractive effect at temperatures within ±15K of T_c, which demonstrates the role of such decoration at elevated temperatures. When a laser beam is incident upon $Ba_2NaNb_5O_{15}$ at T_c with polarization along the polar axis of the crystal, there is a time delay of ~6s after which a phase-matched cone of light is emitted at an initial angle of order 10°; this angle continuously decreases to ~4° as exposure continues. The scattered light arises, we believe, from charged defects driven along the polar axis of the crystal by the incident beam; they require ~6s at input powers of ~130mW to diffuse far enough to form a grating such that phase matched two-wave mixing can occur. Rather surprisingly, the scattered light is polarized in the same (polar) direction as the incident beam (if $E_{inc}^1 \perp$ polar axis, no effect is observed); hence, birefringence plays no obvious role in the phase matching. Such a phase matched photorefractive effect involving charged defects at ~850K is surprising; because at such elevated temperatures charge relaxation is expected to be very fast (Odoulov,[13] this volume and private communication).

V. SUMMARY

Quantitative measurements of dynamic central modes have been made at T_I and T_c in $Ba_2NaNb_5O_{15}$. The width at T_I is in good agreement with that predicted earlier from phonon velocities and linewidths. It implies diffusivities of APBs (known from electron microscopy to be highly mobile near T_I in this material) of order $1cm^2/s$, which is faster than thermal diffusion of entropy fluctuations, in accord with a model calculation. The role of charged defects at high temperatures is shown in two-wave photorefractive mixing experiments at T_c=846K. But the connection between APB diffusion and diffusion of the charged ions that may pin these APBs[14] remains to be demonstrated. In this context it is interesting to note the recent demonstration[15] that 2MeV irradiation will stabilize the IC phase of $Ba_2NaNb_5O_{15}$ at room temperature.

An important question to address is the physical origin of the dynamic central mode(s) reported here.

The central mode shown in Figs. 1,2 is probably related to incommensurate dynamics such as diffusion of anti-phase boundaries. In $Ba_2NaNb_5O_{15}$ these APBs have been observed directly from 293K to 573K by electron microscopy[16] and "as the temperature is raised, the discommensurations become very mobile and their number quickly increases."

In agreement with this, the ratio of the intensity of the central mode I_{cm} measured arbitrarily at 15GHz to that of the background I_b varies from 0.5:1 just below T_{lock} = 543K to a maximum of 1.4 at T = 561K (in the middle of the IC phase) and drops rapidly to 0.5 again by T=594K and to zero by 614K.

This central mode is different from that which characterizes the relaxation in C_{11} above T_I (Ref. 7). It may be the same as that inferred indirectly from a Landau-Khalatnikov analysis of C_{11} below T_I (Ref. 6); hypothesized as due to kink diffusion; a term Dq^2 inferred before[6] with D=$(1.3 \pm 0.3)cm^2$/s and $\tilde{q}$= $2\pi\sqrt{2}$ n w_c with n=2.26 and λ_c =514.5nm yields a predicted width of

$$(2\pi\tau)^{-1} = (112 \pm 25)/2\pi \text{ GHz}$$
$$= 18 \pm 4 \text{ GHz}$$

in excellent agreement with the observed value of 20 GHz in Table I.

The weak divergence in intensity and the temperature independent width of 20-30 GHz are similar to both the dynamic central modes in incommensurate $BaMnF_4$[17] and to those[18] in $Pb_5Ge_3O_{11}$ (which is not incommensurate) near T_c.

REFERENCES

1. G. Errandonea, M. Hebbache, and F. Bonnouvrier, Brillouin Scattering

Study of the Elastic Properties of Incommensurate Barium Sodium Niobate, <u>Phys. Rev.</u> B32:1691 (1985); J. Schneck, Contribution a l'etude des Transitions de Phases dans le Niobate de Baryum et de Sodium, These de Doctorat d'etat es Sciences Physiques, Universite de Paris (1982); J. C. Toledano, Present State of the Studies of Incommensurate Phases in Insulators, Centre National d'Etudes des Telecommunication (Bagneux, France), Report #576 (unpublished).

2. S. M. Lindsay, S. Burgess, and I. W. Shepherd, Correction of Brillouin Linewidths Measured by Multipass Fabry-Perot Spectroscopy, <u>Applied Optics</u>, 16:1404 (1977); S. M. Lindsay, M. W. Anderson, and J. R. Sandercock, Construction and Alignment of a High Performance Multipass Vernier Tandem Fabry-Perot Interferometer, <u>Rev. Sci. Instrum.</u> 52:1478 (1981).

3. R. S. Katiyar, J. F. Ryan, and J. F. Scott, Proton-Phonon Coupling in CsH_2AsO_4 and KH_2AsO_4, <u>Phys. Rev.</u> B4:2635 (1971); P. A. Fleury and P. D. Lazay, Acoustic-Soft-Optic Mode Interactions in Ferroelectric $BaTiO_3$, <u>Phys. Rev. Lett.</u> 26:1331 (1971); I. J. Fritz, R. L. Reese, E. M. Brody, C. M. Wilson and H. Z. Cummins, Light Scattering Studies of the Soft Optic and Acoustic Modes of KH_2PO_4 and KD_2PO_4, <u>in</u>: "Light Scattering in Solids," M. Balkanski, ed., Flammarion Sciences, Paris (1971), p.415.

4. V. A. Golovko and A. P. Levanyuk, Light Scattering and the Dispersion of Susceptibilities in an Incommensurate Phase, <u>Zh. Eksp. Teor. Fiz.</u> 31:2296 (1981) [<u>Sov. Phys. - JETP</u> 54:1217B (1981)].

5. N. I. Lebedev, A. P. Levanyuk, and A. S. Sigov, Is Phason Visible in Optics? (this volume).

6. P. W. Young and J. F. Scott, Brillouin Spectroscopy of Acoustic Phonon Dispersion in $Ba_2NaNb_5O_{15}$ at its Incommmensurate Transition, <u>Ferroelectrics</u> 52:35 (1983); P. W. Young and J. F. Scott, Brillouin Spectroscopy of the Incommensurate - Commensurate Transition in Barium Sodium Niobate, <u>Phase Transitions</u> 6:175 (1986).

7. G. Errandonea, M. Hebbache, and F. Bonnouvrier, Brillouin Scattering Study of the Elastic Properties of Incommensurate Barium Sodium Niobate, <u>Phys. Rev.</u> B32:169 (1985).

8. G. Errandonea, H. Savory, and J. Schneck, Critical Narrowing of a Central Peak at the Ferroelectric Transition of $Ba_2NaNb_5O_{15}$, <u>Ferroelectrics</u> 55:19 (1984).

9. G. N. Hassold, P. D. Beale, J. F. Dreitlein, and J. F. Scott, Dynamics of the Two-Dimensional Axial Third-Nearest-Neighbor Ising Model, <u>Phys. Rev.</u> B33:3581 (1986).

10. J. F. Scott, Thermal Memory and Diffusion in Incommensurate $Ba_2NaNb_5O_{15}$, *Bull. Am. Phys. Soc.* 32:706 (1987).

11. N. V. Kukhtarev, V. B. Markov, S. G. Odoulov, M. S. Soskin and V. L. Vinetskii, Holographic Storage in Electrooptic Crystals, *Ferroelectrics* 22:949 (1979).

12. R. W. Hellwarth, Theory and Observation of Electron-Hole Competition in the Photorefractive Effect, (this volume).

13. S. G. Odoulov and M. Soskin, Coherent Oscillation due to Vectorial Four-Wave Mixing in Photorefractive Crystals, (this volume).

14. D. J. Srolovitz and J. F. Scott, Clock-Model Description of Incommensurate Ferroelectric Films and of Nematic Liquid Crystal Films, *Phys. Rev.* B34:1815 (1986).

15. S. Barre et al. (to be published in *Phase Transitions* 9:225 1987 - title unavailable): C. Manolikas, J. Schneck, J. C. Toledano, J. M. Kiat, and G. Calvarin, Transmission Electron Microscopy Observation of the Memory Effect through the Pattern of Discommensurations in Barium Sodium Niobate, *Phys. Rev.* B35:8884 (1987).

16. G. Van Tendeloo, J. Landuyt, and S. Amelinckx, Electron Microscopy of Incommensurate Structures, in: "Incommensurate Crystals, Liquid Crystals, and Quasi-crystals," J. F. Scott and N. A. Clark, ed., Plenum, New York (1987), p. 71.

17. K. B. Lyons, R. N. Bhatt, T. J. Negran, and H. J. Guggenheim, Incommensurate Structural Phase Transition in $BaMnF_4$: Light Scattering from Phasons, *Phys. Rev.* B25:1791 (1982); K. B. Lyons, T. J. Negran, and H. J. Guggenheim, Anisotropic Order Parameter Phase Fluctuations: $BaMnF_4$, *J. Phys.* C13:L415 (1980).

18. K. B. Lyons and P. A. Fleury, Lgiht-Scattering Investigation of the Ferroelectric Transition in Lead Germanate, *Phys. Rev.* B17:2403 (1978).

ARE PHASONS OBSERVABLE IN OPTICS?

N. I. Lebedev, A. P. Levanyuk and A. S. Sigov

Institute of Crystallography,
USSR Academy of Sciences, Moscow, USSR

ABSTRACT

It is shown that the existence of the phason gap caused by point defects dramatically affects the integral intensity of light scattering from phasons. From the estimates, the possibility of observation of such scattering seems to be highly problematic even for a nominally pure crystal that contains defects of concentration $N \sim 10^{18}$ cm^{-3}.

1. Introduction

In many cases the incommensurate phase structure[1] may be represented as a long-wave periodic inhomogeneity of one or several parameters (e.g. polarization) "built into" a crystal, the period of such an inhomogeneity being, generally speaking, incommensurable with the period of the basic structure.

Any homogeneous shift of this frozen-in wave evidently can't change the energy of the system, i.e. the stiffness with respect to such shifts is equal to zero. Consequently there exist in an incommensurate phase one or several (according to the number of independent directions of modulation) branches of excitation with the frequencies reaching zero at $\vec{k}$ = 0, i.e. the gapless, "acoustic-like" excitations. They are usually called phasons because the homogeneous shift of the modulation wave is equivalent to a change in its phase ϕ.

It is essential that the phason damping coefficient does not tend to zero at k = 0, i.e. the long-wave phason is always overdamped unlike the well-known acoustic phonon. Consequently, the phason line in the light scattering spectrum is a central peak with a width proportional to q^2 ($\vec{q}$ is the scattering wavevector), the width of the order $\omega_{at} q^2 / K_{at}^2 \sim 10^6 \mathrm{s}^{-1}$ being fairly small ($\omega_{at} \sim 10^{13} \mathrm{s}^{-1}$ is the characteristic atomic frequency, and K_{at} is a wavevector of the order of the reciprocal lattice vector).[2]

For the total (integrated over all frequencies) fluctuation in the phase one has $\phi_q^2 \propto q^{-2}$. But the integral intensity of the phason line depends on q weakly as the phason activity in light scattering is determined in fact analogously to the acoustic phonon activity:

$$\Delta \varepsilon_{ik} \propto \frac{\partial \phi}{\partial x}, \qquad\qquad (1)$$

where ε_{ik} is the dielectric constant at the light wave frequency. Experimental attempts to observe the phason peak have proved a failure for the present (see[1], Ch. 7). One may consider the strong changes in the phason characteristics under the influence of defects to be one of the reasons for the failure.

Indeed, in the presence of defects the phason appears not to be a gapless excitation[3-9], and one can express the integral intensity of light scattering from the phason in the form

$$I \propto q^2/(g + q^2), \qquad\qquad (2)$$

where g is the gap for the phason. For temperatures far from the transition to the incommensurate phase $(T = T_i)$ and for defects, determined by the "forces" of the atomic order of magnitude, the value of g may be estimated as

$$g \sim K_{at}^2 \, N/N_{at}.$$

For nominally pure crystals $N/N_{at} \sim 10^{-3}$ to 10^{-4}, thus $g \gg q^2$ i.e. the integral intensity of the phason light scattering in a real crystal is less by several orders of magnitude than its value for an ideal crystal. Despite its preliminary character this estimate shows that even for a nominally pure crystal the phason gap may be substantial and sufficient for making the scattering from the phason practically unobservable. Naturally a special consideration is required if one needs more precise estimates and the dependence of the gap on temperature and other parameters. The necessary treatment has been performed[10] and the following report is devoted to summarizing the results obtained mainly in paper.[10] In the next section of the report we touch upon theoretical estimations for the phason gap.

In section 3 one finds the formulation of the general approach to the problem of a gap in the case of so-called strong pinning, the approach being true for any degenerate system and the physical meaning of the constants in the effective Hamiltonian being not concretized. In section 4 we elucidate the temperature dependences and orders of magnitudes of the constants for the incommensurate phase of the type specific for dielectrics. In section 5 the intensities of light scattering from the phason in ideal and in real crystals are compared.

2.   Short Summary of Theoretical Results

The treatment of the phason gap in a crystal with defects is an extremely complicated theoretical problem so it is not surprising that the results of various authors frequently disagree with each other and contain inherent contradictions.

In thoery there are distinguished two regimes of the interaction of defects with the phase ϕ of the order parameter: the regimes of strong and weak pinning. For the first regime the value of the phase in the point of localization of the defect is defined only by characteristics of the given defect. For the second regime the local value of ϕ is defined by the whole defect ensemble rather than by the given defect. The results of the present report refer to the case of strong pinning.

As to weak pinning it seems to us that reliable results are absent for that case. It was shown by Efetov and Larkin[11] that the static stiffness for the phason tends to zero as $\vec{k} \to 0$ for crystals with defects as well, this conclusion being obtained as the result of consideration of the whole series of the perturbation theory. This conclusion seems unacceptable from the physical point of view since only the case $T = 0$ has been considered (strictly speaking the phason gap is always absent at $T \neq 0$). The nontrivial reason for such a discrepancy was elucidated by Fisher[12]: the perturbation theory methods appeared to be inadequate the formulated problem even with inclusion of all terms of the perturbation series. The fact is that when averaging the expressions of the perturbation theory over the defect positions one takes into account both metastable and unstable states.

This difficulty may be avoided only in the case of strong pinning, when fixing the phase values at the defects (recall that the local phases are defined only by the corresponding defect) allows to discriminate the metastable state. Analogous considerations were used in paper,[8] however the paer[8] can't be considered statisfactory either by the final result or by the method used. The contribution of defects was assumed to be additive, but that is absolutely groundless for incommensurate structures: the phase perturbation decreases as $1/r$ when moving away from the isolated defect. In the expression for the phason gap obtained in paper[8] a certain characteristic length was assumed to be of the order of magnitude of the lattice constant. At the same time we shall show below (see also[10]) that this length considerably exceeds the lattice constant and depends strongly on temperature.

A consistent treatment also leads to a formulation of the condition of strong pinning. For the three-dimensional system this condition has been adduced without deduction in paper,[5] while the deduction given in paper[8] yields another condition.

3. The phason gap in a degenerate system

In the long-wave approximation the effective Hamiltonian of a degenerate system with defects has the form (cf., e.g.,[5,11])

$$H = \int d\vec{r} \; \{ \frac{C}{2} (\nabla\phi)^2 - \sum_i V_1 \delta(\vec{r}-\vec{r}_i)\cos(\phi_i-\psi_i)-h_{ext}\phi \}. \qquad (3)$$

The first term in this equation is referred to the ideal system, V_1 and ψ_i are the amplitude and phase of the "field" of the i-th defect. Eq. (3) is valid not only for all types of degenerate systems, but also for essentially any types of defects in these systems.

In the limiting case of infinitely strong pinning ($V_1 \to \infty$) the values of ϕ at the defects coincide with the phases of a random field:

$$\phi_i = \psi_i + 2\pi n_i. \qquad (4)$$

Various sets of n_i correspond just to various metastable states. The transitions between the states may be realized in two ways: by jumpwise change of the phase at the defect (at $T \neq 0$) and by displacement of the topological defects, specifically dislocations[13] (even at $T = 0$), see.[3] We shall neglect these processes, i.e. we shall find virtually the susceptibility for high enough frequencies. Unfortunately, we failed to estimate the relevant characteristic frequencies.

Having fixed the set of n_i, i.e. having chosen a certain metastable state one can use perturbation theory to calculate the generalized susceptibility describing the change of ϕ under application of the field h_{ext}. One finds[10]

$$\chi(q) = (g + q^2)^{-1}, \qquad (5)$$

where $g = 4\pi C r_0 N$, N is the concentration of defects, and r_0 is the range of the "smearing" of the δ-function that one has to introduce when performing the calculations. To reveal the physical meaning of r_0 one has to abandon the long-wave approximation at the expense of the loss of universality of the treatment.

Eq. (5) continues to be valid also for finite albeit large V_I. The corrections to $\chi(\vec{q})$ connected to the finiteness of V_I may also be obtained with the use of perturbation theory. The condition of smallness of the corrections is the condition of strong pinning:

$$\frac{V_I}{4\pi C r_0} \gg 1. \qquad (6)$$

4. <u>Incommensurate phase in crystals of the potassium selenate type</u>

Let us estimate the parameters C, V_I, r_0 of the long-wave approximation with the use of the model that describes the numerous family of incommensurate dielectrics of the potassium selenate type. The incommensurate phase may be represented here as a spatial inhomogeneity of a certain commensurate phase. In the simplest case the effective Hamiltonian as a function of the order parameter of the implied normal-commensurate transition has the form:

$$H = \int d\vec{r} \left[\frac{A}{2} \rho^2 + \frac{B}{4} \rho^4 + B'\rho^6\cos6\tilde{\phi} + r\rho^2 \frac{\partial\tilde{\phi}}{\partial x} + \frac{D}{2} (\nabla\rho)^2 + \frac{D}{2} \rho^2(\nabla\tilde{\phi})^2 \right], \qquad (7)$$

where $\eta_I = \rho\cos\tilde{\phi}$, $\eta_2 = \rho\sin\tilde{\phi}$. The presence of the Lifshitz invariant $r\rho^2\frac{\partial\tilde{\phi}}{\partial x}$ shows that the second-order normal-commensurate phase transition is impossible and with decrease of the coefficient A formation of the incommensurate phase takes place. The term with coefficient B' is relevant to the incommensurate-commensurate phase transition. We shall focus on the temperature region close to the normal-incommensurate phase transition and thus we may neglect this term due to smallness of ρ in the region.

Let us assume that the defects create the local field with respect to one of the order parameter components, say η_I. Then

$$\Delta H = \sum_i h_1\eta_1(\vec{r}_i) = - \int d\vec{r} \sum_i h_1\delta(\vec{r}-\vec{r}_i)\rho\cos\tilde{\phi}. \qquad (8)$$

The variation of the Hamiltonian (7), (8) leads to two Euler equations for ρ and $\tilde{\phi}$. It is essential that the perturbation of ρ due to a defect falls off exponentially at large distances (because the "stiffness" A is finite) and the variable $\tilde{\phi}$ is related to the phase ϕ of the preceeding section:

$$\phi = \tilde{\phi} + K_0 x, \quad K_0 = r/D. \qquad (9)$$

In addition, for the phases of the random field (Eq. (3)), one obtains

$$\psi_i = K_0 x_i. \tag{10}$$

To determine the parameters C, V_I, r_0 one may consider an isolated defect. The consideration leads to the relations:

$$C = D\rho^2_\infty, \tag{11}$$

$$r_0 = h_I/4\pi D\rho_\infty, \tag{12}$$

$$V_I = h^2_I/4\pi Dd. \tag{13}$$

Here ρ_∞ is the amplitude of the incommensurate modulation in the ideal crystal, $\rho^2_\infty = (-A+r^2/D)/B \propto |T-T_i|$, and d has the order of magnitude of the lattice constant. Using Eqs. (11-13) and (6) one sees that the condition of strong pinning given by Eq. (6) coincides with that of ref.[4]. It means, in particular, that the "random-field" type defects always produce strong pinning in the vicinity of T_i.

One can treat analogously defects of the "random anisotropy" type as well. Here

$$\Delta H = \sum_i A_I n_I^2(\vec{r}_i) = \frac{1}{2} \int d\vec{r} \sum_i A_I \rho^2 \delta(\vec{r}-\vec{r}_i) \ (1+\cos 2\tilde{\phi}). \tag{14}$$

It is found that the strong pinning regime is of extremely small probability in this case.

5. Light Scattering from Phasons

It is natural to expect that not too close to T_i the frequencies of transitions between the metastable states are very low and the observable integral intensity of the light scattering from phasons is determined by the susceptibility given by Eq. (5). In the simplest case the coupling of the dielectric constant at the frequencies of light and the fluctuations of the phase has the form:[2]

$$\varepsilon = \varepsilon_0 + a\rho^2 \frac{\partial\phi}{\partial x}, \tag{15}$$

whence it follows that

$$\langle|\varepsilon(\vec{k})|^2\rangle = a^2\rho^4_\infty k^2 T \ (Nh_1\rho_\infty + D\rho^2_\infty k^2)^{-1}. \tag{16}$$

Eq. (16) differs from the analogous expression for the light scattering intensity in an ideal crystal by the presence of the first term in the denominator. As a result the integral intensity proves to be δ times smaller than in the ideal crystal, δ being given by:

$$\delta \approx \frac{Nh_I}{D\rho_\infty K^2} = \frac{4\pi Nd}{K^2} \ (\frac{\rho_{at}}{\rho_\infty}) \ (\frac{h_I}{h_{at}}), \tag{17}$$

where $h_{Iat} = 4\pi D^{3/2} B^{-1/2}$ and $\rho_{at} = D^{1/2}d \ B^{-1/2}$ are "atomic" (maximum possible) values of h_I and ρ. At $N = 10^{18}$ cm^{-3} (a so-called "undoped" crystal) and $h_I/h_{Iat} = 0.1$, $\rho_\infty/\rho_{at} = 10^{-2}$, $d = 10^{-8}$ to 10^{-7} cm, $K = 10^{-5}$ cm^{-1}, and one obtains $\delta = 10^2$ to 10^3 and, correspondingly, $\delta \sim 1$ at $N = 10^{15}$–10^{16} cm^{-3}. As far as we know incommensurate crystals of this level of perfection are vitually unavailable now.

Let us emphasize that defects are not expected to suppress the neutron scattering from phasons so dramatically because the values of k are much greater for the scattering. In the case of strong pinning the phason should be detected as a vibration with the gap given by Eq. (5), the value of the gap increasing as temperature T_i is approached.

References

1. Incommensurate Phases in Dielectrics - Fundamentals, ed. by R. Blinc and A. P. Levanyuk, (North-Holland, Amsterdam (1986)).
2. V. A. Golovko and A. P. Levanyuk, Zh. Eksp. Teor. Fiz. 81:2296 (1981) (Sov. Phys. JETP.)
3. L. P. Gor'kov, Pis'ma Zh. Eksp. Teor. Fiz. 25:384 (1977) (Sov. Phys. JETP Lett.)
4. H. Fukuyama and P. A. Lee, Phys. Rev. B 17:535 (1978).
5. P. A. Lee and T. M. Rice, Phys. Rev. B 19:3070 (1979).
6. M. V. Feigelman, Zh. Eksp. Teor. Fiz. 79:1095 (1980).
7. G. Grüner, Phys. Rep. 119:117 (1985).
8. S. Abe, J. Phys. Soc. Japan 54:3494 (1985).
9. K. Maki, Phys. Rev. B 33:2852 (1986).
10. N. I. Lebedev, A. P. Kevanyuk, and A. S. Sigov, Zh. Eksp. Teor. Fiz. 92:248 (1987).
11. K. B. Efetov and A. I. Larkin, Zh. Eksp. Teor. Fiz. 72:2350 (1977)
12. D. S. Fisher, Phys. Rev. B 31:7233 (1985).
13. G. E. Volovik and V. N. Mineev, Zh. Eksp. Teor. Fiz. 72:2256 (1977).

LASER SPECTROSCOPY OF FAST KNO$_3$ FERROELECTRIC SWITCHES[*]

J. F. Scott, B. Pouligny,[+] and Zhang Ming-sheng[⊥]

Dept. of Physics
University of Colorado
Boulder, CO 80309-0390

ABSTRACT

Raman spectroscopy of sub-micron ferroelectric films of potassium nitrate reveal qualitative changes from the bulk, which are interpreted in terms of surface electric field effects. Analogous effects have been reported by Farhi and Moch in thiourea. Possible relation to space charge effects is discussed.

INTRODUCTION

The motivation for spectroscopy of ferroelectric thin films is twofold: First, recent theories of finite size effects[1,2] (see especially the very recent review by Binder[3]) suggest that such phenomena are observable for films of thickness of order 100 nm, easily obtained by thermal evaporation or sputtering; and second, ferroelectric nonvolatile memories[4-9] provide a device application of sufficient commercial impact to fuel considerable research on the topic.

In the past the experimental observation of true size effects was usually obscured by poor stoichiometry of thin sputtered films.[10] For example, because BaO is more volatile than rutile, the deposition of barium titanate via BaO + TiO$_2$ → BaTiO$_3$ usually produces Ba$_x$Ti$_y$O$_3$ with x/y

----

* Work supported in part by ARO contract number DAAL03-86-K-0053.
+ Permanent address: CEA/CESTA, Le Barp, France; work at CU supported by French Ministry for Defense
⊥ Permanent address: Inst. Materials Analysis, Univ. Nanjing

$\neq 1$, even if the starting target material is "off-loaded" to compensate the resulting film for the different volatilities. This produces a film of mixed composition, including nominally "perovskite" regions, pyroclore regions, and amorphous regions. There is no hope in measuring subtle shifts in Curie temperature with film thickness, $T_c(d)$, in such films to test finite size effect theories.

By comparison, 70nm thick films of ferroelectric KNO_3 can be evaporated with stoichiometry exceeding 98%. That is because KNO_3 evaporates without dissociating. Finite size effects also seem observable in TGS (triglycine sulfate) films[11,12] and in small particles of lead germanate.[13]

THEORY

Following work by Mills,[14] Lubensky and Rubin[1] and subsequently Tilley and Zeks[2] have developed a theory of ferroelectric thin films which shows that the polarization P_s near the surface may in general be greater than or less than that of the interior. In the former case, $T_c(d)$ increases as the film is made thinner. Their theories ignore depolarization effects, which have however been treated by Binder.[15] The theories are characterized by a single parameter δ, which is an extrapolation length; large negative δ implies a surface layer that is slightly super-polarized compared to the bulk, and small negative δ (i.e., of order the correlation length ξ) implies an abrupt surface polarization (leading to a surface phase transition[16] above T_c bulk). The parameter δ is phenomenological, but in a microscopic theory it would depend upon the work function of the metal electrodes used to measure P_s and on the Fermi level in the ferroelectric. There is an analogy[17] between this situation and that in superconducting thin films, where de Gennes showed the connection to microscopic parameters.

These theories all suggest that polarization $P_s(z)$ is a function of distance z in from the surface in a film, as shown in Fig.1. In this figure δ is taken as large and negative and depolarization effects are neglected. Such models predict that the phonon Raman spectra of ferroelectric films will differ from those of the bulk, because such spectra have LO/TO splittings and intensities that depend upon[18-20] internal electric fields E_i (recall $E_i = 4\pi P_s$). In the section below we examine such spectra for a 0.26 μm KNO_3 film.

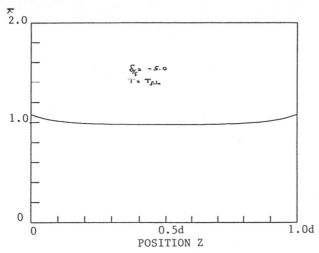

Fig. 1. $P_s(z)$ vs. z, the distance from the film surface, for large negative δ, from Ref. 16. $T=T_c$ (exactly at the first-order transitions).

RAMAN SPECTRA

Fig. 2 compares Γ_3 spectra near $100 cm^{-1}$ for ferroelectric KNO_3 bulk[21] and 0.26μm films.[22] The film spectra resolve the two TO/LO pairs and show an overall shift from $\sim 120 cm^{-1}$ to an average near $105\ cm^{-1}$. We interpret this as due to effective surface fields in the film. Note, however, that the spectra do not agree with what would be predicted from Fig. 1, in that sharp lines corresponding to a precise value of E_i - not a distribution $E_i(z)$ - are observed. This is rather similar to the data of Farhi and Moch[23] on thiourea, where they found increased resolution of TO/LO pairs and strong intensity changes after $10^5 V/m$ electric fields were applied. These changes <u>remained</u> after the applied voltages were turned off! They interpreted their results as due to large fields at domain walls that were frozen in due to charged defects. It seems likely to us that they arise from very specific domain walls, those in space charge regions near the electrodes.[24] In any event, the film spectra of our submicron KNO_3 films are qualitatively different from the bulk.

Fig. 3 shows the empirical dependence of upper and lower transition temperatures in ferroelectric KNO_3 upon film thickness d. T is plotted versus 1/d because T_c is predicted to be linear[1] in 1/d in some theories and regimes. Unfortunately the KNO_3 T_c date also depend upon processing details. The upper (solid) curve with $T_c(d)$ reaching $\sim 190^o C$ is for thermally evaporated films with flash evaporated gold electrodes; the

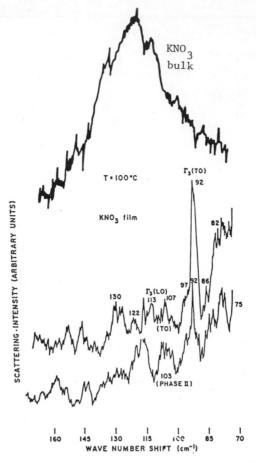

Fig. 2. Raman spectra of ferroelectric KNO_3,
0.26μm film and bulk.

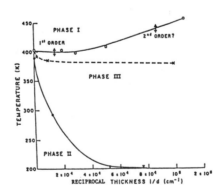

Fig. 3. Phase diagram for KNO_3: T vs. 1/d.

lower (dashed) curve with T_c(d) ~120°C is from General Motors [25] for Al electrodes and slip-cast films. Such data are not highly reproducible, depending probably on other processing details (e.g., substrate temperature?) as well.

Note that there appears to be a finite thickness in KNO_3 at which the FE/PE transition goes from first-order (bulk) to second-order (thin-films). Such a tricritical point does not occur in our model:[16] the transition remains first order for all thicknesses d. although the jump in P_s becomes smaller and smaller as d decreases. so that the transition would <u>appear</u> continuous for sufficiently thin films.

SUMMARY

Although a quantitative theory is still lacking. KNO_3 film data suggest that finite size effects can indeed be observed in submicron films via laser Raman spectroscopy. and that surface electric fields are the dominant perturbations of the bulk spectra. (Stress at film interfaces and grain boundaries can be ruled out from the lack of pressure shifts or splittings in the film spectra and the insensitivity to annealing.) The data suggest a reasonable homogeneous electric field over a significant volume fraction. either at domain walls. grain boundaries. or space charge regions near the electrodes. It appears possible to test finite thickness theories in ferroelectric KNO_3.

REFERENCES

1. T. C. Lubensky and M. H. Rubin. Critical Phenomena in Semi-Infinite Systems: II. Mean-Field Theory. <u>Phys. Rev.</u> B12:3885 (1975).

2. D. R. Tilley and B. Zeks. Landau Theory of Phase Transitions in Thick Films, <u>Sol. St. Commun.</u> 49:823 (1984).

3. K. Binder. Finite Size Effects on Phase Transitions, <u>Ferroelectrics</u> 73:48 (1987).

4. R. B. Godfrey. J. F. Scott. H. B. Meadows. M. Golabi. C. Araujo. and L. McMillan. Analysis of Electrical Switching in Sub-Micron Thin Films, <u>Ferroelectrics Letters</u> 5:167 (1986).

5. C. Araujo. J. F. Scott. R. B. Godfrey. and L. McMillan. Analysis of Switching Transients in KNO_3 Ferroelectric Memories. <u>Appl. Phys. Lett.</u> 48:1439 (1986).

6. M. Sayer. "Fabrication and Application of Multi-Component Piezoelectric Thin Films." Proc. 6th Int. Sym. Appl. Ferroelectrics. IEEE. New York. (1986). p.560.

7. M. H. Francombe. Ferroelectric Films and Their Device Applications. <u>Ferroelectrics</u> 3:199 (1972).

8. J. F. Scott, R. B. Godfrey, C. Araujo, L. D. McMillan, H. B. Meadows, and M. Golabi, "Device Characteristics of Ferroelectric Ceramic KNO_3 Thin-Film Raw Memories," Proc. 6th Int. Sym. Appl. Ferroelectrics, IEEE, New York, (1986), p.569; J. F. Scott and C. A. Araujo, The Physics of Ferroelectric Memories, in: "Molecular Electronics", M. Borissov, ed., World Scientific Pub. Co., Singapore (1987), p.209; C. A. Araujo and J. F. Scott, A Novel High Speed Nonvolatile Memory Based on a Low Coercivity Ferroelectric Thin Film, Ibid, p.216.

9. S. Y. Wu, M. H. Francombe, and W. J. Takei, Domain Switching Effects in Epitaxial Films of Ferroelectric Bismuth Titanate, Ferroelectrics 10:209 (1967).

10. Yu. Ya. Tomashpolsky, "Ferroelectric Thin Films," Radio and Communication Publishing Co., Moscow, (1984) (in Russian).

11. A. Hadni, R. Thomas, S. Ungar, and X. Gerbaux, Drastic Modifications of Electrical Properties of Ferroelectric Crystal Plates with Thickness: The Case of Triglycine Sulfate, Ferroelectrics 47:201 (1983).

12. A. Hadni and R. Thomas, High Electric Fields and Surface Layers in Very Thin Single Crystal Plates of Triglycine Sulfate, Ferroelectrics 59:221 (1984).

13. A. M. Glass, K. Nassau and J. W. Shrever, Evolution of Ferroelectricity in Ultrafine-Grained $Pb_5Ge_3O_{11}$ Crystallized from the Glass, J. Appl. Phys. 48, 5213 (1977).

14. D. L. Mills, Surface Effects in Magnetic Crystals near the Ordering Temperature, Phys. Rev. B3:3887 (1971).

15. K. Binder, Surface Effects on Phase Transitions in Ferroelectircs and Antiferroelectrics, Ferroelectrics, 35:99 (1981).

16. J. F. Scott, H. M. Duiker, P. D. Beale, Properties of Ceramic KNO_3 Thin-Film Memories, Physica B/C (in press, 1987).

17. D. R. Tilley, private communication.

18. R. Loudon, Theory of the First-Order Raman Effect in Crystals, Proc. Roy. Soc. A275:223 (1968).

19. E. Burstein, S. Ushioda, A. Pinczuk, and J. F. Scott, Raman Scattering by Polaritons in Polyatomic Crystals, in: "Light Scattering Spectra of Solids," G. B. Wright, ed., Springer, New York (1969), p.43.

20. P. A. Fleury and J. M. Worlock, Electric-Field-Induced Raman Scattering in $SrTiO_3$ and $KTaO_3$, Phys. Rev. 174:613 (1968); E. Anastassakis and E. Burstein, Electric-Field-Induced Infrared Absorption and Raman Scattering in Diamond, Phys. Rev. B2:1952

(1970); L. J. Brillson and E. Burstein, Surface Electric-Field-Induced Raman Scattering in PbTe and SnTe, Phys. Rev. Lett. 27:808 (1971).

21. M. Balkanski, M. K. Teng, and M. Nusimovici, Raman Scattering in KNO_3 Phases I, II, and III, Phys. Rev. 176:1098 (1968).

22. J. F. Scott, Ming-sheng Zhang, Raman Spectroscopy of Submicron KNO_3 Films, Phys. Rev. B35:4044 (1987).

23. R. Farhi and P. Moch, Raman and Dielectric Susceptibility Studies in Thiourea under Electric Fields, J. Phys. C18:925 (1985).

24. R. Braunstein and K. Barner, Space Charge Injection into a Dipolar Glass, Sol. St. Commun. 33:941 (1980).

25. N. W. Schubring, R. A. Dork, and J. P. Nolta, Ferroelectric and Other Properties of Polycrystalline Potassium Nitrate Films, in: "Ferroelectricity," E. F. Weller, ed., Elsevier, Amsterdam (1967), p.269.

INELASTIC LIGHT SCATTERING BY MAGNONS IN

ANTIFERROMAGNETIC EuTe

S. O. Demokritov, N. M. Kreines and V. I. Kudinov

Institute for Physical Problems, USSR Academy of Sciences
Moscow, USSR

1.   INTRODUCTION

The Brillouin-Mandel'stam scattering (BMS) of light is one of a few experimental methods which makes it possible to study excitations in magnetic solids directly.

Up to now the spin-wave spectra in some ferro-, ferri-, and antiferromagnetic materials have been studied using BMS.[1-4] For a long time experiments on one magnon scattering have revealed contributions from magnetooptic effects (MOe) arising from the large spin-orbital interaction in magnetic materials and which are therefore relativistic in a nature. Among these effects are the Faraday effect (FE), linear magnetic birefringence (LMB), and linear and circular dichroism. Recently a new mechanism of one-magnon scattering, which is of an exchange nature, has been observed.[5] MOE - isotropic magnetic refraction (IMR) which describes the dependence of the refractive index on the magnetization of the crystal corresponds to this mechanism.

In the present work we will give results of BMS by thermal magnons belonging to both branches of the excitation spectrum of "easy-plane" antiferromagnet EuTe. The experiments were carried out at $T \approx 2$ K. The field dependence of magnon frequencies was studied over a broad interval of magnetic fields (0 - 80 kOe) corresponding to the entire domain of existence of the antiferromagnetic phase. The investigation of the magnon frequency dependence on wave vector direction makes it possible to reveal the contribution of the magnetic-dipole interaction into the spectrum of excitations. The considerable softing of one mode near the spin-flip transition is observed. The field dependence of intensity of BMS by thermal magnons is measured. Contributions of different MOE are separated and good quantative agreement of experimental data and results of calculations have been achieved.

2.   Samples and Technique

Europium chalcogenides have the cubic structure of the NaCl type, and the symmetry is described by the space group O_h^5. The magnetic properties of the pure compounds are governed by the strongly localized moment of rare-earth ion Eu^{2+} (electron configuration $4f^7$) whose spin S = 7/2 and orbital angular momentum L = 0. Below T_N = 9.8 K[6] EuTe becomes an easy-plane two-sublattice antiferromagnet where the spins lie in the (111) plane. The corresponding effective anisotropy field H_A is ≈ 10 kOe. Because a cubic crystal has four equivalent (111) planes, below T_N it breaks up into antiferromagnetic T-domains. The external magnetic field cannot bring about a transition of the sample to a single-domain state. In the (111) easy plane there is a weak interplane anisotropy H_a which aligns the spins in the (112) direction. This anisotropy corresponds to an effective field ~ 10 Oe. EuTe undergoes a transition from the antiferromagnetic phase to the ferromagnetic phase (a spin-flip transition) in an external magnetic field H_C = $2H_E$ = 72 kOe (at T = 0 K).[7-11]

The EuTe crystals are relatively transparent to the wavelength of light, λ = 632.8 nm, used by us in the experiment: the penetration depth is ~ 200 μm. At this wavelength of light, EuTe in the ordered state has strong magnetooptical effects: the Faraday effect and the magnetic isotropic refraction. At T = 2 K the Faraday effect is ~ 2 x 10^5 deg/cm (or $n_+ - n_-$ = 0.07).[12] The magnetic isotropic refraction accounts for the quadratic dependence of the magnetic part of the refractive index on the magnetization of the crystal. At T = 2 a change in the refractive index, Δn, of EuTe in the saturated states is ~ 0.05 according to the estimates given in.[5] Both of these magnetooptical effects (and also the magnetic birefringence, a much weaker effect) account for a relatively higher intensity of light scattering by spin waves in EuTe.[11,13] As the spectral instrument we used a five-pass Fabry-Perot interferometer (manufactured by Burleigh) with a contrast > 10^8. The general schematic diagram of the experimental apparatus and the specific details of the measurement procedure were reported in.[1,11]

The experiments were carried out in the backward-scattering geometry with the use of magnons with a wave vector q = 310^5 cm^{-1}. We use two scattering geometry q $\parallel$ M and q $\perp$ M. The magnetic field was applied along the (100) axis. At this orientation all T-domains are in the equivalent position with respect to the field. Since the excitation spectrum of an easy-plane antiferromagnet depends on the angle of inclination of the magnetic field relative to the (111) "easy" plane, the magnon spectra of the different domains are the same for the experimental geometry used by us.

Fig. 1 is a trace of the scattering spectrum at a constant magnetic field. The strong peak with a zero frequency shift (the principal line) is attributable to the elastic scattering of light by the crystal defects. The displaced satellites correspond to the inelastic scattering of light by magnons. From such a spectrum we can measure the frequency of a magnon taking part in a scattering process with a good precision (~ 1%). But we have no opportunity to measure the field dependence of the scattering intensity with precision better than ± 50%. That is why we should turn to measuring the relative intensity. As a reference point we use BMS by longitudinal phonons in TeO_2 propagating along C_4 axis. It is anomalously intense.[14] In one experiment we have simultaneously observed both light scattering by magnons in EuTe and by phonons in TeO_2. A trace of such a spectrum is represented in Fig. 2. The ratio of satellite intensities is measured. Using such a measurement procedure the field dependence of BMS by magnons in EuTe is investigated. (We suggest BMS by phonons is independent of the field.) The precision of this method is about 5%. In these experiments we use the first geometry only when the wave vector of a magnon is along the magnetic field and the magnetization. The measurements were carried out at T ≈ 2 K. For this purpose we used an optical helium cryostat with a horizontal superconducting solenoid.

288

3. Experimental Results

The spectrum of light scattered in EuTe is shown in Fig. 1. As can be seen the magnons belonging to each branch of the spin-wave spectrum of the easy-plane antiferromagnet were detected by the light scattering method. The satellites with a shifted frequency $\nu_1 = \pm (50 \pm 0.5)$ GHz (S_1 and AS_1) correspond to the scattering by magnons belonging to the low-frequency branch of the spectrum and the satellites with $\nu_2 = \pm (94 \pm 0.5)$ GHz (S_2 and AS_2) correspond to the scattering by magnons belonging to the high-frequency branch. From such spectra we can get values of the magnon frequencies for both branches at given magnetic field and the direction of wave vector q relative to the magnetization. The dependence of magnon frequency in each branch on the internal magnetic field at q $\parallel$ M and

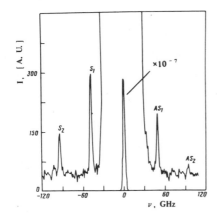

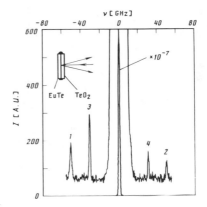

Fig. 1. The spectrum of light scattered in EuTe: T = 2 K, q = 3×10^5 cm^{-1}. q $\parallel$ [010] $\perp$M, H = 25 kOe.

Fig. 2. The spectrum of light scattered in EuTe and TeO$_2$: T = 2 K, H = 45 kOe, 1,2 - magnons, 3,4 - phonons.

q $\perp$ M is given in Fig. 3. In our experimental conditions we can realize only the first scattering geometry (q $\parallel$ M) at strong magnetic field (H $\geq$ 50 kOe.) The experiment showed that intensity of scattering by upper branch magnons at q $\parallel$ M rapidly drops with magnetic field increase. So we have observed BMS only by low-frequency magnons near spin-flip transition. Fig. 3 shows that there exists considerable softing of the magnon branch. Each branch of the spectrum has a gap in the absence of a magnetic field. The gap in the low-frequency branch, $\Delta_1 = \sqrt{2H_E H_a}$, which is caused by the interplane anisotropy H_a and the gap in the high-frequency branch, $\Delta 2 = \sqrt{2H_E H_A}$, which is caused by the easy-plane anisotropy H_A. We can detect the difference between the frequencies of the magnons propagating along and perpendicular to the magnetization of the crystal.

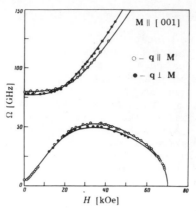

Fig. 3.   Spin-Wave Spectrum. $q = 3 \times 10^5$ cm^{-1}.

In the spectrogram shown in Fig. 3, we see that the intensity of the Stokes satellites (S) is different from the intensity of the anti-Stokes satellites (AS) for each magnon branch. We assume that this difference stems from the different probabilities of the absorption and emission of a magnon by a photon during scattering under conditions where $h\nu_{magn} \sim kT$. It is due to the statistics of the magnons, which are the bosons.

The field dependence of BMS intensity at $q \parallel M$ for the magnons of low frequency branch is represented in Fig. 4. The scattering is seen to be intense in the weak magnetic field. But then when the field is increased, the intensity drops appreciably and begins to increase only at $H > 10$ kOe. Such dependence $I(H)$ is in agreement with the result of our calculations carried out in.[15]

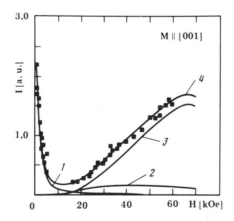

Fig. 4.   Scattered light intensity versus the magnetic field for magnons of low-frequency branch. 1,2,3 – calculated contributions of linear linear magnetic birefringence, Faraday effect and d–f exchange mechanism. 4 – summary.

5.   Analysis of the Observed Magnon Spectrum

Let us turn to consideration of the $\nu(H)$ dependence. The estimation showed the contribution of the dispertion term is smaller than our experimental error in the case of a magnon with $q \sim 10^5$ cm^{-1} in the scattering process. Therefore we can neglect the dispersion in our discussion. The calculation of the magnon spectrum, taking into account the magnetic-dipole interaction, was carried out on the basis of the macroscopic Landau-Lifshitz equations. Results are shown in Fig. 3 in the solid curve for both branches for q $\parallel$ [001] $\parallel$ M and q $\parallel$ [010] $\perp$ M. The value of the exchange field H_E was taken from[9] and values of the effective fields H_A and H_a were determined from the best fit of the experimental data: $H_A = 10.2 \pm 0.2$ kOe, $H_a = 15 \pm 1$ Oe.

6.   Field Dependence of BMS Intensity. Comparison of Experiment with Theory

To analyze our experimental results the intensity of one-magnon light scattering for easy plane antiferromagnet has been calculated. These calculations are carried out on the basis of the macroscopic Landau-Lifshitz equations by analogy with.[16] We consider the situation when the field has an arbitrary value and direction.

Suppose we introduce the extinction coefficient h as a result of total intensity of light scattered in a unit volume into an elementary solid angle dô with scattered frequency between ω^1 and $\omega^1 + d\Omega$ divided by the flux density of the incident light. According to[17] the differential extinction coefficient in the long wave limit is defined by the formula:

$$\frac{dh}{d\hat{o}d\Omega} = \frac{\omega^4}{32\pi^3 C^4} \int_{-\infty}^{+\infty} < \varepsilon_{ij}(0)\varepsilon_{ij}(t)>e^{i\Omega t}dt \tag{1}$$

here ω, ω^1 - frequencies of the incident and scattered light, $\Omega = 2\pi\nu = \omega - \omega^1$. ε_{ij} - the permittivity tensor. We write out the magnetic part of the tensor ε_{ij}:

$$\varepsilon_{ij} (M,L) = if_{ijk} \mu_k + g_{ijkn} \mu_k\mu_n + a\delta_{ij}\mu^2 \tag{2}$$

The first term here describes FE, the second term is responsible for LMB in L. The last term is isotropic magnetic refraction (IMR). This formula relates fluctuations of L and M with fluctuations of ε_{ij}. Let us take M = M_{st} + m(t), L = L_{st} + l(t). Using (2) we can obtain the extinction coefficient for light scattering by magnons in accordance with:[15]

$$\frac{dh}{d\Omega d\hat{o}} = \frac{\omega^4}{32\pi^3 C^4} \{f^2 \frac{<(km)^2>}{k^2} \Omega + g^2 L^2_{st} <l^2> \Omega + 4a^2 <(\mu_{st}-m)^2\Omega\} \tag{3}$$

The third term in (3) is due to IMR. Its microscopic origin is in the exchange of an electron in an excited d-state with the localized electron in an f-state (so called d-f-exchange.) We see that this term occurs when a spin system has a longitudinal vibration of the magnetization m||(t) and has a static value M_{st}. This can occur in a strongly canted antiferromagnet. If the antiferromagnet is compensated (H = 0, M = 0) then IMR makes a contribution only into the two-magnon scattering.[2-4] In the isotropic ferromagnet the mechanism involved can cause neither one-magnon, nor two-magnon scattering. Formulas (2) and (3) show scattering responsible for the exchange mechanism having its distinctive peculiarities: a) it is isotropic, i.e. the intensity is independent of the direction of K (light wave vector); b) scattering takes place without changing the light polarization.

The exchange contribution into one-magnon scattering has been experimentally observed by the authors when studying BMS by excited and thermal magnons in EuTe.[5,15]

In conclusion we underline that d-f-exchange causes one-magnon scattering in canted structures even without the spin-orbital coupling.

As has been mentioned we have measured the dependence of the scattering intensity on the applied magnetic field (see Fig. 4). The comparison of the experimental data with calculations shows that we achieve the best agreement taking into account all MOE.

In a weak field, BMS by low-frequency magnons is mostly due to LMB. The value of this MOE is rather small in EuTe (ψ_{LMB} = 150 rad/cm) but huge fluctuations of L lead to appreciable intensity of the scattering in weak fields. This intensity decreases ($I \sim H^{-2}$) as the magnetic field is increased and at H > 5 kOe LMB becomes negligible (see Fig. 4). For the low-frequency branch the FE contribution is not significant, but for the upper branch this effect plays the main role.

In strong magnetic fields (H > 20 kOe) the IMR contribution (caused by d-f-exchange) is principal. The above mentioned value is isotropic, i.e. is independent of K direction; that is what we in fact experimentally have observed.

The calculated contributions of each MOe and total result are represented in Fig. 4 by the solid curves. The values of FE and IMR are taken from,[5] the constant of LMB and the general scale are determined from the best fit. As can be seen we have succeeded in separating the contribution from each MOE and in getting good quantitative agreement of the experimental data with the theory.

References

1. A. S. Borovik-Romanov, N. M. Kreines, Phys. Rep. 81:5 (1982).
2. V. S. L'vov, Zh. Eksp. Teor. Fiz. 53:163 (1967).
3. T. J. Moriya, J. Phys. Soc. Japan 23:190 (1967).
4. P. A. Fleury, R. London, Phys. Rev. 166:514 (1968).
5. S. O. Demokritov, N. M. Kreines, V. I. Kudinov, JETP Lett. 41:46 (1985).
6. W. R. Johanson, P. C. Mc-Collum, Phys. Rev. B 22:2435 (1980).
7. G. Will, S. J. Pickort, H. A. Alperin, R. J. Nathans, J. Phys. Chem. Sol. 24:1679 (1963).
8. W. Battles, G. E. Everett, Phys. Rev. B 1:3021 (1970).
9. N. F. Oliveira, S. Foner, J. Shapira, T. B. Reed, Phys. Rev. B 5:2634 (1972).
10. P. K. Streit, G. E. Everett, Phys. Rev. B 21:169 (1980).
11. A. S. Borovik-Romanov, S. O. Demokritov, N. M. Kreines, V. I. Kudinov, Sov. Phys. JETP 61:801 (1985).
12. J. Schoenes, P. Wachter, Physica 86-88:125 (1977).
13. S. O. Demokritov, N. M. Kreines, V. I. Kudinov, JETP Lett. 43:403 (1986).
14. V. V. Lemanov, G. A. Smolenski, Usp. Fiz. Nauk 108:465 (1972).
15. S. O. Demokritov, N. M. Kreines, V. I. Kudinov, Zh. Eksp. Teor. Fiz. 92:689 (1987).
16. A. I. Akhiezer, Yu. L. Bolotin, Zh. Eksp. Teor Fiz. 53:267 (1967).
17. L. D. Landau, E. M. Lifshitz, Electrodynamics of Continuous Media, Moscow, Nauka (1982).

SPECTROSCOPY OF MnO_4^- and MnO_4^{2-} CENTERS IN ALKALI HALIDE CRYSTALS

T. I. Maximova and A. M. Mintairov

A. F. Ioffe Physico-Technical Institute, USSR Academy of
Sciences, 194021, Leningrad, USSR

ABSTRACT

Complex molecular impurities in alkali halide crystals possessing a
series of clearly pronounced local modes represent an appropriate subject
for studying the electron-phonon interaction of impurity centers. Of
particular interest are impurity centers with the impurity electron
coupled comparatively weakly to the lattice vibrations and where the
impurity absorption band has a distinct vibronic structure. The resonant
Raman scattering (RRS) method is a promising technique for investigating
the electron-phonon interaction.

I. FIRST ORDER RRS SPECTRUM AND MULTI-PHONON RRS WINGS IN MnO_4^- DOPED KBr AND RbBr

Polarized spectra were obtained on oriented crystals in z(yy)x and
z(yz)x geometries where x, y, z are the four-fold crystal axes.

We have studied the temperature behavior of RRS spectra of MnO_4^- in
RbBr excited with a 5145 Å line which falls within the structure of a
vibronic band (see inset to Fig. 1). One readily sees in the spectrum at
all temperatures equidistant series of narrow lines repeating with a
period of 850 cm^{-1} and representing multiphonon RRS progressions due to
the overtones and combination modes of the MnO_4^- intramolecular vibrations $\nu_1(A_1)$, $\nu_3(F_2)$, and $\nu_4(F_2)$.

At 2K and 100K additional broad bands (of width ~250 cm^{-1} at 2K)
repeating with the same period are seen at the base of the most intense
lines (Fig. 1). As the temperature increases, these bands drop in
intensity and spread out. Frequency dependence measurements have shown
that this part of the spectrum also corresponds to resonant Raman scattering and is due to the phonon wings (PW) of MnO_4^- impurity ion local
vibrations.

The vibrations ν_1, $2\nu_1$, ν_4 in RbBr (MnO_4^-) excited by the lines at
5145, 4965 and 4880 Å , respectively, exhibit the strongest PW.
Polarized RRS spectra obtained in the region of these vibrations at 100K
are presented in Fig. 2. The PW of the totally symmetrical vibrations

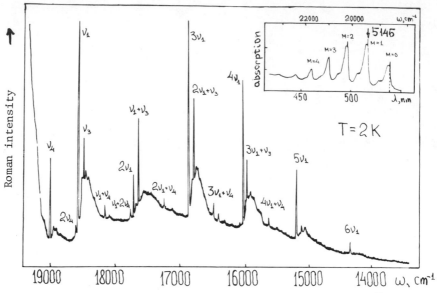

Fig. 1. Low temperature RRS spectrum of RbBr (MnO$_4^-$).

The $^1A_1 \rightarrow {}^1T_2 (1t_1 \rightarrow 2e)$ vibronic band is shown in the inset.

ν_1 and $2\nu_1$ are seen to have the same structure, differing from that of the ν_4 wing. At the same time, the intensity distribution in the ν_4 wing obtained in the z(yz)x polarization practically coincides with that observed in the ν_1 and $2\nu_1$ PW in the z(yy)x polarization. In KBr (MnO$_4^-$) we have studied the region of the vibration ν_4 and $2\nu_1$ excited at $\lambda = 5145$ and 5017 Å, respectively. The PW of ν_4 (Fig. 3) reveals clearly, apart from the maximum at 50 cm^{-1} observed in both polarizations, strong narrow peaks lying at 92 cm^{-1} in z(yy)x and at 165 cm^{-1} in the z(yz)x polarization. For the ν_1 and $2\nu_1$ vibrations, the peak's polarization becomes reversed. The difference in the PW polarization of the local modes of different symmetry can be explained if we recall that by selection rules the PW polarization in the RRS spectrum is determined by the direct product of the representations corresponding to the symmetry types of the local vibrations and of the phonon projected density. Projected densities of the same symmetry in combination with the totally symmetrical and non-totally symmetrical vibration should manifest themselves in PW in different polarizations. Note that the diagonal projected densities (A$_1$+E) manifest themselves in the same polarization as the local vibration, and the nondiagonal ones (F$_2$), in the polarization forbidden for this vibration, just as is observed in our experiments.

It was of interest to investigate the structure of the frequency region in the scattering spectrum of the impurity crystal adjoining directly the excitation line. As is well known, the first order spectrum in alkali halides is forbidden by selection rules, only second order RS being observed in pure crystals.

Figure 4 presents polarized spectra of RbBr (MnO$_4^-$) and RbBr in the region 20 to 300 cm^{-1} when excited at $\lambda = 5145$Å. At 300K, the spectra of a pure and a doped crystal practically coincide, a second order RS spectrum characteristic of pure RbBr being observed in both[1]. At liquid

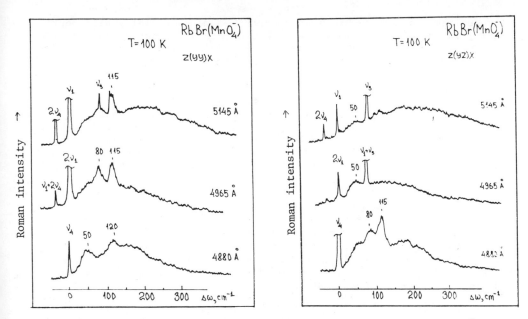

Fig. 2. RRS spectrum in the region of phonon wings of $V_1(A_1)$, $2V_1(A_1)$, $V_4(F_2)$ vibrations in RbBr (MnO_4^-) at T = 100K.

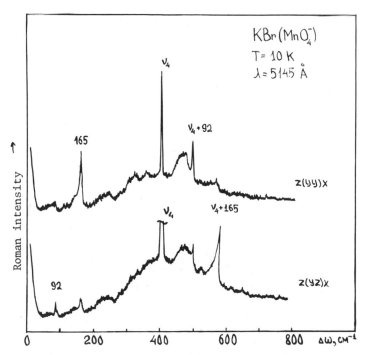

Fig. 3. Quasilocal modes in polarized RRS spectrum of KBr(MnO_4^-) at T = 10K (λ = 5145Å).

nitrogen temperature the observed spectra become essentially different; namely, in the pure crystal one sees a considerably weakened second order spectrum with characteristic peaks at 770 and 190 cm^{-1}; while in the doped crystal at this temperature the spectrum consists of two intense maxima at 80 and 115 cm^{-1} and a number of overlapping bands in the region of 150 to 300 cm^{-1}. Obviously, the spectrum observed in the above crystals at low temperatures in the region of the crystal vibrational frequencies of the doped crystal is actually a one-phonon scattering spectrum induced by the impurity. The narrow peaks at the boundaries of the vibrational branches in KBr(MnO$_4$) (Fig. 3) have been interpreted as quasi-local modes[5,6] One readily sees that the structure of the low-temperature doped crystal spectrum coincides with that of the MnO$_4^-$ local mode PW described earlier and shown in Fig. 2.

A study of the temperature behavior of the low frequency region in the RRS spectrum permits one to observe subsequently the manifestation in scattering of "pure" and "impurity-affected" parts of a doped crystal. The second order scattering intensity falls off with temperature according to the well known T^2 law, whereas the intensity of the one-phonon spectrum of a doped crystal grows rapidly with decreasing temperature, thus revealing its resonant nature.

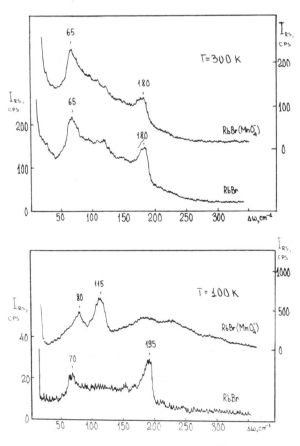

Fig. 4. Temperature dependence of RbBr(MnO$_4^-$) and RbBr low frequency Raman spectra in z(yy)x geometry, excited by light with λ = 5145Å.

When excited in the ZP line ($\omega_o = \Omega_e$) in a weakly absorbing medium, the scattering intensity can be written as

$$I(\omega) \sim \frac{1}{\Gamma^2(0)} \frac{\Gamma^+(\Omega_e - \omega)}{(\Omega_e - \omega)^2 + \Gamma^2(\Omega_e - \omega)} \tag{1}$$

where $\Gamma(\Omega_e - \omega) = \Gamma^+(\Omega_e - \omega) + \Gamma^-(\Omega_e - \omega) = \sum_k \xi_k^2 [\delta(\Omega_e - \omega - \omega_k)(1 + \langle n_k \rangle) +$

$+ \delta(\Omega_e - \omega + \omega_k)\langle n_k \rangle]$, $\Gamma(0) = \Gamma(\omega_o - \Omega_e)|_{\omega_o = \Omega_e}$ is the relaxation at the ZP line

frequency, ξ_k^2 is the electron phonon interaction constant for the k-th

phonon, $\langle n_k \rangle$ is a temperature factor. The one phonon scattering

intensity from lattice phonons in the resonance case is seen to be proportional to T^{-3}, i.e. it should increase dramatically with decreasing temperature.

Thus in the temperature range from 300 to 100K one may expect the intensity of one-phonon resonant impurity scattering to increase 20-30 times. This agrees qualitatively with experiment.

II. VIBRONIC MIXING in MnO_4^- IMPURITY CENTERS

We have studied the frequency dependence of the scattering intensity of the triply degenerate vibration ν_3, as well as of the combination tones $n\nu_1 + \nu_3$. The values obtained for the ν_3 vibration for different excitation frequencies are presented in Fig. 5 together with data for ν_1. The non-totally symmetrical vibration ν_3 is seen to reveal as strong a frequency dependence as that of the totally symmetrical one. This dependence implies that the vibration ν_3 is resonantly coupled with the excited electron state $^1T_2$ (Jahn-Teller effect). The relatively low intensity of the scattering lines involving ν_3 suggests that the interaction of the $^1T_2$ electron state with this vibration is much weaker than that with the totally symmetrical one.

In RRS excitation profile (REP) calculations for the non-totally symmetrical vibration ν_3 in RbBr (MnO_4^-) and KBr(MnO_4^-) we used the fact that the coupling of ν_3 with the electron state $^1T_2$ is weak. In the case of a weak Jahn-Teller (JT) interaction, perturbation theory can be used for its description. It can be shown that in cases where the Condon approximation is valid the expression for the scattering intensity from a non-totally symmetrical vibration reduces to a form similar to that for a totally symmetrical one[2]:

$$I(\omega_o) = C\omega_o \omega_o^3 |M_{eg}^\alpha \xi_{F_2} M_{ge}^\beta|^2 |\Phi(\omega_o) - \Phi(\omega_o - \omega_{F_2}^e)|^2 , \tag{2}$$

where $\omega_{F_2}^e$ is the non-totally symmetrical vibration frequency in an

excited 1_{T_2} state, and $\xi_{F_2}^2$ is a dimensionless constant of the $T_2 \times F_2$

interaction. The value of ξ used in the calculations is 0.1 for KBr and 0.2 for RbBr. Figure 5 reveals a noticeable disagreement between the theoretical and experimental data.

Obviously, to account for the observed frequency dependence of the scattering intensity one should take into consideration vibronic mixing

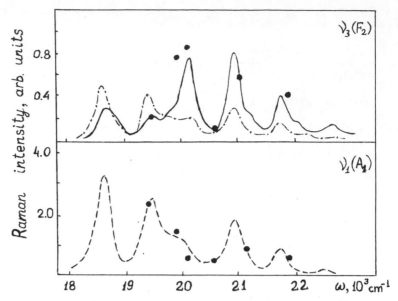

Fig. 5. REP of $\nu_3(F_2)$ and $\nu_1(A_1)$ vibrations of $RbBr(MnO_4^-)$ at T=100K. Circles— experiment, dashed curve — calculation in Franck—Condon approximation. Dash—dotted curve —calculation taking account of weak JT coupling. Solid curve — calculation with account of small HT correction to JT coupling.

of the state $^1T_2$ with other excited electron states of the MnO_4^- ion, i.e. the Herzberg—Teller (HT) interaction. The existence of this mixing is suggested also by the weak violation of the selection rules for the vibrations ν_3 and ν_4 observed by us. The HT interaction results in a dependence of the matrix element of the M_{eg} electron transition on vibrational coordinates Q, i.e. in a deviation from the Condon approximation. To describe properly the scattering involving one degenerate phonon, it is sufficient to include the linear dependence of M_{eg} on nuclear displacements: $M_{eg}(Q) = M_{eg}(0)(1+mQ)$ where m is the HT coupling constant. The expression for the one-phonon scattering intensity in this case can be written as[3,4]:

$$I(\omega_o) = C\omega_o\omega^3|M^\alpha_{eg}\xi_{F_2}M^\beta_{ge}|^2|\Phi(\omega_o)(1+C_n)-\phi(\omega_o-\omega^e_{F_2}(1-C_n)|^2 , \qquad (3)$$

where $C_n = m/\xi_{F_2}$.

We used this expression to calculate the REP for the vibration taking into account the linear HT coupling. In the calculations, the value of C_n was varied to obtain the best fit between the theory and experiment. The best fit was found to be for $C_n = -0.2$.

Thus, the case of MnO_4^- illustrates that studying the electron-phonon interaction by RRS spectroscopy offers a possibility of obtaining detailed information on the mechanisms of interaction of non-totally symmetrical vibrations and on the nature of vibronic mixing.

III. RRS EXCITATION PROFILES FOR KI DOPED WITH MnO_4^- and MnO_4^{2-}.

REP for four orders were measured with a dye laser for $KI(MnO_4^-)$ over the range 16600–18500 cm^{-1} covering the first two vibronic maxima of the absorption band $^1A_1 \rightarrow {}^1T_2 (1t_1 \rightarrow 2e)$. The experimental REP are presented in Fig. 6 showing also the impurity absorption band. One readily sees that the vibration reveals a pronounced resonance in the scattering cross section whose maximum lies at 17250 cm^{-1} and, thus, is slightly shifted toward low frequencies from the absorption maximum M = 0 (Table I).

TABLE 1

Model parameters of MnO_4^- and MnO_4^{2-} centers in KI extracted from REP and optical absorption data (T = 100K).

Parameter	MnO_4^-	MnO_4^{2-}
Ω_e, cm^{-1}	17310	16000
ω^e, cm^{-1}	760	750
γ, cm^{-1}	300	450
ξ^2	1.46	1.23

A similar procedure is observed for the RRS excitation profile of the MnO_4^{2-} ion in KI (Fig. 6). Here the excitation region (15800–17500 cm^{-1}) of the RRS spectra shifts somewhat toward low frequencies because of the maxima of the band $^2E \rightarrow {}^2T_2 (1t_1 \rightarrow 2e)$ lying at lower frequencies. The vibration has apparently two resonances with maxima at 16550 and 17100 cm^{-1}. The overtones $2\nu_1$ and $3\nu_1$ have one resonance each with maxima at 16350 and 17600 cm^{-1}, respectively. Thus the position and width of the resonances in the excitation profiles varies with RRS order and differs substantially from that of the maxima in the vibronic structure of the optical absorption spectrum, suggesting a substantial role of interference effects in resonant scattering.

The small step in frequency scanning used and the broad frequency range covering the most intense part of the absorption band permit using the measured REP to check the validity of the "standard" assumptions and to determine the electron-phonon coupling constant for the totally symmetrical vibration (the Stokes losses). The excitation profiles were calculated by the expressions of Ref. 2 using the experimental data on the optical absorption spectrum. The experimental spectrum was given numerically at 200 frequency points with the background absorption subtracted properly. A comparison of the experimental and calculated excitation profiles (Fig. 6) gives good agreement in the shape of REP in both cases, thus establishing the validity of the standard assumptions for ν_1. The scaling coefficients obtained in the excitation profile comparison were used to determine the Stokes losses for the totally symmetrical vibration (Table I).

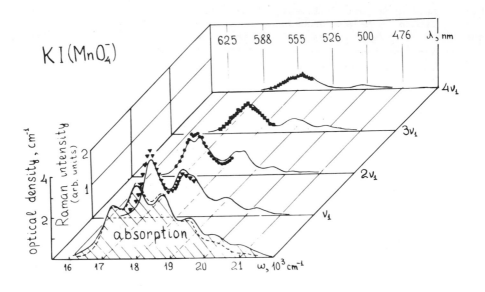

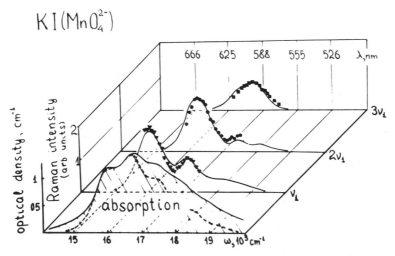

Fig. 6. Experimental and calculated (solid lines) REP for $n\nu_1(A_1)$ (n = 1-4) vibrations of MnO_4^- and MnO_4^{2-} centers in KI(T = 100K). In the foreground the measured (solid curve) and calculated [one oscillator model (dashed curve)] impurity absorption bands are given.

Thus our studies have shown that the data obtained from REP can be used to advantage in determining the mechanisms and evaluating quantitatively the constants of the electron-phonon interaction of impurity centers. This is particularly important in the case of weakly pronounced optical absorption spectra where obtaining such information from optical absorption meets with difficulties.

REFERENCES

1. J. E. Potts, C. T. Walker, and I. R. Nair, Phys. Rev. B8, 2756 (1979).
2. T. I. Maximova and A. M. Mintairov, Sov. Phys. Solid State 28 827 (1986).
3. I. Tehver, Opt. Commun. 38 279 (1981).
4. T. I. Maximova and A. M. Mintairov, Sov. Phys. Solid State 29, 1422 (1987).
5. L. A. Rebane, E. G. Blumberg, T. A. Fimberg, Pis´ma v. Zh. eksp. Teor Fiz., 44, 339 (1986).
6. T. I. Maximova and A. M. Mintairov, Fiz. Tverd. Tela 30, (1988)

THEORY OF TRANSIENT OPTICAL RESPONSE AND PULSE PROPAGATION IN BOUNDED SPATIALLY DISPERSIVE MEDIA.*

Spiros V. Branis, K. Arya, and Joseph L. Birman

Physics Department, City College of CUNY
Convent Ave. at 138th Street
NY, NY 10031

ABSTRACT.

The reflection of a finite duration optical pulse from a semi-infinite nonlocal medium at normal incidence for various ABC's, and at oblique incidence for Pekar's ABC, is investigated theoretically. We have obtained explicit expressions for the amplitude of the transient reflected field (local and nonlocal portions) and evaluated them numerically for different ABC's for normal incidence, and for different polarizations for oblique incidence. The effects of spatial dispersion on the reflected transients associated with the light pulse are important for laser frequency at the vicinity of an exciton-polariton resonance. For various ABC's, we find quantitative differences in the magnitude; these can be used to analyse different ABC's experimentally. The time decay of the transients shows oscillatory behavior in the time domain. In the case of oblique incidence, using p-polarization as well as increasing the angle of incidence enhances the spatial dispersion effects and oscillatory response of the medium after the pulse dies out.

INTRODUCTION.

There are several reasons for continuing interest in the nonlinear optical response of spatially dispersive media, such as semiconductors in the exciton-dominated optical region. Two of the principal matters are: the resolution of the ABC problem, and the prediction and measurement of novel optical effects, such as transient (pulse) propagation [1]. In this paper we present new results for transient reflectivity with special emphasis on the implication for helping to resolve the ABC problem. Especially of interest are our new results predicting time-oscillatory decay of the reflected pulse, and the dependence of the pulse on the polarization of the light, and on the ABC. We also remark on predictions of novel precursor effects which can be carried out with pico- and femto-second experimental capabilities currently available.

The domain of exciton-polariton spectroscopy investigates the electrodynamic and optical consequences of the finite mass exciton effects leading to spatial dispersion. Nonlocality occurs when the exciting laser frequency ω_0 is in the region of the exciton resonance ω_t. Steady-state optical spectroscopy includes reflectivity and transmissivity measurements (elastic scattering), and resonant Brillouin (inelastic) scattering. We recall that, because of the extra waves which propagate in nonlocal media, additional boundary conditions (ABC) are required, in order to completely determine the reflectivity function (Fresnel theory) [2].

Pekar's ABC [3] is based on the ansatz that the exciton polarization vanishes at the vacuum-medium boundary:

$$P_{exc}(\vec{r}, t)|_{\Sigma} = 0 \quad or \quad (n_1^2 - \varepsilon_0)E_1 + (n_2^2 - \varepsilon_0)E_2 = 0 \tag{1}$$

where n_1, n_2 are the refractive indices and E_j, $j = 1, 2$ are the electric field amplitudes for the two transverse propagating modes $n_j = \dfrac{k_j}{k_0}$, ε_0 is the dielectric constant of the background.

Hopfield and Thomas [4] and Kiselev [5] proposed a general ABC of the form:

$$P_{exc}(\vec{r}, t) + \gamma \frac{c}{\omega} \frac{\partial P_{exc}(\vec{r}, t)}{\partial z} |_{\Sigma} = 0 \tag{2}$$

or $$(1 + i\gamma n_1)(n_1^2 - \varepsilon_0)E_1 + (1 + i\gamma n_2)(n_2^2 - \varepsilon_0)E_2 = 0$$

where γ is a phenomenological constant (or function of frequency) which takes into account the flow of polarization on the surface of the medium.

In the seventies Sein and Birman [6], Maradudin and Mills [7], Agrawal et al. [8] independently proposed a third ABC:

$$i k_+ P_{exc}(\vec{r}, t) + \frac{\partial P_{exc}(\vec{r}, t)}{\partial z} |_{\Sigma} = 0 \tag{3}$$

or $$\frac{E_1}{n_1 - n_+} + \frac{E_2}{n_2 - n_+} = 0$$

with $$k_+^2 = n_+^2 \frac{\omega^2}{c^2} = \frac{m^*}{\hbar \omega_t}(\omega^2 - \omega_t^2 + i\omega\Gamma) \tag{4}$$

where m^* is the effective exciton polariton mass, ω_t is the transverse exciton frequency, Γ is a phenomenological damping factor and n_+ is the refractive index for the bare exciton.

An ABC was proposed by Ting, Frankel, and Birman [9], according to which the flow of polarization vanishes on the surface:

$$\frac{\partial P_{exc}(\vec{r}, t)}{\partial z} |_{\Sigma} = 0; \quad or \quad n_1(n_1^2 - \varepsilon_0)E_1 + n_2(n_2^2 - \varepsilon_0)E_2 = 0 \tag{5}$$

TIME DOMAIN-TRANSIENT RESPONSES.

(A). Reflectivity

Transient optical experiments are another tool by which the optical consequences of spatial dispersion can be investigated. In particular, transient reflectivity offers the possibility of separating local and nonlocal components, and investigating the influence of different ABC's. Thus one may decide the correct one while determining the parameters including ω_t and ω_l in a transient experiment. Transient optical reflectivity from semi-infinite local media was first analysed in detail by Elert [10]. For the case of a nonlocal medium for normal incidence the transient optical reflectivity was analysed recently by Agrawal et al. [11].

(i) The case of normal incidence for different ABC's [12].

Consider a semi-infinite nonlocal medium occupying the half space z>L, with z<L being vacuum (see Fig.1). A detector is placed at z=0. We assume that a normally incident laser pulse corresponds to a linearly polarized, monochromatic, plane wave -field. For a square pulse of unit intensity and duration T at frequency ω_0, the incident electric field at the plane z=0 is given by:

$$E_I(0,\ t) = \sin(\omega_0 t)\,[\Theta(t) - \Theta(t - T)] \tag{6}$$

where $\Theta(t)$ is the Heaviside step function.

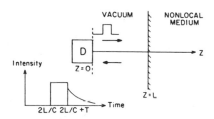

FIG. 1. Schematic illustration of the geometry and the notation used. At time $t = 0$ a square pulse of duration T is emitted from the plane $z = 0$ where a detector is place to monitor the reflectivity. For $2L/c < t < 2L/c + T$ the steady-state signal and for $\tau = t - 2L/c - T \geq 0$ the transient signal is detected.

The dielectric function for the semi-infinite medium using the "dielectric approximation" takes the form:

$$\varepsilon(\vec{k},\ \omega) = \varepsilon_0 + 4\pi\chi(\vec{k},\ \omega) = \varepsilon_0 + \frac{4\pi\alpha_0\omega_t^2}{\omega_t^2 - \omega^2 - i\omega\Gamma + (\frac{\hbar\omega_t}{m^*})\,k^2} \tag{7}$$

where ε_0 is the background dielectric constant, α_0 is the oscillator strength, m^* is the effective exciton mass, ω_t is the transverse exciton-polariton frequency, c is the velocity of light and Γ is a phenomenological damping factor, which recently [12] was shown to be frequency dependent. In the present work we take Γ constant. The Fresnel amplitude reflection coefficient is given in the following form:

$$\rho(\omega) = \frac{1 - \bar{n}(\omega)}{1 + \bar{n}(\omega)} \tag{8}$$

where $\bar{n}(\omega)$ is the effective refractive index of the nonlocal medium which for different ABC's has the following forms:

Pekar: $\qquad \bar{n}(\omega) = \dfrac{n_1 n_2 + \varepsilon_0}{n_1 + n_2}$

Sein-Birman: $\qquad \bar{n}(\omega) = n_1 + n_2 - n_+$

Ting et al.: $\qquad \bar{n}(\omega) = \dfrac{n_1 n_2 (n_1 + n_2)}{n_1^2 + n_2^2 + n_1 n_2 - \varepsilon_0}$

Kiselev: $\qquad \bar{n}(\omega) = \dfrac{n_1 n_2 + \varepsilon_0 + i\gamma\, n_1 n_2 (n_1 + n_2)}{n_1 + n_2 + i\gamma\, (n_1^2 + n_2^2 + n_1 n_2 - \varepsilon_0)} \tag{9}$

and n_1, n_2 are the solutions of the implicit relation:

$$n_j = +[\varepsilon(k_j, \omega)]^{\frac{1}{2}}, \quad k_j = n_j \frac{\omega}{c} \tag{10}$$

while n_+ is the refractive index of the uncoupled exciton.

The reflected field in the time domain is obtained as a superposition of the reflected components and can be written as a frequency integral:

$$E_R(0, t, \omega_0) = \mathrm{Lim}\ \mathrm{Re}\ \frac{1}{2\pi} \int_{-\infty}^{+\infty} \frac{d\omega\, \rho(\omega)}{\omega_0 - \omega - i\eta} \{1 - e^{i(\omega - \omega_0)T}\} e^{i\omega(t - \frac{2L}{c})} \tag{11}$$
$$\eta \to 0$$

where the time delay $\frac{2L}{c}$ corresponds to a round trip between z=0 and z=L.

The reflected field can be separated into steady-state (pole contribution) and transient parts (branch points contribution). The steady-state part gives rise to a reflected pulse (of duration T), whose leading and trailing edges contain the rapidly time-varying transient contributions. The integral in Eq. 11 is evaluated in the complex ω plane, using the method of contour integration and asymptotic approximation. For $t < \frac{2L}{c}$, $E_R(0, t < \frac{2L}{c}) = 0$, as required by causality. For $t > \frac{2L}{c}$ the evaluation of the frequency integral is lengthy, although straightforward (algebraic details wil be presented elsewhere) [14]. The time-dependent reflectivity consists of three parts:

$$E_R(0, t) = \mathrm{Re}\,[E_s(t, \delta) + E_{NL}(t, \delta) + E_L(t, \delta)] \tag{12}$$

where $E_s(t, \delta)$ is the signal reflectivity for $\frac{2L}{c} \leq t \leq \frac{2L}{c} + T$, $E_L(t, \delta)$ is the local transient and $E_{NL}(t, \delta)$ is the nonlocal transient part. Although $E_L(t, \delta)$ and $E_{NL}(t, \delta)$ are both affected by spatial dispersion, $E_{NL}(t, \delta)$ arises solely from spatial dispersion and vanishes as $\delta \to 0$ (where $\delta = [\frac{\hbar \omega_t}{m^* c^2}]^{\frac{1}{2}}$ or $m^* \to \infty$). In the strict limit of $\delta \to 0$, $E_s(t, \delta)$ and $E_L(t, \delta)$ reduce to the results previously obtained by Elert (local optics).

A careful numerical and analytical study of these terms, shows interesting and novel effects for the signal and transient part, when the laser frequency is tuned to resonance $\omega_t < \omega_0 < \omega_l$, for all ABC's. In particular, we plot in Fig. 2 the signal reflectivity $R_0 = |R_0(\omega_0)|^2$ as a function of ω_0 / ω_t in the vicinity of exciton-polariton resonance for parameters appropriate to a CdS crystal, for all ABC's. Signal reflectivity for the case of local medium $\delta \to 0$ is shown by a solid line for comparison. We note that the main effect of spatial dispersion is to reduce the reflectivity maximum for all ABC's.

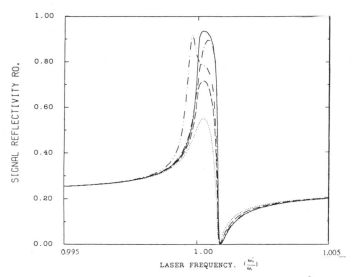

Fig. 2   Resonance enhancement of steady-state reflectivity $R_0 = |R_0(\omega_0)|^2$ which persists for $\frac{2L}{c} \leq t \leq \frac{2L}{c} + T$. The parameters are chosen appropriate to a CdS crystal with $\varepsilon_0 = 8$, $\hbar \omega_t = 2.55eV$, $\Gamma = 0.1275meV$, $\beta^2 = 4\pi\alpha_0 = 0.0125$, $m^* = 0.9m_e$ and $\gamma = 0.1$.

For comparison each curve is represented by an ABC as follows:

Local Optics   ——————

Pekar   ····················

Sein, and Birman   - - - - - - - - - -

Ting, Frankel, and Birman   -.-.-.-.-.-.-.-.-.-

Kiselev   -..-..-..-..-..-..-

If the pulse duration T is longer than the effective time during which transients contribute significantly, leading and trailing edge transients will not interfere and can be considered separately. For T>few psecs, leading edge transients would die out before the trailing edge of the pulse arrives, so experimentally it may be more convenient to look for transients for $t > (\frac{2L}{c} + T)$. The total transient reflectivity has the following form now:

$$E_T(\tau) = E_{NL}(\tau) + E_L(\tau) \qquad (13)$$

where: $\tau = (t - \frac{2L}{c} - T) > 0$.

We evaluated the total transient reflectivity for parameters appropriate to CdS. In Fig. 3 , we plot the time behaviour of total transient reflectivity at resonance $\omega_0 = \omega_t$. To be specially noted is the predicted oscillatory transient in the decay of the reflected pulse. We believe this is due to time-dependent dephasing of the emission from Lorentz dipoles which comprise the medium. Transient oscillations decay in few psecs; the local part is stronger and dominates but it decays faster than the nonlocal part. Various ABC curves show quantitative differences in the time domain. For fixed time $\tau=0.1$psec around the vicinity of ω_t , we plot total reflectivity versus ω_0 / ω_t in Fig. 4. Spatial dispersion reduces the total amplitude for all ABC's, while the local part still dominates. The peak of the local part appears at ω_l while the nonlocal peak is at ω_t . An interesting feature of Fig. 3 is the double-peaked spectrum of the total transient reflectivity for Pekar's ABC. The two peaks occur at ω_l and ω_t , which are the longitudinal and transverse exciton-resonance frequencies, respectively.

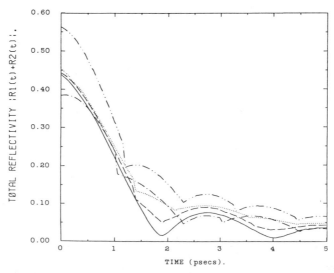

Fig. 3   Total transient reflectivity for a CdS crystal for various ABC's in the time domain at $\omega_0 = \omega_t$. The parameters appropriate to CdS crystals are those of Fig.2.

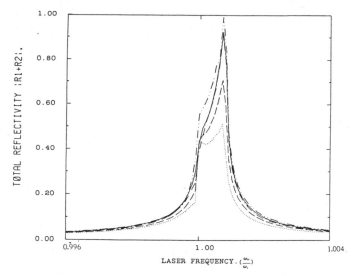

Fig.4
Total transient reflectivity resonance enhancement for a CdS crystal for various ABC's at fixed time $\tau=0.1$psec as a function of ω_0 in the vicinity of ω_t.

(ii) Oblique incidence using Pekar's ABC.

Experiments are often designed for oblique incidence. This experimental set up is more suitable for measuring reflectivity at an angle relative to the normal. The new factor that enters our analysis now is the vector polarization of the incident light.

We consider again a semi-infinite medium nonlocal medium for z>Lcosθ with z<Lcosθ being vacuum. Three modes of propagation travel in the nonlocal medium: two transverse and one longitudinal.

The generalized Pekar ABC for any angle is given by the following formula [15]:

$$\vec{P}_{exc}(\vec{r},\ t)_{\Sigma} = 0 \quad or \quad (n_1^2 - \varepsilon_0)\vec{E}_1 + (n_2^2 - \varepsilon_0)\vec{E}_2 - \varepsilon_0\vec{E}_L = 0 \qquad (14)$$

We investigate two cases: s-polarization and p-polarization. The formulas for amplitude reflection coefficient and effective refractive index of refraction for both polarizations are given by the following formulas [16]:

s-polarization: $\qquad \rho_s(\omega,\ \theta) = \dfrac{1 - n_s^*(\omega,\ \theta)}{1 + n_s^*(\omega,\ \theta)}$ $\qquad\qquad\qquad\qquad$ (15)

where: $\qquad n_s^*(\omega,\ \theta) = \dfrac{w_1(n_2^2 - \varepsilon_0) - w_2(n_1^2 - \varepsilon_0)}{(n_2^2 - n_1^2)\cos\ \theta}$

p-polarization: $\rho_p(\omega,\,\theta) = -\dfrac{1 - n_p^*(\omega,\,\theta)}{1 + n_p^*(\omega,\,\theta)}$ (16)

where:

$$n_p^*(\omega,\,\theta) = \frac{\varepsilon_0 \cos\,\theta\,[\varepsilon_0 \sin^2\,\theta\,(n_1^2 - n_2^2) + w_3[n_2^2(n_1^2 - \varepsilon_0)w_1 - n_1^2(n_2^2 - \varepsilon_0)w_2]]}{[\varepsilon_0(n_1^2 - n_2^2)w_1 w_2 w_3 + \sin^2\,\theta\,[n_2^2(n_1^2 - \varepsilon_0)w_2 - n_1^2(n_2^2 - \varepsilon_0)w_1]]}$$

while $w_i = \sqrt{n_i^2 - \sin^2\theta}$, $i = 1, 2, 3$. For $\theta \to 0$, we obtain Pekar's ABC result for normal incidence (see Eq. 9).

The reflected field (see Eq. 11) for s and p polarized light is angle and polarization dependent now:

$$E_R(0,\,t,\,\omega_0,\,\theta) = Lim \ \ Re \ \frac{1}{2\pi} \int_{-\infty}^{+\infty} \frac{d\,\omega\,\rho(\omega,\,\theta)}{\omega_0 - \omega - i\eta} \{1 - e^{i(\omega - \omega_0)T}\} e^{i\omega(t - \frac{2L}{c})} \quad (17)$$
$$\eta \to 0$$

where $\rho(\omega,\,\theta)$ is either $\rho_p(\omega,\,\theta)$ or $\rho_s(\omega,\,\theta)$. s and p-polarizations have a common pole at $\omega = \omega_0 - i\,\eta$ but their branch points are different and contribute in a different way to the local and nonlocal parts. Again we investigate for time $t > (\frac{2L}{c} + T)$ after the pulse dies out, while for $\delta \to 0$ the nonlocal part $E_{NL}(t,\,\delta,\,\theta) \to 0$ and the local one reduces to an angle-dependence Elert's result (local optics).

We evaluate the total transient reflectivity for parameters appropriate to CdS crystals. In Fig. 5, we plot the time behaviour of total transient reflectivity for p-polarization at resonance $\omega_0 = \omega_t$. Transient oscillations decay in few psecs; oscillatory behavior becomes more distinct for small values of damping factor Γ (low temperatures, pure material). Various angle dependent curves show quantitative differences in the time domain; for p-polarization the total transient reflectivity is enhanced for increasing angles. In opposite fashion, s-polarization decreases for large angles. For fixed time τ=0.1psec around the vicinity of ω_t , we plot total reflectivity versus ω_0 / ω_t for p-polarization in Fig. 6. The peak due to the local part appears at ω_l while that due to the nonlocal part peaks at ω_t . An interesting feature of Fig. 6 is the double -peak spectrum of the total transient reflectivity for p-polarization which is enhanced for increasing angles. For s-polarization, the double peak becomes weaker for the nonlocal part for increasing angles. The two peaks occur at ω_l and ω_t , the longitudinal and transverse exciton-resonance frequencies, respectively. Experiments for large angles should show time decay and frequency dependence of the transient part quite clearly, as well as giving the means of a possible measurement of ω_t and ω_l .

(B). Precursors and Gaussian Pulse Propagation.

To complete our presentation of novel effects expected in transient regime for spatially dispersive media we recall some previous predictions from our group's work on precursors. Following the general approach of Sommerfeld and Brillouin, consider the amplitude f(z,t) of the electromagnetic wave received at point z, time t>0, when an incident sharp-front truncated laser $f(0^-,\,t)$ impinges on the plane boundary (z=0) at t=0. Take the semi-infinite spatially dispersive medium to occupy z>0. Then, with

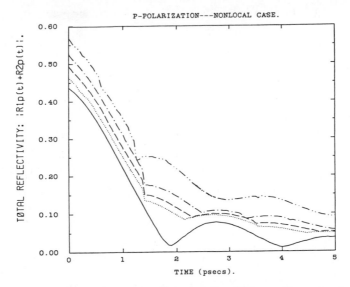

Fig.5   Total transient reflectivity (p-polarization) for a CdS crystal for various incident angles in the time domain at $\omega_0 = \omega_t$. All curves used Pekar's ABC.

For comparison each curve represents an incident angle as follows:

Local Optics ($\theta=0°$)   ——————

     Pekar   ($\theta=0°$)   ·················

              ($\theta=30°$)   - - - - - - - - -

              ($\theta=45°$)   -.-.-.-.-.-.-.-.-

              ($\theta=60°$)   -..-..-..-..-..-

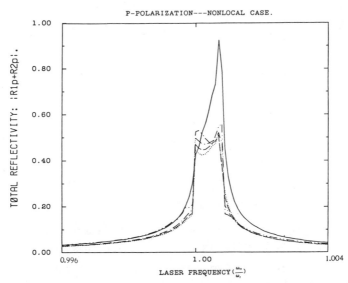

Fig.6   Total transient reflectivity (p-polarization) resonance enhancement for a CdS crystal for various incident angles at fixed time τ=0.1psec as a function of ω_0 in the vicinity of ω_t. Pekar's ABC was used.

$$f(0^-, t) = \sin(\omega_0 t) \, \Theta(t) \qquad\qquad (18)$$

we have:

$$f(z, t) = \mathrm{Re} \, \frac{1}{2\pi} \int_{-\infty}^{+\infty} \sum_{j=1}^{2} \frac{A_j(\omega) \exp\{i[k_j(\omega)z - \omega t)]\}}{(\omega - \omega_0)} \, d\omega \qquad (19)$$

The sum on j is over the two propagating polariton modes, and the coefficients are the coupling coefficients $A_1(\omega)$, $A_2(\omega)$ for upper and lower polaritons which are determined by the Maxwell plus Additional Boundary Conditions. Analysing the complex ω integral by standard complex variable methods we find the results for the "precursor regime":

$$f(z, t) = 0 \qquad\qquad \text{for} \qquad t < \frac{z}{c}$$

$$f(z, t) = \sum_{j=1}^{2} f_\alpha(z, t) \, \Theta(t - t_\alpha) \qquad \text{for} \quad \frac{z}{c} < t < \frac{z}{v_{SIG}(\omega_0)} \qquad (20)$$

The observed non-zero amplitude consists of the superposition of three partially overlapping packets. The new result is that an Exciton Precursor will appear, between the Sommerfeld (fast) and Brillouin (slow) Precursors. The precursor scenario procedes the "Signal Arrival", and is achieved in a few picoseconds. Finally the main signal arrives. In Fig. 7 the Analytical expressions have been obtained for the shapes and delay-times of the three contributions in the nonlocal case in the original paper of Frankel and Birman [17].

We conclude by pointing out that the time evolution propagation (and distortion) of a Gaussian pulse in spatially dispersive media has also been investigated theoretically and shows some quite interesting features. Taking an initial Gaussian pulse entering the medium with form:

$$f(z, t) = \tau \, (2\pi)^{-\frac{1}{2}} \int_{-\infty}^{+\infty} d\omega \, \exp[-i\omega t] \sum_{j=1}^{2} f_j(z, \omega) \qquad (21)$$

where:

$$f_j(z, \omega) = A_j(\omega) e^{i k_j(\omega) z} \, e^{-(\omega - \varpi)^2 \tau^2 / 2}$$

where $k_j(\omega)$ is the upper or lower branch dispersion, ϖ the center frequency of the Gaussian, τ^{-1} the pulse half-width and $A_j(\omega)$ is the same as in Eq. 19; we find that if $\Gamma\tau \gg 1$ or $\Gamma\tau = 1$ the pulse propagates with little distortion and remains essentially Gaussian, while the peak propagates at velocity close to the classical group velocity. For $\Gamma\tau \ll 1$ distortion occurs. A very interesting cross-over in the power spectrum occurs from lower to upper branch, as the frequency varies through resonance. In the above, Γ is the exciton damping coefficient in the dielectric function $\varepsilon(\vec{k}, \omega)$, which is taken as a constant. The cases of distorted pulse propagation have been discussed in detail in the original paper to which the reader is referred for details.

Propagation of Gaussian pulses in disordered dielectric media has recently been of interest in connection with studies of photon localization [18].

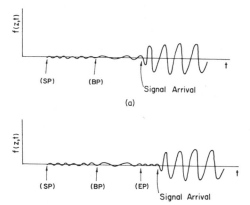

Fig.7 (a) Representation of the signal precursors for the case of no spatial dispersion.

(b) Representation of signal precursors including spatial dispersion effects.

In this case there are three precursors instead of two as in (a).

SP, Sommerfeld precursor arrival time: $\sqrt{\varepsilon_0}\left(\frac{z}{c}\right)$

BP, Brillouin precursor arrival time: $\sqrt{\varepsilon_0}\left(\frac{z}{c}\right)\left(1 + \frac{4\pi\alpha_0}{\varepsilon_0}\right)^{\frac{1}{2}}$

EP, Excitonic precursor arrival time: $\frac{z}{\delta}$

* This work was supported in part by FRAP-PSC CUNY and NASC.

REFERENCES.

1. A review of various optical transient effects connected with spatial dispersion is given in "Transient and Pulse Propagation in Linear Spatially Dispersive Media" by A. Puri and J.L. Birman, Ch. 22 of "Semiconductors Probed by Ultrafast Laser Spectroscopy", Vol. II, p.331, ed. by R.R. Alfano, (Acad. Press, NY, 1984).
2. A review of the ABC problem for spatial dispersive medium is given in "Electrodynamic and Nonlocal Optical Effects Mediated By Exciton Polaritons", by J.L. Birman, Ch. 2 of "Excitons", ed. by E.I. Rashba and M.D. Sturge (North-Holland Publishing Company, NY, 1982).
3. S.I. Pekar, Sov. Phys. JETP 6, 785 (1958).
4. J.J. Hopfield, D.G. Thomas, Phys. Rev. 132, 563 (1963).
5. V.A. Kiselev, Sov. Phys.-Sol. State 15, 2338 (1974).
6. J.J. Sein, J.L. Birman, Phys. Rev. B6, 2482 (1972); J.J. Sein, Ph.D. thesis, NYU (1969).
7. A.A. Maradudin, D.L. Mills, Phys. Rev. B1, 2787 (1973).
8. G.P. Agrawal, D.N. Pattanayak, and E. Wolf, Phys. Rev. Lett. 27, 1022 (1971); Phys. Rev. B10, 1477 (1974).
9. C.S. Ting, M. Frankel, and J.L. Birman, Sol. State Comm. 17, 1285 (1975).
10. D. Elert, Ann. Phys. (Leipzig) 7, 65 (1930).
11. G.P. Agrawal, et al., Phys. Rev. B25, 2715 (1982).
12. T. Shigenari, X.Z. Lu, and H.Z. Cummins, Phys. Rev. B30, 1962 (1984); M. Dagenais, W.F. Sharfin, Phys. Rev. Lett, 58, 1776 (1987).
13. A preliminary report "Transient Optical Reflectivity from Bounded Nonlocal Media" by S.V. Branis, K. Arya, and J.L. Birman was published in "Ultrafast Lasers Probe

Phenomena in Bulk and Microstructure Semiconductors", SPIE Proceedings, ed. R.R. Alfano, Vol. 793, (1987).

14. S.V. Branis, K. Arya, and J.L. Birman (to be published).

15. E.L. Ivchenko, "Spatial Dispersion Effects in the Exciton Resonance Region", Ch. 4 of "Excitons", ed. by E.I. Rashba and M.D. Sturge (North-Holland Publishing Company, NY, 1982).

16. T. Skettrup, Phys. Status Solidi (b)60, 695 (1973); I. Broser et al., Phys. Status Solidi (b)90, 77 (1978); J.S. Nkoma, J. Phys.: Sol. State, C16, 3713 (1983).

17. M. Frankel, J.L. Birman, Phys. Rev. A15, 2000 (1977).

18. G.H. Watson Jr., P.A. Fleury, and S.L. McCall, Phys. Rev. Lett. 58, 945 (1987).

PICOSECOND DYNAMICS OF EXCITONIC POLARITONS

IN THE BOTTLENECK REGION

J. Aaviksoo, A. Freiberg, J. Lippmaa and T. Reinot
Institute of Physics, Estonian SSR Academy of Science
202400 Tartu, USSR

1.   INTRODUCTION

Ultrashort light pulses have made possible the direct observation of complicated relaxation processes of material excitations in solids. The dynamics of this relaxation for photon-coupled excitations is reflected in the transient resonant secondary emission (RSE) spectrum. Strong wavelength dependence of RSE time evolution in the resonant region makes it necessary to use simultaneous spectral and temporal resolution. In the pioneering work of Toyozawa[1] it was pointed out that energy relaxation of excitons coupled to photons (polaritons) is not restricted to the exciton band, but can continue below the band bottom through the so called "polariton bottleneck." It was also shown that this bottleneck plays a crucial role in the formation of the RSE spectrum, as the relaxation is slowed down considerably near the bottleneck. The formation of the RSE spectrum in the polariton framework is the result of the relaxation of initially excited polaritons through multiple scattering by phonons and other defects of an ideal lattice (surfaces, impurities, etc.). The polariton emission has several characteristics features: i) the lineshape of the polariton emission reflects the quasistationary distribution (QSD) of polaritons in the sample (if the exciton lifetime is long compared to the characteristic timescale of relaxation processes), ii) two maxima appear in the spectrum due to the upper and lower polariton branches, iii) the spectral shape of the emission band depends on the crystal size and temperature.

The first time-resolved polariton luminescence experiment was carried out for CdS,[2] and established clearly the bottleneck for the relaxation of excitonic polaritons. A subnanosecond resolution experiment in anthracene revealed frequency-dependent decay of polaritons.[3] Recent picosecond experiments in semiconductors CuBr,[4] ZnTe,[5] CdSe[6] have allowed one to follow directly the formation of the polariton distribution near the bottleneck and its relaxation to still lower energies. Another interesting feature concerning the polariton dynamics is the complex temporal behaviour of the "Raman-like" line.[5] Investigation of this line using picosecond temporal resolution reveals new aspects of the Raman/luminescence distinction problem in the time domain.

We have tried to follow the kinetic energy relaxation processes in the polariton bottleneck region in two widely investigated model crystals - molecular crystal anthracene and direct gap semiconductor CdS.

2.   Experimental

Thin samples which were grown by sublimation, were immersed in liquid He and pumped down to 2K.   Observations were carried out in reflection goemetry perpendicular to the excitation.   As an excitation source, a mode-locked synchronously pumped dye laser was used, which yielded a train of 3 ps pulses. Secondary emission was analyzed by a 0.4 m double monochromator in the subtractive dispersion mount to avoid excess broadening of the time response. A synchroscan streak camera was used to obtain temporal resolution (dynamic range > 100, resolution 10 ps).   Steady-state spectra were recorded on the same setup with a standard photon counting apparatus.

3.   Polariton Dynamics in Anthracene Crystals

The emission spectrum of anthracene crystals at low temperatures has a well resolved structure consisting of 4-6 cm^{-1} broad phonon replicas of the pure exciton line (ν_0 = 25097 cm^{-1}) due to internal asymmetric shape that reflects the QSD of excitons (polaritons) at the band bottom. Besides the latter, as a result of scattering processes on acoustic phonons, a secondary broad distribution of polaritons is formed below the resonance frequency, that appears as a broad (20-40 cm^{-1}) emission band in the spectrum.[3] The kinetic behaviour of these emission bands directly reflects the formation and decay of the QSD at the band bottom and in the long wavelength region of the polariton bottleneck, respectively.   Our picosecond measurements show[7] that the narrow QSD is formed in 30 ps at 2 K and it decays in 980 ps as a result of scattering via acoustic phonons, that brings the polaritons below the bottleneck. The rise and decay times of the broad emission band depend on wavelength. The characteristic timescales are 100 ps and 1400 ps, respectively.   An important conclusion from the observed kinetics is that the broad distribution of polaritons tens of wavenumbers below the resonance frequency, can survive after the main distribution at the band bottom has disappeared.   It is clear that these fast polaritons below ν_0 contribute essentially to energy transfer in the crystal.

The dynamics of the polaritons immediately below ν_0 is best revealed under resonant excitation of these states.   In a time-of-flight experiment[8,9] the group velocities of polaritons were determined and a 10^4 - 10^5 times slowing down was observed which almost equalized the velocities of light and sound in the crystal. We have followed the kinetics of the "Raman-like" lines, that appear as distinct maxima in the secondary emission spectrum and shift together with excitation.   In a series of experiments, the excitation frequency was tuned below resonance and the time profiles of the 81 cm^{-1} librational line was recorded together with the time-of-flight profiles in the forward scattering geometry (Fig. 1).   The preresonant scattering process results in a complicated time behaviour of the "Raman-like" line, which can crudely be described as consisting of a "fast" and "slow" components. Both of these depend essentially on the detuning of the excitation. Off resonance, the fast component is more intense and its time profile follows closely that of the exciting pulse, whereas the slow component is weak and has a decay time which is long, but less than 1 ns. Approaching the resonance, the slow component, with a decay time of about 1-3 ns, strengthens and at a detuning of about 20 cm^{-1}, it governs the time profile. This should be compared to the corresponding time-of-flight profiles of the excitation pulses recorded under same experimental conditions. the latter consist of a leading pulse at zero delay, corresponding to the a-polarized component of the initial pulse and one or two weaker pulses which correspond to single or triple transits of the b-polarized component through the 18 μ m thick crystal. From this comparison a straightforward suggestion can be made: the group velocity of polaritons plays a significant role in the temporal behaviour of secondary emission. It is best illustrated in the central part of the Fig. 1. Every time a part of the exciting pulse leaves the crystal, a sudden drop in the scattering intensity is detected. Further, an important observation concerns the effect of temperature on the time profile of the scattering line. Increasing temperature enchances the slow component as shown in Fig. 1C. At even higher temperatures the fast component is quenched and only the slow component can be observed. These two components can be attributed to different channels that populate final polariton states - a single scattering by an optical phonon and multiple scattering by acoustic phonons. Close to the resonance the time evolution of a "Raman-like" line, however, does not provide the basis for the distinction between scattering and luminescence.

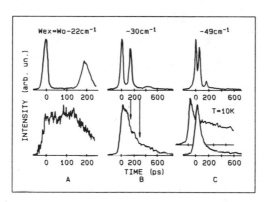

Fig. 1. Time-of-flight profiles of the exciting light pulses (upper curves) and the corresponding time profiles (lower curves) of the "81 cm^{-1} line" in the forward scattering geometry in an anthracene crystal (d = 18 μ m). The excitation frequency is: (A) 22 cm^{-1}, (B) 30 cm^{-1} and (C) 49 cm^{-1} below the exciton resonance. T = 1.8 K (A,B), T > 2 K (C).

4.   Relaxation of Excitonic Polaritons in CdS

The low temperature free exciton spectrum of CdS consists of a broad emission band at the exciton transition (E^A_T = 2.552 eV) and its LO phonon replicas. The main emission band has a typical double-peaked structure due to two polariton branches[10] (Fig. 2). We measured the decay times of polariton emission for exciton kinetic energies up to 10 meV. The decay is very fast (< 10 ps) in the high energy region ($E > E^A_T + 2$ meV), indicating that the kinetic energy relaxation due to LA phonons is effective.[11] The emission kinetics at the bottleneck region is more complicated - we see a rise time < 200 ps and a decay, which can be approximated with two exponentials. This can be explained as the formation and subsequent decay of the QSD of excitons above and at the bottleneck. The dependence of relaxation probabilities W on exciton kinetic energy is plotted in Fig. 2. The straight line shows the deformation potential approximation and corresponds to $E_C - E_V = 2.7$ eV, where E_C and E_V are the deformation potentials for the conduction and valence bands respectively.[11] The actual relationship is not linear, and therefore other relaxation mechanisms must exist in the high energy region.

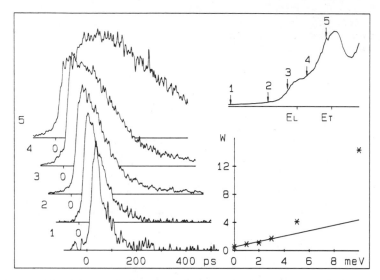

Fig. 2.   Decay kinetics of free exciton emission in CdS (left, a steady-state luminescence showing the energies of the kinetic curves (upper right) and the dependence of the scattering probability $(W*10^{-10})$ on exciton kinetic energy.

The emission kinetics of the zero-phonon band reflects the dynamics of exciton distribution in the surface layer due to the polariton effect. On the other hand the kinetics of the phonon replicas reflect the dynamics of QSD in the crystal bulk. This explains the observed difference between the emission kinetics of the zero-phonon band and the LO bands. We therefore measured the 2LO line, which involves the whole Brillouin zone of excitons. The observed rise time ($\approx$ 100 ps) describes the formation of QSD above the bottleneck and the decay time (440 ps) is determined by the trapping rate of free excitons into bound states.

Detailed analysis of time-resolved bound exciton emission shows, that these states are populated mainly by QSD of excitons at the free exciton band bottom. Under this assumption we can say that: i) bound excitons at neutral donors have a radiative decay time of 580 ps and radiative energy transfer plays an important role in the kinetics of these states, ii) bound excitons at neutral acceptors have a decay time of 1 ns and a rise time of 100 ps. The latter is most probably due to a complicated mechanism for populating of these states through I_{1B} intermediate states, which lie immediately below the E^A_T.[12]

We have also measured the decay times of two weak emission lines at 2.575 and 2.569 eV corresponding to the A(n=2) and B(n=1) excitons respectively. It is interesting to note that the lower B_1 exciton decays more than three times faster (τ = 12 ps) than the higher lying A_2 exciton (τ = 40 ps).

5. Conclusions

Our picosecond time-resolved measurements of RSE kinetics in anthracene and CdS allow us to make the following conclusions:

1. The emission kinetics depends strongly on wavelength and has, as a rule, nonexponential behaviour.

2. An initial ultrafast stage (<10 ps) of relaxation can be distinguished, that populates the polariton states around the bottleneck uniformly according to the density-of-states function.

3. The formation time of the quasistationary distribution of excitons at the band bottom is 30 ps for anthracene and 100 ps for CdS.

4. Group velocity dispersion near the bottleneck plays the most important role in polariton dynamics.

5. In anthracene, a secondary, long-lived (<1 ns) distribution of polaritons exists below the bottleneck. These polaritons contribute essentially to the observed excitonic energy transfer in the crystal.

6. Energy transfer to localized states in CdS is mediated by the quasistationary distribution of excitons near the band bottom.

7. An adequate theoretical model must account for the inhomogenous spatial distribution of polaritons.

8. A spectrochronographic analysis of the secondary emission of polaritons reveals new aspects of the scattering/luminescence distinction problem - the fast scattering vs. slow luminescence approach appears inadequate.

References

1. Y. Toyozawa, Prog. Theor. Suppl. 12:111 (1959).
2. U. Heium and P. Wiesner, Phys. Rev. Lett. 30:1205 (1975); P. Wiesner and U. Heim, Phys. Rev. B 11:3071 (1975).
3. M. D. Galanin, E. N. Myasnikov, Sh. D. Khan-Magometova, Mol. Cryst. Liquid Cryst. 57:119 (1980).
4. Y. Masumoto, S. Shionoya, J. Phys. Soc. Japan 51:181 (1982).
5. T. Kushida, T. Kinoshita, F. Ueno, T. Ohtsuki, ibid. 52:1838 (1983).
6. Y. Masumoto and S. Shionoya, Phys. Rev. B 2 30:1076 (1984).
7. J. Aaviksoo, A. Freiberg, T. Reinot, S. Savikhin, J. Lumin. 35:267 (1986).
8. J. Aaviksoo, J. Lippmaa, A. Freiberg, A. Anijalg, Solid State Commun. 49:115 (1984).
9. N. A. Vidmont, A. A. Maksimov, I. I. Tartakovskii, JETP Lett. 37:689 (1983).
10. E. Gross, S. Permogorov, V. Travnikov, A. Selkin, Solid State Commun. 10 11:1071 (1972).
11. Y. Toyozawa, Progr. Theor. Phys. 20:53 (1958).
12. R. Planel, A. Bonnot, C. Benoit a la Guillaume, Phys. Stat. Sol. 1 58:251 (1973).

POLARITON WAVES NEAR THE THRESHOLD FOR STIMULATED SCATTERING

N. A. Gippius, L. V. Keldysh, and S. G. Tikhodeev
Academy of Sciences of the USSR
P.N. Lebedev Physical Institute, Moscow, and
Institute for General Physics, Moscow

1.   We investigated the propagation in a crystal of a polariton wave whose amplitude is close to the threshold for stimulated Mandelstam-Brillouin scattering.   In this situation, the interaction of the initial wave with the scattered polariton and the phonon noise becomes appreciable.   Correlations between scattered polaritons and phonons emitted during this scattering lead to the formation of mixed phonon-polariton ("phonoriton") modes.[1] Near threshold the decay rate of one of these mixed modes vanishes, and the number of quanta in the mode grows. Subsequent backscattering of these quanta in the forward direction leads to the development of intense fluctuations consisting of correlated polariton pairs, whose frequencies and propagation directions are close to those of the initial wave.[2] This significantly modifies the polariton spectrum: a pseudogap arises with frequency close to that of the initial wave.

2. Formal investigation of the physical picture described above was carried out within the framework of the diagram technique for noneequilibrium processes,[3,4] using Green's functions in the matrix form

$$G = \begin{bmatrix} 0 & G^a \\ G^r & F \end{bmatrix}$$

Here, the retarded G^r and advanced G^a functions describe the response of the system to the external field while the correlation function F describes fluctuations.   In particular, the imaginary part of G^r is connected with the density of states

$$W(\varepsilon) = \int \left| \text{Im } G^r(\vec{p},\varepsilon) \right| d^3\vec{p}/(2\pi)^3,$$

and the quantity

$$N(\varepsilon) = \int N(\vec{p},\varepsilon) d^3\vec{p}/(2\pi)^3 = \int \{iF/2 + \text{Im}G^r\} d^3\vec{p}/(2\pi)^3$$

is the spectral intensity of noise. In thermodynamic equilibrium F is uniquely realted to G^r by the fluctuation-dissipation theorem, but under conditions far from equilibrium as considered here, it becomes a fully autonomous nontrivial physical characteristic of the system. As in the Feynman diagram technique, we obtain Dyson-type equations, which connect functions G with free functions G_0 and self-energy (polarization) operators Σ.

As was noted above, a characteristic feature of this system is the onset of the correlation of the scattered polaritons with emitted phonons, and also of the scattered polaritons with one another. From the formal standpoint, this coherence is accounted for in a natural way by the appearance of the so-called anomalous Green's functions, similar to those introduced by Belyaev[5,6] and by Gor'kov[7,6] in superfluidity and superconductivity theories, respectively. This substantially alters the character of the equations that describe the spectral distribution of the scattered particles. The usual kinetic equation for an incoherent strongly nonequilibrium many-particle system is replaced by equations similar to the equation for the density matrix of a two-level system in a resonant external field well known in quantum radiophysics. In these equations anomalous Green's functions take the part of the off-diagonal matrix elements. Before writing out these equations, let us comment on the types of particles they describe.

The lower polariton branch with its dispersion law $\varepsilon = \varepsilon(\vec{p})$ is shown in Fig. 1. The mode $p_0 = (\vec{p}_0, \varepsilon_0)$ corresponds to the coherent initial wave. The matrix element for interaction with phonons grows with momentum transfer, and is maximal for backward scattering into the modes $p_\pm = (\vec{p}_\pm, \varepsilon_\pm)$, where

$$\varepsilon_\pm \equiv \varepsilon(\vec{p}_\pm) = \varepsilon(\vec{p}_0) \mp u \, |\vec{p}_\pm - \vec{p}_0|,$$

u is the speed of sound, the upper sign and the lower sign refer to the anti-Stokes and Stokes scattering, respectively. As a result, the above-mentioned mixing of polaritons with phonons is greatest for frequency and momentum around p_+. Forward-rescattered polaritons produce fluctuations of the initial wave in some region close to p_0.

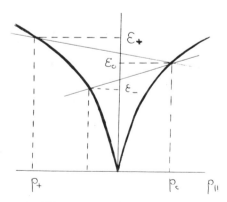

Fig. 1. Polariton branch (1); along with the spectra of absorbed (2); and emitted (3) phonons.

The equations describing the system are shown in Fig. 2, where the Green's functions of back-scattered polaritons G_{pol}, of phonons G_{ph}, and of forward-rescattered polaritons D are denoted by double lines (see Fig. 3a), the free-particle functions by single lines (Fig. 3b). The interaction (scattering) is denoted by vertices (Fig. 3c) which couple Green's functions of different types. The coherent initial wave is accounted for as in superfluidity theory[5] by anomalous vertices $\Phi = \Phi^* = \mu(2n_0)^{1/2}$ (Fig. 3d), where $n_0 = I/n\epsilon_0 c_0$ is the spatial density of the coherent-mode polaritons, I is the intensity of the initial wave, μ the matrix element for phonon-polariton interaction. The anomalous function G_X (-p,p) (Fig. 3e) accounts for the coherency between backward-scattered polaritons and emitted phonons, D_X (-p,p) (Fig. 3f) between forward-rescattered polaritons. The simplest diagrams for the functions G_X and D_X are shown in in Fig. 3g-3h. The polarization operators $\Sigma_{pol,ph,02,20}$, and $\Gamma_{11,02,20}$ are denoted in Fig. 2 by circles; some diagrams for them are shown in Fig. 4.

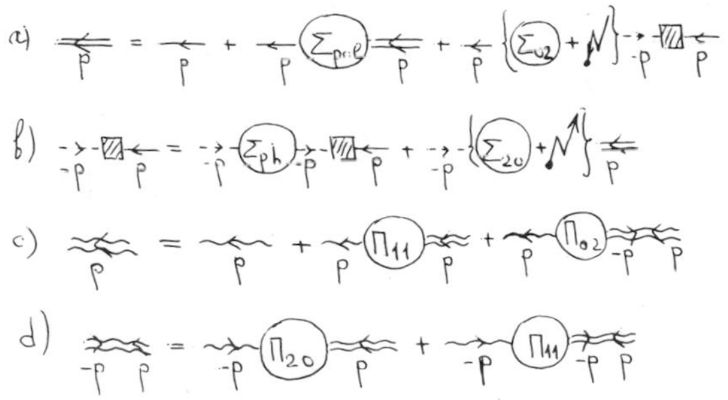

Fig. 2. Equations for Green's functions G_{pol} (a); G_X (b); D (c); D_X (d).

3. In the simplest approximation (which we call the τ-approximation) $\Sigma_{20,02} = \Pi_{20,02} = 0$ and the operators $\Sigma_{pol,ph}, \Pi_{11}$ take the standard form, imposed by the fluctuation -dissipation theorem

$$\Sigma_\alpha, \ \Pi_{11} = \begin{bmatrix} -2i\gamma_\alpha(1 - 2N_\alpha) & -i\gamma_\alpha \\ i\gamma_\alpha & 0 \end{bmatrix}$$

where $\alpha = $ pol,ph; γ_α are the inverse lifetimes, and $N_\alpha = [\exp(\hbar\omega_\alpha/k_BT) - 1]^{-1}$ the occupation numbers for polaritons and phonons in thermal equilibrium.

323

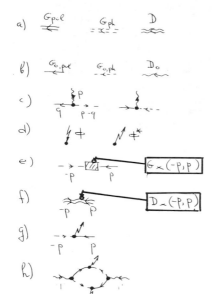

Fig. 3. The graphic notation for Fig. 2. (For details see the text). In Fig. 2-4 p means $k-p_\pm$ and $k-p_0$ for backward-scattered and forward-scattered polaritons, $k-p_0 + p_\pm$ for phonons and k is momentum and frequency of corresponding particle.

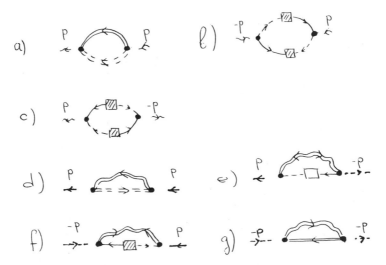

Fig. 4. One-loop diagrams for the polarization operators $\Pi_{11}(P)$ (a), $\Pi_{20}(-p,p)$ (b); $\Pi_{02}(p,-p)$ (c); $\Sigma_{pol}(p)$ (d); $\Sigma_{02}(p,-p)$ (e); $\Sigma_{20}(-p,p)$ (f); $\Sigma_{ph}(-p)$ (g).

In the τ-approximation the equations shown in Fig. 2 can be solved explicitly.[2] The poles of G^r_{pol}-functions yield the "phonoriton" excitation spectrum

$$\varepsilon_{1,2}(\vec{p}) = \frac{1}{2} \{\varepsilon_{\pm}(\vec{p}) - \omega_{\pm}(-\vec{p}) - i\Gamma + (-)[(\varepsilon_{\pm}(\vec{p}) + \omega_{\pm}(-\vec{p}) - i\gamma)^2 \pm 4\phi^2]^{1/2}\}$$

where $\Gamma = \gamma_{pol} + \gamma_{ph}$; $\gamma = \gamma_{pol} - \gamma_{ph}$; $\varepsilon_{\pm}(\vec{p}) = \varepsilon(\vec{p} + \vec{p}_{\pm}) - \varepsilon_{\pm}$; $\omega_{\pm} = \pm u|\vec{p} + \vec{p}_o - \vec{p}_{\pm}| + \varepsilon_{\pm} - \varepsilon_o$.

Anti-Stokes scattering was analysed in Ref. 1, in which it was shown that a gap in the polariton and phonon spectra arises as the intensity of the intial wave increases. A typical dependence of the polariton density of states $|\text{Im } G^r_{pol}|$ on ε and $p_{||}$ at $p_{\perp} = 0$ is shown in Fig. 5. Some evidence of phonoriton modification of the spectrum in CdS was obtained experimentally.[8]

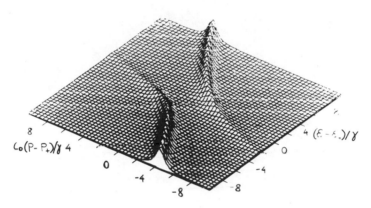

Fig. 5. A typical dependence of Im G^r_{pol} on ε, $p_{||}$ in the region close to the anti-Stokes resonance $(p_{||} = \vec{p} \, \vec{p}_o / |\vec{p}_o|)$.

A different picture takes place in the case of Stokes scattering (see Fig. 6). In sufficiently strong fields at $I \sim I_c$, the sign of the damping of one of the phonoriton modes is reversed. The quantity $I_c = \gamma_{pol}\gamma_{ph}/^2\mu^2\hbar\varepsilon_o c_o$ is the threshold for stimulated Mandelstam-Brillouin scattering.

4. Of special interest is the behavior of the system near the threshold; for $I = I_c (1-\lambda)$, $0 < \lambda \ll 1$. The damping of one of the phonoriton modes becomes small in the vicinity of $\vec{p}$:

$$\text{Im}\,\varepsilon_2(\vec{p}) \sim \gamma(\lambda + a\vec{p}^2), \text{ where } \gamma = \gamma_{pol}\,\gamma_{ph}\Gamma^{-1}; \; a \sim (c_o\Gamma^{-1})^2.$$

As a result the number of quanta in the mode grows and the feedback effect of this noise on the propagation of the initial wave becomes appreciable. A self-consistent approximation which takes into account one-loop diagrams for the polarization operators (see Fig. 4) was developed in Ref. 2. For simplicity a one-dimensional system was

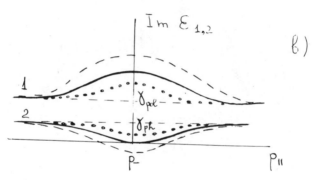

Fig. 6. (a) the dependence of $\mathrm{Re}\,\varepsilon_1(p_{||})$ (curve 1) and $\mathrm{Re}\,\varepsilon_2$ (curve 2); (b) the same for $\mathrm{Im}\,\varepsilon_1(p_{|1})$ (curve 1) and $\mathrm{Im}\,\varepsilon_2(p_{||})$ (curve 2). The dependence of $\mathrm{Im}\,\varepsilon_{1,2}$ shown in Fig. 6b by dotted line corresponds to $I<I_c$, while by dashed line - to $I>I_c$.

investigated. Green's functions for forward-rescattered polaritons have the form

$$D(p) = 2^{-1}[D_1(p)+D_2(p)], \quad D_X(-p,p)=2^{-1}(D_1-D_2),$$

where

$$D_{1,2} = \begin{bmatrix} 0 & (\varepsilon-c_0\vec{p} - \Pi^a{}_{1,2})^{-1} \\ c.c. & \Delta_{1,2}\,|\varepsilon-c_0\vec{p} - \Pi^a{}_{1,2}|^{-2} \end{bmatrix}$$

The polarizations operators $\Pi_{a1,2}$, $\Delta_{1,2}$ are large in the vicinity of $\varepsilon = o, \vec{p} = 0$. In particular, for $|\varepsilon| \lesssim \lambda\gamma$, $|\vec{p}| \lesssim \lambda^{1/2} \gamma/c_0$; $\Pi^a{}_1 \sim \Delta_1 \sim i\lambda^{-1/2}\gamma$; $\Pi^a{}_2 \sim i\lambda^{-3/2}\gamma$, $\Delta_2 \sim -i\lambda^{-5/2}\gamma$. As a result a pseudogap arises in the forward-rescattered polariton density of states: $|\text{Im } D^r(0,0)| \sim \lambda^{1/2}\gamma^{-1}$. It can be shown that the width of the gap is of the order of $\lambda^{-1/4}\gamma$. The physical reason for a pseudogap formation is the following. We have seen above that a pseudogap arises in the region of anti-Stokes backward scattering. The same phenomenon takes place in our case, because the forward-rescattering of the Stokes backscattered polaritons is the anti-Stokes process.

Typical dependence of the densities of states $|\text{Im } D^r(\varepsilon,\vec{p})|$ and $W(\varepsilon) = \int |\text{Im } D^r(\varepsilon,\vec{p})d\vec{p}|2\pi$ for different values of λ are shown in Fig. 7, 8. Close to the threshold narrow spikes in the density of states appear located at $\vec{p}_m \sim 0$, $\varepsilon_m - A \gamma \lambda^{-1/4}$ (where $A \sim 1$), along the edges of the pseudogap. The spikes are caused eventually by the gap in the framework of sum rules. The behavior of the spectral intensities of the forward-scattered polaritons is illustrated in Fig. 9, 10.

$$\lambda = 10^{-2}$$

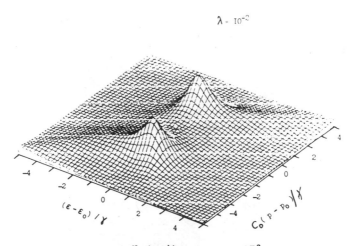

Fig. 7. The dependence of $\text{Im } D^r (\varepsilon,\vec{p})$ for $\lambda = 10^{-2}$.

The results given were obtained in the one-loop approximation for the polarization operators. In Ref. 2 it was shown that close to the threshold higher-order diagrams are of the same order of magnitude as the one-loop diagrams. It is natural to assume that the exact polarization operators stay large at $p = 0$ and the conclusion about the formation of the pseudogap and the spikes remains valid. Then, only the critical exponents will change in the asymptotic formulae cited above, as in the theory of phase transitions.

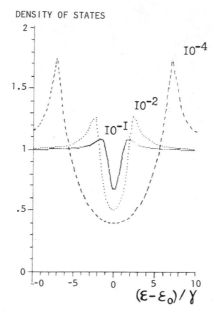

Fig. 8. The dependence of $W(\varepsilon)$ for $\lambda = 10^{-1}$, 10^{-2}, 10^{-4}.

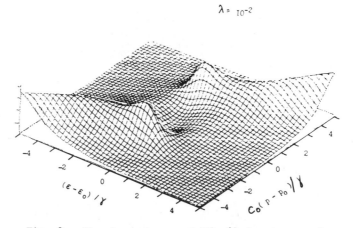

Fig. 9. The dependence of $N(\varepsilon, \vec{p})$ for $\lambda = 10^{-2}$.

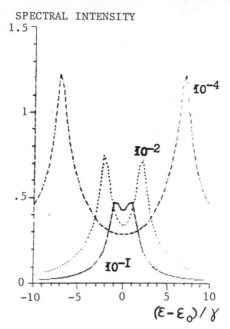

SPECTRAL INTENSITY

Fig. 10. The dependence of $N(\epsilon)$ for $\lambda = 10^{-1}$, 10^{-2}, 10^{-4}.

References

1. A. L. Ivanov, L. V. Keldysh, Sov. Phys. JETP 57:234 (1983).
2. L. V. Keldysh, S. G. Tikhodeev, ibid. 63:1086 (1986).
3. L. V. Keldysh, ibid. 20:1018 (1964).
4. E. M. Lifshitz and L. P. Pitaevskii, Physical Kinetics, Pergamon, Chap. 10 (1981).
5. S. T. Belyaev, Sov. Phys. JETP 7:289, 299 (1958).
6. A. A. Abrikosov, L. P. Gor'kov, I. E. Dzyaloshinskii, Quantum Field Theoretical Methods in Statistical Physics, Pergamon, (1965).
7. L. P. Gor'kov, Sov. Phys. JETP 9:1364 (1959).
8. G. S. Vygovskii, et al., Zh. Eksp. Teor. Fiz. Pis'ma 49:134 (1985).

QUANTUM FLUCTUATIONS AND STATISTICAL

PROPERTIES OF INTENSE POLARITON WAVES

S. A. Moskalenko, A. H. Rotaru, and Yu.M. Shvera

Institute of Applied Physics, Academy of Sciences of the
Moldavian SSR, Kishinev, USSR

INTRODUCTION

The dynamics of the maintaining order in complex non-linear systems has recently been attracting great interest. The ordering process is related to the cooperative behaviour of the system. The appearance of order is due to some selective instability and to the competition between different possibilities and constituent parts of the system. Consideration of fluctuations in such a system is very important, especially in the vicinity of the order-disorder phase transition.[1]

The possibility of Bose-Einstein condensation of excitons, biexcitons or polaritons induced by resonant laser radiation is the topic of many articles.[2-7] In particular, theoretical investigation of the properties of the coherent polariton wave induced by an intense light beam and its interaction with phonons was discussed in a number of papers.[6,7] The main peculiarities of the polariton-phonon system, excited in a resonant way are related to establishing the phonoriton energy spectrum and the existence of a threshold concentration of coherent polaritons: above such a concentration induced resonant Mandelstam-Brillouin scattering takes place.

In our paper we deal with similar properties of the coherent polariton wave in the case when polariton-polariton interaction is taken into account instead of polariton-phonon one. There are also changes in the energy spectrum of the incoherent polariton modes. They were investigated for the first time.[8] We pay attention not to the changes of the energy spectrum, but to the influence of the quantum and thermal fluctuation on establishing the coherent polariton mode in the presence of an external coherent driving field. The Fokker-Planck equation (FPE), which describes the coherent mode in a selfconsistent way, will be obtained. Coefficients in the FPE depend on the mean values of the occupation numbers of the incoherent polaritons, which in their turn depend on the mean number of coherent polaritons. The first attempt in this direction was initiated by the work.[9] But in this work only the deterministic approximation was considered without taking into account the diffusion terms in FPE which have the same order value as the drift ones. The selfconsistent character of the problem was not discussed either.

The Master-Equation for the Reduced Density Matrix and the Kinetic Equations

The Hamiltonian of the system contains only the creation and annihilation operators $a^+_{\vec{k}}$, $a_{\vec{k}}$ of the lower polariton branch

$$H = \sum_{\vec{k}} \hbar\omega(\vec{k})\, a^+_{\vec{k}} a_{\vec{k}} + i\,[d_p E_0 e^{-i\omega_L t} a^+_{\vec{k}_0}$$

(1)

$$-(dpE_0)^* e^{i\omega_L t}\, a_{\vec{k}_0} + \frac{1}{2V} \sum_{\vec{p}\vec{q}\vec{k}} g a^+_{\vec{p}} a^+_{\vec{q}} a_{\vec{q}+\vec{k}} a_{\vec{p}-\vec{k}}$$

The first term represents free polaritons, the second one - the coherent pumping and the third one - the polariton-polariton interaction which leads to scattering within the same branch. For simplicity it was assumed that exciton-exciton interaction constant g and the dipole momentum d_p do not depend on wave vectors. $\omega(\vec{k})$ represents the lower polariton dispersion curve, whereas E_0 and ω_L are amplitude and frequency of the external coherent laser field. V is the volume of the system. The laser field induces a macroscopically large occupation number of a single particle state with wave vector $\vec{k}_0$ of the Bose-Einstein condensation type. Further this state will be denoted as a coherent polariton state. The operators $a^+_{\vec{k}_0}$, $a_{\vec{k}_0}$ of the coherent mode can be considered proportional to $\sqrt{V}$, whereas the other operators $a^+_{\vec{k}_0+\vec{k}}$ with $\vec{k}$ = 0 have a quantum nature and their mean occupation numbers $\bar{n}_{\vec{k}+\vec{k}}$ = $\langle a^+_{\vec{k}_0+\vec{k}} a_{\vec{k}_0 \vec{k}} \rangle$ are of the order of I. The polariton system can be devided into two parts. One of them consists only of coherent polaritons with wave vector $\vec{k}_0$. Another one is formed in incoherent polaritons with wave vector $\vec{k}_0+\vec{k}$. They appear due to the processes of real loss of quasiparticles form the condensate. The density operator $\hat{\rho}(0)$ at time t =0 can be represented in a factorized way (II)

$$\hat{\rho}(0) = \hat{\rho}_c(0)\,\hat{\rho}_{inc}(0); \quad \hat{\rho}_c(0) \quad \hat{\rho}_{\vec{k}_0}(0); \quad \hat{\rho}_{inc}(0) \quad \hat{\rho}_{\{\vec{k}_0+\vec{k}\}}(0)$$

(2)

This approximation together with the assumption of the Markovian character of the processes involved form the basis of the derivation of the master-equation for the reduced density operator of the coherent mode. On the same basis the kinetic equations for the average values $\bar{n}_{\vec{k}_0+\vec{k}}$ of the incoherent polaritons were obtained. In agreement with the assumption of Markovian character of the processes involved we can extract the density operator from the integrand due to its slow variation in time. As a result we have obtained the following master equation for the reduced density operator of the coherent polaritons in a rotating reference system

$$\frac{\partial \hat{\rho}_c}{\partial t} = -i\left[(\omega(\vec{k}_0) - \omega_L)\,\hat{N}_0 + i\left(\frac{d_p E_0}{\hbar}\right) a^+_0 - i\left(\frac{d_p E_0}{h}\right)^* a^+_0, \hat{\rho}_c\right] -$$

$$- i\left\{\frac{\ell}{\hbar}\,[\hat{N}_0, \hat{\rho}_c] + \frac{g}{2\hbar V}\,[\hat{B}, \hat{\rho}_c] + \frac{f_1}{2V}\,([\hat{A}^+, \hat{A}\hat{\rho}_c]-\right.$$

$$\left. - [\hat{\rho}_c \hat{A}^+, \hat{A}]) + \frac{g_1}{2V}\,([\hat{\rho}_c \hat{A}, \hat{A}^+] - [\hat{A}, \hat{A}^+ \hat{\rho}_c]) + \right.$$

$$+ \; P_1([a_o^+, a_o \hat{\rho}_C] - [\hat{\rho}_C a_o^+, a_o] + g_1 \; ([\hat{\rho}_C a_o, a_o^+] -$$

$$- \; [a_o, a_o^+ \hat{\rho}_C])\} - \{\frac{f_2}{2V} \; ([\hat{A}^+, \hat{A}\hat{\rho}_C] + [\hat{\rho}_C \hat{A}^+, \hat{A}]) +$$

$$+ \; \frac{g_2}{2V} \; ([\hat{A}, \hat{A}^+ \hat{\rho}_C] + [\hat{\rho}_C \hat{A}, \hat{A}^+] + p_2 \; [a_o^+, a_o \hat{\rho}_C] +$$

$$+ \; [\hat{\rho}_C a_o^+, a_o]) + q_2 \; ([a_o, a_o^+ \hat{\rho}_C] + [\hat{\rho}_C a_o, a_o^+])\} \qquad (3)$$

Here we have introduced the symbols

$$a^+_{\vec{k}_0} = a_o, \quad a_{\vec{k}_0} = a_o, \quad \hat{N}_o = a^+_o a_o, \quad \hat{A} = a_o a_o,$$

$$\hat{B} = a^+_o a^+_o a_o a_o, \quad \hat{C} = a_o a_o a^+_o a^+_o$$

The coefficients, ℓ, f_i, g_i, p_i, q_i, where $i = I, 2$, depend on the average values of the occupation numbers $\bar{n}_{\vec{k}_0 + \vec{k}}$. They are defined as

$$\bar{n}_{\vec{k}_0 + \vec{k}} \quad \bar{n}_{\vec{k}} = S_p \; (a^+_{\vec{k}_0 + \vec{k}} \; a_{\vec{k} + \vec{k}}) \qquad (4)$$

and obey the following kinetic equations

$$\frac{\partial \bar{n}_{\vec{k}}}{\partial t} = \frac{2 \pi g^2}{\hbar^2 V^2} \delta \; (\Omega(\vec{k}) \; [<B>(1 + \bar{n}_{\vec{k}}) \; (1 + \bar{n}_{-\vec{k}} -$$

$$- \; <C> \bar{n}_{\vec{k}} \bar{n}_{-\vec{k}}] + \frac{4 \pi g^2}{\hbar^2 V^2} \sum_{\vec{q}} \delta \; (\Theta(\vec{k}, \vec{q})) \; [<1 + \hat{N}_o> \; \times$$

$$\bar{n}_{-\vec{q}} \bar{n}_{q \rightarrow + \vec{k}} \; (1 + \bar{n}_{\vec{k}}) - <\hat{N}_o> \bar{n}_{\vec{k}} \; (1 + \bar{n}_{-\vec{k} + \vec{q}}) \; (1 + \bar{n}_{-\vec{q}})] +$$

$$+ \; \frac{8 \pi g^2}{\hbar^2 V^2} \sum_{\vec{p}} \delta \; (\Theta(\vec{p}, -\vec{k})) \; [<\hat{N}_o>n_{p \rightarrow}(1 + \bar{n}_{\vec{k}}) \; (1 + \bar{n}_{\vec{p} - \vec{k}}) -$$

$$- \; <1 + \hat{N}_o> \bar{n}_{\vec{k}} \bar{N}_{\vec{p} - \vec{k}} \; (1 + \bar{n}_{\vec{k}})] \qquad (5)$$

The symbol $< \; >$ denotes the average of the coherent mode operators. The frequencies Ω $(\vec{k})$ and Θ $(\vec{p}\vec{q})$ are the following

$$\Omega(\vec{k}) = 2\omega(\vec{k}_0) - \omega(\vec{k}_0 + \vec{k}) - \omega(\vec{k}_0 - \vec{k})$$

$$\Theta(\vec{p}, \vec{q}) = \omega(\vec{k}_0) + \omega(\vec{k}_0 + \vec{p}) - \omega(\vec{k}_0 - \vec{q}) - \omega(\vec{k}_0 + \vec{p} + \vec{q}) \qquad (6)$$

They show explicitly that in the scattering processes either two coherent and two incoherent polaritons or one coherent and three incoherent quasiparticles take part. The steady-state value of the incoherent polariton distribution function can be found only if the singular $\delta(\Omega(\vec{k}))$-function standing outside of the collision integral is substituted by a slow varying Lorentz-type function. The justification for such a substitution is related to the existence of the reservoir in addition to the polariton-polariton interaction. The reservoir can be formed of impurities, and crystal surfaces and leads to finite life time for the polaritons. Practically it means that the introduction of an additional damping rate in the kinetic equations goes side by side with δ-function smoothing. Another restriction of our solution is related to the assumption that incoherent polariton occupation numbers are less than 1. Under these conditions we can neglect the scattering integrals containing the products of two and three average numbers of incoherent polaritons because these terms are smaller than the ones containing only one value $\bar{n}_{\vec{k}}$. The steady-state solution has the form

$$\bar{n}_{\vec{k}} = \frac{1}{2} \frac{b}{n^2_c(\vec{k})-b} \frac{\gamma^2(\vec{k})}{\Omega^2(\vec{k})+\gamma^2(\vec{k})}$$

$$n_c^2(\vec{k}) = (\hbar\gamma(\kappa))^2/g^2 \tag{7}$$

A very important feature of this solution is the existence of the threshold for the average $b = V^{-2} \langle \hat{B} \rangle$, which must not exceed the minimal value from among a set of values $n_c^2(\vec{k})$. For simplicity we neglect the dependence of $n_c(\vec{k})$ on the wave vector $\vec{k}$ and employ the n_c at the point $\vec{k} = 0$. The threshold concentration n_c is inversely proportional to the interaction constant g and directly proportional to the damping rate γ. This result is similar to the one obtained for the first time in the case of polariton-phonon interaction.[7] We confine ourselves to the region $K_0 < K_0^*$ where K_0^* corresponds to the point with a minimal polariton group velocity on the dispersion curve. Namely in the region $K_0 < K_0^*$ the real appearance of the polariton from the condensate mode is possible. The nonequilibrium distribution function (7) represents two overlapping Lorentz-type curves symmetrically located on each side of the vicinity of the point K_0. The more complicated scattering processes will give rise to an asymmetrical behaviour of $\bar{n}_{\vec{k}}$. With the help of (7) we obtain the steady-state expressions for the coefficients ℓ, f_i, g_i, p_i, q_i. On the one hand, they determine the master-equation and the averages of the coherent mode operators. On the other hand, the coefficients in their turn depend on the averages mentioned above. Thus, the problem has to be solved selfconsistently.

The Fokker-Planck Equation

The mean-field approximation is probably not sufficient in the case of a comparatively low density of coherent polaritons. But even in its framework there appears a new feature of the master equation solution and of the average values calculated on this basis. To avoid the mean field approximation we have to search for a more exact solution of the master equation. Such solution can be obtained in the generalized P-representation.[10] The quantum and thermal fluctuations were included via FPE, and its solution was found for the generalized P-function. In contrast to them the coefficients in the FPE as well as the ones in the master-equation depend self-consistently on the concentration of coherent polaritons. The generalized P-representation has the form[10]

$$\hat{\rho}_c(t) = \int dx \int d\beta P(\alpha,\beta,t) |\alpha\rangle \langle\beta^*|/\langle\beta^*|\alpha\rangle \tag{8}$$

Here $|\alpha\rangle$ and $|\beta\rangle$ are coherent states; α and β are considered as independent complex variables. With the help of (8) we find the FPE as follows:

$$\frac{\partial P(\xi,\eta,t)}{\partial t} = [\frac{\partial}{\partial\xi} (k\xi + 2\chi\xi^2\eta - (\frac{d_p\varepsilon_0}{h})) -$$

$$- \chi \frac{\partial_2}{\partial\xi^2} \xi^2 + \frac{\partial}{\partial\eta} (k^*\eta + 2\chi^*\eta^2\xi - (\frac{d_p\varepsilon_0}{h})^*) -$$

$$- \chi^* \frac{\partial^2}{\partial\eta^2} \eta^2] P(\xi,\eta,t) \tag{9}$$

Here the higher order derivatives were neglected and the following notation was used

$$\xi = \alpha \sqrt{\frac{a_{ex}^3}{V}}, \quad \eta = \beta \sqrt{\frac{a_{ex}^3}{V}}, \quad \epsilon_0 = E_0 \sqrt{\frac{a_{ex}^3}{V}},$$

$$k = i\,\Delta\bar{\omega} + \Theta, \quad \chi = i\mu + \delta, \quad \Delta\omega = \omega(k_0) - \omega_L,$$

$$\Delta\bar{\omega} = \Delta\omega + \frac{\ell}{\hbar} + p_I - q_I, \quad \delta = \frac{1}{2a_{ex}^3}(f_2 - g_2),$$

$$\mu = \frac{g}{ha_{ex}^3} + \frac{1}{2a_{ex}^3}(f_1 - g_1), \quad \Theta = \frac{\gamma}{2} + P_2 - q_2 \qquad (10)$$

One can see that along with the diagonal diffusion terms there are also nondiagonal ones. But the coefficient in the latter case depend only on the thermal fluctuation and are less than coefficients of the diagonal terms as long as the mean numbers of incoherent polaritons are small. This fact permits us to neglect nondiagonal terms and to employ the steady-state solution of FPE[10]

$$P(\xi,\eta) = \xi^{c-2}\eta^{d-2} \exp\left[\frac{z}{\xi} + \frac{z^*}{\eta} + 2\xi\eta\right] \qquad (11)$$

where

$$c = (\frac{k}{\chi}), \quad d = (\frac{k}{\chi})^*, \quad z = (\frac{dp\epsilon_0}{h\chi}). \qquad (12)$$

The average values of coherent polariton operators of the type $a^+_0{}^p a_0{}^q$ are expressed in terms of generalized hypergeometric series ${}_0F_2(c,d,x)$ as follows[10]

$$\langle a^+_0{}^p a_0{}^q \rangle = (\frac{V}{a_{ex}^3})^{(\frac{p+q}{2})} z^q z^{*p} \frac{{}_0F_2(c+q,d+p,2|z|^2)\Gamma(c)\Gamma(d)}{{}_0F_2(c,d,2|z|^2)\Gamma(c+q)\Gamma(d+p)} \qquad (13)$$

Further investigation can only be done in some particular cases. We chose the model of exciton polariton common to crystals of CdS type. Taking into account the large values of the constant χ due to the big factor a^{-3}_{ex}, we confined ourselves to the case $|C|^2 < 1$ and $2|Z|^2 < 1$. With this restriction some simplifications can be made. They permit us to obtain a selfconsistent system of algebraic nonlinear equations for the values $|C|^2$, $|Z|^2$ and $b < n_c^2$. The coefficients of these equations can be expressed by use of exciton and polariton parameters such as the Rudberg constant, reduced mass, radius and transverse exciton frequency. To deal with the polariton dispersion function $\omega(\vec{K}_0 + \vec{k})$ we employ a series expansion in the neighborhood of the point $\vec{K}_0$, where $|\vec{k}|$ is less than K_0. The system obtained solves in principle the problem. Numerical estimation shows that the incoherent polariton spatial density n_I is small in comparison with n_0.

In conclusion we should like to point out the intrinsic relation of this model to the theory of a single mode semiconductor quantum generator.[12]

References

1. Nonlinear waves. Selforganisation, edited by A. V. Gaponov-Grehov and M. I. Rabinovich Nauka, Moscow, (1983).
2. S. A. Moskalenko, Fiz. Tverd. Tela 4:276 (1962).
3. L. V. Keldysh, in Problems of Theoretical Physics, Nauka, Moscow, (1972), p. 433-444.
4. V. F. Elesin and Yu. V. Kopaev, Zh. Eksp. Fiz. 63:1447 (1972).
5. N. Peyghambarian, L. L. Chase, and A. Mysyrowicz, Phys. Rev. B 27:2325 (1983).
6. A. L. Ivanov, L. V. Keldysh, Zh. Eksp. Teor. Fiz. 84:404 (1983).
7. L. V. Keldysh and S. G. Tihodeev, Zh. Eksp. Teor. Fiz. 91:78 (1986).
8. M. I. Shmigliuj and V. N. Pitey, Ukr. Fiz. Zh. 30:56 (1985); ibid. 31:1670 (1986).
9. M. L. Steyn-Ross and C. W. Gardiner, Phys. Rev. A 27:310 (1983).
10. P. D. Drummond and D. F. Walls, J. Phys. A 13:725 (1980).
11. F. Arecchi, M. Scully, H. Haken, and W. Weidlich, Quantum Fluctuations of Laser Radiation, Mir, Moscow (1974).
12. V. F. Elesin and E. B. Levchenko, Izv. Vuzov. Radiofiz. 22:130 (1979).

TRANSPORT AND QUANTUM STATISTICS OF EXCITONS IN Cu_2O

J. P. Wolfe

Department of Physics and
Materials Research Laboratory
University of Illinois at Urbana-Champaign
Urbana, Illinois   61801

Since the pioneering studies of Gross and others in the 1950's, the semiconductor cuprous oxide (Cu_2O) has provided a classic example of a Wannier exciton. Spectroscopic studies such as optical absorption, luminescence and two-photon absorption have established the hydrogen-like energy levels and the existence of ortho- and paraexcitons. Also, recent experiments have indicated that Cu_2O is a good candidate for demonstrating Bose-Einstein condensation of excitons. Towards this end, we have conducted a variety of experiments to determine the thermodynamic and transport properties of ortho- and paraexcitons in Cu_2O. In naturally grown crystals, the paraexciton lifetime exceeds a microsecond, and its diffusivity and drift mobility can be measured by time-resolved luminescence imaging. In this way, the low-temperature regime of deformation-potential scattering has been measured. Under intense photoexcitation, the orthoexcitons display quantum statistics. Time-resolved spectroscopy and imaging reveals the evolution of the basic thermodynamic parameters: density, temperature and volume of the excitonic gas. The conditions of quantum saturation have been reached and the system is observed to follow the phase boundary for Bose-Einstein condensation, $n = CT^{3/2}$, for over an order-of-magnitude in gas density.

In this talk, I will summarize the important results on the transport and quantum statistics of excitons that we have obtained at the University of Illinois. These experiments were performed in collaboration with D. P. Trauernicht, A. Mysyrowicz and D. Snoke.

The key idea in these studies is that the exciton is not simply a localized excited state of the crystal, but a mobile entity which moves diffusively through the crystal until its constituents--the electron and hole--recombine. The recombination event provides a characteristic photon which contains information about the location of the exciton in the crystal and its kinetic energy. By time- and space-resolving the luminescence spectrum of excitons, the experimenter gains a detailed picture of the thermodynamics of this unique particle.

In the usual direct-gap semiconductors such as GaAs, the excitons have lifetimes of order a nanosecond or less, and after being created by a photon of high energy they cannot completely cool to the lattice temperature within their short lifetime. Cuprous oxide, however, is an unusual direct-gap semiconductor. Transitions between the electron and hole bands are parity forbidden, and thus the excitonic lifetimes can be much longer than a few nanoseconds. Thus, the excitons in Cu_2O provide a unique system in which macroscopic transport processes and equilibrium thermodynamics can be observed.

One interesting complication is that the exciton in Cu_2O can occur in either ortho (S = 1 like) or para (S = 0 like) states, which are separated by an exchange interaction. The paraexciton state lies lowest and its recombination luminescence is much weaker due to the spin selection rule. Consequently, the paraexciton also has a long intrinsic lifetime; A. Mysyrowicz et al.,[1] measured paraexciton lifetimes of several micro-seconds in naturally grown crystals of Cu_2O. The orthoexciton lumine-scence is not spin-forbidden, so it is much easier to detect; however, the orthoexciton is observed to have a much shorter lifetime because it down-converts rather quickly into paraexcitons. The ortho-exciton lifetime is typically 10 nanoseconds, but this depends upon temperature and density of the excitonic gas. In pure crystals, the luminescence spectrum of both orthoexcitons and paraexcitons produced by moderate CW excitation exhibit a Maxwell-Boltzmann distribution of kinetic energies characterized by a temperature approximately equal to that of the lattice.

While the direct (no-phonon) recombination of the paraexcitons is highly forbidden, the application of external stress to the crystal breaks some of the symmetry and makes this luminescence line observable.[2] The emergence of the paraexciton direct luminescence line is shown the data[3] of Figure 1. These <u>direct</u> recombination lines from ortho- and para-

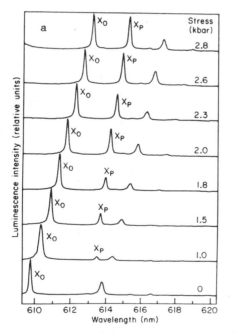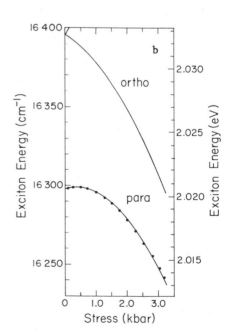

Figure 1: (a) Luminescence spectra as a function of applied stress recorded at T = 2.0 K. X_0 is direct recombination of lowest strain-split orthoexcitons. X_p is direct recombination of paraexcitons. (Reference 3) (b) Energy of the orthoexciton and paraexciton as a function of applied < 100 > uniaxial stress σ. (Reference 7.)

excitons are quite sharp because only excitons with wavevector equal to the photon wavevector are observed. As we shall see, it is only the phonon-assisted lines which display the Maxwell-Boltzmann statistics of excitonic gas. The stress-induced shift of the excitonic lines is due in part to the lowering of the semiconductors band-gap, but the nonlinear shift with stress arises from the coupling between the ortho- and para-excitons due to the exchange interaction, as discussed in Refs. 4 and 5.

The stress-dependence of the exciton energy provides a useful way of applying a force to the exciton, even though it is a neutral particle. An energy gradient is set up in the crystal by producing an inhomogenous stress. This strain gradient technique was originally used by Tamor and Wolfe[6] to measure the drift mobility of excitons in silicon. Trauernicht[7] applied a similar technique to Cu_2O and his results are shown in Fig. 2.

Part (a) of Fig. 2 shows the approximate equipotential surfaces set up by the inhomogeneous strain. (A rounded plunger is pressed on the top of the sample to produce a Hertzian contact stress.) The excitons are created resonantly at a single point in the crystal by directly pumping

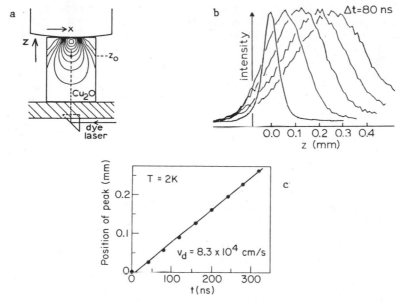

Figure 2: (a) stressing arrangement used to provide the motive force for excitons in Cu_2O. The curves approximate lines of constant exciton energy. A maximum of sheer stress is created inside the sample, providing a minimum in energy to which the excitons are attracted. Excitons are created at position z_0 by resonant excitation of the direct orthoexciton line. (b) Time-resolved spatial profiles of paraexciton luminescence shows the drift due to the strain gradient. (c) Position of the peak of the exciton distribution as a function of time. The slope yields the drift velocity v_d. From a callibration of the motive force, F, and this v_d, one determines the mobility $\mu = v_d/F = 1.7 \times 10^6$ $cm^2/eV \cdot s$ for this temperature in applied stress. (Reference 7)

the orthoexciton absorption line with a tunable dye laser. Orthoexcitons are created at the point in the crystal where their stress-shifted energy exactly matches the energy of the dye-laser photons. The orthoexcitons rapidly down-convert to paraexcitons, which are detected by their direct recombination. By mapping out the wavelength of the paraexciton luminescence line as a function of position in the crystal, one obtains the energy gradient or local force on the exciton. The paraexcitons are observed to drift in this force field, and by time-resolving their motion, it is possible to measure their mobility.

Using a pulsed dye laser and time-resolved photon counting, the spatial profiles of the exciton gas can be obtained as shown in part (b) of Fig. 2. This measurement is a little tricky, because the wavelength of the luminescence line shifts with spatial position. Thus, a computer is used to continuously control the grating of the spectrometer as the image of the crystal is scanned across the entrance slit of the spectrometer to obtain the spatial profiles.

A clear picture of the excitonic motion is obtained at increasing time intervals of 80 nanoseconds, as shown in the Figure. Initially, the spatial distribution of the excitonic gas is very narrow because the excitons are created in a very localized region. As time evolves, the spatial distribution shifts its center of mass due to drift, and it also broadens due to diffusion of the excitons. The displacment of the center of mass as a function of time gives the drift velocity of the excitons, as shown in part (c) of the Figure.

The mobility of the excitons, which is simply defined as their drift velocity divided by the motive force, $\mu = v_d/F$, is found to be extremely high. From standard drift theory, the exciton mobility is directly related to the particle scattering time and its mass, $\mu = \tau/m$. At $T = 1.2$ K, the paraexciton mobility is observed to be $\mu \gtrsim 2 \times 10^7$ cm^2/eV $\cdot$ s. Assuming an excitonic mass of approximately three times the free electron mass,[8] this implies a mean scattering time $\tau \gtrsim 20$ ns and a mean free path between scattering events of about 70 µm. It is remarkable that this paraexciton, with a Bohr radius of about 7 Å, travels ballistically for 70 micrometers before scattering. Of course, the neutrality of this particle and the low temperatures are key elements in this observation. Still, it is amazing that impurities or defects wouldn't shorten this path length.

At this point one must ask what is the physical process for the scattering of excitons. Their density is relatively low, so the scattering between excitons is relatively infrequent. The temperature dependence of the exciton mobility holds the answer to this question. As seen in Figure 3, the paraexciton mobility is very temperature dependent. This strong temperature dependence is indicative of excitons scattering from acoustic phonons. Usually, it is expected that electron-phonon scattering yields a mobility which varies as $T^{-3/2}$, according to deformation-potential theory.[9] This is actually a result using the high-temperature approximation, whereby the energy of the excitons is much larger than that of the scattered phonons. The excitonic mobility has a stronger temperature dependence than $T^{-3/2}$, and Trauernicht and Wolfe have shown that this is a result of deformation potential theory in the low temperature regime. Basically, when the average thermal velocity of the exciton becomes less than the sound velocity in the crystal, it is impossible for the exciton to lose energy by emitting an acoustic phonon. Of course, there is a distribution of kinetic energies according to Maxwell-Boltzmann statistics, but it turns out that at these low temperatures, the peak of the distribution occurs at a velocity which is close to the minimum velocity for longitudinal phonon emission. Thus, the scattering time and mobility are expected to rise rapidly with decreasing temperature, as is observed in the experiments.

Another unexpected observation in the measurement of paraexciton transport is that the mobilities depend upon the local stress in the crystal. As can be seen from the figure, the mobilities decrease with

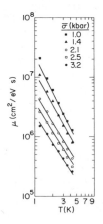

Figure 3: Temperature dependence of the measured drift mobility of paraexcitons. The five sets of data are for different applied stresses. The lowest curve is approximately $T^{-3/2}$. The symbols represent experimental data and the solid lines are theory, as discussed in Reference 7.

increasing stress. The explanation of this effect, which is not observed in other crystals such as silicon, is quite interesting. At low stress, symmetry dictates that these spherically symmetric electrons and holes comprising the exciton interact only with longitudinal phonons. But this symmetry restriction is relaxed as stress is applied to the crystal, permitting excitons to interact with transverse phonons. The transverse phonons have four times smaller velocity than longitudinal, and thus the minimum velocity for phonon emission by the excitons is much smaller for transverse phonons. This explains both the reduction in the mobility with stress, and the change in the temperature dependence to nearly a $T^{-3/2}$ form at high stress. Trauernicht has considered this problem in some quantitative detail and, with very few parameters, he has obtained the solid curves in Figure 3, which match the data very well.[7]

Complementary to these drift-mobility measurements of paraexcitons in Cu_2O, the diffusion constant of these excitons has been measured under zero stress conditions by time-resolved luminescence imaging.[7] The measured diffusion constants are in good accord with the mobilities measured by the strain gradient technique. Extremely high diffusion constants, $D \tilde{=} 1000 \text{ cm}^2/\text{s}$ at 1.2 K, are observed. In effect, these highly mobile paraexcitons in Cu_2O are able to diffuse millimeter distances in their lifetime of a few microseconds.

In addition to the transport properties of excitons, the thermo-dynamics of the excitonic gas in Cu_2O is being studied extensively. As previously mentioned, the excitons at low densities exhibit a Maxwell-Boltzmann distribution in kinetic energies. This fact is demonstrated clearly in Figure 4.[10] Both direct (S_d) and phonon-assisted (S_p) orthoexciton luminescence lines are shown in this figure, for two different bath temperatures. The gas density is fairly low in this case, because CW laser excitation is used. ($\lambda = 5145$ Å, which is absorbed within two microns of the crystal surface.) Thus, at low gas densities the excitons behave according to classical statistical mechanics.

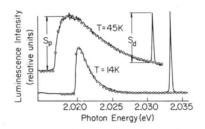

Figure 4: Orthoexciton luminescence spectra for two temperatures under very low power CW laser excitation. Open circles are a fit of the phonon-assisted luminescence line to a Maxwell-Boltzmann distribution with the given temperatures. (Reference 10.)

The exciton, by virtue of its integral spin is a boson, and at high densities the excitonic gas should exhibit Bose-Einstein statistics. In particular, it has been recognized for some time that excitons might undergo Bose-Einstein condensation. The critical density for Bose-Einstein condensation is given by

$$n_c = 2.612 \text{ g } (m/2\pi\hbar^2)^{3/2}(k_BT)^{3/2} = CT^{3/2}$$

where m is the particle mass and g is the spin degeneracy. According to Einstein, once the population of the excited states reaches this saturation condition, further addition of particles at constant T and V must be accommodated by the ground state--Bose-Einstein condensation (BEC).

The exciton has several desirable properties in this respect: 1) its mass is much smaller than that of an atom or molecule, thus requiring lower n_c for a given T; 2) the excitonic gas density can be controlled over a wide range by changing the optical excitation level; and 3) in many cases the exciton gas emits a recombination luminescence which directly reflects the kinetic energy distribution in the gas. Degenerate BE statistics have been reported for excitons in Ge[11] and by excitons in CuCl.[12] In the latter case, a stable macroscopic occupation of low-energy states was observed by resonant pumping of these states above a critical threshold in excitation power. Hulin, et al.,[13] have reported a gradual transition from classical to degenerate Bose statistics for orthoexcitons in Cu_2O generated by a nonresonant laser. The system appeared to approach $n_c(T)$ asymptotically.

Experiments performed at Illinois in collaboration with D. Snoke and A. Mysyrowicz have recently succeeded in producing a saturated quantum gas of orthoexcitons in Cu_2O.[10] The experiments employed time- and space-resolved photoluminescence techniques. The crystal is excited by intense 10-ns pulses from a cavity-dumped argon-ion laser (5145 Å). Time-resolved photon-counting techniques show the spectrum of phonon-assisted ortho-exciton luminescence during and after the laser pulse. The spectra are the solid lines in Figure 5. Both the density and the temperature of the

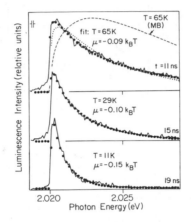

Figure 5: Time-resolved spectra of the orthoexciton phonon-assisted luminescence in Cu_2O showing quantum-statistical nature of this gas. Maximum absorbed power $\simeq 5 \times 10^6$ W/cm^2. Spectra are collected from the center of the excitation spot, integrated over a 1-ns time interval, and normalized to the same heights. Solid circles indicate best fits for a constant-potential Bose-Einstein distribution. The dotted line is a least-squares fit to the first spectrum, imposing a lower degeneracy $\mu = -0.3k_BT$. (The corresponding T is 43 K.) The dashed curve is a classical ($\mu < -2k_BT$) distribution for the same T as the first spectrum.

excitonic gas evolve in time during the pulse. The maximum density of excitons produced in this case is above 10^{19} cm^{-3}. The gas particles at such densities reach a quasi-equilibrium on a time scale much shorter than a nanosecond due to their rapid collisions.

Indeed, the spectra can be fit assuming equilibrium thermodynamics. The solid dots represent predicted energy distributions assuming Bose-Einstein statistics. Two parameters determine this distribution, temperature T and chemical potential μ. As can be seen from the fits at three different times during the laser pulse, the determined chemical potential is well within the regime for quantum statistics, $\mu \gtrsim -k_BT$. For comparison, the dashed line shows a Maxwell-Boltzmann distribution at the same temperature as the fit temperature. Clearly a highly degenerate gas of excitons has been produced under these conditions; however, no direct evidence for Bose-Einstein condensation has yet been observed, for example, a spike at the low-energy edge of the spectrum.

The principal result of this work is plotted in Figure 6, which shows the density of the gas as a function of its temperature. The density is

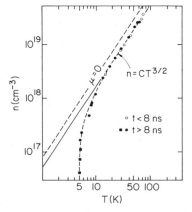

Figure 6: Density vs. temperature from the data of Fig. 5, showing Bose-Einstein saturation of the exciton gas. Circles are densities deduced from spectral fits at 1-ns intervals, where density is increasing for $t < 8$ ns and decreasing for $t > 8$ ns. Squares are densities deduced from I/V, where I is the luminescence intensity and V is the measured volume of the exciton cloud.

obtained from the fitted chemical potential using a standard Bose-Einstein distribution function and an $E^{1/2}$ density-of-states. The points on this curve correspond to successive times in the pulsed experiment. We see that as the gas density rises, its temperature rises in precisely such a way to keep the system on the $n = CT^{3/2}$ curve. In effect, the gas has evolved in such a way to map out the phase boundary for Bose-Einstein condensation.

It should be noted at this point that the chemical potential and temperature derived from the spectra must represent average values within the gas. Thus, the maximum local density in the gas must be higher than that determined from the fits. We believe that this is why the gas follows a $\mu = -0.1\ k_B T$ curve rather than $\mu = 0$. The principle question remaining is that if some portion of the gas has reached the critical density, why doesn't it condense? Several factors may act to limit the k = 0 population and cause the system to follow the BEC phase boundary. fundamental limitation may be that the gas simply does not have time on this 10 nanosecond time scale to actually condense into the k = 0 state, which requires emission of acoustic phonons. There are also some technical difficulties which may prevent excitonic condensation in these particular experiments. First of all, since the gas is created very near the excitation surface, there is a large gradient in gas density into the crystal, and a gradient in chemical potential acts as an expansive force on the particles. Thus particles with $\mu \stackrel{\sim}{<} 0$ are quickly pushed into lower density regions of the crystal. Also, the lifetime of low-energy particles should be considerably shorter than that of the more highly excited particles due to the direct (no-phonon) recombination mechanism. Compared to a classical gas, a degenerate system has a much larger fraction of particles near k = 0.

Whether these difficulties can be overcome remains to be seen. The present experiments, however, have provided a substantial new perspective on the problem of exciton condensation. The quantum statistics have significantly modified the time evolution of the particle temperature. It will be quite interesting to study how this process occurs microscopically, presumably using higher time-resolution. The principle result in the present experiments is the demonstration of the saturated gas condition predicted by Einstein and the corresponding phase boundary, $n = CT^{3/2}$.

In summary, the transport and thermodynamic experiments discussed in this paper demonstrate vividly that the exciton is a highly mobile particle which interacts with the lattice phonons and displays the statistical properties of atomic particles. Due to the light mass of this particle, basic phenomena in quantum thermodynamics can be studied at reasonable temperatures. Fundamental questions still remain about the transport and kinetics of excitons in Cu_2O.

This research was supported in part by grants AFOSR 84-0384 and NSF-DMR-85-21444, and by a NATO travel grant.

REFERENCES

1. A. Mysyrowicz, D. Hulin, and A. Antonetti, Phys. Rev. Lett. 43:1123 (1979).
2. F. I. Kreingold and V. L. Makarov, Fiz. Tekh. Poloprovodn 8:1475 (1974) Sov. Phys. Semicond. 8:962 (1975)].
3. A. Mysyrowicz, D. P. Trauernicht, J. P. Wolfe, H.-R. Trebin, Phys. Rev. B 27:2562 (1983).
4. H.-R. Trebin, H. Z. Cummins, and J. L. Birman, Phys. Rev. B 23:597 (1981).
5. R. G. Waters, F. H. Pollak, R. H. Bruce, and H. Z. Cummins, Phys. Rev. B 21:1665 (1980).
6. M. A. Tamor and J. P. Wolfe, Phys. Rev. Lett., 44:1703 (1980); Phys. Rev. B 26:5743 (1982).
7. D. P. Trauernicht and J. P. Wolfe, Phys. Rev. B 33:8506 (1986).
8. P. Y. Yu and Y. R. Shen, Phys. Rev. Lett. 32, 939 (1974).
9. J. Bardeen and W. Shockley, Phys. Rev. 80:72 (1950).
10. D. Snoke, J. P. Wolfe and A. Mysyrowicz, Phys. Rev. Lett. 59:827 (1987).
11. V. B. Timofeev et al., p. 327 in: "Proceedings of the Sixteenth International Conference on Physics of Semiconductors, Montpellier, France, 1982," M. Averous, ed., North-Holland, Amsterdam (1983).
12. N. Peyghambarian, L. L. Chase, and A. Mysyrowicz, Phys. Rev. B 27:2325 (1983).
13. N. Hulin, A. Mysyrowicz, and C. Benoît a la Guillaume, Phys. Rev. Lett. 45:1970 (1980).

STUDY OF LOCALIZED EXCITONS IN SEMICONDUCTOR SOLID SOLUTIONS

BY SELECTIVE EXCITATION

S. Permogorov and A. Reznitsky

A.F. Ioffe Physical-Technical Institute
USSR Academy of Sciences
194021 Leningrad, USSR

INTRODUCTION

Semiconductor solid solutions are good model substances for the study of electronic spectra and energy migration processes in weakly disordered solids. Random distribution of different atomic species over the regular crystal sites leads to the tailing of otherwise sharp energy band edges which in turn favours the localization of excitons or carriers. In this work we discuss the results of an experimental study of localized excitons in $CdS_{1-x}Se_x$ and $ZnSe_{1-x}Te_x$ solid solutions with substitution in the anion sub-lattice by selective excitation with monochromatic light.

A microscopic mechanism of exciton localization depends on the difference in the electronic properties of the atomic components forming the solution. Two limiting situations which were treated theoretically are represented in Fig. 1. For the case when the perturbation produced by the isolated atom of the other component is small, exciton localization takes place in the potential wells produced by long range concentration fluctuations.[1] The local levels of the exciton states within the wells form the inhomogeneously broadened tail of the exciton band.[x] In the density of states spectrum of this type, a mobility edge (EM) separating the states of localized and delocalized excitons should exist.

In the opposite case of strong difference in the electronic properties of two components exciton localization by small clusters takes place in dilute solutions.[2] As a result, the isolated levels corresponding to different types of clusters appear in the forbidden gap. However, with an increase of solution concentration these levels will broaden and merge with the band edge due to the intercluster interaction. In solutions with comparable concentration of the components, no isolated cluster can exist and the electronic spectrum attains qualitative similarity to the case of exciton localization via concentration fluctuations with the mobility edge somewhere near the band bottom.

The experimental characterization of the exciton mobility edge is of special interest. The Anderson model[3] which was developed for particles with infinite lifetime is not applicable to excitons in the real semiconductor solid solutions. First of all, excitons have a finite lifetime, limited at least by the radiative recombination. Secondly, strong interaction with the phonon system causes fast and irreversible relaxation of exciton energy. The situation is further complicated by the dependence of the internal exciton parameters (radius of internal motion and transition moment) on its energy. As a criterion of exciton localization the ratio of the occupation time for the given state to the total exciton lifetime can be used. This quantity is very small for free excitons and should be unity for completely localized states. As in the Anderson model, the position of the exciton mobility edge comes from the interplay of energy dependences of density of states and transfer probabilities.

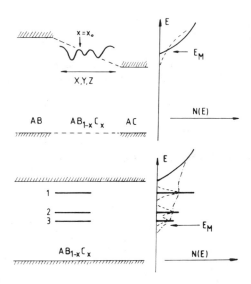

Fig. 1. Models for exciton localization and the exciton density of states N(E) spectra in semiconductor solid solutions.

The process of exciton energy relaxation is shown on Fig. 2. At sufficiently high energies the density of states is high and exciton motion in solid solutions is qualitatively similar to that in ideal semiconductors except for the appearance of additional exciton scattering by fluctuations of potential. In this region exciton migration is practically unlimited and is assisted by fast energy relaxation due to phonon emission. As the energy is lowered this process slows down due to the decrease of the density of final states with lower energies available for exciton relaxation. At the same time the exciton motion changes to the tunnelling between those spatially confined states which can be considered as a starting stage of exciton localization. If the decrease of exciton density of states with energy is smooth the spectral extent of this tunnelling region can be sufficiently wide for experimental detection. Finally, the characteristic migration time exceeds the radiative lifetime, and the exciton becomes completely localized. It should be noted, however, that the transformation of exciton motion from the regime of strong scattering by potential fluctuations to tunnelling migration with an energy dependent rate is quite gradual. As a result one can expect that in a réal semiconductor solutions the exciton mobility edge will not appear as an infinitely sharp threshold predicted by the Anderson model, but as an extended region of characteristic changes in exciton dynamical properties.

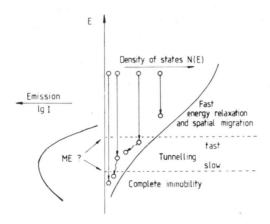

Fig. 2. Energy relaxation and emission of light by excitons in semiconductor solid solutions.

Exciton dynamics is clearly reflected in the radiative recombination spectra. Emission of light during the early stages of energy relaxation forms hot luminescence which has a very small intensity due to the high relaxation rate. A substantial increase of the emission intensity can be expected for the region where the energy relaxation is slower, i.e. in the region of the exciton mobility edge. The maximum position for the exciton emission band is given by the compromise between the growing occupation time and the decreasing density of states, taking into account the decrease of the exciton oscillator strength with the increase of localization energy.[4]

The model of exciton relaxation and recombination discussed above is in good qualitative agreement with the experimental results. In $CdS_{1-x}Se_x$ solid solutions the luminescence of localized excitons was observed as a broad and very intense band in the immediate vicinity of the absorption edge. A high intensity of emission reflects the fact, that exciton localization supresses the energy transfer to the centers of nonradiative recombination and hence increase the quantum efficiency of radiative recombination. With the help of resonant monochromatic excitation it has been shown that the emission band is inhomogeneously broadened.[5,6] The emission spectra of the selectively excited exciton states are in good qualitative agreement with the model of exciton localization by concentration fluctuations. The dependence of emission spectral shape on the excitation energy, as well as the frequency dependence of the degree of polarization of exciton emission at selective excitation[7] were used to obtain the position of the exciton mobility edge with an accuracy up to 1 meV.

Relaxation processes and the spectral migration of energy are evidenced by the distribution of the decay times across the emission band.[8] Fig. 3 shows the results of the study of emission decay times in $CdS_{1-x}Se_x$ with picosecond time-resolution. In the region of the band maximum the decay times are of the order of several nanoseconds which seem to be reasonable for values of the radiative times for localized excitons. At the high energy side of the band the decay times decrease down to several picoseconds in parallel with the decrease of the stationary emission intensity by 6 orders of magnitude. Such fast decay times reflect fast exciton relaxation above the mobility edge. A tunnelling mechanism of exciton migration shows up in the nonexponential character of emission decay all over the observation region. For deeper localized states the intensity rise can be also clearly detected in the emission kinetics.[8]

$ZnSe_{1-x}Te_x$ solid solutions are an example of a system where the exciton localization takes place on small clusters. Fig. 4 compares the emission spectrum of pure ZnSe with the spectrum of a sample with 1% Te content. The latter spectrum consists of a broad and intense band with a complex structure,[9] corresponding to the exciton states, localized by Te clusters of different types.[10] At selective excitation the same sample shows a pronounced phonon structure of the emission spectrum (Fig. 5). This structure is in good correspondence with the singularities of the density of states in ZnSe phonon spectrum; however, the LO peak has exceptionally high intensity. The phonon structure of the selectively excited emission of localized excitons in ZnSe:Te is in agreement with the model proposed in,[10] according to which the holes are tightly localized at Te clusters whereas the electrons move around the holes on large coulombic orbits. As a consequence both short-range deformation and long-range Frohlich electron-phonon interaction can be observed in selectively excited emission.

Simultaneous participation of different cluster types in exciton localization is proved by the excitation spectrum of the 1LO line intensity. As can be seen from Fig. 6, such an excitation spectrum for the sample with 1% of Te shows two distinct maxima (black points) corresponding to two Te clusters with different exciton localization energies.[11]

By increasing the concentration in $ZnSe_{1-x}Te_x$ one can observe the above-mentioned transition from cluster localization of excitons to localization by concentration fluctuations. Fig. 7 shows the concentration dependence of the $ZnSe_{1-x}Te_x$ luminescence spectra. It can be seen that at Te concentrations higher than 30% the emission spectra of localized excitons in $ZnSe_{1-x}Te_x$ are very similar to those of $CdS_{1-x}Se_x$.[10]

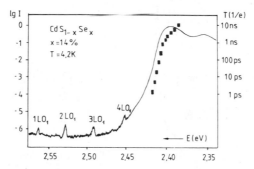

Fig. 3. Distribution of decay times (black points) across the emission band of localized excitons in $CdS_{1-x}Se_x$ (x = 0.14). Excitation energy is 2.460 eV.

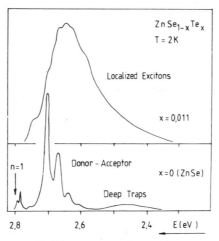

Fig. 4. Comparison of luminescence spectra of pure ZnSe and ZnSe:Te solid solutions with 1% Te content.

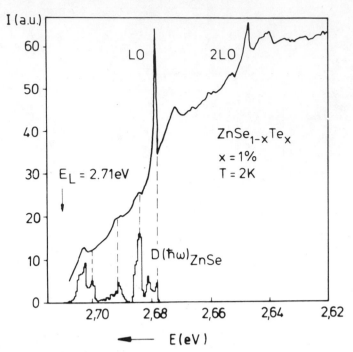

Fig. 5. Part of ZnSe$_{1-x}$Te$_x$ emission spectrum at selective excitation of localized excitons with E$_L$ = 2.710 eV.

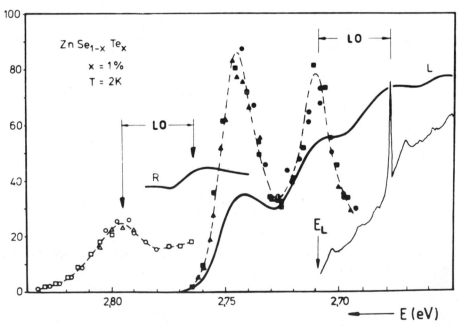

Fig. 6. Excitation spectrum of 1LO line intensity (black point) in ZnSe$_{1-x}$Te$_x$ sample with x = 1%.

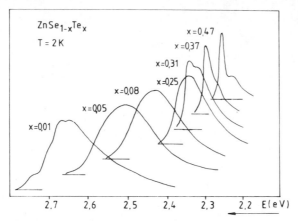

Fig. 7. Concentration dependence of localized exciton luminescence in ZnSe$_{1-x}$Te$_x$.

References

1. S. D. Baranovskii, A. L. Efros, Sov. Phys. Semicond., 12:1328 (1978).
2. M. A. Ivanov, Yu. G. Pogorelov, Zn. Eksp. Teor. Fiz., 72:2198 (1977).
3. P. W. Anderson, Phys. Rev., 109:1492 (1958); N. F. Mott, Philos. Mag., 18:835 (1966).
4. E. I. Rashba, G. E. Gurgenishvili, Sov. Phys. Solid State, 4:759 (1962).
5. S. Permogorov, A. Reznitskii, S. Verbin, G. O. Müller, P. Flogel, M. Nikiforova, Phys. Stat. Sol., (b) 113:589 (1982).
6. E. Cohen, M. D. Sturge, Phys. Rev., B25:3828 (1982).
7. S. Permogorov, V. Lysenko, S. Verbin, A. Reznitsky, Sol. St. Commun., 47:5 (1983).
8. J. Aaviksoo, J. Lippmaa, S. Permogorov, A. Reznitsky, P. Lavallar, K. Gurdon, Zn. Eksp. Teor. Fiz. Pis'ma, 45:391 (1987).
9. A. Reznitsky, S. Permogorov, S. Verbin, A. Naumov, Yu. Korostelin, V. Novozhilov, S. Prokoviev, Sol. St. Commun., 52:13 (1984).
10. A. Yu. Naumov, S. A. Permogorov, A. N. Reznitsky, V. J. Zhulai, V. A. Novozhilov, G. T. Petrovsky, Fiz. Tverd. Tela, 29:377 (1987).
11. S. Permogorov, A. Reznitsky, A. Naumov, H. Stolz, W. von der Osten, Proc. Int. Conf. Luminescence, Beijing (1987).

LUMINESCENCE OF BOUND EXCITONS IN THE QUASI-TWO-DIMENSIONAL TlGaS$_2$ CRYSTALS

G. I. Abutalibov[1], V. F. Agekyan[2], K. R. Allakhverdiyev[1], and E. U. Salayev[1]

[1] Institute of Physics of the Azerbaijan SSR Academy of Sciences, 370143, 33, pr. Narimanova, Baku. [2] Institute of Physics of Leningrad State University, 198904, 1, Ulyanovskaya ul. St. Petergof, Leningrad.

ABSTRACT

The results of an investigation of photoluminescence and magnetic field effects in the layered crystal TlGaS$_2$ at T = 2K are reported.

The TlGaS$_2$ luminescence spectrum contains a number of narrow lines in the energy interval 2.035 ÷ 2.085 eV. The exciton absorption peak near the direct transition edge in TlGaS$_2$ has the energy 2.60 eV, and indirect transitions begin 0.15 eV below it. Hence, the emission region in question is 0.4 eV away from the absorption edge and is interpreted as luminescence of deep bound excitons. Investigation of the luminescence spectrum and its energy dependence, selective excitation, and Raman scattering establishes that two short-wave lines are due to zero phonon transitions, and the complex adjacent structure results from electron-phonon processes. The relative intensities of the lines depend on the temperature and are reproduced from sample to sample. Hence, the emission relates to one center with a lifetime τ = 250 ns common for all structures.

Strong complex splitting of all lines has been observed in a magnetic field H. Each line is split into 6 or 8 components depending on the orientation of H in the crystalline layer (H⊥c). In the case when H∥c, the splitting is 1.5 times smaller than for H⊥c. The magnetic splitting is identical for all lines, and symmetry of the intensities relative to the central component occurs even when the energy distance between the end components exceeds $k_B T$ by an order of magnitude. Thus the common level of the two zero-phonon electron transitions is the lower one, which possesses a strong magnetic splitting.

INTRODUCTION

Recently the ternary semiconductor compounds of the TlM^{III}X$_2^{VI}$-type (M^{III}-In,Ga; X - S, Se) have been synthesized and their crystallographic structure has been reported[1]. The TlGaS$_2$ layered crystals have mono-clinic symmetry with the cell period a ≈ b = 1.04 nm, c = 1.52 nm,

$\angle ab = 100°$ [2]. The crystal layers perpendicular to the c-axis are binary, each sublayer constituting a sequence of Ga_4S_{10} pyramids constructed of four GaS_4 tetrahedrons. Two sublayers are turned towards each other by the vertices of pyramids and are arranged with respect to each other in such a way as to form the trigonal voids, where the Tl atoms are located.

The $TlGaS_2$ crystals investigated in this work were grown by the Bridgman-Stockbarger technique. At room temperature they possess high resistance and high photosensitivity in the region of the fundamental absorption edge.

1. TlGaS$_2$ Absorption Spectrum

A sharp absorption peak due to free excitons (Fig. 1) is observed on the edge of the direct interband transition at T = 2K. This structure is observed up to room temperature. The temperature shift of the absorption edge in $TlGaS_2$ is anomalous; up to 50K the exciton peak energy is little affected, however at higher temperatures it increases (Fig. 1, inset). Assuming that the Rydberg constant of the exciton series R_{ex}, is changed little with temperature, one can determine the variation of the forbidden

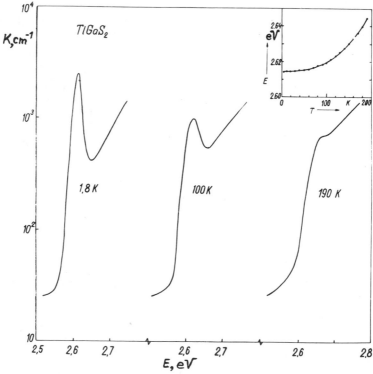

Fig. 1. Exciton structure on the direct absorption edge of $TlGaS_2$. Inset – temperature dependence of the exciton peak energy.

gap with temperature for direct transition, $E_g^{(n)}$ and describe it by the dependence

$$E_g^{(n)}(T) = E_g^{(n)}(0) + \frac{\alpha T^2}{T+\theta} \, ,$$

where the Debye temperature $\theta = 60K$, and $\alpha = 1.10^{-4}$. The analysis of the absorption edge shape allows an estimate of $R_{ex} = 0.02$ eV. In this case $E_g^{(n)}(\theta) = 2.62$ eV; the exciton radius $a_{ex} = 3$ nm; the reduced mass $\mu = 0.1$ m (the dielectric constant of $TlGaS_2$ is equal to 6.7). In the other $TlM^{III}X_2^{IV}$ crystals the forbidden gap decreases with increasing temperature.

2. Raman Scattering

Due to the low symmetry of the crystal and the large number of atoms in the unit cell, z = 16, the Raman scattering spectrum in $TlGaS_2$ is rich. Up to now, the Raman scattering and lattice reflectance spectra have been investigated only for phonons with energies above 40 cm^{-1} (3,4); therefore we investigated the Raman scattering at T = 300K in the region of low-frequency phonons, essential for the layered structures (Fig. 2 and Table 1). We observed phonon lines with low energies equal to 4, 6 and 9 cm^{-1} (see inset in Fig. 2), which might correspond to the excitation of the bending vibration in $TlGaS_2$ crystal layers.

TABLE 1

Phonon symmetry	Phonon energy cm^{-1}	Principal Polarization	Relative intensity in c-polarization
	4	?	10^{-3}
	6	?	10^{-2}
	9	nonpolarized	10^{-2}
$E_g^{(1)}$	22.5	nonpolarized	1
$E_g^{(2)}$	28.5	nonpolarized	0.15
$A_{1g}^{(1)}$	44	E⊥c	0.5
$B_{1g}^{(1)}$	68	"	0.3
$B_{2g}^{(1)}$	75.5	"	0.2
$B_{2g}^{(2)}$	112	"	0.4
$E_g^{(3)}$	120	"	0.05
$B_{1g}^{(2)}$	151	"	0.02
$A_{1g}^{(2)}$	185.5	"	0.2

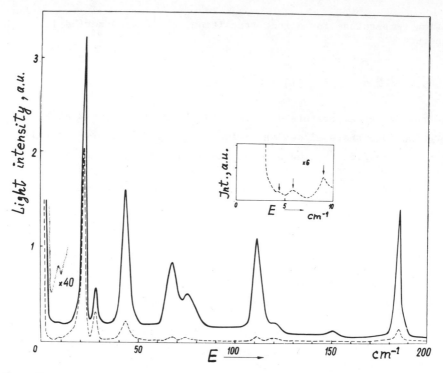

Fig. 2. Raman spectrum of TlGaS$_2$ at T = 300 K for E⊥c (solid line) and E∥c (dashed line) polarizations. The excitation wave length λ_e = 632.8 nm.

3. TlGaS$_2$ Luminescence.

In some series of TlGaS$_2$ crystals at band-to-band excitation, one observes unusual low-temperature luminescence consisting, as is seen from Fig. 3a, of numerous narrow lines[5,6]. The indirect absorption edge in TlGaS$_2$ is located 0.15 eV below the direct one; thus the emission energy is 0.4 eV below E$_g$. In different samples all the luminescence lines have the same relative intensities and should be attributed to one impurity center or structural defect. There is no doubt that the line I (Fig. 3a) is the zero-phonon line. All the other lines, except for line II, are to be interpreted as electronic transitions with simultaneous phonon emission. As will be seen from the following experimental data, line II corresponds to the zero phonon transition. Comprehensive decoding of the luminescence phonon structure is hindered by the richness of the TlGaS$_2$ phonon spectrum and the closeness of the value of LO-TO splitting and the distance between the lines I and II ($\Delta E = 5$ cm^{-1}). Moreover, the majority of the data on phonons is obtained at 300K. Even so, groups of intense phonon replicas are observed in the luminescence at distances about 20, 85, 120, 150, 180 and 330 cm^{-1} which agree with the data on the "strong" phonons in the TlGaS$_2$ lattice spectrum. The phonon spectrum of luminescence breaks at about 400 cm^{-1} from the zero phonon lines. This coincides with the energy of highest-energy phonon in TlGaS$_2$.

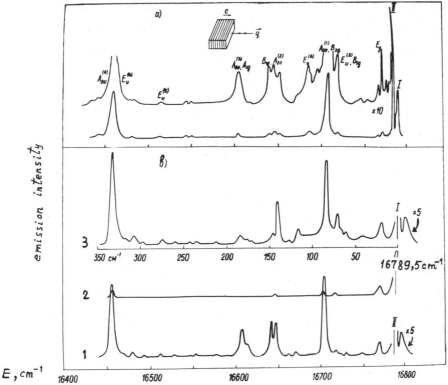

Fig. 3. Luminescence spectrum of $TlGaS_2$ at $T = 2K$. "a" -band-to-
band excitation $\left[h\nu_e > E_g^{(n)}\right]$. Lines I and II are the zero-phonon
electronic transitions with energies E_I and E_{II}; other lines are
their phonon replicas. "b" -spectra of selective excitation
with $h\nu_B = E_{II}(1)$, $E_I > h\nu_B > E_{II}(2)$, $h\nu_B = E_I(3)$. Phonon indexes
are given according to Table 3 from Ref. (3), and correspond to the
tetragonal model of the $TlGaS_2$ crystal with the only type of layer
stacking; q is the photon wave vector.

As the temperature increases from 2K, the enhancement of the luminescence of the phonon lines with respect to the zero phonon ones occurs
involving the activation of the long-wave wing of the line I, probably
corresponding to acoustic phonon emission. At $T = 15K$ the phonon lines
are increased relatively by an order of magnitude. At further temperature increase they are relatively reattenuated, and the spectrum structure is broadened (Fig. 4). As in the $TlGaS_2$ Raman scattering spectrum
one observes phonons with the energies close to ΔE, the question arises
whether line II, like the others, is the phonon replica of line I. To
clarify the nature of line II we made an experiment on the selective
excitation of $TlGaS_2$ into the luminescence region by means of a dye laser
at $T = 2K$. When the excitation energy $h\nu_e$ coincides with the energies
of lines I and II, E_I and E_{II}, there occurs a resonant enhancement of the
phonon spectrum, while the excitation with $h\nu_e > E_I$, $h\nu_e < E_{II}$ and

$E_{II} < h\nu_e < E_I$ even in the close vicinity of E_I and E_{II} produces a phonon spectrum which does not differ from the nonresonant one (Fig. 3b). Such an acute selectivity of the excitation of the phonon component points to the fact that line II, as well as line I is a zero phonon line. It should be mentioned that the relative intensities of some phonon components at $\nu_e = E_I$ and $\nu_e = E_{II}$ strongly differ. This must be due to the concrete symmetry of two zero phonon transitions.

4. Magnetic Splitting of Luminescence Lines.

Magnetic splitting is strong, complex, and identical for lines I and II (Fig. 5,6). If the field $H\perp c$, the magnitude and structure of the splitting is determined by the position of H in the $TlGaS_2$ crystallographic layer. When the direction of H is close to the a- and b-axis in a low field, triplet splitting with g-factor equal to 2 takes place; in high field the central component of the triplet splits into a quarter, the side components - into doublets (Fig. 5a, 6a). When H is directed along the angle bisector between the a- and b-axis, first the lines split into doublets, then upon increase of H each component of the doublet splits into three components (Fig. 5b,6b). Hence, depending on the orientation of H in the crystal layer, lines I and II split into 6 and 8 components. In the case of $H\parallel c$ (Fig. 5c) splitting in low field is into a triplet; the magnitude of the splitting is 1.5 as small as that

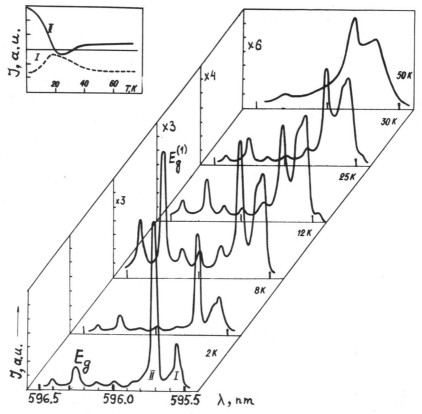

Fig. 4. Short-wavelength region of luminescence spectrum of $TlGaS_2$ at different temperatures. Inset - the temperature dependence of relative intensities of lines I and II (solid line) and phonon replica $E_g^{(1)}$ (dashed line) normalized to the intensity of line I.

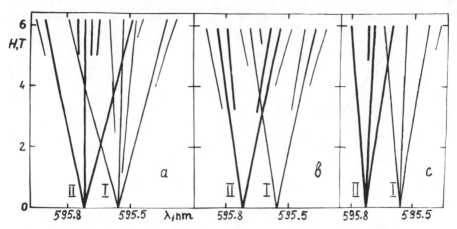

Fig. 5. The scheme of magnetic splitting for lines I and II.
"a" - H‖a or H‖b; "b" - H is parallel to the angle bisector between
the a- and b-axis.

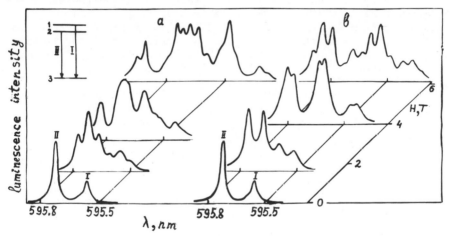

Fig. 6. Magnetic splitting of the zero-phonon lines in the
luminescence of TlGaS$_2$, T = 2K, H⊥c. Field direction in the layer
for "a" and "b" corresponds to Fig. 5a and Fig. 5b.

with H‖a or H‖b. In the maximum field H = 6 T one can see an additional
splitting of the central component. It was verified that the long-wave
lines of phonon replicas split like lines I and II.

The identity of the complex magnetic splitting of lines I and II
shows that they have at least one level of the corresponding transitions
in common. It is noteworthy that even for splitting values much greater
than k_BT the intensities of the central and side components of line II
(or line I) are approximately equal; hence the splitting picture of each
line is symmetrical (Fig. 6). Therefore, one can conclude that the lower
level of the two electronic transitions is the common level, possessing a
strong magnetic splitting. The luminescence decay time is equal to 250
ns, so that in the system of the Zeeman sublevels of the upper state an
equilibrium population is established.

Let us analyze the temperature variations in the relative intensities of lines I and II. The upper levels of transitions 1 and 2 are
split by an interval equal to 5 cm^{-1} (Fig. 6), and in the equilibrium
state it follows from the intensity relation at T = 2K that the

oscillator strength for the transition (1-3) is higher by an order of magnitude than that for (2-3). Heating of the crystal from 2 to 4K was expected to have relatively amplified the shortwave line I by a factor of 6 times; however, in fact the intensity for lines I and II in heating from 2 to 4K remained practically invariable, and the intensities became equalized only at about 20K (Fig. 3, inset). The temperature dependence is explained by the fact that an increase in T leads to the mixing of the wave functions for levels 1 and 2 through the mechanism of electron-phonon interaction. In this case the transition oscillator strength (2-3) increases, and at $k_B T \gg \Delta E$ both the equality of populations for levels 1 and 2 and of the oscillator strengths of the (1-3) and (2-3) transitions are attained.

5. Electron Spin Resonance in $TlGaS_2$.

The ESR spectra have been investigated at the frequency of 9.48 GHz, T = 4 - 300K. A great number of signals with g-factor equal to 0.7 - 5 are observed in the low temperature region. The narrow ESR lines can be divided into two groups. The signals with g = 3.7 - 5 do not shift significantly during the transition from $H \perp c$ to $H \parallel c$. On the other hand, some narrow signals exhibit an appreciable anisotropy. This is true for the lines with g = 2; 1.5 and 1.4: When H deviates somewhat from the crystallographic layer plane, one observes a notable shift of lines, which is associated with decreases in the g-factor to 1.75, 1.25, and 1.20, respectively.

One can see that two ESR signals have a distinct ultrafine structure. The distance between the components of ultrafine structure depends upon the direction of H. Such behavior points to the fact that these ESR signals refer to structural defects.

The ESR spectral lines have a complex temperature dependence, and they are observed in the narrow interval from 5.5 to 8K.

It should be noted that the strong and narrow signals with g = 2 ($H \perp c$) and g = 1.3 ($H \parallel c$) correspond to the g-factors of triplet splitting in a weak field of the luminescence lines I and II for $H \perp c$ and $H \parallel c$. Regarding the additional splitting of luminescence triplets or doublets which takes place at high field, g = 0.3 - 0.5, we failed to observe the ESR transitions between the respective sublevels due to the experimental limitation g_{min} = 0.7. Thus, the results of the investigation of Raman scattering, selective excitation of luminescence, magnetic splitting, and ESR indicate that low temperature luminescence in $TlGaS_2$ is generated by two zero phonon transitions from upper levels, separated by an interval equal to 5 cm^{-1}, to a common lower level with strong magnetic splitting, and their numerous phonon replicas. The luminescence spectrum seems to correspond to a $TlGaS_2$ structural lattice defect formed under certain crystal growth conditions. This emission can be described in terms of a deep bound exciton or lattice defect excited via the exciton transfer of energy.

To gain information on the nature of the center and its electron levels, to compare the Zeeman splitting with ESR, a comprehensive investigation of ESR and its temperature behavior in $TlGaS_2$ and its analogs are needed.

REFERENCES

1. D. Muller and H. Hahn, Z. anorg. und allg. Chem. 438 258 (1978).
2. S. G. Abdullaeva, S. G. Abdinbekov, and G. G. Gusseinov, DAN AzSSR 36 23 (1980).
3. N. M. Gasanly, A. F. Goncharov, N. M. Melnik, A. S. Ragimov, and V. I. Tagirov, phys. stat. sol. (b)116, 427 (1983).
4. R. A. Aliev, K. P. Allakhverdiev, R. M. Sardarly, and V. E. Steinschreiber, phys. stat. sol. (b)112, K153 (1982).
5. G. I. Abutalibov, A. A. Aliev, L. S. Larionkina, I. K. Neiman-zade, and E. U. Salayev, Fiz. Tverd. Tela 26 1221 (1984).
6. G. I. Abutalibov, V. F. Agekyan, A. A. Aliev, E. U. Salayev, and Yu. A. Stepanov, Fiz. Tekhn. Poluprov. 19 351 (1985).

OPTICALLY DENSE ACTIVE MEDIA OF SOLID-STATE LASERS

A.A. Danilov, A.M. Prokhorov and I.A. Scherbakov

Institute of General Physics, Academy of Sciences of the USSR
Vavilov Street 38, Moscow USSR

INTRODUCTION

The design of solid-state high-power lasers for the solution of problems of laser technology and for investigations of radiation interaction with matter is now one of the more active trends in applied quantum electronics. Papers are now available concerning CW solid-state YAG:Nd^{3+} lasers in which the use of consecutive location of several active elements in one resonator makes it possible to obtain power generation up to I kW at a characteristic radiation divergence of about 30 mrad.[1,2] The most interesting, however, are investigations in the field of pulsed-periodic high-power lasers based on wave-guide active elements.[3,4] The main features of such lasers are as follows:[5-10] the angular density of laser emission is virtually independent of optical pumping power and the gain coefficient in the bulk of the active volume is, in principle inhomogeneous. Although the results of investigations published, for example, in[11-15] seem to be rather promising they reveal some difficulties in the design of such high-power lasers. The planar structure of the active wave-guide element highly impedes the design of an efficient pumping element, so that in the currently used lasers the energy efficiency is 2-3 orders of magnitude lower compared to ordinary pumping elements employed cfor cylindrical active elements. Since the wave-guide emission regime is obtained with the aid of a pair of parallel planes, polished by a mirror-like finish, it is difficult to sustain a wave-guide regime for such active media as neodymium-doped glasses in contact with a cooling liquid because of partial etching of the polished surfaces. The cooling of the wave-guide active elements via static gas layers[16] does not provide quite an effective heat removal and leads to the heating of the active element and, consequently, to the worsening of the emission characteristics.

It is known that the efficiency of Nd-glass lasers operating in the pulsed-periodic regime can be markedly increased by using chromium and neodymium doped rare-earth scandium garnets as the active media.[17] In these media the rate of excitation energy transfer from donors to acceptors (i.e. from chromium to neodymium ions) is determined only by the acceptor concentration and virtually does not depend on the donor concentration.[17] This enables one to vary the chromium concentration considerably in the matrix of the active medium thereby obtaining high values of the spectral averaged absorption coefficient $\bar{a}$. The efficiency of utilization of the radiation of the pulsed pumping lamp, having a quasigrey spectrum, is determined by high $\bar{a}$ values. The spectrally averaged absorption coefficient in a spectral band of 400-950 nm versus concentration of the doped impurities is represented in Fig. I. It can be seen, that chromium ions contribute greatly to the absorption of pumping radiation. The characteristic value of $\bar{a}$ for YAG:Nd/C_{Nd} = 1.5 x 10^{20} cm^{-3} / is 0.25 cm^{-1}, for neodymium doped phosphate glasses (even for increased neodymium concentration) it is not over 0.5 cm^{-1}, while for GSGG:Cr,NdC_{Cr} = C_{Nd} = 2 x 10^{20} cm^{-3} (crystal) is equal to 4.5 cm^{-1}. The waveguide GSGG:Cr,Nd active element plate with a thickness of 1 cm absorbs 99% of normally incident pumping radiation per one transmission, for YAG:Nd the fraction of the absorbed energy is 22%, for phosphate glasses - it is not greater than 40%. Moreover, rare-earth scandium garnets have a large refractive index and are inert with respect to the ordinary cooling liquids. All this makes these crystals promising as active media of solid-states lasers with wave-guide active elements. For such lasers, it is, probably, possible to obtain high energy efficiency even with a reflector-type pumping element, keeping at the same time, the wave-guide regime stability in a wide range of pumping power with direct cooling of the active element. However, a relatively large fraction of absorbed pumping energy creates great heat release in the bulk of active element; then relative to the needed heat conductivity the rare-earth scandium garnets are at a disadvantage in relation to YAG.[18]

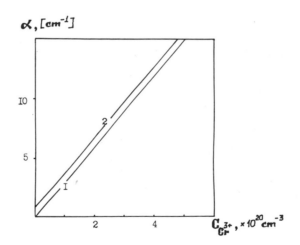

Fig. 1. Spectrally averaged absorption coefficient of optically dense active media - GSGG:Cr, Nd crystals doping concentration; 1 - noedymium concentration is equal to zero; 2 - neodymium concentration is equal to 5 x 10^{20} cm^{-3}.

So, using these crystals for high-power lasers was traditionally considered to be either impossible or requiring a considerable decrease of their optical density due to decreasing of chromium concentration. The stationary heat field investigations in an optically dense active medium plate placed into a pumping element showed that under suitable conditions thermooptical inhomogeneities could be smoothed.[19] The main condition for static "STON-" or smoothing effects of thermo-optical inhomogeneities with respect to higher-power optical pumping is the presence of spectrally grey absorption of the optical pumping at high α values. For all existing active media only chromium and neodymium doped rare-earth scandium garnets can satisfy this condition. The heat-release distribution in such a plate with a thickness of $2a_S$ can be accurately described by the following relation:

$$Q_T(x) \approx \frac{\kappa P_\rho}{\nu_S} \bar{a}a_S ch(\bar{a}x)/sh(\bar{a}a_S) \qquad (1)$$

where P_ρ is the power of optical pumping, κ is the fraction of this power which changes into heat, ν_S is the active medium bulk. The stationary distribution of the temperature drop is the following:

$$\Delta T_S(x) \approx \frac{\kappa P_\rho a_S^2}{\lambda_T \nu_S} \frac{ch(\bar{a}a_S) - ch(\bar{a}x)}{\bar{a}a_S sh(\bar{a}a_S)} \qquad (2)$$

where λ_T is the plate thermal conductivity $\bar{a}a_S$ is the plate optical density. Analysing relation (2) we can see, that at large optical density the maximum temperature drop decreases in inverse proportion to the optical density of the active element. The difference $\overline{\Delta T_S}(x) - \Delta T_S(x)$ important in calculation of thermo-elastic stresses in the plate,[20]

$$\Delta T_S(x) \approx \frac{1}{2a_S} \int_{-a_S}^{+a_S} \Delta T_S(x)dx,$$

behaves in the same way.

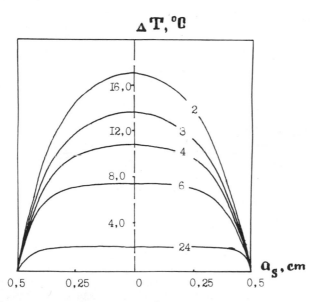

Fig. 2. Maximum temperature drop in GSGG Cr, Nd plate versus optically dense active media (digitals in the break of curves) $P_\rho = 2.5$ kW, bulk of active medium - 40 cm³.

The temperature drop distribution in a plate at different values of optical density are shown in Fig. 2. In our work[19] we estimated the limiting values of the averaged high-power optical pumping for an optically dense active medium plate:

$$P_\rho{}^{lim} \approx \frac{\lambda_T}{\kappa} \; \frac{2(\bar{a}a_S)^2 sh(\bar{a}a_S)}{\bar{a}a_S ch(\bar{a}a_S) - sh(\bar{a}a_S)} \; \frac{\sigma_S(1-\nu)}{\beta_T E} \; \frac{b_S L_S}{a_S} \tag{3}$$

where σ_S is the limiting value of the cleavage stress, E is the Young modulus, ν is the Poisson coefficient, β_T is the temperature expansion coefficient, b_S is the width, L_S is the length of the plate. From the well-known expression for $P_\rho limit$[21] this inequality is distinguished by a multiplier $\phi(\bar{a}a_S) = 2(\bar{a}a_S)^2 sh(\bar{a}a_S)/(\bar{a}a_S sh\bar{a}a_S - sh(\bar{a}a_S))$,[21] - which only depends on the active medium optical density. For the case of slow optical density the multiplier is equal to 6.0 while it can increase asymptotically to a double value. The pumping power $P_\rho{}^{lim}$ versus optical density for the wave-guide active element with a fixed GSGG:Cr, Nd geometry is represented in Fig. 3.

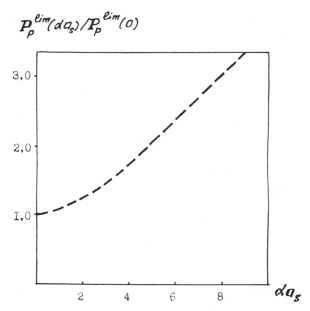

Fig. 3.   The relative averaged optical pumping power corresponding to the damaged threshold of the wave-guide active element versus optical density.

With increase of the optical density the inhomogeneity of absorbed pumping energy distribution grows, i.e. the gain coefficient in the bulk of the active element becomes inhomogeneous. The most efficient combination of the static "STON-effect" with slightly pronounced inhomogeneous gain in the bulk active element can be achieved by using the wave-guide active element.   Thus, high-power pulsed-periodic lasers with energetically effective optical dense active media could be designed on the basis of the wave-guide active elements.   A combination of the static "STON-effect" with a high absorption probability of the active media is very important. In particular, a solid-state GSGG:Cr, Nd laser with the wave-guide active element of dimension 1 x 4 x 15 cm^3 at chromium concentration 3 x 10^{20} can be irradiated with mean power of the order of 2.5 kW under limiting pumping power of 60 kW.

We have examined the feasibilities of optically dense active media in the with solid-state lasers in the case of wave-guide GSGG:Cr, Nd active element with dimensions of $0.5 \times 0.5 \times 10$ cm^{-3} (neodymium ion concentration was 3×10^{20} cm^{-3}). The optical density of the active element and its geometrical sizes were chosen so as to provide limiting pumping regimes at relatively low levels of the mean power. The pumping element of the laser design was cylindrical which allows us to compare the wave-guide and cylindrical active elements equally. A theoretical damage threshold with respect to the optical pumping power was estimated according relation (3) and was 4.2 kW. The laser irradiation was 75-80 W at the absolute efficiency 2% and pulse repetion rate of 35 Hz. Then, in the whole region of pumping a divergence change of the radiation emitted was not greater than 1.5 mrad; the damage threshold measured experimentally with respect to pumping was 4.2 kW, which fully agrees with the theoretical results mentioned above. It is additionally well to note that the active element was cooled due to direct contact with the cooling liquids. As a whole the experimental results confirm a promising use of optically dense active media for the pulsed-periodic high-power lasers with the wave-guide active elements.

An optically dense medium exhibits, along with the static "STON-effect", the dynamic "STON-effet" which lies in the onset of a quasistationary thermal regime in the active medium (under pulse periodic pumping). In this case the thermal flux is directed on the average, outwardly while during each pumping pulse non-uniform heat release forms a counterflux directed into the active medium. For optically dense active media this flux can be quite noticeable, and a situation is possible, when during each pumping pulse the outward thermal flux is partially compensated or even changes its direction at a certain instant following the beginning of the next pumping pulse. The modulus of the thermal flux directed in quasistationary regime from the bulk of the active medium is equal to:

$$q_{st} \approx \frac{\kappa P \rho_{as}}{\nu_s} \; sh(\bar{a}x)/sh(\bar{a}a_s) \tag{4}$$

where $P_\rho = E_p.f$ - is the optical pumping power, E_p - is the pumping pulse energy, f - is the pulse repetition rate of pumping. For a counterflux which is formed in an active medium by means each pumping pulse one can write:

$$q_{dym} \approx - \lambda_T \frac{E_p}{c\rho\nu_s} \; \bar{a}^2 a_s sh(\bar{a}x)/sh(\bar{a}a_s) \tag{5}$$

where c - is the thermal capacity, ρ - is the active medium density. Assuming the pumping pulse to be rectangular the dynamics of thermal flux in the optically dense active media per each pulse of pumping is described by the following expression:

$$q_{sum} \approx \frac{\kappa P \rho a_s}{\nu_s} \frac{sh(\bar{a}x)}{sh(\bar{a}a_s)} \left[1 - \frac{\lambda_T}{c\rho f} \; \bar{a}^{-2} \; \frac{\tau}{\tau_u} \right] \tag{6}$$

where τ_u is the pumping pulse duration, τ is the time of the current. The parameter $D = \frac{\lambda_T}{c\rho t} \; \bar{a}^2$ characterizes a dynamic contribution of nonequilibrium heat release to a total quasistationary regime. If $D < 1$ the total flux does not vanish (no autocompensation occurs); if $D \approx 1$ autocompensation occurs just at the finish of the next pumping pulse; if $D > 1$ autocompensation occurs at the moment of $\tau \approx D^{-1}\tau_u$ from the onset of the next pumping pulse. Full compensation of thermooptical inhomogeneities corresponds to the moment of autocompensation which can be controlled while vaying the pulse repetition rate f. As follows just from the parameters this effect is feasible only in optically dense active media for the pulse periodic regime of the solid-state laser. The form of the parameter D for YSGG:Cr, Nd and GSAG:Cr, Nd crystals versus chromium concentration is represented in Table I. The dynamic "STON-effect" is of a special interest for Q-switched regime. In this regime the moment of dynamic compensation of thermooptical inhomogeneities and the moment corresponding to the inverse population in each pumping pulse can be made coincident by appropriately choosing the pumping pulse repetition rate under specified thermal characteristics of the active medium and the absorption coefficient $\bar{\alpha}$. Switching the resonator Q-factor on just at that moment, one can markedly improve the divergence and the mode composition of the emitted radiation.

Table I,[22]

Cr concentration	$1\times10^{20}cm^{-3}$	$2\times10^{20}cm^{-3}$	$3\times10^{20}cm^{-3}$	$4\times10^{20}cm^{-3}$
Absorption coeff.	$3cm^{-1}$	$6cm^{-1}$	$9cm^{-1}$	$12cm^{-1}$
D YSGG:Cr, Nd	0.22/f	0.86/f	1.95/f	3.46/f
D GSAG:Cr, Nd	0.41/f	1.62/f	3.65/f	6.5/f

We have controlled the dynamic "STON-effect" with the help of a solid-state laser having a cylindrical active element (7.2 x 80 mm) for chromium and neodymium doped concentration of 2×10^{20} cm^{-3} in a matrix of rare-earth scandium garnet YSGG. The irradiation of the laser was 120 W with absolute efficiency of 5%, pulse repetition rate of pumping - 15 Hz; with the dynamic "STON-effect" divergence of radiation emitted was not greater than 10 mrad, whereas without influence of the "STON-effect" it should be 25-30 mrad. The angular intensity of radiation emitted versus averaged pumping power at different pulse repetition rates are represented in Fig. 4a. The divergences of radiation emitted versus averaged pumping power at different pulse repetition rate of pumping are represented in Fig. 4b.[23]

The dynamic "STON-effect" also makes it possible to apply an unexpectedly high resonator stability at the high pumping energies. This aspect of the dynamic "STON-effect" has enabled us to use the solid-state laser on the basis of cylindrical YSGG:Cr, Nd active element (doping concentration is 3×10^{20} cm^{-3}) with a record energy efficiency; the differential efficiency obtained at the pulse repetition rate of pumping of ~ 2 Hz was 11.5%.

Thus, on the basis of an optically dense active medium one can design the solid-state lasers, operating in a Q-switched regime with a sub-diffraction divergence of the radiation emitted. This is feasible by coordinating the choice of the pulse repetition rate of pumping and the moment of switching on the Q-factor within the framework of the "STON-effect" described. This effect also causes smoothing of the thermooptical inhomogeneities in the solid-state active elements under the pulse-periodic regime of their operation in case the pulse repetition rates are not too large.

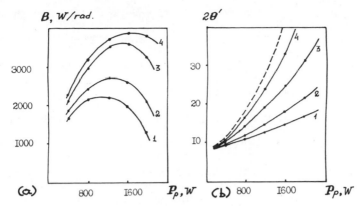

Fig. 4. a) Angular intensity of optically dense active medium solid-state layers versus averaged pumping power at the different pulse repetition rates; 1 - 5 Hz; 2 - 10,Hz; 3 - 20 Hz; 4 - 35 Hz.
b) The divergence of optically dense active medium solid-state laser radiation versus averaged pumping power at the different pulse repetition rates.

In conclusion, it is well once more to emphasize, that optically dense active media probably allow us to solve the problem of further increasing the solid-state laser efficiency under simultaneous increase of the averaged emission power. The "STON-effect" described relatively to solid-state lasers means that a search for new sensitizers helping the further increase of the optical density is not restricted, but expands the energy feasibilities of the the solid-state lasers. We hope that solid-state lasers with wave-guide active elements using optically dense active media will solve the problem of the generation of high-power laser sources.

References

1. Ye Biging Proc. of the Intern. Conf. and School, "Laser and Applications," Bucharest, 1:190, (1982).
2. Y. Yamada, S. Yoshida, S. Ishida, Y. Fujmori, K. Ishikawa, Proc. of the CLEO-87, R WOZ., p.190.
3. W. S. Martin, J. P. Chevnoch, USA Patent, - 3.633.126, (1972).
4. A. L. Mikaelyan, V. V. D'yachenko, Pis'ma v ZHETF 16:25 (1972).
5. A. L. Mikaelyan, V. V. D'yachenko, Kvantovaya Electronika 1:4 (1974).
6. J. M. Eggleston, T. J. Kane, J. Unternahrer, R. L. Byer, Opt. Lett. 9:405 (1982).
7. T. J. Kane, R. C. Echardt, R. L. Byer, IEEE J. Quant. Electron. QE-19:1351 (1983).
8. J. M. Eggleston, T. J. Kane, K. Kuhn, J. Unternahrer, R. B. Byer, IEEE J. Quant. Electron. QE-20:289 (1984).
9. T. J. Kane, J. M. Eggleston, R. L. Byer, IEEE J. Quant. Electron. QE-21:1195 (1985).
10. S. Basu, T. J. Kane, R. L. Byer, IEEE J. Quant. Electron. QE-22:2052 (1986).
11. G. F. Albrecht, J. M. Eggleston, J. J. Wing, IEEE J. Quant. Electron. QE-22:2099 (1986).
12. J. M. Eggleston, G. F. Albrecht, R. A. Petr, J. F. Zumdieck, IEEE J. Quant. Electron. QE-22:2092 (1986).
13. T. J. Kane, W. J. Kozlovsky, R. L. Byer, Opt. Lett. 11:216 (1986).
14. D. A. Rockwell, Proc. of the CLEO-87, R. WB, 1:114.

15. K. Maeda, H. Hayakawa, T. Ishikawa, T. Yokoyama, Proc. of the CLEO-87, R.WB, 3:116.
16. M. Reed, K. Kuhn, J. Unternahren, R. L. Byer, IEEE J. Quant. Electron. QE-21:412 (1985).
17. E. V. Zharikov, V. V. Osiko, A. M. Prokhorov, I. A. Scherbakov, Izv. AN SSSR, ser. fiz. 48:1330 (1984).
18. V. F. Kitaeva, E. V. Zharikov, I. L. Chisyi, Phys. Stat. Sol. 92(a):475 (1985).
19. A. A. Danilov, V. V. Osiko, A. M. Prokhorov, I. A. Scherbakov, Preprint IOFAN 23M (1987).
20. V. L. Indenbom, I. M. Cil'verstova, Yu. I. Sirotin, Kristallograph 1:599 (1956).
21. J. E. Marion, J. Appl. Phys. 60:69 (1986).
22. A. L. Denisov, E. V. Zharikov, A. I. Zagumyenny, S. P. Kalitin, M. A. Nogin, V. G. Ostroumov, V. A. Smirnov, I. T. Sorokina, I. A. Scherbakov, Preprint IOFAN 350M (1986).
23. A. A. Danilov, E. V. Zharikov, Yu. D. Zavartsev, M. Yu. Nikol'sky, P. A. Studenikin, I. A. Scherbakov, Preprint IOFAN 160M (1987).

NEGATIVE ABSOLUTE ELECTRICAL CONDUCTIVITY OF OPTICALLY

EXCITED RUBY: MICROSCOPIC NATURE OF THE PHENOMENON

S.A. Basun, A.A. Kaplyanskii and S. P. Feofilov

Ioffe Physical Technical Institute
USSR Academy of Sciences
194021 Leningrad, USSR

1. INTRODUCTION

Ruby, $Al_2O_3:Cr^{3+}$, is clearly an insulator, its electrical conductivity becoming finite only under optical excitation. Recently, the unusual photoelectrical properties of ruby have become a subject of considerable interest. A study has been carried out[1] at a low (~ 10 K) temperature on crystals of concentrated ruby of the effect of intense Ar laser irradiation in the region of the broad U and Y absorption bands. A strong uniform electric field E_S ~ 10^6 V/cm directed along the trigonal C axis was found to appear gradually and reach saturation in the excited region of the crystal: the field persisted after the pumping was turned off and the sample warmed up to room temperature.

The photoinduced field in ruby was established[2] to have a complex spatial structure consisting of regions (domains) with equal but oppositely directed electric fields ($\pm E_S$). The phenomenological theory[3] connected the formation of photoinduced field domains with the electrical instability of optically excited ruby against small fluctuations of the electric field. This instability comes from the anomalous character of the dependence of photocurrent $j\|$ on the electric field parallel to the C axis. It was assumed[3] that the $j\|$(E) relation has an N-shaped feature near zero (see Fig. 1). Namely, at fields $-E_S<E<+E_S$, the absolute electrical conductivity is negative and the photocurrent flows against the field, while in the interval $-E_t<E+E_t$ the differential conductivity is also negative. The current flow against the field for $|E|<|E_S|$ was phenomenologically explained as due to the current being actually a bulk photovoltaic current which is induced in centrosymmetric ruby crystals by the electric field.

The N-shaped $j\|$(E) dependence was directly observed experimentally[4,5] in measurements of the I-V characteristic of the stationary photocurrent in ruby, $j\|$(E_0), in the presence of an external field $E_0\|$C produced in ruby plates (d = 0.1 + 0.2 mm thick) by applying a voltage U (E_0 = U/d) (see inset in Fig. 1). As seen from Fig. 1, at fields $E_0<E_S$ the current $j\|$ is opposed to the field and reverses sign at $E_0 = E_S$ (= 475 kV/cm). The measured $j\|$ is essentially a stationary current flowing indefinitely through j a homogeneous sample. Thus the measurements of $j\|$ fully confirm the main assumption of the phenomenological theory[3,6] of the formation of induced domains containing strong electric fields in ruby.

It should be pointed out that the N-shaped anomaly in $j\|(E_0)$ resulting in an electrical instability of optically excited ruby occurs within a critically bounded interval of conditions, namely, at chromium concentrations $c>c_c = 0.15\%$,[1,7] temperatures $T<T_c = 150$ K,[2,5] and pumping densities $P\leq10$ kW/cm^2.[1,5] In fields $E_0|C$, the I-V characteristic of of ruby always has⁴ a shape which is close-to-ohmic (Fig. 1).

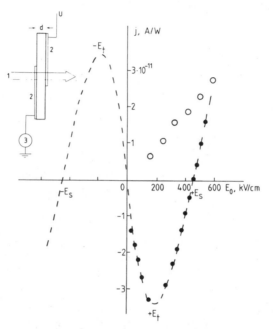

Fig. 1. Schematic representation of the N-shaped field dependence of photocurrent, $j\|(E)$, for $E\|C$ (dashed line) and experimental I-V characteristics of the photocurrent j in external fields $E_0\|C$ (o) and $E_0\|C$ (·), $T = 77$ K. Inset: schematic of measurements; 1 - laser beam (514.5 nm); 2 - transparent electrodes; 3 - electrometric amplifier.

The present paper reports on recent kinetic and spectral measurements of the photocurrent which provide an insight into the microscopic nature of the absolute negative electrical conductivity of optically excited ruby in fields $\|C$.

2.   Kinetics and Spectrum of Photocurrent: Experimental Data

2.1   Photocurrent Kinetics

The photocurrent kinetics was measured[8] using the scheme of Fig. 1 (see inset). The crystal was excited with rectangular pulses by properly chopping the Ar-laser beam (Fig. 2a). The large pulse duration, with long off intervals, permitted the current to reach saturation. Fig. 2b-d shows the measured time dependence of the photoecurrent for three values of the applied external field $E_0\|C$ corresponding to the positive ($E_0>E_S$), zero ($E_0 = E_S$) ($E_0<E_S$) values of the stationary current j_{st} observed under

constant excitation in the I-V characteristics (Fig. 1).

With the pumping switched on (t = t_1), a fast (<10 μs) pulse of positive current appears along the field. Next the current falls off on a time scale of a few ms down to the stationary value j_{st}, which may be positive (Fig. 2b), zero (Fig. 2c) or negative (Fig. 2d) depending on E_0. In the latter case the photocurrent reverses polarity in time.

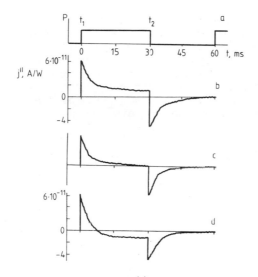

Fig. 2. Kinetics of photocurrent $j_{||}(t)$ for $E_0 || C$. (a) Ar 514.5 nm pulses; (b)-(d) photocurrent $j_{||}(t)$ for E_0 = 550 kV/cm, 450 kV/cm, 280 kV/cm, respectively. The critical field in the sample E_S = 450 kV/cm, chromium concentration 0.5 wt.% Cr_2O_3.

When the pumping is switched off (t = t_2), the photocurrent drops rapidly in all three cases (<10 μs) by an amount equal to the amplitude of the positive current observed when the pumping is switched on. As a result, the current at t = t_2 always becomes negative. In the case $E_0 > E_S$, Fig. 2b, the current even reverses polarity. Next the amplitude of the negative current falls to zero on the time scale of a few ms which corresponds to the zero current in the dark.

The complex kinetics of the photocurrent implies the existence of two contributions to the stationary photocurrent, namely, one "positive" (along the field) and one "negative" (oppositive to it) the contributions of opposite signs having different kinetics. The "positive" contribution has fast kinetics (<10 μs) as revealed in fast growth and decay of the photocurrent as the pumping is turned on and off. The "negative" contribution exhibits ms-scale kinetics manifesting itself in a slow growth and falloff of the current amplitude when the optical pumping is turned on and off.

2.2   Photocurrent Spectral Response

The above results were obtained with the ruby excited by discrete Ar laser lines which fall in the region of broad absorption bands corresponding to transitions to the highly excited states of the Cr^{3+} ions, $^4A_2 \rightarrow {}^4T_2$ (U-band) and $^4A_2 \rightarrow {}^4T_1$ (Y-band). These states undergo fast relaxation to the lowest metastable state ^2E, therefore it was natural to assume that direct "resonant" excitation of the Cr^{3+} ions via $^4A_2 \rightarrow {}^2E$ in the R_1, R_2 absorption lines of ruby would otherwise produce a photocurrent.

In the recent experiments[9] carried out using the scheme of Fig. 1 (inset) using a dye laser "resonant" excitation of ^2E-states in the R lines $(^4A_2 \rightarrow {}^2E)$ was found to generate photocurrent.

Fig. 3a presents the spectral response, $j_{\perp R}(\nu)$, of the "resonant" photocurrent in the region of the R line in an external field $E_0 \perp C$ (in this case, the $J_\perp(E_0)$ dependence does not have anomalies under "nonresondant" excitation in the U-band, Fig. 1). The photocurrent spectral response $j_{\perp R}(\nu)$ (solid line) is presented together with the $R(\nu)$ lineshape (dashed line): its width which is of the order of 1 cm^{-1}, is due to inhomogeneous broadening. The photocurrent spectrum $j_{\perp R}(\nu)$ agrees, on the whole, with the $R(\nu)$ absorption profile while being broader than the latter and stronger in the wings of the R line.

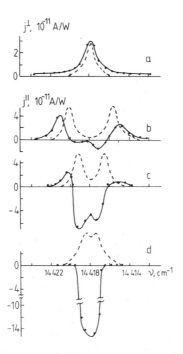

Fig. 3.   Spectral response of stationary resonant photocurrent under R line excitation; T = 77 K, chromium concentration 0.5 wt.% Cr_2O_3. (a) E_0 C; E_0 = 420 kV/cm; (b)-(d) $E_0||C$; E_0 = 640 kV/cm, 370 kV/cm, 210 kV/cm, respectively.

Fig. 3b-d shows the spectral responses of the resonant photocurrent $j||_R(\nu)$ in fields $E_0||C$ (solid line). As is known,[10] in fields $E_0||C$ the R lines, just as the other bands in the ruby spectrum, become Stark-split into symmetrical doublets, their width depending linearly on the field $|E_0|$ (dashed line). This doublet "pseudo-Stark" splitting is due to the linear Stark shift of transition frequencies in the Cr^{3+} ions which occupy two (A and B) Al^{3+} sites differing in the direction of the polar axis of their site symmetry (C_3) in the centrosymmetrical (D_{3d}) lattice of Al_2O_3. This Stark shift at the two sites is equal in magnitude and opposite in sign. As seen from Fig. 3b-d, the photocurrent reverses its polarity as the excitation frequency is scanned over the pseudo-Stark R-doublet profile. The photocurrent is positive in the outer doublet wings and negative (i.e. directed against the field) in the inner wings.

3. On the Microscopic Nature of the Phenomenon

We believe the possibility of charge transfer in photoexcited ruby to be due to the presence in the ruby lattice, also of a relatively small number of chromium ions in another, anomalous, charge state Cr^{2+} and/or Cr^{4+}[11] besides Cr^{3+} ions. When in the dark, these chromium states are stable, the corresponding "excess" charges being fixed. However, when the Cr^{3+} ions residing close to an excess charge carrier (Cr^{3+} ions residing close to an excess charge carrier (Cr^{2+}, Cr^{4+}) become optically excited in metastable ^2E states, ion charge exchange becomes possible involving spatial transfer of excess charge from the charge carrier (CC) to the excited Cr^{3+} ion:

$$Cr^{2+}(or\ Cr^{4+}) + Cr^{3+}(^2E) \rightarrow (Cr^{3+}(^4A_2) + Cr^{2+}(or\ Cr^{4+}) \qquad (1)$$

In the absence of an electric field, photostimulated charge transfer in randomly directed, the resultant current being zero. However, when a field is applied, the charge exchange (1) becomes oriented producing a photocurrent.

It is also very important to take into consideration the effect of the CC Coulomb field, $\vec{E} = \dfrac{e\vec{r}}{\varepsilon r^3}$, where $\varepsilon = 11.3$ is the dielectric constant of corundum Al_2O_3, on the Cr^{3+} ion environment which participates in the charge exchange under excitation in ^2E-state. For mean distances $r \approx 20$ Å in concentrated ruby the field E_C is high, its $z(||C)$ component, E^Z_C, producing a pseudo-Stark shift of the $^4A_2 \rightarrow {}^2E$ transition frequencies of the Cr^{3+} ions close to the CC toward the wings of the inhomogeneous R lineshape. In particular, excitation in the R line wings of these "perturbed" ions which are active in the charge exchange accounts, for the experimentally observed manifestation of the resonant photocurrent primarily in the wings of the R lineshape in the photocurrent spectra $j\perp_R(\nu)$ (Fig. 3a) and $j''_R(\nu)$ (Fig. 3b-d).

In an external field $E_0\perp C$ which does not produce a Stark shift of the Cr^{3+} levels, the Cr^{3+} ions are excited in the R-line with equal probability (i.e. symmetrically) on all sides of the CC. In this case one observes (Fig. 3a) a positive photocurrent due to direct action of the field E_0 on the charge exchange probabilities (1) resulting in a field-aligned charge transfer, which is typical of the hopping electrical conduction over impurities.

However, in an external field $E_0||C$ which can produce a pseudo-Stark spectral line splitting a radically different photocurrent mechanism becomes possible which is, in principle, similar to the mechanism of a "bulk photovoltaic" current. Indeed, in a field $E_0 \| C$, the resultant field E_0 plus E^Z_C turn out to be different for the Cr^{3+} ions residing on the right and on the left of a CC. On one side of the CC the fields E_0 and E^Z_C are parallel while on the other they are antiparallel, thus effectively canceling one another (Fig. 4). Therefore the width of the pseudo-Stark R-doublet splitting for the "right" Cr^{3+} ions (in Fig. 4) is broader, and for the "left" ones, narrower than the splitting of the main R line shape determined by E_0. Hence, by spectrally selective optical excitation of the crystal in the outer (inner) R doublet wings one can achieve spatially selective excitation of the Cr^{3+} ions lying on one or the other side of the CC. Under these conditions photostimulated charge transfer from the CC is possible only in the direction where excited Cr^{3+} ions are present, i.e. a "photovoltaic current" determined by the right-left asymmetry of excitation appears. As follows from purely geometrical considerations,[9,12] irrespective of the CC sign (Cr^{2+} or Cr^{4+}) excitation in the outer wings of the R doublet produces always a positive, and in the inner wings, a negative (opposite to the field) photovoltaic current. It is this situation that is observed experimentally for $E_0 \| C$ in the spectral response of the R photocurrent (Fig. 3b-d).

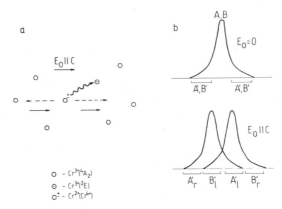

Fig. 4. (a) Schematic of photostimulated charge transfer in ruby lattice. Wavy arrow: chromium ion charge transfer; straight dashed arrows: excess (positive) charge carrier field; solid arrows: applied field $E_0 \| C$. (b) Inhomogeneously broadened R lineshape. Brackets below identify transition frequencies in "affected" (A',B') Cr^{3+} ions residing on the right (r) and left (1) of the excess (positive) charge carrier.

Thus experiments with frequency-selective excitation of photocurrent in the R line (Subsection 2.2, Fig. 3) show that the postulated[3] photovoltaic nature of the photocurrent against the field for $E_0 \| C$ originates on the microscopic level from the polar asymmetry of the excitation to ^2E of the Cr^{3+} ions residing close to CC and participating actively in charge exchange. Why, however, does an asymmetry arise with sign corresponding to a current flowing against the field for $E_0 \| C$, $E_0 < E_S$, and under "nonresonant" Cr^{3+} excitation to ^2E "from above" via the broad U-band (Section 1, Fig. 1) when the spectral selectively factor for Cr^{3+} excitation is negligible because of the large band width?

The answer to this question which was obtained recently[8] is as follows. The "perturbed" Cr^{3+} ions located close to CC may become optically excited to ^2E not only directly by absorbing a photon in the U band with a subsequent $^4T_2 \rightarrow {}^2E$ relaxation, but also via resonant nonradiative energy transfer[13,14] to these ions of ^2E excitation from the "main" inhomogeneously broadened Cr^{3+} ensemble which is excited efficiently to ^2E through the U band. The spectral distribution of ^2E excitations is obviously described by the doublet R lineshape (dashed line in Fig. 3b-d). Under incomplete resolution of the pseudo-Stark A, B doublet for $E_0 < E_S$, the maximum of one component (A or B) in its central part overlaps strongly with the wings of the other (B or A) doublet component. As a result, the Cr^{3+} ions residing in the inner doublet wings and "affected" by the proximity to CC will be energetically in resonance with a larger number of excited Cr^{3+} ions than in the case with the ions in the outer wings of the doublet. This is what accounts for the more efficient $A \rightarrow B$, $B \rightarrow A$ resonant nonradiative transfer of ^2E excitation to the Cr^{3+} ions in the inner wings and hence provides (see Fig. 3b-d) the asymmetry corresponding to a negative current.

The conclusion about two mechanisms of excitation of Cr^{3+} ions residing close to CC and of the decisive role of the resonant nonradiative ^2E excitation transfer in the appearance of absolute negative conductivity in ruby in fields $E_0 || C$ ($E_0 < E_S$) when exciting in the U band is confirmed convincingly by measurements of the complex kinetics of the photocurrent (Subsection 2.1, Fig. 2). The positive contribution to the photocurrent exhibiting fast (<10 μs) kinetics is the "photoconduction current." It originates from direct excitation of the "pertubed" Cr^{3+} ions which accounts for the symmetrical excitation of the Cr^{3+} ions near CC and results in a positive current due to the field action on the charge transfer probabilities in favor of charge transfer along the field (the kinetics of this contribution demonstrates fast, <10 μs, relaxation of the ^2E excitation of the Cr^{3+} ions residing close to CC). The negative contribution to the photocurrent with nonexponential ms-scale kinetics is an "induced photovoltaic current." It comes from the excitation of "affected" Cr^{3+} ions close to CC via resonant nonradiative ^2E-energy transfer from the bulk of the Cr^{3+} ions. In this process, the polar asymmetry of excitation appears due to $A \vec{\leftrightarrow} B$ energy transfer which results in a negative (against the E_0 field) sign of the photovoltaic current. The role of resonant nonradiative $A \vec{\leftrightarrow} B$ excitation energy transfer is persuasively confirmed by the long (ms scale) times typical of the negative current kinetics agreeing with the direct measurements of the ^2E excitation $A \vec{\leftrightarrow} B$ transfer times in ruby,[13,14] as well as by the nonexponential character of this kinetics. One readily sees that to produce the excitation asymmetry leading to $j || < 0$ it is essential to have an overlap of the A, B components of the pseudo-Stark R doublet which decreases with increasing E_0. This contributes to the negative absolute conductivity is limited by the fields $|E| < |E_S|$ and affects the N-shaped behaviour of the field dependence of the photocurrent $j || (E)$ (Fig. 1).*

---

*Obviously enough, the $A \vec{\leftrightarrow} B$ tranfer of ^2E excitations occurs only in the "resonant" R excitation of the photocurrent, where it affects the amplitudes of the negative and positive photocurrents in the $j \perp {}_R(\nu)$ spectrum for different E_0 (Fig. 3b-d).

Thus the results of the present work shed light on the role of some microscopic processes in the appearance of absolute negative electrical conductivity in ruby. This phenomenon is seen to arise as a result of combined action of a number of effects well known to exist in ruby, such as the pseudo-Stark R line splitting due to the polarity of distinct crystallographic sites of Cr^{3+} ions, the existence of anomalous Cr charge states and resonant nonradiative excitation energy transfer. The elucidation of the microscopic mechanism of the photoinduced charge transfer between chromium ions (1) remains an unsolved problem.

References

1. P. F. Liao , A. M. Glass, L. M. Humphrey, Phys. Rev. B22:2276 (1980).
2. S. A. Basun, A. A. Kaplyanskii, S. P. Feofilov, Pisma Zh. Eksp. Teor. Fiz. 37:492 (1983).
3. M. I. Dyakonov, Pisma Zh. Eksp. Teor. Fiz. 39:158 (1984).
4. S. A. Basun, A. A. Kaplyanskii, S. P. Feofilov, A. S. Furman, Pisma Zh. Eksp. Teor. Fiz. 39:161 (1984).
5. S. A. Basun, A. A. Kaplyanskii, S. P. Feofilov, Zh. Eksp. Teor. Fiz. 87:2047 (1984).
6. M. I. Dyakonov, A. S. Furman, Zh. Eksp. Teor. Fiz. 87:2063 (1984).
7. S. A. Basun, A. A. Kaplyanskii, S. P. Feofilov, Fiz. Tverd. Tela 28:929 (1986).
8. S. A. Basun, A. A. Kaplyanskii, S. P. Feofilov, Fiz. Tverd. Tela 29:1284 (1987).
9. S. A. Basun, A. A. Kaplyanskii, S. P. Feofilov, Pisma Zh. Eksp. Teor. Fiz. 43:344 (1986).
10. W. Kaiser, S. Sugano, D. L. Wood, Phys. Rev. Lett. 6:605 (1961).
11. G. E. Arkhangelskii, Z. L. Morgenshtern, V. B. Neustruev, Phys. Status Solidi 22:289 (1967).
12. A. A. Kaplyanskii, S. A. Basun, S. P. Feofilov, in: "Zero-Phonon Lines and Spectral Holeburning in Spectroscopy and Photochemsitry," eds. O. Sild, K. Haller, Springer-Verlag, 1987.
13. P. E. Jessop, A. Szabo, Phys. Rev. Lett. 45:1712 (1980).
14. S. Chu, H. M. Gibbs, S. L. McCall, A. Passner, Phys. Rev. Lett. 45:1715 (1980).

THEORY AND OBSERVATION OF ELECTRON–HOLE COMPETITION IN THE PHOTOREFRACTIVE EFFECT

Robert W. Hellwarth

University of Southern California
Department of Physics
Los Angeles, CA 90089–0484

An extension of Kukhtarev's single–carrier model of photorefractive charge transport to allow simultaneous photo–excitation of both holes and electrons improves agreement with a wide variety of data in many crystals. This indicates that electron–hole competition plays a much wider role than previously suspected. This competition does not however reduce the advantages of using moving gratings to enhance grating strength.

INTRODUCTION

A reversible refractive index grating can be formed in photorefractive crystals by two overlapping monochromatic laser beams. This makes photorefractive crystals attractive for applications such as dynamic holography and optical phase conjugation. Many such crystals are now available, the most widely used of which are p–type $BaTiO_3$ (p–BTO) and n–type $Bi_{12}SiO_{20}$ (n–BSO). The physical process involved is a light–induced redistribution of charges among deep traps. The resulting space–charge field then modulates the refractive index of the materials by means of the electrooptic (Pockels) effect. Two models have been used to describe the behavior of these materials, the band–conduction model[1-3] and the hopping model.[4] Until recently these models assumed that a single type of carrier was involved in the charge transport process. For the approximately dozen samples of the above materials studied using these models,[4-9] a single–carrier model, especially that of Kukhtarev, gave fair quantitative agreement with the various optical measurements, yielding in the process the important impurity parameters (deep–trap density, quantum efficiency, and carrier diffusion length) for predicting all photorefractive effects slower than carrier–trap recombination. Nevertheless, one $BaTiO_3$ crystal showed behavior clearly at variance with these models and, was called "anomalous".[6] Also most crystals behaved as if their electro–optic coefficients were smaller than those measured by standard methods[6-9]. These discrepancies were resolved by a simple extension of the Kukhtarev model which incorporates simultaneous hole and electron excitation from the same deep–trap levels, i.e., "bipolar" behavior.[10,11] This model[10-12] also shows improved agreement with grating amplitude and decay measurements, for all crystals examined to date.[10,11,13] That is, it seems likely at this point that some observable electron–hole competition will be found in any of the presently available photorefractive samples. This new model explains the observation of Kamshilin and Petrov[14] that n–BSO is bipolar on a 1 micron scale, whereas Hou, et.al.,[15] and Kostyuk, et.al.,[16] found hole photoconductivity to be at least an order of magnitude smaller than electron photoconductivity on a millimeter scale. Also as Orlowski and Kratzig demonstrated in $LiNbO_3$, hole–electron competition (bipolarity) reduces the magnitude of the photorefractive effect.[17] It is this reduction that has almost certainly been responsible for the apparent lowering of the electro–optic coefficient mentioned above, rather than 180^0 domain structure or other crystal imperfections.

In this paper we review the new bipolar model of photorefractive grating formation, extending it to the case where the optical writing beams have different frequencies. We summarize experimental values for both hole and electron excitation cross–sections and diffusion lengths in n–BSO which have been inferred from data using this model. We also show that the moving grating written in a bipolar crystal by beams of different frequencies has a much larger amplitude than when the beams have equal frequencies, similarly as is the case for a single carrier.

THEORY

The bipolar model we describe assumes, as have previous treatments,[1-4] that N_D per unit volume of a single species of dopant (e.g., Fe) creates an impurity band that is partially filled at temperature T. A number N_D^i per unit volume of ionized dopants (e.g., Fe^{3+}) are centers for electron recombination and hole photoexcitation. The remaining $(N_D - N_D^i)$ m^{-3} of the un–ionized dopants (e.g., Fe^{2+}) are centers for hole recombination and electron photoexcitation. In accordance with the experimental conditions, we assume that electron and hole current densities j_e and j_h, as well as the internal field plus the applied electric field E, are parallel to the crystal c axis along which all variations in optical intensity I occur. The crystal is assumed to be uniaxial. Introducing the densities of free holes and electrons n_e, n_h; their charge $\pm e$; their excitation cross sections s_e, s_h; recombination coefficients γ_e, γ_h; and mobilities μ_e, μ_h (>0), we can write the following basic equations (in SI units), which are a simple extension of those of Kukhtarev[1]: the rate equation for ionized dopants

$$\partial N_D^i/\partial t = s_e I (N_D - N_D^i) - \gamma_e n_e N_D^i - s_h I N_D^i + \gamma_h n_h (N_D - N_D^i) , \qquad (1)$$

the current densities for electrons and holes (k_B is Boltzmann's constant)

$$j_e = en_e \mu_e E + \mu_e k_B T \nabla n_e , \qquad (2)$$

$$j_h = en_h \mu_h E - \mu_h k_B T \nabla n_h , \qquad (3)$$

the continuity equations for electrons and holes

$$\partial n_e/\partial t = \nabla j_e/e + s_e I (N_D - N_D^i) - \gamma_e n_e N_D^i , \qquad (4)$$

$$\partial n_h/\partial t = -\nabla j_h/e + s_h I N_D^i - \gamma_h n_h (N_D - N_D^i) , \qquad (5)$$

and the Poisson equation

$$\nabla E = -(e/\epsilon)(n_e + N_A - N_D^i - n_h) . \qquad (6)$$

The density of compensative acceptors N_A equals the density of ionized photorefractive dopants in the dark. The dielectric constant ϵ, parallel to the crystal axis, equals the relative dielectric constant times that of free space.

We assume an intensity distribution $I = I_0 \{1 + \mathcal{R}e[m \exp(i\phi)]\}$, where $\phi = kz - \Omega t$ ($k>0$), and the complex modulation index is m. The spatial variation of all other quantities is therefore described in the same way by a constant zeroth-order term plus a small sinusoidal first-order term:

$$N_D^i(t,z) = N_0(t) + \mathscr{Re}[N_1(t)\exp(i\phi)] \; ,$$

$$n_e(t,z) = n_{e0}(t) + \mathscr{Re}[n_{e1}(t)\exp(i\phi)] \; ,$$

$$n_h(t,z) = n_{h0}(t) + \mathscr{Re}[n_{h1}(t)\exp(i\phi)] \; ,$$

$$j_e(t,z) = j_{e0}(t) + \mathscr{Re}[j_{e1}(t)\exp(i\phi)] \; ,$$

$$j_h(t,z) = j_{h0}(t) + \mathscr{Re}[j_{h1}(t)\exp(i\phi)] \; ,$$

$$E(z,t) = E_0 + \mathscr{Re}[E_1(t)\exp(i\phi)] \; , \tag{7}$$

where E_0 is the externally applied electric field.

Substitution of the expressions of Eqs. (7) into Eqs. (1)–(6) yields equations in which we are able to neglect all terms of quadratic or higher order in the $\exp(i\phi)$ in Eqs. (7). This is because we assume that the magnitude $|m|$ of the optical modulation is much less than unity. We also use the so-called quasi-cw illumination approximation in which it is assumed that the time scale for establishing or erasing of a photorefractive grating is much longer than either free-carrier lifetime. This means that we can neglect the time derivatives in Eqs. (4) and (5). Furthermore, the illumination is assumed to be so small that (a) the contribution of free-carrier densities to the spatially modulated component of Eqs. (7) is negligible, (b) $|n_{e1}| \ll n_{e0}$, and (c) $|n_{e0}| \ll N_A$, and similarly for hole densities. This requires the satisfaction of two inequalities, one for each carrier, of the form

$$I_0 \ll h\nu k_i^2 \epsilon k_B T / \tau \alpha_P e^2 \; , \tag{8}$$

where $h\nu$ is the photon energy, τ is the carrier lifetime, and α_P is the optical absorption caused by excitation of the carrier. Here k_i equals k, k_0, K_e or K_h (as defined below) whichever is least. We also assume that the beam frequency difference Ω is much less than the electron or hole recombination rates $\gamma_e N_A$ or $\gamma_h (N_D - N_A)$, as relaxing this condition does not give accurate formulae without relaxing the condition (8).

The zeroth-order free-carrier densities are then

$$n_{e0} = \frac{s_e I_0 (N_D - N_0)}{\gamma_e N_0} \; , \qquad n_{h0} = \frac{s_h I_0 N_0}{\gamma_h (N_D - N_0)} \; , \tag{9}$$

and we obtain a first-order differential equation for the time evolution of the space–charge field $E_1(t)$:

$$-\frac{E_1(t)}{dt} = E_1(t) \left[\frac{n_{e0} e\mu_e}{\epsilon} \frac{1 + k(k - iV)/k_0^2}{1 + k(k - iV)/K_e^2} \right.$$

$$\left. + \frac{n_{h0} e\mu_h}{\epsilon} \frac{1 + k(k + iV)/k_0^2}{1 + k(k + iV)/K_h^2} - i\Omega \right]$$

$$+ im \frac{k_B T}{e} \left[\frac{n_{e0} e\mu_e}{\epsilon} \frac{k - iV}{1 + k(k - iV)/K_e^2} \right.$$

$$\left. - \frac{n_{h0} e\mu_h}{\epsilon} \frac{k + iV}{1 + k(k + iV)/K_h^2} \right], \qquad (10)$$

with

$$K_e^{-2} \equiv \frac{k_B T \mu_e}{e\gamma_e N_0}, \qquad\qquad K_h^{-2} = \frac{k_B T \mu_h}{e\gamma_h (N_D - N_0)}, \qquad (11)$$

$$k_0^2 \equiv \frac{e^2 N_0 (N_D - N_0)}{\epsilon k_B T N_D}, \qquad V = eE_0/k_B T . \qquad (12)$$

K_e^{-1} and K_h^{-1} are average distances traveled by the two kinds of charge carrier between excitation and recombination, k_0 is the inverse Debye screening length, and E_0 is the externally applied electric field.

When there is no voltage applied to the crystal and $\Omega = 0$, the amplitude of the steady–state space–charge field is, From Eq. (10),

$$E_{SC} = -im \frac{k_B T}{e} \frac{k}{1 + k^2/k_0^2} \xi(k) \, , \qquad (13)$$

where

$$\xi(k) \equiv (1 - C)/(1 + C) \, , \qquad (14)$$

in which

$$C \equiv \frac{\alpha_h (k^2 + K_e^2)}{\alpha_e (k^2 + K_h^2)} \, , \qquad (15)$$

where α_h and α_e are the contribution of hole–and electron–excitation to the total absorption coefficient α cm^{-1} of the material. In the single–charge–carrier case, $\xi(k) = 1$ for electrons and $\xi(k) = -1$ for holes. The expressions for the space–charge field derived by Orlowski and Kratzig[17] and by Klein and Valley agree, as expected, with Eq.(13) in the limit of small k and m.

EXPERIMENT

A reversal of the sign of E_{SC} as k was increased was reported for one "anomalous" BTO crystal by Klein and Valley.[6] The relation (13) fits very well the "anomalous" data for this crystal, with reasonable values for $S(\equiv \alpha_h/\alpha_e)$, K_e, K_h, and k_0. Also unadjusted values for the unclamped[6] electro–optic coefficients measured by standard methods were used.

The photo–induced grating decay rates and two–beam coupling coefficients recently have been measured as a function of k for four different n–BSO samples. In each case the two curves agreed within experimental error with the functional forms predicted by (10) and (13) with the impurity parameters summarized in Table 1.[13] The quantum efficiencies ϕ listed for the four crystals in Table 1 are defined as the fraction of the optical absorption which arises from the excitation of holes or electrons: $\phi \equiv (\alpha_e + \alpha_h)/\alpha$. By comparing (1) with (10), one sees that the quantity $\alpha_e + \alpha_h$ can be obtained from the observed asymptotic decay rate at large k. To prepare Table 1, we have used only the unadjusted value at 515 nm of the electrooptic coefficient ($r_{41} = 4.5 \pm 0.1$ pm/V) as measured by precise electro–optic techniques.[19] The electron–hole competition among these crystals is seen to be similar, although no care was taken in selecting them. We have recently observed competition of the minority carrier of similar magnitude in a typical p–BTO sample used in phase–conjugation experiments. These experiences have lead us to suppose that many if not all photorefractive crystals which have undergone no special treatment exhibit electron–hole competition which reduces, although not greatly, the photorefractive efficiency of the crystal.

The enhancement of the photorefractive grating amplitude E_1, by imposing a moving intensity grating (having velocity $v = \Omega/k$ m/sec) simultaneously with a constant applied electric field E_0, has often been considered in theory.[4,20-24] Experimental observations of this case have been studied commonly in n–BSO and for K_e^2, $K_h^2 \ll Vk$, $k^2 \ll k_0^2$,[22-24] at least if our measurements showing $K_e \sim K_h$ are correct. The steady state solution of (10) is simply

$$eE_1 \sim - imk_BT(1-S)/(A+iB) \qquad (16)$$

where

$$A \equiv k(1+S)\,(k^2+V^2)^{-1} \qquad (17)$$

and

$$B \equiv V(1-S)\,(k^2+V^2)^{-1} - k\Omega/\Gamma_{e0}K_e^2 \ . \qquad (18)$$

Here $\Gamma_{e0} \equiv n_{e0}\,e\mu_e/\epsilon$ is the inverse photoconductive dielectric relaxation time for electrons only. The peak enhancement of two–beam coupling occurs for given V when $|\mathrm{Im}\,E_1|$ is largest. This occurs when Ω is adjusted to make B nearly zero, giving in the regime defined above,

$$\max_{\Omega} |\mathrm{Im}\,eE_1| \sim k_BT\,(1-S)(1+S)^{-1}\,(k^2+V^2)/k \ . \qquad (19)$$

In practice, values of V approaching k_0 can be obtained before electrical breakdown occurs, and the enhancement over the value given in (13) is of order (k_0^2/k^2) which can be as large as 10^3 in BSO when k becomes as small as K_e and K_h. (From Table 1 and Eqn. (12) one sees that $k_0 \sim 10^5$ cm^{-1}.) Such enhancement has been observed.[24] We predict from (19) that the enhancement factor for this regime will not be found to be affected by electron–hole competition, although the absolute magnitude of E_1 is affected.

Table 1. Electron and hole parameters of four typical crystals of n-type $Bi_{12}SiO_{20}$, as determined from the measured dependences on grating wavevector k of their grating-decay and two-beam coupling constants.[a] The electron-hole competition parameter $\xi(k)$, the inverse electron and hole diffusion lengths K_e and K_h, the effective trap density N_E, the ratio of hole to electron excitation rates s, the measured absorption α at 515 nm, and the photorefractive quantum efficiency ϕ are all derived as described in the text.

Crystal	$\xi(\infty)$	$\xi(0)$	$K_e^{-1}(\mu m)$	$K_h^{-1}(\mu m)$	$N_E(10^{16}cm^{-3})$	S (%)	$\alpha (cm^{-1})$	ϕ
Sumitomo	0.73	--	5.8+2.3	>3	0.86+-0.08	15.9+-1.4	1.57+-0.16	1.01+-0.17
Crystal Tech.1	0.61	0.93	1.39+-0.16	0.55+-0.08	1.01+-0.18	24.2+-2.7	1.56+-0.16	1.64+-0.62
Crystal Tech.2	0.66	--	>4.5	>1.5	1.23+-0.20	20.3+-2.2	1.15+-0.1	1.63+-0.35
Crystal Tech.3	0.70	0.92	4.0+-0.7	3.0+-1.5	1.45+-0.16	17.7+-2.4	1.33+-0.1	0.91+-0.18

[a]Reference 13.

SUMMARY

In conclusion, we have found that a simple extension of Kukhtarev's equations (governing photorefractive charge transport and trapping) to include both hole and electron excitation produces much better agreement with the observed functional forms of grating decay and two-beam coupling coefficient considered as functions of the grating wavevector. Fitting the theoretical expressions to data has yielded both electron and hole excitation rates and drift lengths in a number of samples of $BaTiO_3$ and $Bi_{12}SiO_{20}$. We show that electron–hole competition does not reduce the enhancement in grating amplitude that is produced by an optical beam frequency difference (moving grating) which is properly matched to the value of a constant applied electric field.

REFERENCES

1. N.V. Kukhtarev, Sov. Tech. Phys. Lett. $\underline{2}$, 438 (1976).
2. N.V. Kukhtarev, V.B. Markov, S.G. Odulov, M.S. Soskin, and V.L. Vinetskii, Ferroelectrics $\underline{22}$, 949 (1979).
3. G.C. Valley, IEEE J. Quantum Electron. $\underline{QE-19}$, 1637 (1983).
4. J. Feinberg, D. Heiman, A.R. Tanguay, Jr., and R.W. Hellwarth, J. Appl. Phys $\underline{51}$, 1297 (1980).
5. R.A. Mullen and R.W. Hellwarth, J. Appl. Phys. $\underline{58}$, 40 (1985).
6. M.B. Klein and G.C. Valley, J. Appl. Phys. $\underline{57}$, 4901 (1985).
7. S. Ducharme and J. Feinberg, J. Opt. Soc. Am. B $\underline{3}$. 283 (1986).
8. J.M.C. Jonathan, R.W. Hellwarth, and G. Roosen, IEEE J. Quantum Electron. $\underline{QE-22}$, 1936 (1986).
9. G. Pauliat, J.M.C. Jonathan, M. Allain, J.C. Launay and G. Roosen, Opt. Comm. $\underline{59}$, 266 (1986).
10. F.P. Strohkendl, J.M.C. Jonathan, and R.W. Hellwarth, Opt. Lett. $\underline{11}$, 312 (1986).
11. F.P. Strohkendl and R.W. Hellwarth, J. Appl. Phys., September 1987).
12. G.C. Valley, J. Appl. Phys. $\underline{59}$, 3363 (1986).
13. F.P. Strohkendl, P. Tayebati and R.W. Hellwarth, Tech. Digest Series $\underline{17}$, 32 (Opt. Soc. Am. 1987) and to be published.
14. A.A. Kamshilin, M.P. Petrov, Sov. Phys. Solid State $\underline{23}$, 1811 (1981).
15. L. Hou, R.B. Lauer and R.E. Aldrich, J. Appl. Phys., $\underline{44}$, 2652 (1973).
16. B. Kh. Kostyuk, A. Yu. Kudzin, and G. Kh. Sokolyanskii, Sov. Phys. Solid State $\underline{22}$, 1429 (1980).
17. R. Orlowski, and E. Kratzig, Solid State Commun. $\underline{27}$, 1351 (1978).
18. P. Gunther, Phys. Rev. $\underline{93}$, 200 (1983).
19. A.R. Tanguay, Jr., Ph.D. Thesis, Yale University, 1977.
20. G. Valley, J. Opt. Soc. Am. $\underline{B1}$, 868 (1984).
21. Ph. Refregier, L. Solymar, H. Rajbenbach, and J.P. Huignard, Electron. Lett. $\underline{20}$, 656 (1984).
22. J.P. Huignard and A. Marrakshi, Optics Commun. $\underline{38}$, 249 (1981).
23. S.I. Stepanov, K. Kolikov, and M. Petrov, Optics Commun. $\underline{44}$, 19 (1982).
24. H. Rajbenbach, J.P. Huignard, and B. Loiseaux, Optics Commun. $\underline{48}$, 247 (1983).

COHERENT OPTICAL OSCILLATION DUE TO VECTORIAL FOUR-WAVE MIXING IN

PHOTOREFRACTIVE CRYSTALS

S. Odoulov and M. Soskin

Institute of Physics, Academy of Sciences of the Ukrainian
SSR, 252 650, Kiev, USSR

ABSTRACT

Photorefractive coherent oscillators[1,2] attract much attention
because of their unusual properties and the possibility of their
practical application in interferometry, laser beam conjugation and
cleanup, etc. The majority of experiments (see the review article[3])
have been conducted with barium titanate or similar crystals where the
nonlocal nonlinear response is governed by the diffusion of the
photoexcited charge carriers[4].

In this paper another kind of photorefractive nonlinearity, nonlocal
as well, which arises due to excitation of circular photovoltaic currents
in Fe- or Cu-doped $LiNbO_3$ or $LiTaO_3$ crystals, and coherent oscillators
based on this nonlinearity, are reported.

PHOTOVOLTAIC NONLINEARITY OF PHOTOREFRACTIVE CRYSTALS

The appearance of a steady-state electric current in a short-
circuited sample illuminated by a light wave (photovoltaic effect[6,7])
is one of the most effective charge transport mechanisms in doped photorefractive crystals. In the pioneering studies by A. M. Glass, only the
current in the direction of the spontaneous polarization of $LiNbO_3$ and
$LiTaO_3$ was detected, the current density being proportional to the
incident light intensity I and sample absorptivity α,

$$j = k\alpha I, \tag{1}$$

where k is the Glass constant.

More recent theoretical studies (see the review article[7]) predicted the appearance of a photovoltaic current also along the transverse
polar axes. It is the photovoltaic tensor $\overset{\leftrightarrow}{\beta}$ that indicates the possible
directions of current propagation and the conditions of its excitation

$$j_i = \beta_{ijk} E_j E_k^*, \quad \beta_{ijk} = \beta_{ikj}^*, \tag{2}$$

where $\vec{E}$ is the electric field of the light wave.

For 3m point group crystals this tensor has the form

$$
\beta_{ijk} =
\begin{vmatrix}
0 & -\beta_{22} & \beta_{31} \\
0 & \beta_{22} & \beta_{31} \\
0 & 0 & \beta_{33} \\
0 & \beta_{15} & 0 \\
-\beta_{15} & 0 & 0 \\
-\beta_{22} & 0 & 0
\end{vmatrix}
\cdot
\tag{3}
$$

The standard notations for indices, viz $11 \to 1$, $22 \to 2$, $33 \to 3$ and $31 \to 5$, are used here.

All components of $\vec{\vec{\beta}}$ (3), with only one exception, β_{15}, are real. The component $\beta_{15} = \beta_{15}^{s} + i\beta_{15}^{a}$ contains the antisymmetric part, β^{a} (7,8). Its corresponding current can be excited only by a circularly polarized wave[9,10]. That is why, following the suggestion of the authors of[9,10], we will call them "circular" photovoltaic currents. The currents corresponding to all real components can be excited by light with linear polarization, and hence they are called "linear" currents.[7]

The excitation of the linear currents and the resulting space charge formation lead to the appearance of the local nonlinear response, whereas the circular currents feature the nonlocal nonlinear response.[5]

A nonvanishing off-diagonal component β_{15} makes possible anisotropic holographic recording, i.e. grating recording by orthogonally polarized waves. The field

$$
\vec{E} = \hat{e}_o A_o \exp(i\vec{k}_o \cdot \vec{r}) + \hat{e}_e A_e \exp(i\vec{k}_e \cdot \vec{r}),
\tag{4}
$$

representing ordinary (o) and extraordinary (e) waves intersecting in a plane perpendicular to the spontaneous polarization direction excites a spatially-oscillating current

$$
\vec{j} = 2\hat{e}_o |A_o A_e| [\beta_{15}^{s} \cos(\vec{K} \cdot \vec{r}) + \beta_{15}^{a} \sin(\vec{K} \cdot \vec{r})],
\tag{5}
$$

where $\vec{K} = \vec{k}_o - \vec{k}_e$ is the current grating wave vector. The first term in (5) determines the recording of an "unshifted" grating (local response), and the second one, of a "shifted" grating (nonlocal response)[11].

The photovoltaic current (5) results in the formation of a space charge grating, the space charge field modulating the crystal permittivity. Because crystals belonging to the point group 3m have a nonzero Pockels coefficient r_{51}, anisotropic diffraction (with a 90° rotation of the polarization of the diffracted wave with respect to the polarization of the reconstructing wave) is possible[12,13].

The shifted grating ensures a steady-state coupling of two interacting beams. The output intensities of the beams become

$$I_{o,e}(\ell) = [I_o(0) + I_e(0)][1 + m^{\pm 1}\exp(\pm\Gamma\ell)]^{-1}, \tag{6}$$

where $\Gamma = 2\pi n_o^2 n_e^2 r_{51}\beta_{15}^a(\hat{s}\cdot\hat{e}_o)^2/\kappa\lambda\cos\theta$ is the gain[14], $\hat{s} = \vec{K}/|\vec{K}|$ is the grating unit vector, $m = I_o(0)/I_e(0)$, the wavelength is λ, and the angle 2θ is that of the beam intersection inside the sample.

Note some distinguishing features of this interaction as opposed to those in "diffusion type" photorefractive crystals: (i) the coupling direction is independent of the orientation of the polar axes; energy is always transferred from the ordinary wave to the extraordinary one for crystals with $r_{51}\beta_{15}^a < 0$, and _vice versa_; (ii) the gain Γ is proportional to β_{15}^a, i.e. energy coupling is possible only in doped samples and Γ can be controlled by the dopant concentration; (iii) the angular dependence of the gain is determined by the crystal birefringence; Γ drops rapidly to zero at low spatial frequencies of the recorded grating.

The backward wave vectorial four-wave mixing due to this type of nonlinearity has been treated in previous papers[14-15]. The set of equations for the complex amplitudes of the interacting waves for this particular nonlinearity differs from that known for the diffusion-type nonlinearity. This dissimilarity reflects a difference in the physical situations: two transmission gratings recorded by the counterpropagating pairs of waves are here strictly spatially superposed (also in phase !). As a result some unusual features appear: (i) the highest efficiency (greatest transmittivity of the signal wave and greatest reflectivity of the phase conjugate wave) correspond to equal intensities of the pump waves; (ii) in the undepleted pump approximation both signal transmittivity and phase conjugate reflectivity become infinite for a finite coupling strength ($\Gamma\ell = 4$ for equal intensities of the pump waves), which indicates the threshold of parametric mirrorless oscillation; pump intensity imbalance only increases the oscillation threshold; (iii) for any given coupling strength $\Gamma\ell < 4$ the nonlinearity under consideration ensures the greatest phase conjugate reflectivity when compared with other types of nonlinearity (Fig. 1).

Unusual properties of the photovoltaic nonlinearity itself and its unusual manifestation in four-wave mixing processes influence the properties of coherent oscillators[14-19]. Oscillation was obtained in the various cavity configurations shown in Fig. 2. Some of them, such as the unidirectional ring oscillator with a single pump beam (Fig. 2a), the oscillator with a phase-conjugate mirror (Fig. 2b), or the oscillator with a linear cavity (Fig. 2c), have already been utilized to produce oscillation in "diffusion-type" photorefractive materials[2,3]. The oscillators based on the forward parametric four-wave mixing (Figs. 2d,e,f)[18] or mirrorless oscillation (Fig. 2g)[19] cannot be obtained in principle with nonlinearity of the diffusion-type.

It should be noted, however, that some oscillator configurations successfully used with diffusion-type media (semilinear open cavity[20] or "cat-conjugator"[21]) are impossible with photovoltaic-type nonlinear media.

BACKWARD-WAVE VECTORIAL FOUR-WAVE MIXING OSCILLATOR WITH A LINEAR CAVITY[15-17]

By putting a LiNbO$_3$:Fe sample inside a cavity (Fig. 2c) and by illuminating it with a single pump wave (ordinary wave in the sample) we can achieve oscillation. Two counterpropagating oscillating (extraordinary) beams 1 and 2 arise in the cavity and an additional (ordinary) wave 4, phase conjugate to the pump wave 3 appears.

In a 3 mm thick sample, where flat input and output faces served as the cavity mirrors (with reflectance $R_1 = R_2 \simeq 0.16$), the output intensity of the forward oscillation beam 1 was $\simeq 35\%$ of the incident pump intensity. Thus, this oscillator can efficiently convert a pump beam with a complex wavefront: the o-polarization of the pump is simultaneously changed into the e-polarization of the oscillation beam.

To optimize this oscillator as a phase conjugate mirror, the reflectance of both mirrors should be as close to unity as possible ($R_1 = R_2 = 1$). In contrast to the similar device with diffusion-type media, where the achievable phase conjugate reflectivity R_{pc} is limited only by the reflectance of the back mirror, in the case under consideration R_{pc} cannot exceed the reflectance of the more transparent of the two cavity mirrors[15].

FORWARD FOUR-WAVE VECTORIAL MIXING AND OSCILLATORS

The difference in phase velocities of the ordinary and extraordinary waves in a birefringent photorefractive crystal allows a forward parametric mixing with the phase matching condition

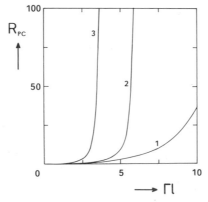

Fig. 1. Dependence of phase conjugate reflectivity (R_{pc}) on the coupling strength ($4\pi \, \Delta nl/\lambda$) for (1) diffusion type photorefractive nonlinearity, (2) third order local nonlinearity, and (3) circular photovoltaic nonlinearity. All curves correspond to optimized pump intensity ratios.

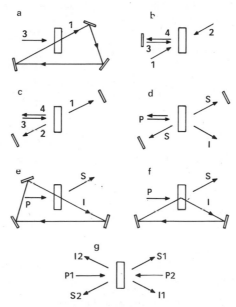

Fig. 2. Optical oscillators with the gain due to photovoltaic photorefractive nonlinearity.

$$2\vec{k}_p^e = \vec{k}_s^o + \vec{k}_i^o \; , \tag{7}$$

where p,s, and i denote pump, signal, and idler waves, respectively. Orthogonally polarized waves s and p create a dynamic volume grating with wave vector $\vec{K} = \vec{k}_p - \vec{k}_s$; and a third, idler, wave i appears due to diffraction of the pump wave by this grating. The idler wave together with the pump wave also records the grating with the same wave vector $\vec{K}$ but with a somewhat different phase angle[22].

This process is similar to the forward parametric mixing of waves with the same polarization in a medium with a positive third order non-linearity, as discussed in Ref. 23. The difference is that the matching angle, determined by the birefringence, can be $\simeq 6.5°$ for LiTaO$_3$ and $\simeq 42°$ for LiNbO$_3$ crystals, whereas for isotropic media[23] it cannot exceed 1° even if the nonlinearity is rather large.

In the undepleted pump approximation the output intensities of the signal and the idler waves are

$$I_s(\ell) = I_s(0)\exp(\Gamma\ell)\cosh^2(\ell\Gamma^2+\Omega^2)^{1/2}/2),$$

$$I_i(\ell) = I_s(0)\exp(\Gamma\ell)\sinh^2(\ell\Gamma^2+\Omega^2)^{1/2}/2), \tag{8}$$

where Γ denotes the gain, and $\Omega = n_o^2 n_e^2 r_{51}\beta_{15}^s(\hat{s}\cdot\hat{e}_o)^2/\lambda\kappa\cos\theta)$.

It is evident that both signal wave transmittance $T_s = I_s(\ell)/I_s(0)$ and conversion efficiency $T_i = I_i(\ell)/I_s(0)$ may exceed unity both in a local response medium ($|\Omega|\neq 0$), and in media with purely nonlocal response ($\Gamma > 0$, $\Omega = 0$). This makes oscillation possible with the set-ups shown in Figs. 2d–f.

Such parameteric oscillation was achieved in LiNbO$_3$:Cu crystals where an ordinary signal wave is amplified due to both the direct two-beam interaction with the pump beam and to the forward four-wave mixing[18]. The parametric nature of the amplification is unambiguously confirmed by the oscillation in the cavity of Fig. 2f, where the feedback was established by returning part of the output oscillating beam in the direction of the symmetric idler wave at the input edge of the crystal (the notations "signal" and "idler" waves are used in the oscillation regime only to distinguish between two waves; oscillation is self-starting, i.e. it develops without any input signal wave).

At saturation the oscillation intensity exhibits low frequency modulation $\simeq 10^{-2}\sec^{-1}$, the beat frequency increasing linearly with the pump intensity. The most probable cause of such a modulation is the rather large local response of the crystal used. The spectral profile of the gain for $\Omega \neq 0$ becomes asymmetric and its maximum is shifted with respect to the pump frequency.

MIRRORLESS OSCILLATION DUE TO FORWARD/BACKWARD MIXING

Theory[14,15] predicts mirrorless oscillation for vectorial backward four-wave mixing, but the directions of the oscillation beams are not defined. For the forward four-wave mixing the signal wave transmittivity increases infinitely only for infinite coupling strength (see (8)). It has been shown that in some special cases the superposition of both parametric processes can unambiguously indicate the direction of the oscillation beams in the sample. The oscillation appears most easily when both pump waves are slightly misaligned in the plane containing the c-axis of the crystal.

Mirrorless oscillation was obtained in $LiNbO_3$:Cu crystals[19] with CW pump waves from an Ar^+ laser (0.51 μm). The oscillation was not very sensitive to variation of the pump intensity ratio: it was observed for $I_{p1}/I_{p2} \simeq 1:30$.

SPECTROSCOPIC APPLICATIONS

Studies of the amplification and oscillation due to vectorial four-wave mixing provides information about the signs, the ratios, and the absolute values of the symmetric and antisymmetric parts of the β_{15} component of the photovoltaic tensor[15,24,26]. By comparing the data for $LiNbO_3$:Fe, $LiNbO_3$:Cu, and $LiTaO_3$:Cu we arrive at the conclusion that the specific features of the photovoltaic effect are primarily determined by the impurity center rather than by the crystal matrix.

REFERENCES

1. J. Feinberg and R. Hellwarth, Opt. Lett. 5, 519 (1980).
2. J. O. White, M. Cronin-Golomb, B. Fisher, and A. Yariv, Appl. Phys. Lett. 40, 450 (1982).
3. S. K. Kwong, M. Cronin-Golomb, and A. Yariv, IEEE J. Quant. Electron. 22, 1508 (1986).
4. N. Kukhtarev, V. Markov, S. Odoulov, M. Soskin, and V. Vinetski, Ferroelectrics 22, 949 (1979).
5. B. Sturman, Kvant. Electr. 7, 483 (1980) [Sov. J. Quant. Electr. 10, 276 (1980)].
6. A. M. Glass, D. von der Linde, and T. K. Negran, Appl. Phys. Lett. 25, 233 (1974).
7. V. Belinitcher and B. Sturman, Usp. Fiz. Nauk. 130, 415 (1980) [Sov. Phys. Usp. 23, 199 (1980)].
8. V. Belinitcher, Phys. Lett. 66A, 213 (1978)
9. E. Ivchenko and G. Pikus, Pis´ma v Zh. Eksp. Teor. Fiz. 27, 640 (1978) [Sov. Phys. JETP Lett. 27, 604 (1978)].
10. V. Asnin, A. Bakun, A. Danishevski, E. Ivchenko, G. Pikus, and A. Rogatchev, Sol. State Commun. 30, 565 (1979).
11. S. Odoulov, Pis´ma v Zh. Eksp. Teor. Fiz. 35, 10 (1982) [Sov. Phys. JETP Lett. 35, 10 (1982)].
12. S. Stepanov, M. Petrov, and A. Kamshilin, Pis´ma v Zh. Tekh. Fiz. 3, 849 (1977) [Sov. Techn. Phys. Lett. 3, 345 (1977)].
13. M. Petrov, S. Stepanov, and A. Khomenko, Photosensitive Electrooptic Media in Holography and Optical Information Processing (Nauka, Leningrad, 1983) (in Russian).
14. S. Odoulov and B. Sturman, Zh. Eksp. Teor. Fiz. 92, 2016 (1987) [Sov. Phys. JETP, 92, nº 6, 1987].

15. A. Novikov, S. Odoulov, O. Oleinik, and B. Sturman, Ferroelectrics <u>66</u>, Special Issue on Electrooptic Materials, 1986.
16. S. Odoulov and M. Soskin, Pis´ma v Zh. Eksp. Teor. Fiz. <u>37</u>, 243 (1983) [Sov. Phys. JETP Lett. <u>37</u>, 289 (1983)].
17. S. Odoulov, Kvant. Elektr. <u>11</u>, 529 (1984) [Sov. J. Quant. Electron. <u>14</u>, 360 (1984)].
18. A. Novikov, S. Odoulov, and M. Soskin, Dokl. Acad. Nauk SSSR <u>295</u>, n⁰ 3, (1987) [Sov. Phys. Dokl. <u>32</u>, 000 (1987)].
19. A. Novikov, V. Obukhovski, S. Odoulov, and B. Sturman, Pis´ma v Zh. Eksp. Teor. Fiz. <u>44</u>, 418 (1986) [Sov. Phys. JETP Lett. <u>44</u>, 538 (1986)].
20. M. Cronin-Golomb, B. Fisher, J. O. White, and A. Yariv, Appl. Phys. Lett. <u>41</u>, 689 (1982).
21. J. Feinberg, Opt. Lett. <u>7</u>, 486 (1982).
22. S. Odoulov, K. Balabaev, and I. Kiseleva, Opt. Lett. <u>10</u>, 31 (1985).
23. R. Y. Chaio, P. L. Kelley, and E. Garmire, Phys. Rev. Lett. <u>17</u>, 1158 (1966).
24. I. Kiseleva, V. Obukhovski, and S. Odoulov, Fiz. Tverd. Tela <u>28</u>, 2975 (1986) [Sov. Phys. Solid State <u>28</u>, 1673 (1986)].
25. S. Odoulov, Izv. Akad. Nauk SSSR, ser. Fiz. <u>50</u>, 670 (1986) [Bull. Acad. Sc. USSR (Allerton Press) <u>50</u>, 000 (1986)].

STIMULATED PHOTOREFRACTIVE SCATTERING

AND OPTICAL PHASE CONJUGATION

Jack Feinberg

Department of Physics
University of Southern California
University Park
Los Angeles, CA 90089-0484

INTRODUCTION

If a single laser beam is incident on a photorefractive crystal of barium titanate, a phase-conjugate beam can emerge.[1] Nature seems to have a proclivity for producing the phase-conjugate beam, and here I will discuss some of our recent experimental attempts to understand this phenomenon. This study is motivated by a search for a new kind of optical computer, one that can take entire patterns, or images, and transform them to other images. We already know how to do this on a limited scale; for example, we know how to use four-wave mixing to produce the correlation pattern between two images.[2] However, what I'd like to discuss here is somewhat different, in that it involves a stimulated process.

SINGLE-BEAM PHASE CONJUGATORS

Stimulated processes rely on the parametric amplification of light in a nonlinear material. A well-known example is stimulated Brillouin scattering[3] (SBS), in which an intense image-bearing optical beam of amplitude $Re\{E exp(ikx-iwt)\}$ focused into a liquid can, in a few nanoseconds, produce a new beam whose amplitude is the phase-conjugate replica $Re\{E^* exp(-ikx-iwt)\}$ of the original image. It is not obvious why a process as complicated as stimulated Brillouin scattering should produce a an output beam that, mathematically, is so easily described. A number of papers have shown that under some conditions (notably when the image is sufficiently complex), the phase-conjugate beam has twice the gain of any other beam, and so it grows more quickly from the noise and dominates the competition.[4,5]

Stimulated photorefractive scattering (SPS) resembles stimulated Brillouin scattering (SBS) in that a single beam of light causes a new beam of light to grow from noise and become spatially structured. Both SPS and SBS have a natural propensity

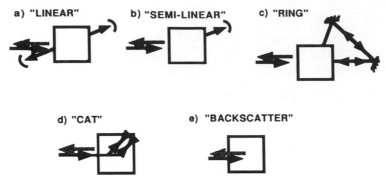

Figure 1) Various geometries for producing a phase-conjugate wave using stimulated photorefractive scattering with one input beam. a) The "LINEAR" has a linear resonator cavity around the crystal; b) the "SEMILINEAR" has only one external mirror; c) the "RING" feeds the transmitted beam back into the crystal; d) the "CAT" uses internal reflection at the crystal faces; e) the "BACKSCATTER" is similar to the geometry used in SBS.

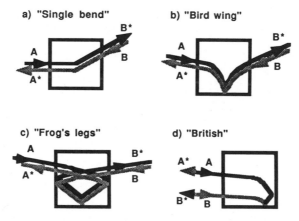

Fig.2. Various "double-beam" self-pumped phase conjugators. The two input beams A and B need not be coherent with each other.

for producing the phase-conjugate replica of an optical input beam. However, SPS requires a factor of about 10^6 less optical intensity than SBS. In addition, the time it takes for the phase-conjugate beam to form in SPS is determined by the intensity of the incident beam, and the formation time can be made conveniently slow (a few seconds with incident intensity ~ milliwatt/cm^2) or fast (nanoseconds with megawatts/cm^2 intensity). The detailed formation of the phase-conjugate beam can be observed by eye and has been recorded on movie film, and the interactions of the incoming and outgoing optical beams in the photorefractive crystal can be studied using a microscope.

Figure 1 shows a number of different geometries in which SPS can efficiently generate a phase-conjugate wave.[1,6-8] In all of these geometries, a single incoming beam creates gain for one or more beams. A common feature of all of these phase conjugators is that the incoming beam is deflected by one or more volume holograms. These holograms grow from noise. The hologram that has the most gain is the one that produces the phase-conjugate of the input beam.[8] In the "backscatter" geometry of Fig.1e (which is identical to the geometry used in SBS) there is one reflection hologram, and it backscatters the incident beam.[6] In the ring geometry of Fig. 1c, it is a transmission hologram that grows from noise and dominates the scattering.[8] This process will be described in more detail later.

DOUBLE-BEAM PHASE CONJUGATORS

There is another group of phase conjugators, called "double-beam phase conjugators," which use SPS but require *two* input beams, as shown in Fig. 2. The two input beams do not have to be coherent with each other. In some geometries the two beams do not even have to have the same wavelength. If the two incident beams contain different images, then the phase-conjugate of image A is observed to propagate back along the direction beam A, and likewise for image B.

The conjugators shown in Fig. 2 all share a common principle: each beam sets up a photorefractive "waveguide" inside the crystal, which then alters both of the beams. Of all the waveguides that could spontaneously form in the crystal, there is one waveguide that will have twice the gain of the others. This particular waveguide is the one that alters beam A so that it travels through the waveguide to emerge as B* (the conjugate of B). This same waveguide alters beam B (which enters the other end of the waveguide) so that it emerges as A*. Note that beam A acts as the reading beam to generate beam B*, so that if beam A is blocked, beam B* instantly disappears. If beam B is blocked, beam B* fades only slowly.

The conjugators shown in Fig. 2 differ only in the number of internal reflections needed inside the crystal to complete the waveguide: a) no reflections are needed in the "single bend" geometry,[9,13] b) one reflection at the crystal c-face in the "bird wing" geometry,[10] c) two reflections at the crystal a-faces in the "frog legs" geometry,[11] and d) two reflections at an a-face and a c-face in the "British" geometry.[12] Experimentally it is found that the conjugators having longer beam paths in the crystal (Fig. 2b-2d) produce the highest-quality phase conjugate

Fig. 3. Photograph of a "bird"wing conjugator in full flight. (Left): The two beams enter through prisms into opposite "a" faces of a barium titanate crystal. The beams bend and "find" each other on the bottom face of the crystal. (Right): Detail of the bottom face of the crystal. The +c-axis direction points from the top to the bottom in these photographs.

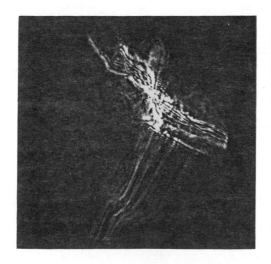

Figure 4. Image quality obtained with the "cat" conjugator. The conjugator can correct for a distorter placed in the path of the image. (Left): distorted image. (Right): phase-conjugated image.

images.[13],[14] The increased path length allows the waveguides a longer interaction region over which to subtly alter the incoming waves and transform them into the outgoing waves. The "single bend" conjugator of Fig. 2a can also produce high-resolution images if the two incident beams are almost counterpropagating to begin with, so that the relatively short waveguide does not have to alter the direction of the beams too much. The single waveguiding region of this conjugator makes it the most tolerant of a wavelength mismatch between the two incident beams, and it has been used to demonstrate Gaussian beam steering using input beams that differ by 30% in wavelength.[15]

Figure 3 is a photograph of the laser beams inside the "bird wing" double-beam phase conjugator of Fig. 2b. The incident Gaussian beams are observed to curve soon after they enter the crystal. This curving is caused by a holographic waveguide spontaneously generated in the crystal by the presence of the two input beams, which are not coherent with each other. After curving, the beams are seen to form stable filaments of light which propagate in a straight line, and reflect off the crystal face at the bottom of the right photograph. (The coupling prisms seen in the left photograph are used only to obtain a wider range of incident beam angles, and are not needed in general.)

Figure 4 shows the image quality obtainable using the aptly named "Cat" conjugator[1] of Fig. 1d. The phase conjugator can correct for the severe distortion (left) by double-passing the image back through the distorter (right). Figure 5 shows images of a resolution chart obtained using the "bird wing" conjugator of Fig. 2b. Line spacings of about 35 lines per mm can be resolved. (The resolution is limited by the collecting optics, and not by the conjugators; image resolution exceeding 1000 lines/mm has been demonstrated using four-wave mixing in a photorefractive crystal.)[16]

An argument showing why stimulated scattering produces a phase-conjugate wave is as follows. First consider a single beam entering a poled photorefractive crystal. The beam will be scattered from defects and other impurities in the crystal. Each randomly scattered beam interferes with the incident beam to form a photorefractive hologram, which scatters the beam even more. There are a multitude of such scattered beams, each with its own hologram. The end result is that the energy of the incident beam is diverted into a broad fan of light[17] shown in Fig. 6. The fan appears on one side of the incident beam, as determined by the direction of the positive c-axis of the crystal.

Now consider the simplest double-beam phase conjugator, the "single-bend" conjugator[9] of Fig. 2a, shown schematically in Fig. 7. Each of the two incident light beams will produce its own fan. However, if one of the holograms causing beam "A's" fan happens to coincide with a hologram producing beam "B's" fan, then that hologram will grow a factor of 2 faster than the other holograms, and it will dominate. In fact, there *is* one such hologram that is Bragg-matched to both of the incident beams, and it is exactly the correct hologram to convert beam "A" into the phase-conjugate of beam "B", and vice versa. Note that beams "A" and "B" are, in general, not coherent with each other, and so do not interfere and write a

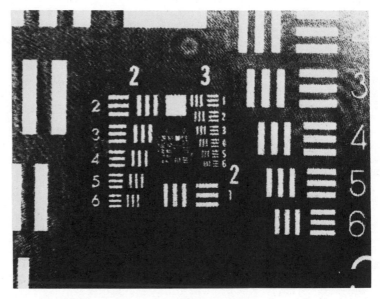

Figure 5. Image of a resolution chart obtained with the "bird-wing" conjugator.

Figure 6. Beam fanning in barium titanate. A single beam
causes a fan of light to emerge on one side of a barium
titanate crystal, seen here as the small rectangle inside
the cuvette. (top): no fanning (ordinary polarization).
(bottom): fanning is seen above the main beam
(extraordinary polarization).

hologram directly. Instead, beam B *reinforces* a hologram that is written by beam A and by beam A's scattering.

In general, the hologram with the most photorefractive gain will dominate over the others. Which of the various geometries shown in Fig. 2 will dominate is determined by the incident angles of the input beams, by the distance of the beams from the crystal faces, and by the orientation of the crystal. The various double-beam phase conjugators shown in Fig. 2 vary only in the number and the location of the holograms in the crystal. For example, in the "bird-wing" conjugator, there are two holograms located where the beams are seen to bend sharply. In reality, each hologram is a series of holograms spread over a finite region, so that the beam is not simply Bragg-deflected off of a single hologram but is waveguided by a series of holograms.

FOOLING A PHASE CONJUGATOR: COMPUTING NEW PATTERNS

Even though nature likes the phase-conjugate beam, she can be fooled into generating a different output mode. Consider the ring phase conjugator shown in Fig 1c. This device will only generate the phase-conjugate replica of the input beam if the optical path length is the same for light propagating around the ring in the clockwise and the counter-clockwise directions.[18] If this condition is not satisfied, then the time-reversal symmetry of the ring is broken, and the device can produce a non-phase-conjugate output mode.[19]

Figure 8 shows the ring conjugator now with a Faraday rotator (sandwiched between quarter-wave plates) inserted into the ring to produce a phase length difference for light propagating in different directions around the ring. The magnitude of the phase difference $\Delta\phi$ can be varied by altering the magnetic field applied along the length of the Faraday glass (Hoya FR-5).

Figure 9 shows the result of varying the nonreciprocal phase shift $\Delta\phi$ when the input beam is a TEM_{00} Gaussian mode: for small asymmetry the output mode is the phase-conjugate TEM_{00} Gaussian mode, as shown in the left photograph in Fig. 9, but with a small (~1 Hz) frequency shift imparted to the output beam. However, as the phase shift $\Delta\phi$ is increased, the phase-conjugate mode becomes unstable, and a new output mode appears. The old mode and the new mode beat together and cause the combined output pattern to "dance." As the phase shift $\Delta\phi$ is increased further, the new mode dominates, as seen in Fig. 9(center). A further increase in $\Delta\phi$ causes new modes to appear, Fig. 9(right), until $\Delta\phi$ approaches 2π and the cycle repeats itself.

An explanation for the frequency shifts and the dancing modes is as follows.[19] In Fig. 8, many holograms, both stationary and moving, are generated by stimulated photorefractive scattering (beam fanning) of the input beam inside a $BaTiO_3$ crystal. One of these holograms will deflect the input beam counterclockwise around the ring and back into

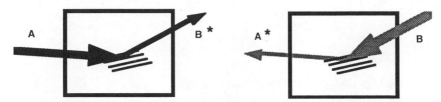

Figure 7. A double-beam phase conjugator: (Left): beam A reads and reinforces the grating and emerges as beam B*. (Right): beam B reads and reinforces the same grating and emerges as beam A*.

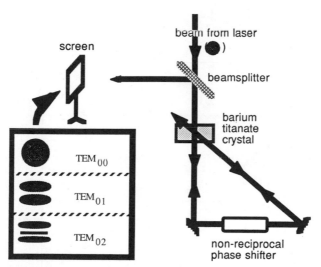

Figure 8. Ring phase conjugator: For a TEM_{00} input mode the output mode will choose between the phase-conjugate mode and higher-order modes depending on the magnitude of the nonreciprocal phase shift $\Delta\phi$ in the ring.

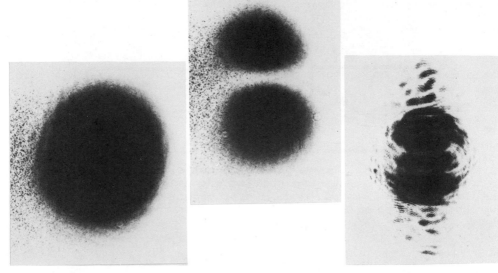

Figure 9. Output modes obtained with the altered ring phase conjugator. Left: phase-conjugate mode with no ring asymmetry ($\Delta\phi=0$). Right: higher order modes obtained by increasing the ring asymmetry ($\Delta\phi\neq0$).

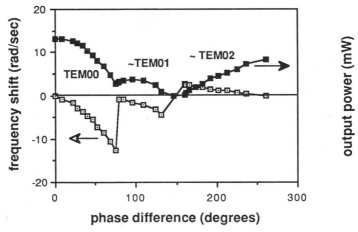

Figure 10. Frequency shift vs. phase nonreciprocity $\Delta\phi$ in the ring conjugator. Note the interplay between the mode shape and the mode frequency.

the crystal. The undeflected portion of the input beam will travel clockwise around the ring and back into the crystal. These two beams, which have travelled around the ring in opposite directions, will interfere in the crystal and write a new photorefractive hologram. If the optical path length is the same for clockwise and counterclockwise propagation around the ring, then this new hologram will reinforce a stationary hologram. However, if there is a nonreciprocal phase delay $\Delta\phi$ in the ring (caused by a Faraday cell, for example), then the hologram must translate in order to be reinforced. This moving hologram Doppler-shifts the output beam and gives it a small frequency shift. For large $\Delta\phi$, the speed required for hologram reinforcement exceeds the response time of the photorefractive crystal, and the phase-conjugate output intensity begins to drop. However, a new hologram, which can partially compensate for the ring asymmetry, can spring up and be reinforced. This new hologram travels more slowly and does *not* produce a phase-conjugate beam; instead it makes the modes *different* for the two different directions of propagation around the ring. This new output mode will have a different propagation constant k, which will compensate for the extra phase delay $\Delta\phi$ caused by the Faraday cell. Figure 10 shows a plot of the measured frequency shift of the output beams versus the strength of the nonreciprocal phase shift $\Delta\phi$ of the ring. The interplay between frequency shift and mode pattern can be clearly seen. The crystal has two degrees of freedom for its self-generated hologram: it can alter either the hologram's shape or the hologram's velocity, and it picks the combination that maximizes the hologram's gain.

In the future, a general purpose optical computer may function by converting a complicated input pattern into a complicated output pattern. I have shown an example of a stimulated process that does just that: it can be made to compute the phase-conjugate replica of the input pattern, or it can be made to produce a different output pattern. Nature has provided us with a gift: stimulated effects which have a propensity for computing the phase-conjugate wave. Our task is to use this gift to design a more general-purpose optical computer.

REFERENCES

1) J. Feinberg, "Self-pumped, continuous-wave phase conjugator using internal reflection," Opt. Lett. **7**, 486 (1982).
2) T.R. O'Meara, D.M. Pepper, and J.O. White, "Applications of nonlinear optical phase conjugation," in: "Optical Phase Conjugation," R.A. Fisher, ed., Academic Press, New York, 1983.
3) B.Ya. Zel'dovich, V.I. Popovichev, V.V. Ragul'skii, and F.S. Faizullov, "Connection between the wave fronts of the reflected and exciting light in stimulated Mandel'shtam–Brillouin scattering," Zh. Eksp. Teor. Fiz. Pis'ma Red **15**, 160 (1972) [English transl.: Sov. Phys. JETP **15**, 109 (1972)].

4) R.W. Hellwarth, "Phase conjugation by stimulated backscattering," in: "Optical Phase Conjugation," R.A. Fisher, ed., Academic Press, New York, 1983.

5) B. Ya. Zel'dovich, N.F. Pilipetskii, and V.V Shkunov, "Experimental investiagion of wave-front reversal under stimulated scattering," in: "Optical Phase Conjugation," R.A. Fisher, ed., Academic Press, New York, 1983.

6) T.Y. Chang and R.W. Hellwarth, "Optical phase conjugation by backscattering in barium titanate," Opt. Lett. **10**, 408 (1985).

7) J.O. White, M. Cronin-Golomb, B. Fischer, and A. Yariv, "Coherent oscillation by self-induced gratings in the photorefractive crystal $BaTiO_3$," Appl. Phys. Lett. **40**, 450 (1982).

8) M. Cronin-Golomb, B. Fischer, J.O. White, and A. Yariv, "Passive phase conjugate mirror based on self-induced oscillation in an optical ring cavity," Appl. Phys. Lett. **42**, 919 (1983).

9) S. Weiss, S. Sternklar, and B. Fischer, "Double phase-conjugate mirror: analysis, demonstration and applications," Opt. Lett. **12**, 114 (1986).

10) M.D. Ewbank, "New mechanism for photorefractive phase conjugation using incoherent beams," submitted to Opt. Lett. (1987).

11) J. Feinberg and M.D. Ewbank, unpublished.

12) A.M.C. Smout and R.W. Eason, "Analysis of mutually incoherent beam coupling in $BaTiO_3$," Opt. Lett. **12**, 498 (1987).

13) M.D. Ewbank, "Incoherent beams sharing photorefractive holograms," OSA topical meeting on photorefractive materials, effects, and devices, Los Angeles, California, August, 1987.

14) R.S. Cudney and J. Feinberg, unpublished.

15) B. Fischer and S. Sternklar, "Self Bragg-matched beam steering using the double color pumped photorefractive oscillator," Appl. Phys. Lett. **51**, 74 (1987).

16) M.D. Levenson, "High-resolution imaging by wave-front conjugation," Opt. Lett. **5**, 182 (1980); M.D. Levenson, K.M. Johnson, V.C. Hanchett, and K. Chiang, "Projection photolithography by wave-front conjugation," J. Opt. Soc. Am. **71**, 737 (1981).

17) V.V. Voronov, I.R. Dorosh, Yu. S. Kuz'minov, and N.V. Tkachenko, "Photoinduced light scattering in cerium-doped barium strontium niobate crystals," Sov. J. Quantum Electron. **10**, 1346 (1980); J. Feinberg, "Asymmetric self-defocusing of an optical beam from the photorefractive effect," J. Opt. Soc. Am. **72**, 46 (1982).

18) B. Fischer and S. Sternklar, "New optical gyroscope based on the ring passive phase conjugator," Appl. Phys. Lett. **47**, 1 (1985).

19) J.P. Jiang and J. Feinberg, "Dancing modes and frequency shifts in a phase conjugator," Opt. Lett. **12**, 266 (1987).

DEPHASING OF IMPURITIES IN ORGANIC GLASSES BETWEEN 0.04 and 1.5K VIA

SPECTRAL HOLE PHOTOBURNING

A. A. Gorokhovskii, V. Kh. Korrovits, V. V. Palm, and M. A. Trummal

Institute of Physics, Academy of Sciences of the Estonian SSR

202400 Tartu, USSR

ABSTRACT

Zero-phonon lines of impurity spectra in glasses reveal an anomalous (as compared to crystals) low-temperature dependence of the homogeneous width[1,2]. To study the optical dephasing in organic glasses the method of photoburning of stable holes in spectra was used[3]. The pure electronic S_0-S_1 transition of H_2-octaethylporphine (OEP) in the matrices of amorphous polymers – polystyrene (PS) and polymethylmetacrylate (PMMA) – was studied. To obtain superlow temperatures an optical $^3He/^4He$ dilution refrigerator was used.

For OEP-PS at T = 0.05K the holewidths δ are 26 Mhz for 620.1 nm and 34 MHz for 621.8 nm, which exceed the lifetime-limited value δ_0 = 10.8 Mhz. The temperature dependences $\delta(T)$ have two crossovers; at $T_1 \simeq 0.1K$ and T_2 = 0.2-0.4K, and can be approximated by $\delta(T) \sim T^n$ with n = 1.5 ÷ 1.8 for T < T_1, n = 0.6 – 0.8 for T_1 < T < T_2 and n = 1.2 ÷ 1.3 for T > T_2 [4]. The shortening of the measurement time from 1 s to 10^{-5} at 1.5K causes a notable decrease of the holewidths.

For OEP-PMMA at T = 0.04K the holewidths are δ = 16 - 18 Mhz, which is close to the lifetime-limited value, δ_0 = 15 MHz; the temperature dependences have crossovers at T_1 = 0.8 K; the powers n = 2 ÷ 3 at T < T_1 and n = 1.1 at T > T_1.

I. INTRODUCTION

In 1971 it was found[1] that for glasses in the range 0.1-1K the heat capacity is linear (C~T) and the thermal conductivity is quadratic with T (Q~T^2), instead of C, Q~T^3 predicted within the Debye model and observed for crystals. Subsequently a number of experiments have been carried out (see for example Refs. 2,3), confirming the results of Ref. 1. To explain these phenomena a number of theoretical models have been proposed, the best-recognized and experimentally-confirmed of which Refs. (4, 5) assumes the presence of groups of atoms and other structural elements in glasses, which are capable of tunneling between two nearly

411

equivalent equilibrium configurations and have therefore effectively two energy levels – the so-called two-level systems (TLS).

Such anomalous low-temperature properties of glasses (in comparison with crystals) have attracted the attention of spectroscopists. The difficulties due to large ($\Delta \approx 100-1000$ cm^{-1}) inhomogeneous broadening of the impurity spectra in glasses as a result of their structural disorder have been overcome with the help of the methods of selective spectroscopy, photoburning of persistent spectral holes (PSH)[6-9] and luminescence line narrowing (LLN)[7]. The vibronic spectra of impurities, obtained by these methods were in basic agreement with those predicted by theory[10] and observed in experiments on crystals: an intensive and narrow zero-phonon line (ZPL) accompanied by a broad phonon wing. Anomalies were found in finer but fundamental details, such as homogeneous ZPL widths and their temperature dependences, characterizing the time τ_2 and the mechanisms for phase relaxation of vibronic transitions: $\Gamma = (\pi\tau_2)^{-1} = (2\pi\tau_1)^{-1} + (\pi\tau_2^-)^{-1}$. Here τ_1 is the energy relaxation time and τ_2^-, a "pure" dephasing time. This anomalous behavior was first reported for inorganic systems[11] and then for organic systems[12,13] and later in a number of other works[14,15]). The following common features are observed:

1) In the liquid helium temperature range the homogeneous ZPL widths, Γ, exceed substantially (by 1-3 orders of magnitude) the values of Γ observed for the same impurities in crystalline matrices[12,16] and approach the limiting value $\Gamma_0 = (2\pi\tau_1)^{-1}$ determined for optical transitions solely by the lifetime τ_1 of the electronic state[10].

2) At temperatures considerably below the Debye temperature, weak temperature dependences of the homogeneous ZPL width are observed, $\Gamma(T) \sim T^n$ with $0.5 < n < 2.6$. In some cases the broadening law changes in the 0.1-1K range[16,17], contrary to both the $\sim T^7$ dependence predicted by the adiabatic theory for crystals in the case of interaction with acoustical phonons[10,18], and the $\sim e^{-\hbar\Omega/kT}$ behavior expected in the interaction with a pseudolocal vibration of frequency $\Omega > kT/\hbar$ [18].

Because the anomalous thermal properties of glasses become most apparent for T < 1K, the study of the mechanism of ZPL broadening in this temperature range is of great interest. The first results of such experiments were presented in Ref. (17).

II. PHOTOBURNING OF SPECTRAL HOLES (PSH)

in case of PSH, monochromatic radiation excites a subset of centers within the inhomogeneously broadened band whose ZPL frequencies are distributed in correspondence with the homogeneous absorption spectrum. Due to the finite probability for the excited states to convert to a state with a changed ZPL frequency (another electronic state, a photochemical transformation, etc.), in the inhomogeneous distribution of the centers over ZPL frequency, a spectral hole is formed. The lifetime of the hole is determined by the lifetime of the photoproduct, which may last for days or longer[6,9]. The hole can be registered in the absorption or luminescence spectra. The shape of the hole obtained may be rather complicated[6-9]. However, owing to the high peak intensity of the ZPL, at the excitation frequency a narrow feature, the "zero-phonon" hole, can be discriminated within it, which corresponds to the absence of centers that have absorbed the pumping radiation through their ZPL. In view of the long lifetime of the hole, the times of burning and recording of the hole spectrum may considerably exceed the lifetime τ_1 of the electronic state. This allows the burning and recording processes to be separated and ensures a preference for PSH over LLN in recording homogeneous ZPL spectra: the difficulties of the separation of the luminescence from the scattered laser radiation in resonance measurements have been eliminated. On the other hand, an important new time appears – the measurement time τ_m. The times of burning, delay and recording of the hole usually make $\tau_m = 10^{-2}-10^3$ sec. If during τ_m the matrix structure remains unchanged (which evidently holds for crystals in thermodynamic equilibrium), then, under certain, easily fulfilled conditions[6], the width of the "zero-phonon" hole is related to the ZPL width by the simple relation $\delta = 2\Gamma$.

In the case of glasses, whose structure is thermodynamically meta-stable, during the time τ_m, relaxation of the structure to a new more stable state is possible. In the course of such structural relaxation, the surroundings of impurity centers undergo random changes, which cause corresponding random changes of ZPL frequencies. This phenomenon is called spectral diffusion. It leads to the change of the holeshape in time – i.e. it broadens and becomes more shallow[19,20]. With the lowering of temperature thermal motions are gradually frozen out. At liquid He temperatures the main relaxation processes arise from tunneling motions, including the motions of the matrix elements which form TLS.

In accordance with theoretical models[4,5] the TLS display a wide variety of the relaxation times τ_Γ. An estimate of the minimal relaxation time for quartz at T=1K gives $\tau_{\Gamma min} \simeq 10^{-9}$ sec, for polystyrene, $\tau_{\Gamma min} \simeq 10^{-12}$ sec. A comparison to the optical lifetime of the allowed transition $\tau_1 \simeq 10^{-8}$ sec indicates that transitions in TLS can perform a fast ($\tau_\Gamma < \tau_1$) as well as slow ($\tau_\Gamma > \tau_1$) random modulation of the ZPL frequency. In the case of fast modulation spectral diffusion contributes to the homogeneous ZPL width. The slow motions lead to the spectral diffusion which adds a quantity $\delta_{sp\ dif}$ to the inhomogeneous broadening of the hole. Thus, in the case of glasses, the hole spectrum carries information about both the fast processes of the homogeneous ZPL broadening and the slow processes of spectral diffusion which broaden the hole inhomogeneously. The holewidth is $\delta = 2\Gamma + \delta_{sp\ dif}$.

The separation of the contributions from the homogeneous broadening and from slow spectral diffusion is possible by measuring the homogeneous ZPL widths by independent fast ($\tau_m < \tau_1$) methods (LLN, photon echo, fast hole burning) and by comparing the results with those of PSH[19,21,22].

III. EXPERIMENT

Holes in the spectra of the purely electronic S_1-S_0 transition of the H_2-octaethylporphin (OEP) impurity in polystyrene (PS) and poly-methylmetacrylate (PMMA) matrices, and H_2-monoazaetioporphyrin IV (MAEP) impurity in a PS matrix were studied. The burning of holes in the spectra of these impurities is connected with a photochemical reaction of the reorientation of intra-layer protons[23]. The samples were prepared by block polymerization of the impurity-implanted solution of the monomer and they had impurity concentration of $10^{-4} \div 10^{-3}$ mole/l. The hole was burned with a single-frequency laser (jitter $\simeq 5$ MHz) and was registered

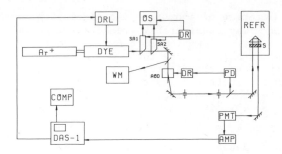

Fig. 1. Experimental setup. DYE – single-frequency ring dye laser
CR-699-21 pumped by Ar$^+$-laser; SA1, SA2 – spectrum analyzers; WM –
wavemeter; AOD – acousto-optic deflector, DR – driver and PD –
photodiode form the power stabilizing system; REFR – ^{3}He/^{4}He
dilution refrigerator; S – sample; PMP – photomultiplier; DAS-1 –
multichannel analyzer; COMP – computer.

with a multichannel analyzer in the transmission spectrum by scanning the
laser through the burning region and by accumulating the signal
(see Fig. 1).

To attain temperatures 0.04-2K, an optical ^{3}He/^{4}He dilution
refrigerator was used. The incident laser beam was directed vertically,
passed through the object which was placed into the mixing chamber,
reflected back from the prism, and passed through the crystal for the
second time before exiting the cryostat into a photomultiplier. The
experimental set-up has been described in more detail in[17,24].

IV. RESULTS

The holes for OEP-PS were burned in the long-wavelength side of the
0-0 absorption band (λ_{max} = 618.5 nm, $\Delta \simeq$ 150 cm^{-1}) at λ_1 = 620.1 nm and
λ_2 = 621.8 nm. The holewidths at T = 0.05K were δ_1 = 26.0 ± 1.5 MHz and
Δ_2 = 34 ± 1.5 for λ_1 and λ_2, respectively, which considerably exceed the
limiting holewidth as determined by the lifetime of the S_1 state,
τ_1 = 17.5 ± 1 nsec: δ_0 = $2\Gamma_0$ = $(\pi\tau_1)^{-1}$ =18 MHz.* τ_1 depends weakly on
temperature in the range of interest. Therefore the lifetime contribu-
tion to the holewidth can be excluded by analyzing the difference
$(\delta(T)-\delta_0)$. These dependences are presented on a double logarithmic scale
in Fig. 2a. The broadening follows complicated laws with two slope
changes (crossovers) at T $\simeq$ 0.1 and T $\simeq$ 0.2K, slightly different for
λ_1 and λ_2. These slopes are given as approximations to separate parts of

---

*In the results published earlier[17], an underestimated value
δ_0 = 10.8 MHz was used.

the power law $(\delta(T)-\delta_o) = T^n$. In the range $T > 0K$ $n = 2.4$ and 2.6 (± 0.5), at $T > 0.2K$ $n = 1.07$ and 1.22 (± 0.05), at $T = 0.1 \div 0.2K$ $n = 0.75$ and 0.5 (± 0.1) for λ_1 and λ_2, respectively. Extrapolation of $\delta(T)$ to $T \to 0$ indicates that the holewidths approach the limiting value only at $T = 0.01-0.03K$.

In the case of OEP-PMMA, the holes were burned at $\lambda_1 = 6.185$ nm and $\lambda_2 = 6.16$ nm. The widths obtained and their temperature dependences coincide within the limits of the accuracy of the measurement. At $T = 0.04K$ $\delta = 16 \div 18$ MHz, which is close to the limiting lifetime width $\delta_o = 15$ MHz ($\tau_1 = 21$ nsec). The temperature dependences of $(\delta(T)-\delta_o)$ (Fig. 2b) exhibit one crossover point at $T = 0.1K$, where the broadening law changes from $n = 1.3 \pm 0.5$ to $n = 1.2 \pm 0.05$.

For MAEP-PS, at $T = 0.06K$ the holewidth $\delta = 75$ MHz ($\lambda = 617.3$ nm), which exceeds twice the lifetime limit $\delta_o = 37$ MHz ($\tau_1 = 8.6$ nsec). The temperature dependence $(\delta(T)-\delta_o)$ does not display any slope changes in the range $0.1-2K$, $n = 1.1 \pm 0.05$. Extrapolation to $T \to 0$ indicates that in this system the holewidth approaches the limiting lifetime value only at $T \lesssim 0.01K$.

For OEP-PS, the holeshape was also measured, at $T = 0.05$ and $T = 0.4K$ (Fig. 3). The contours were recorded by multiple accumulation of measurement cycles. It can be seen that at $T = 0.05K$ the hole can be well approximated to the Lorentz curve. At $T = 0.4K$ the hole is broadened on the Stokes side and its shape is slightly asymmetric.

5. DISCUSSION

The fact that at very low temperatures, $T \simeq 0.1K$, the holewidth in all three systems investigated exceeds by several times the limiting lifetime width, approaching it only at $T = 0.01-0.04K$, suggests that the impurity interacts with extremely low-frequency excitations of the amorphous polymer matrix. The thermal properties of polymers in this temperature region are determined by TLS[2,3] with their density of states considerably exceeding the phonon density of states[2]. Consequently, a conclusion can be reached, that the TLS play a significant part in the processes of ZPL broadening, and that their influence is frozen out only in the range $T = 0.01-0.04K$.

The holewidth in the OEP spectrum is considerably (1.5-3 times) larger in PS than it is in PMMA. This result can be understood qualitatively, using the data about TLS from thermal conductivity measurements of these polymers in the range $0.1-1K$[2,3]. The quantity $\gamma^2 p$ (γ – the constant of phonon-TLS interaction, p – TLS density) has a value 1.5 times higher in PS than in PMMA. In the case of MAEP-PS, however, the holewidth at $T > 0.1K$ only slightly exceeds the holewidth in OEP-PS. At $T < 0.1K$ this difference becomes more noticeable, obviously due to the shorter lifetime of the S_1 state in MAEP. A conclusion can be drawn that the holewidth in the temperature region, where $\delta \gg \delta_o$, is determined to a greater extent by the matrix properties than by the properties of the impurity. The latter are more important in the range $T < 0.1K$ where $\delta \gtrsim \delta_o$, and δ depends strongly on decay processes.

The hole is also sensitive to the local environment of the impurity as a result of a slight difference in values and dependences $\delta(T)$ for different burning wavelengths in OEP-PS. However, this effect is weaker

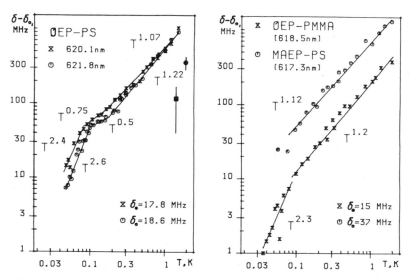

Fig. 2. Temperature dependences of the holewidths $\delta(T) - \delta_o$ in the spectra of OEP-PS (left) and MAEP-PS, OEP-PMMA (right), $\delta_o \triangleq$ lifetime limited holewidths. Point- fast measurements, $\tau_m = 10^{-5}$ s [Refs. (14-22)]; square – stimulated photon echo measurement (more exact data(28)).

than the dependence of the holewidth on the matrix properties. It is evidently connected with the weak dependence of the impurity-TLS interaction on distance, which causes the impurity to be influenced by a great number of TLS within a macroscopic volume of the matrix.

Two regions can be discriminated in the measured dependences $\delta(T)-\delta_o$. In the range $T > 0.1-0.2K$, $(\delta(T)-\delta_o) \sim T^{1.1-1.2}$, which is close to the $T^{1.3}$-law observed for a number of organic[16] as well as inorganic[25] systems. In the low-temperature range, $T < 0.1K$, the broadening follows a steeper relation $(\delta(T)-\delta_o) \sim T^{2.3-2.6}$. Note that such crossover occurs in a temperature region, where $(\delta(T)-\delta_o) \simeq \delta_o$* i.e. when the pure phase part of the holewidths become comparable to the decay width. Such behavior might be connected, in principle, with the incorrect subtraction of the decay width from the whole width; the phase and decay parts are additive only in case of Lorentz shapes for the corresponding spectral distributions. However, since the decay kinetics of the S_1 state for the systems under investigations are accurately exponential, the decay contours are Lorentzian. Since the hole contours are also Lorentzian (as in the case of OEP-PS in Fig. 3a), the procedure under discussion is correct.
Note that on recording the hole in the transmission spectrum, the holeshape in the absence of saturation must be symmetric even in the case of an asymmetric ZPL[6]. The observed asymmetry at $T = 0.4K$ (Fig. 3b) may be due to the manifestation of ZPL asymmetry via saturation and/or the processes of slow spectral diffusion, more probable in the Stokes region.

There are a great number of theoretical works on homogeneous ZPL broadening in glasses (see reviews in Refs. 14,15) which explain the $T^{1.3}$ relation. As a rule, interactions of the impurity with TLS and (one or more) phonons are taken into account. However, only a few of them predict one[26] or two[27] crossovers to a steeper dependence $T^{2.3}$ in the region $T < 0.1-0.2K$, considering only the processes of homogeneous broadening during the time $t < \tau_1$, and neglecting slow processes $\tau_1 < t < \tau_m$. At the same time, fast time-dependent measurements in OEP-PS, depicted in Fig. 2 (square-holewidth measurement, $\tau_m = 10^{-5}$ sec [14,22], circle-stimulated photon echo[28]), indicate a considerable influence of slow spectral diffusion processes on the persistent holewidth, at least at $T = 1.5-2K$. The holewidth is discussed on the basis of the model of spectral diffusion if Ref. 29. Broadening with a $T^{4/3}$ dependence is predicted, for dipole-quadrupole (under the condition $P(E) = const$, $E = $ TLS energy) or dipole-dipole ($P(E) \sim E^{0.3}$) interactions between an impurity and TLS, the main contribution to the broadening being given by TLS with the relaxation time $\tau_2 < \tau_\Gamma < \tau_m$. However, the important case of very low temperatures, when $\tau_2 \simeq \tau_1$ and crossovers have been observed, has not yet been examined theoretically.

Thus, it can be concluded that for polymer matrices: a) in OEP-PMMA the broadening of the hole reaches the decay limit at $T = 0.4K$ and, consequently, processes of slow spectral diffusion are frozen out; for OEP-PS and MAEP-PS the broadening approaches the lifetime limit at

---

*It can be assumed that at $T < 0.1K$, the contribution from slow spectral diffusion in δ is small – the near-lifetime limit value of $\delta(T)$ for OEP-PMMA also indicates this and, consequently, $(\delta(T)-\delta_o)$ describes the processes of pure dephasing.

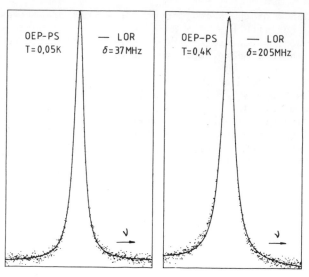

Fig. 3. Holes in the transmission spectrum of OEP-PS at T = 0.05K (left) and T = 0.4K (right); points - experiment, lines - approximation to the Lorentz curve. Measurement conditions for T = 0.05K: intensity 0.5 μW/cm^2, hole-burning time 2 s, scanning rate 0.5 GHz/s, number of accumulated cycles 50.

T < 0.04K; b) the temperature dependences of broadening have one or two crossovers at T = 0.1-0.2K; c) dephasing at T > 0.1K depends to a greater extent on excitations in the matrix volume than on the surroundings of the impurity, at T < 0.1K, on the properties of the impurity.

The authors are grateful to K. Rebane and L. Rebane for discussion, to I. Renge for preparing the samples, and to A. Schulga for providing MAEP.

REFERENCES
1. R. C. Zeller and R. O. Pohl, Phys. Rev., B4, 2029 (1971).
2. R. M. Stephens, Phys. Rev. B8, 2981 (1973).
3. J. E. Graebner, B. Golding, and L. C. Allen, Phys. Rev. B34, 5696 (1986).
4. P. W. Anderson, B. I. Halperin, and C. M. Varma, Phil. Mag., 25, 1 (1972).
5. W. A. Phillips, J. Low Temp. Physics 7, 351 (1972).
6. L. A. Rebane, A. A. Gorokhovskii, and J. V. Kikas, Appl. Phys., B29, 35 (1982).
7. R. I. Personov, in: Spectroscopy and Excitation Dynamics of Condensed Molecular Systems, eds. V. M. Agranovich, R. M. Hochstrasser, (North-Holland, Amsterdam, 1983) p. 333.
8. G. J. Small, in: Spectroscopy and Excitation Dynamics of Condensed Molecular Systems, eds. V. M. Agranovich, R. M. Hochstrasser (North-Holland, Amsterdam, 1983), p. 515.
9. J. Friedrich and D. Haarer, Angewandte Chemie 23, 113 (1984).
10. K. K. Rebane, Impurity Spectra of Solids, (Plenum Press, New York, 1970).
11. P. M. Selzer, D. L. Huber, D. S. Hamilton, W. M. Yen, and M. J. Weber, Phys. Rev. Lett. 36, 813 (1976).
12. A. A. Gorokhovskii and L. A. Rebane, Izv. Akad. Nauk SSSR, er. iz. 44, 859 (1980).
13. A. A. Gorokhovskii, J. V. Kikas, V. V. Palm, and L. A. Rebane, Fiz. Tverd. Tela 23, 1040 (1981).

14. K. K. Rebane and A. A. Gorokhovskii, J. Luminescence 36, 237 (1987).
15. R. M. MacFarlane and R. M. Shelby, J. Luminescence 36, 179 (1987).
16. H. P. H. Thijssen, S. Volker, M. Schmidt, and H. Port, Chem. Phys. Lett. 94, 537 (1983); S. Volker, J. Luminescence 36, 251 (1987).
17. A. A. Gorokhovskii, V. Kh. Korrovits, V. V. Palm, and M. A. Trummal, JETP Lett. 42, 307 (1985); Chem. Phys. Lett. 125, 355 (1986).
18. M. A. Krivoglaz, Fiz. Tverd. Tela 6, 1707 (1964).
19. A. A. Gorokhovskii and V. V. Palm, JETP Lett. 37, 237 (1983).
20. W. Breinl, J. Friedrich, and D. Haarer, Chem. Phys. Lett. 106, 487 (1984).
21. R. M. MacFarlane and R. M. Shelby, Opt. Commun. 45, 46 (1983).
22. K. K. Rebane, Cryst. Latt. Def. and Amorph. Mat. 12, 427 (1985); ESSR, Fiz. Matem., to be published.
23. K. N. Solov´ev, I. E. Zalessky, V. N. Kotlo, and S. F. Shkirman, JETP Lett. 17, 332, 1973.
24. V. Korrovits and M. Trummal, Izv. Akad. Nauk ESSR, Fiz. Matem. 35, 198, 1986.
25. D. L. Huber, M. M. Broer, and B. Golding, Phys. Rev. lett. 52, 2281 (1984).
26. K. Kassner and P. Reineker, Phys. Rev. B35, 828 (1987).
27. I. S. Osad´ko, Zh. Eksp. Teor. Fiz. 90, 1453 (1986).
28. R. K. Kaarli and M. L. Ratsep, in abstracts "Modern methods of laser spectroscopy of molecules in low-temperature media", Tallinn, (1987), p. 39-40.
29. M. A. Krivoglaz, Zh. Eksp. Teor. Fiz. 88, 2171 (1985).

PHOTOBURNING OF PERSISTENT SPECTRAL HOLES, AND SPACE-TIME DOMAIN

HOLOGRAPHY OF ULTRAFAST EVENTS OF NANO- AND PICOSECOND DURATION

K. K. Rebane, R. K. Kaarli, A. K. Rebane, and P. M. Saari

Institute of Physics, Estonian S.S.R. Academy of Science
142 Riia Str., Tartu 202400, USSR

ABSTRACT

Photoburning of persistent spectral holes is not only an effective method to eliminate the inhomogeneous broadening and perform high resolution spectroscopy, but it serves also as a powerful tool to control the absorption coefficient and refractive index of the matter by means of illumination. It opens new horizons for frequency-selective optical memories, holographic data storage and data processing. A short review of the results obtained in the field is presented.

1. INTRODUCTION

Narrow zero-phonon lines (ZPL) of low-temperature solids (see[1] and review papers)[2] is the cornerstone of photoburning of spectral holes in inhomogeneously broadened bands of impurity absorption.

Actually an inhomogeneously-broadened ZPL of impurity molecules in a disordered molecular medium is a wide band comprising many thousands of very sharp (10^{-2}–10^{-4} cm^{-1}), resonances continuously spread along the frequency axis in each micron-size spatial element of the medium. This feature is a good starting point towards applications in the frequency domain for information storage and data processing.

The possibility of photoburning of persistent spectral holes with lifetimes up to year and narrow line widths (10^{-2}–10^{-4} cm^{-1})[3,4] and review papers[5-7] serves as a next step in this direction. As a matter of fact, one can burn not sharp holes by using monochromatic laser excitation, but also burn arbitrary profiles by using illumination of various spectral composition.

The third step is to use the principles of holography including holographic storage, and playback of the time dependence of signal light fields. This is a generalization of the conventional holography to the 4th (time) dimension.

2. HIGH-SELECTIVE "SPECTRAL ENGINEERING" OF OPTICAL PROPERTIES OF LOW-TEMPERATURE SOLIDS

2.1 Homogeneous Zero-Phonon Lines (ZPLs), Phonon Sidebands and Inhomogeneous Broadening

More than a couple of decades ago it had been predicted theoretically[1,2] (and references therein) that impurity absorption and luminescence bands at low temperatures often consist of a very narrow ZPL (an optical analog[8,9] of Mossbauer γ-resonance line) and a wide phonon sideband. At liquid helium temperatures the homogeneous linewidth of ZPL is 10^{-3}-10^{-4} cm^{-1} and its integral intensity is comparable to that of the phonon sideband. As a result, the absorption cross section at the peak of ZPL may be rather high and the transition-frequency to line-width ratio (resonance quality factor Q) reaches values up to 10^8 (even more for forbidden transitions).

However, in most real impurity systems inhomogeneous broadening, while practically not affecting phonon sidebands, masks, more or less, the ZPLs: matrice irregularities spread the frequencies of ZPL of individual impurity centres over a comparatively wide glassy matrices spectral interval of ~ 0.1-1 cm^{-1} in single crystals, up to 10^4 cm^{-1}.

2.2 Photoburning of Persistent Spectral Holes (HB)[3-7]

Under illumination by nearly monochromatic laser light in the inhomogeneous impurity absorption band the impurities whose ZPLs are in resonance with the laser light undergo excitation-deexcitation cycles with a high repetition rate. It is the result of their very large absorption cross sections at the sharp ZPL peak. In the end something changes in the molecule itself or in its surrounding matrix. The changes result in large shifts of the ZPLs' positions (in comparison with the ZPL's linewidth.) In the end the ZPL absorption of a number of molecules is "burned out" and a sharp hole in the inhomogeneous band is created. The changes are fixed by rearrangement of the positions of atoms or parts of molecules in impurities or in the surrounding matrix and can have long lifetimes.

That is why these holes are persistent contrary to the transient spectral holes of saturation created under very strong illumination. For persistent holeburning high light intensities are not needed: It is the dose of illumination that matters.

HB is a tool to control the absorption coefficient and (the closely related to it) refractive index by decent illumination. HB turns the inhomogenous broadening of the sharp ZPLs into a feature useful for studies of the properties of matter and also for technological applications.

It enables one to perform nice high-resolution spectroscopy of solid matter, including disordered systems (glasses, polymers) (see e.g.[6,10]) and molecules, including large molecules like chlorophull (see e.g. [11]).

2.3 HB Applications - Optical Spectral Memory

The broad absorption band with a sharp hole in it represents a narrow-line (10^{-3}-10^{-4} cm^{-1}) wide-aperture spectral filter.

The next possible application is high, and very high, capacity optical spectral memory. We can attach to a diffraction-limited 10^{-8} cm^2 area up to 10^5 bits of information - defining the content of a bit by absence or presence of a hole at a certain frequency in the absorption band. The total information-packing density becomes really high: $10^8 \times 10^5 = 10^{13}$ bits cm^{-2}. If one second is needed to burn a hole, then almost a million years will be required to fill one cm^2 of such a memory. Actually the writing-in time of a single hole may be much shorter - down to nanoseconds.[12,13] Nevertheless, the point by point and frequency by frequency writing-in and reading-out time of the entire memory remains still unreasonably long. The conclusion is that for optical memories with such a high capacity parallel processing is required. This way of thinking brings us to hole-burning with picosecond pulses and time-domain holography.

2.4 Picosecond Pulse Hole-Burning[14,15]

If we illuminate a hole-burning medium with picosecond pulses we shall burn in a very broad hole in comparison with the ZPL's width. The pulses in the first experiments[14] were 2-3 ps long and about 5 cm^{-1} spectral width, i.e. the pulses were well-formed in the sense that the product of their widths in time and frequency domains was close to the Fourier transform-limited value. The profile of the hole fixes the distribution of intensity among the harmonics of the pulse with the accuracy of the ZPLs homogeneous linewidths, in about 10^4 points on the frequency axis. It should be mentioned that the proper version of the theory of time-dependent spectra[16,17] tells us that such a description of the written-in picture is correct provided we look at it not too soon after the excitation was completed, because during the excitation and some minimum period after it the matter is still "hot".[17] But later, when everything has "cooled down," the Fourier-transform intensity spectrum is really stored. That time interval must be longer than the excited electronic state's lifetime. It should be mentioned here also that two-step holeburning opens possibilities of getting holes narrower than the ZPL's linewidth determined limit.[18]

In order to restore the time behaviour of the initial pulse from the hole-burned information we must know also the phases of the Fourier components. Here we come to holography with the main idea to use in addition to the signal pulse, a reference pulse and to store the intensity distribution of their interference picture. The latter stores the phases.

3. TIME- AND SPACE-DOMAIN HOLOGRAPHY

3.1 Photochemically Accumulated Stimulated Photon Echo (PASPE)[14,15]

The main idea centers upon the creation of an interference between two picosecond-scale pulses (pulse 1 - the reference, pulse 2 - the object pulse) hitting the HB medium at different moments of time. Let the delay time of the second pulse t_d be much longer than the duration of the picosecond pulse. It is obvious that the pulses never meet in space, and therefore, there will be no overlap between them and no interference in the conventional sense will occur. Nevertheless the interference in the HB medium can take place. This precious feature is provided by the phase memory of the medium. The first pulse excites the electronic state of the impurity molecules. The wave functions of the electrons "remember" for a while - until the phase relaxation will be completed - the phases of the light wave which gave them the first excitation. If the second pulse hits before the characteristic phase relaxation time has expired ($T_2 \sim 1$ ns - in the experiments,[14,15,19,20]) the interaction with the excited electrons takes place with phase relations preserved correctly enough.

Let the Fourier transform of the signal pulse hitting the HB medium at the moment t = 0 be S(ω). The Fourier transform of a very short (δ-like) reference pulse is exp(-iωt_d). Thus, the spectrum of the field illuminating the medium is given by

$$|S(\omega) + \exp(-i\omega t_d)|^2 = 1 + S(\omega)|^2 + S(\omega)\exp(i\omega t_d) + S^*(\omega)\exp(-i\omega t_d). \quad (1)$$

The first term causes a uniform reduction of absorption over the band, the second one shows the storage of the signal spectrum (with loss of the phase information). The last two terms are the interference terms which store the phases.

So the frequency domain interference results in a specific modulation pattern of intensity distribution between the Fourier components (fixed via HB due to changes in the dielectric permittivity of the HB medium) of the joint event - pulse 1 (reference) + pulse 2 (signal).

Read-out is performed by sending a reference pulse 1 in the same direction (or the opposite one, if we want to create conjugated wave fronts) through the plate of an HB medium with a permittivity (absorption coefficient + refractive index) modulated in the frequency domain. Two response pulses emerge, as predicted by the inverse and causality-corrected Fourier transformation of expression (1), proportional to the transmission of the illuminated sample (the casuality correction takes place automatically, if we calculate the permittivity of the sample after HB, where the density of the impurities is changed in accord with (1)). The first one is the attenuated reference pulse 1 itself, only its shape is changed to some extent due to the second term in (1). The second pulse emerges later, the delay time being exactly the same t_d as was fixed for writing-in. The shape (time-dependene) of that pulse repeats to a good approximation the initial signal pulse (see a review of the experimental results in)[19].

The second (delayed) light pulse presents a new type of coherent optical transient and is called, due to its relation to some versions of stimulated photon echoes,[21-23] photochemically accumulated stimulated photon echo PASPE.) The term "photochemical" underlines the decisive role of the photochemical accumulation of the modulation depth and, consequently, gives rise to the high intensity of the echo signal. In experiments[14,15] about 10^{10} identical pairs of writing-in pulses were applied and echo signal intensities up to 50 per cent of that of the arbitraring weak readout pulse response were achieved.

The following parameters of a HB medium are principally important for time-domain holography: (i) To write down a τ sec long detail of the signal pulse the spectral width of the inhomogeneous absorption band has to be considerably broader than the spectral width τ^{-1} sec^{-1} of the pulse. (ii) The reciprocal value of the ZPL's width, i.e. the phase relaxation time has to exceed the maximum duration of the cycle, i.e. the interval from the leading edge of the first (reference) pulse of the trailing edge of the second (signal) pulse.

3.2 Holography of Time-and-Space Domain Scenes[15,19,20]

The spatial resolution of HB media is not less than that of photographic materials for holography. Moreover, as the resolution is, in principle, determined by the dimensions of impurity centres, HB media are well suitable for holography of spatially-,modulated light fields. As HB media possess an additional dimension - the frequency dimension - the time dependence of the field at very given spatial point can be memorized. So, by making use of such media one can perform time-and-space domain holography - storage and playback of scenes with their whole time dependene (movements of the parts of the object scene, rise and decay and phase modulation of light sources on the scene).

Dynamic holography also deals with nonstationary light fields. In order to avoid a terminological confusion, it should be stressed that a dynamic hologram works in "a real time" without any storage of the temporal behaviour of the field.

A detailed discussion of the procedure and corresponding theory of time-and-space domain holography as well as photographs of holographically restored images obtained in model experiments were presented in.[19]

Impurity oscillators are distributed in a HB medium not only over the spatial and frequency coordinates, but also over the arbitrary orientations of the dipole momentum of the resonant electronic transition. This feature leads to a possibility of recording the polarization characteristics of the field, including the spatial and temporal dependences of the electric vector.[20] Consequently, we can accomplish this holography, which precisely corresponds to the meaning of the term ("complete recording".) The first experiments on storage and playback of picosecond signals with time-dependent polarization have already been carried out.[24]

Further, associative recall of time-and-space domain hlograms provide a means to be able to recognize scenes of ultrafast events by their time-space fragments.[25] This feature opens new possibilities for high-speed optical data processing and all-optical time and space-dependent pattern recognition. Insofar as an associative recall of time and space domain holograms must obey the causality principle, it means also the existence of high-capacity content-addressable spectral memories where information is stored, and may be recalled in the form of causality-connected "causes" and "results".

References

1. K. K. Rebane, Impurity Spectra of Solids, Plenun Press, New York (1970).
2. A. A. Maradudin, Rev. Mod. Phys. 36:417 (1964); R. H. Silsbee, D. B. Fitchen, Rev. Mod. Phys. 36:433 (1964); K. K. Rebane, Zh. Prikl. Spektrosk. 37:906 (1982); L. A. Rebane, Zh. Prikl. Spektrosk. 34:1023 (1981); I. S. Osad'ko, Spectroscopy and Excitation Dynamics of Condensed Molecular Systems, ed. by V. M. Agranovich and R. M. Hochstrasser, 4:437, North-Holland, Amsterdam (1983); M. N. Sapozhikov, Alekseev, Phys. Status Solidi (b) 120:435 (1983).
3. A. A. Gorokhovskii, R. K. Kaarli, L. A. Rebane, JETP Lett. 20:216 (1974); Optics Commun. 16:282 (1976).
4. B. M. Kharlamov, R. I. Personov, L. A. Bykovskaya, Optics Commun. 12: 191 (1974).
5. L. A. Rebane, A. A. Gorokhovskii, J. V., Appl. Phys. B 29:235 (1982).
6. G. Small, Spectroscopy and Excitation Dynamics of Condensed Molecular Systems, ed. by V. M. Agranovich and R. M. Hochstrasser, 4:515, North Holland, Amsterdam (1983).
7. J. Friedrich, D. Haarer, Angew. Chemie 23:113 (1984).
8. E. D. Trifonov, Dokl. Akad. Nauk SSR 147:826 (1962).
9. K. K. Rebane, V. V. Hizhnyakov, Opt. Spektros 14:362 (1963).
10. K. K. Rebane, A. A. Gorokhovskii, J. Luminescence 36:237 (1986).
11. R. A. Avarmaa, K. K. Rebane, Spectrochim. Acta 41A:1365 (1985).

12. K. K. Rebane, J. Luminescence 31/32:744 (1984); Cryst. Latt. Def. and Amorph. Mat. 12:427 (1985).

13. M. Romagnoli, W. E. Moerner, S. M. Schellenberg, M. D. Levenson, G. C. Bjorklund, J. Opt. Soc. Am. B1:341 (1984); S. Winnacker, R. M. Shelby, R. M. Macfarlane, Opt. Lett. 10:350 (1985).

14. A. K. Rebane, R. K. Kaarli, P. M. Saari, Opt. Spektrosk. 55:405 (1983); A. K. Rebane, R. K. Kaarli, P. M. Saari, JETP Lett. 38:383 (1983); A. Rebane, R. Kaarli, P. Saari, A. Anijalg, K. Timpmann, Optics Commun. 47:170 (1983).

15. P. M. Saari, R. K. Kaarli, A. K. Rebane, Kvant. Elektron. 12:627 (1985); A. Rebane and R. Kaarli, Chem. Phys. Lett. 101:317 (1983); A. K. Rebane, R. K. Kaarli and P. M. Saari, J. Molec. Struct. 114:343 (1984); A. Rebane R. Kaarli, P. Saari, Izv. Akad. Nauk Eston. SSR 34:444 (1985); J. Kikas, R. Kaarli and A. Rebane, Opt. Spektrosk. 56:387 (1984); P. Saari and P. A. Rebane, Izv. Akad. Nauk Eston. SSR 33:322 (1984).

16. V. V. Hizhnyakov and I. K. Rebane, Zh. Eksp. Toer. Fiz. 74:885 (1978) (Engl. transl. Sov. Phys. JETP 47:463 (1978)); I. K. Rebane, A. L. Tuul, V. V. Hizhnyakov, Zh. Eksp. Teor. Fiz. 77:1302 (1979) (Engl. transl. Sov. Phys. JETP 50:655 (1979)); P. Saari, Light Scattering in Solids, ed. by J. L. Birman, H. Z. Cummins and K. K. Rebane, p.315, Plenum Press, New York and London (1979).

17. I. Rebane, Proc. of the Acad. Sci. of Estonian SSR 34:438 (1985).

18. I. Rebane, Proc. of the Acad. Sci. of Estonian SSR 35:296 (1986); 35:400 (1986); 36:204 (1987).

19. P. Saari, R. Kaarli and A. Rebane, J. Opt. Soc. Am. B3:527 (1986).

20. P. Saari, Proc. Inst. Phys. Estonian SSR Acad. Sci. 59:157 (1986).

21. E. I. Shtyrkov, V. V. Samartsev, Phys. Stat. Sol. A45:647 (1978).

22. W. H. Hesselink, D. A. Wiersma, J. Chem. Phys. 75:4197 (1981).

23. T. W. Mossberg, Opt. Lett. 7:77 (1982).

24. R. K. Kaarli, P. M. Saari, H. R. Sonajalg, Abstracts of Symposium "Modern Methods of Laser Spectroscopy of Molecules in Low-Temperature Media,"? Tallinn (1987); P. K. Kaarli, P. M. Saari, H. R. Sonajalg, Abstracts of V International Symposium "Ultrafast Phenomena in Spectroscopy," Vilnius (1987).

25. A. K. Rebane, R. K. Kaarli, Proc. of the Acad. Sci. of Estonian SSR 36:208 (1987); A. K. Rebane, Abstracts of Symposium "Modern Methods of Laser Spectroscopy of Molecules in Low-Temperature Media," Tallinn (1987); A. K. Rebane, Abstracts of International Symposium "Ultrafast Phenomena in Spectroscopy," Vilnius (1987); A. K. Rebane (see in present issue).

ASSOCIATIVE SPACE-AND-TIME DOMAIN HOLOGRAPHY OF PICOSECOND LIGHT FIELDS VIA PHOTOCHEMICAL HOLE BURNING

A. Rebane

Institute of Physics, Estonian S.S.R. Academy of Sciences, 142 Riia Str., Tartu, 202400, USSR

ABSTRACT

A new version of space-and-time domain holography in media with photochemical hole burning is proposed in which holographic recall is completed by partial space-time episodes of the ultrafast scene. Associative holographic storage and recall of picosecond light signals is experimentally demonstrated.

I. INTRODUCTION

Combining the spectroscopic method of photochemical spectral hole burning (PHB) in low-temperature impurity media[1,2] with the principles of optical holography makes possible holographic recording and playback of space-time behavior of ultrashort light signals[3-5]. The phenomenon of PHB provides recording and long term storage of the intensity spectrum of the incident object light field with an accuracy to 10^{-3}-10^{-4} cm^{-1}, while the introduction of an additional reference pulse provides the storage and playback of the phases of the quasimonochromatic components of the object light field also.

For the sake of convenience it has earlier been assumed that the role of the reference signal in recording and playback of space-time holograms in PHB media should be performed by special δ-like reference pulses, i.e. pulses with durations very short compared to the duration of the object pulse[4,5]. At first glance one might conclude that in the opposite case, i.e. when the duration of the reference pulse is comparable to or even longer than that of the object pulse, the signal recalled from the hologram would be temporally hopelessly distorted and dispersed.

In recent papers[6,7] a different, so-called associative space-time domain holographic recording and playback of picosecond signals was demonstrated. In this version of space-time holography the role of the reference signals is accomplished by the object pulse itself or by its separate space-time fragments (episodes). Recall of such associative holograms can be done, analogously to an associative recall of ordinary space-domain holograms, by feeding into the hologram one or more fragments of the stored signal. What is important, is that in the present case not only does reproduction of the spatial image occur but

427

also recall of the temporal features of the original signal takes place. In other words, one can reproduce from the hologram the whole event provided a single episode of this event is still available for the readout procedure.

In the present communication an experiment demonstrating holographic associative recall of picosecond light signals in PHB media is described. Future prospects of this approach are discussed.

II. EXPERIMENTAL

The experimental setup used in the present work was similar to that described in[4,5]: laser pulses with 2-3 ps duration and 6 cm^{-1} spectral width were generated with an 82 MHz repetition rate by a picosecond dye laser synchronously pumped by an actively mode-locked argon ion laser. The sample was prepared from polystyrene activated with octaethylporphin molecules at molar concentrations 10^{-3}-10^{-4} and had the dimensions 15 x 15 x 5 mm. The sample was mounted at 2K temperature in a liquid-He cryostat with pass-through optical windows. The inhomogeneously broadened absorption band used for PHB at 620 nm wavelength had a 200 cm^{-1} width. The starting value of the optical thickness of the sample (prior to PHB) was 2.0-2.5.

To record a hologram exposures of about 10 mJcm^{-2} were needed; to perform the readout the exposures were 2-3 orders of magnitude lower. The time interval required to write a hologram depended on the sample illumination conditions and varied from tens to several hundreds of seconds. During this long interval of time the PHB-action of 10^{10}-10^{11} identical light pulses was accumulated in the holographic medium.

Picosecond temporal resolution was provided in the experiment by a synchroscan streak camera. A photographic camera was used to record temporally averaged spatial images.

3. RESULTS AND DISCUSSION

The output beam of the laser was expanded in a telescope and passed through a Michelson echelon (Fig. 1). The optical paths through the neighboring segments of the echelon differed by 34 ps. At the output of the echelon the laser pulse appeared to be divided up into eight pulses. Each of the eight pulses corresponding to a fraction of the input pulse cross-section travelled through a certain segment of the echelon so that the relative delays of the segments were correspondingly 34 ps, 68 ps, 102 ps, etc.

The total duration of the pulse train did not exceed, as is required in space-time holography, the phase relaxation time $T_2 \simeq 500$ ps of the excited electronic state of the impurity centers responsible for PHB.

By inserting into the laser beam various transparencies the spatial outlines of the light pulses could also be tailored, say, in the form of stripes, fragments of printed text, etc.

As mentioned above, in the present experiments there were no special recording or readout reference pulses with wavefronts extending well over the whole area of the hologram plate illuminated by object light field. For that reason in order to write a hologram mutual interference of various parts of the incident object light field had to be arranged by making all the spatial parts of the object pulse coincide over the entire hologram. To do this a mat glass plate was inserted into the laser beam some 15 cm from the incident window of the cryostat. As a result the

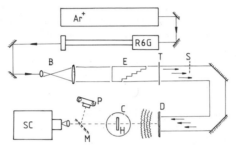

Fig. 1. Schematic of the experimental setup. B – beam expander; E
– Michelson echelon; S – beamstop used to eliminate parts of the
beam during readout of the hologram; D – mat glass beam disperser; C
– cryostat; H – hologram; P – photographic camera; SC – synchroscan
streak camera; M – semitransparent mirror.

laser beam, which at the outset of the echelon comprised a train of co-propagating pulses with spatially non-overlapping wavefronts, was turned at the incident plane of the hologram into a train of pulses with each pulse evenly illuminating the whole area of the sample.

In Fig. 2a is presented a photograph of a spatial image produced by object pulses scattered from the mat glass disperser. This image was observed through the sample and the windows of the cryostat during the writing exposure of the hologram. In Fig. 2b the temporal structure of the same light signal recorded with the streak camera is presented. In order to enhance the visibility of the spatial image a transparency was used to shape the object pulse in such a way that only pulses with delays 0, 34, 68 and 170 ps of the whole echelon output train were present.

To perform the read-out of the hologram certain parts of the object beam were stopped so that they did not reach the mat glass. At the time both spatial (averaged over the time interval of the photographic camera exposure) and temporal images of the light signal emerging from the hologram were recorded.

In Figs. 3a,b a photograph of the spatial image and the streak camera record of the signal recalled from the hologram are presented. The readout of the hologram was carried out with a space time fragment comprising the first three pulses out of the four pulse sequence used to write the hologram. Comparing the data presented in Fig. 2 and Fig. 3 one can conclude that the hologram indeed had the ability to reproduce both the spatial and the temporal structure of the missing part of the object signal, although with a notably lower intensity as compared to the intensity of the passed-through readout fragment.

Similar results were obtained when the readout of the hologram was carried out by different combinations of time-space fragments. At the same time it was established that the missing part of the signal emerges only in the case that at least one of the readout fragments preceded the missing part of the signal. In the opposite case when the readout fragment belonged entirely to the retarded part of the signal no recall was observed. This result corresponds well to the causality properties of space-time holography in PHB media[5].

During the readout of the hologram it was also observed that the greater the number of signals that preceded the missing part of the scene, the more intense and detailed was the signal recalled from the hologram. This feature may be interpreted as a further property of an associative memory to reproduce the details of the stored-in scene depending on the amount of the preliminary information submitted for the readout.

Figures 4a, b display the photographs of an associatively recorded and recalled picosecond space-time signal which had the form of a printed text each line of which had a time duration of 2 ps. The interval between successive lines was 34 ps. By illuminating the hologram with some of the first lines of the text it was possible to read from the recalled image the rest of the text while the streak camera records showed that the temporal features of the signal were reproduced as well.

It should be noted that associative properties of the temporal recall observed in the experiments described above manifested themselves strongly only because different temporal fragments of the recorded object scene had irregular, and distinct structures[7]. In this sense the present experiments are reminiscent of the reproduction of a so-called phantom image observed in conventional holography.

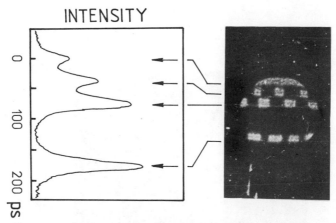

Fig. 2 Photograph of the spatial image (a) and corresponding streak camera record (b) of the object signal used to write the hologram. Correspondence between spatial and temporal fragments is indicated by arrows.

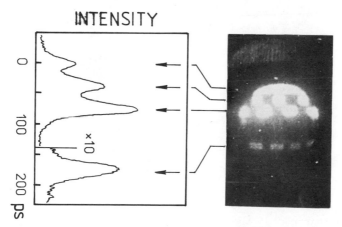

Fig. 3. Photograph of the spatial image (a) and the streak camera record (b) of the signal recalled from the hologram. For the readout fragment see the bright part at the left of the photograph (a) and the three pulses with delays 0, 34 and 68 ps (b).

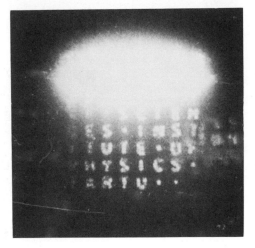

Fig. 4. Photographs of the image of a picosecond space-time signal applied to record a hologram (a) and the image recalled from the recorded hologram (see text).

Associative readout of space-time holograms can also be performed making use only of the time-domain structure of the object signal, provided the different fragments of the object scene do not correlate strongly with each other in the time domain[7]. In this case readout of the hologram requires a fragment which possesses temporal structure coinciding with an episode belonging to the object scene.

As a concluding remark it may be noted that the possibility of holographic temporal recall utilizing fragments of the recorded signal has been already indicated earlier by Gabor[8] and the analogy between holographic recall and some properties of human memory was also pointed out.

ACKNOWLEDGEMENTS

The author is indebted to K. K. Rebane and P. M. Saari for useful discussion of this work and to R. K. Kaarli for taking part in the experiments.

REFERENCES

1. A. A. Gorokhovskii, R. K. Kaarli, and L. A. Rebane, JETP Lett. 20, 7, 474 (1974); B. M. Kharlamov, R. I. Personov, and L. A. Bykovskaya, Optics Commun. 16, 282 (1974).
2. L. A. Rebane, A. A. Gorokhovskii, and J. V. Kikas, Appl. Phys. B29, 234 (1982); J. Friedrich and D. Haarer, Angew. Chem. Int. Ed. Engl. 23, 113, (1984).
3. A. Rebane, R. Kaarli, P. Saari, A. Anijalg, and K. Timpman, Optics Commun. 47, 3, 173 (1983); A. K. Rebane, R. K. Kaarli and P. M. Saari, Optika i spektr. 55, N°3, 405 (1983); A. K. Rebane, R. K. Kaarli, and P. M. Saari, JETP Lett. 38, 383 (1983).
4. A. Rebane, R. Kaarli, and P. Saari, Izv. Akad. Nauk Eston. SSR 34, 444 (1984); 34, 328 (1985); A. Rebane and A. Kaarli, Chem. Phys. Lett. 101, 317 (1983).
5. P. Saari, R. Kaarli, and A. Rebane, J. Opt. Soc. Am. B3, 4, 527 (1986); P. Saari and A. Rebane, Izv. Akad. Nauk Eston. SSR 33, 322 (1984); P. Saari, R. Kaarli, and P. Rebane, Kvant. elektron. 12, 672 (1985).
6. A. Rebane and R. Kaarli, Izv. Akad. Nauk Eston. SSR 36, 208 (1987).
7. A. Rebane, Abstracts of Symposium "Modern methods of laser spectroscopy of molecules in low-temperature media," Tallinn, 1987; A. Rebane, Abstracts of V International Symposium "Ultrafast phenomena in spectroscopy," Vilnius, 1987.
8. D. Gabor, Nature 217, 1288 (1968).

STUDY OF HOT CARRIER DIFFUSION IN SEMICONDUCTORS BY TRANSIENT GRATING

TECHNIQUES

J. Vaitkus, L. Subačius, and K. Jarašiūnas

Vilnius State University and Semiconductor Physics Institute
of the Lithuanian Academy of Sciences, Vilnius, USSR

ABSTRACT

The applicability of transient grating techniques for the study of
nonequilibrium carrier diffusion and drift features has been considered.
An experimental set-up for the investigation of hot carrier diffusion has
been constructed. It combines the advantages of electron heating by an
external electric field and those of diffusion-sensitive decay of the
laser-induced free-carrier grating. Field dependences of hot carrier
ambipolar diffusion coefficients have been determined in Si and GaAs.

INTRODUCTION

The interaction of a powerful laser radiation with a semiconductor
leads to the modulation of the optical properties of the medium in real-
time. A periodic excitation field modulates the permittivity of the
matter and forms a dynamic diffraction grating[1,2]. These nonstationary
light-induced structures are of great interest for active spectroscopy of
condensed matter, for studies of the properties of semiconductors and
defects, for real-time holography, optical phase conjugation, optical
processing, deflection of light, DFB-lasers, etc.[2,3,4]

In strong electric fields the Einstein relation is violated due to
carrier heating. Therefore, when investigating the transition phenomena
in those fields, it is necessary to consider the mobility and the
diffusion coefficient of hot carriers separately. Current electrical
techniques for diffusion coefficient measurements with hot carriers (HC)
entail considerable experimental difficulties, and are applicable, as a
rule, only for high-resistivity samples[5]. In addition, reliable data
about the field dependences of HC diffusion coefficient are lacking for a
number of semiconductors.

This work presents a new contactless technique for the investigation
of HC diffusion. It combines electron-hole plasma heating by external
microwaves or pulses and measurements of the diffusion coefficient from
the decay of light-induced transient gratings of nonequilibrium carriers
(NC). From the changes of light intensity response (diffraction
efficiency of the dynamic grating) in strong electric fields, field
dependences of the coefficients of HC ambipolar diffusion in Si and GaAs
have been determined.

The excitation of a semiconductor by a pulse of light interference field results in a spatial modulation of the electric and optical properties of the medium. The authors of numerous works have shown[1,2,4,6] that the dominating mechanism for the modulation of the optical properties is the electron-hole plasma, if the semiconductor is excited and probed in the region of non-resonant interaction and at excitation levels insufficient to transform the structure of the solid or to heat it. A direct relation of refractive index change Δn with NC concentration N, according to the Drude-Lorenz model, enables one, from the experimentally measured optical characteristics, to determine electric parameters of the semiconductor: NC concentration, lifetime and diffusion coefficient, surface recombination velocity, the perfection of a crystal structure, etc.[1,2,6].

In the linear approximation, when the kinetic coefficients μ, D, τ_R do not depend on the NC concentration, NC dynamics are described by the continuity equation

$$\frac{\partial N(x,t)}{\partial t} = D_a \frac{\partial^2 N(x,t)}{\partial x^2} - \mu E \frac{\partial N(x,t)}{\partial x} - \frac{N(x,t)}{\tau_R} + \alpha I(x)F(t), \tag{1}$$

where μ and D_a are the ambipolar mobility and diffusion coefficient, respectively, τ_R is the NC lifetime, α is the absorption coefficient for the wavelength of the laser radiation, $I(x)$ is the interference field, forming the dynamic grating with period Λ, and $F(t)$ is the temporal shape of a laser pulse with duration τ_L.

The solution of equation (1) in the presence of volume excitation of the semiconductor[7] has a sinusoidal distribution of NC $N \simeq N_0 + \Delta N \cos(2\pi x/\Lambda)$. The dynamics of the modulation depth of the grating ΔN are determined by the characteristic times $\tau_e^{-1} = \tau_R^{-1} + \tau_D^{-1}$ and $\tau_e = \Lambda/\mu E$. Here the time decay of the grating τ_e is governed by recombination and NC diffusion ($\tau_D = \Lambda^2/4\pi^2 D$) and τ_E is connected with ambipolar drift. In the case that $\Delta N = \Delta P \gg N_0, P_0$, or in the absence of electric field, the depth of the grating modulation is determined only by diffusion and recombination. The rate of diffusion decay can be monitored by changing the grating period.

For a thin phase grating with a light induced phase modulation of the probing beam $\Phi = 2\pi \Delta n d/\lambda \leqslant 0.4$, the self-diffraction efficiency is described by the relation:

$$\eta(t) = \frac{I_1(t)}{I_T(t)} \simeq [\Phi(t)/2]^2 = [\frac{\pi d}{\lambda} n_{eh} \Delta N \int_0^t F(t-\xi)e^{-\xi/\tau_e} d\xi]^2, \tag{2}$$

where I_1 is the first order diffraction intensity, I_T is the intensity of the transmitted laser beam, λ is the wavelength of the incident beam, n_{eh} is the modulation coefficient of the refractive index by one electron-hole pair.

In the configuration of self-diffraction, the value of the parameter η will be influenced by the rate of nonequilibrium processes, if τ_e is comparable to the laser pulse duration τ_L. According to a numerical calculation based on Eq. (2) (Fig. 1), the investigation of NC diffusion ought to be carried out in Si and in low-doped GaAs at $\Lambda = 20 \div 30$ μm.

The electric field in a semiconductor can influence the optical parameters by means of the direct electrooptical interaction, by change of the distribution function of the carriers, or by spatial modification of the carrier distribution. The fields used ($E < 5$ kV·cm^{-1}) provided for the last factor be the dominating one. In a monopolar semiconductor ($N \neq P$) additional changes of the grating modulation depth due to NC drift can be expected. Figure 2 presents the results of numerical calculations of the drift of NC packets according to Eqs. (1) and (2).

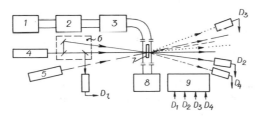

Fig. 1. The efficiency of light self-diffraction vs. grating erasure time due to diffusion (Si) or diffusion and recombination (GaAs).

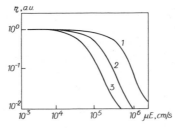

Fig. 2. The calculation of carrier drift influence on grating self-diffraction efficiency for Λ, μm: 50 (1), 40 (2), 30 (3).

In accordance with the calculations, the drift component can be neglected for $\mu E < 10^4$ cm $\cdot$s^{-1} and $\Lambda \simeq 30$ m. Consequently, under the conditions of bipolarity, the influence of an external electric field will mainly be revealed in the heating of the NC plasma with subsequent changes of the ambipolar diffusion.

Thus, the above analysis of the dynamic grating technique points out the possibility of measuring the dependence $D_a(E)$, which is of essential interest for the fundamental and applied physics of hot carriers.

EXPERIMENTAL TECHNIQUE

Samples of single crystals of n-Si (ρ = 100 and 300 $\Omega \cdot$cm) and semi-insulating GaAs have been investigated. For the latter, a lifetime τ_R = 20 ÷ 40 ns and a D_a = 15.5 cm^2/s, determined from the dynamics of the grating decay, do not prevent the diffusion-sensitive erasure of the grating during the action of the laser pulse. All measurements have been

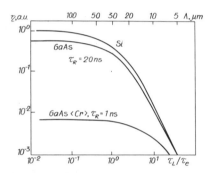

Fig. 3. Experimental setup: microwave generator (1), tuners and power meters (2,3,8), pulsed lasers (4,5), beamsplitter (6), sample (7), and data processing system (9).

carried out at T = 300K. The exciting NC interference field has been formed by placing the sample at the intersection of two beams from a neodymium laser(λ = 1.06 μm, τ_L = 14 ns). NC heating has been accomplished by microwave electric field pulses $\vec{E} = \vec{E}_m \sin\omega t$ ($\omega = 6 \cdot 10^{10}$ Hz). Its orientation relative to the NC gradient in the grating was $\vec{E}_m \perp$gradN and $\vec{E}_m \parallel$gradN. The amplitude of the mw field E_m in the sample was evaluated from the value of the absorbed mw power[8].

Figure 3 shows the experimental set-up. The measured intensities of the exciting I_0, transmitted I_1 and diffracted I_T beams entered the computer controlled data processing system. The data extraction has been presented as I_1, $I_T = f(I_0)$. The measurements have been carried out at relatively low excitation levels corresponding to $\eta \simeq 0.2$ ÷0.3 per cent ($\Delta N \lesssim 10^{17}$ cm^{-3}) for Si and $\eta \simeq 0.1$ per cent for GaAs in order to prevent imbalance of the mw circuit, sample heating, and to keep up the value Φ restricted by Eq.. (2). More detailed information about the technique is given in Refs. 9 and 10.

EXPERIMENTAL RESULTS AND DISCUSSION

A linear increase of NC concentration versus excitation level correlates with a cubic dependence of the self-diffraction intensity[9]. The application of an electric field with E_m < 5 kV/cm changed the self-diffraction intensity, while the transmission was not sensitive to field action. The observed changes in self-diffraction efficiency increased with the decrease of the grating period, due to diffusion-sensitive erasure, in full agreement with the above-mentioned calculations (Fig. 1).

Field dependences $\eta(\vec{E}_m)$ in n-Si for different orientations of the electric field vector are shown in Fig. 4. The observed anisotropy of η points to the essential dependence of the kinetic coefficient on the

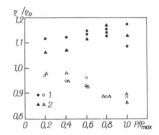

Fig. 4. Self diffraction efficiency vs. mw power in waveguide for Si: $\vec{E}_m \perp$ grad N (open points), $\vec{E}_m \parallel$ grad N (full points); 1 – 100 $\Omega \cdot$cm, 2 – 300 $\Omega \cdot$cm.

orientation of the field $\vec{E}_m$. Assuming that the field dependence of η is determined by a dependence of the diffusion coefficient D_a on $\vec{E}$, the data obtained are in qualitative agreement with the measurements of a decrease of the longitudinal and an increase of the transverse diffusion coefficients of the HC[5].

Figure 5 presents field dependences of the transverse ambipolar diffusion coefficient, calculated according to the relation of D_a with τ_e from Eq. (2); here $\tau_R \gg \tau_e$. This same figure gives calculated field dependences $\langle D_{a\perp} \rangle$ using experimental results for the monopolar diffusion coefficients $D_{n\perp}$, $D_{p\perp}$ and differential mobilities $\tilde{\mu}_n$, $\tilde{\mu}_p$ of electrons and holes obtained by other techniques[5]. The calculation of the bipolar diffusion coefficients has been carried out from the formula

$$D_{a\perp} = (\tilde{\mu}_p D_{n\perp} + \tilde{\mu}_n D_{p\perp})/(\tilde{\mu}_n + \tilde{\mu}_p). \tag{3}$$

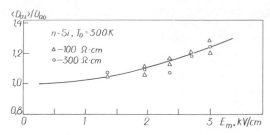

Fig. 5. Field dependence of transverse ambipolar hot carrier diffusion coefficient in Si: points – experiments, line – calculation.

It should be mentioned that the results presented in Fig. 5 are time-averaged over the the period of the mw field, and are related to the instantaneous values of D_a by the integral equation

$$\langle D_{a\perp}(E_m)\rangle = \frac{1}{2\pi} \int_0^{2\pi} D_{\alpha\perp}(E)d(\omega t). \tag{4}$$

The good agreement between the calculated and directly measured transverse ambipolar diffusion coefficient proves that diffusion is the main determining factor of the features of transient gratings in strong electric fields.

The determination of a longitudinal ambipolar diffusion coefficient from the observed increase of efficient grating erasure time for $\vec{E}_m \parallel$ grad N is considerably complicated. In this case the electric field distribution in the sample is inhomogeneous due to an NC concentration gradient and is determined by the carrier dynamics $N(x,t)$. Therefore, along with HC diffusion, determined by the NC gradient, a thermodiffusion of HC may also take place due to inhomogeneous carrier heating. The latter factor would increase the grating modulation depth. Evaluation of the definite contribution of both mechanisms to NC redistribution requires further investigation.

Field dependences $\eta(E_m)$ for $\vec{E}_m \perp$ grad N and the corresponding dependences $\langle D_{a\perp}(E_m)\rangle$ have been obtained in GaAs in a similar way (Fig. 6). In the GaAs samples investigated, the grating erasure is dependent on diffusion as well as on NC recombination. Therefore, the measurements have been carried out at several grating periods to reveal the diffusion process. Assuming that the HC heating by transverse electric fields in GaAs adiabatically follows the mw amplitude[5], the dependence of the instantaneous values of $D_{a\perp}$ on E (Fig. 6b, solid line) has been obtained from the field dependence $\langle D_{a\perp}(E_m)\rangle$ using Eq. (4). According to Eq. (3), the transverse hot electron diffusion coefficient $D_{n\perp}$ has been

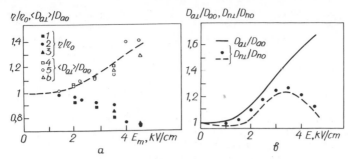

Fig. 6. a. Dependence of η and averaged $\langle D_{a\perp} \rangle$ on mw field amplitude for $\Lambda, \mu m$: 26 (1,3), 31 (2). Points 1,2,4,5 and 3,6 – two different samples of GaAs. b. Field dependence of ambipolar $D_{a\perp}$ and hot electron $D_{n\perp}$ diffusion coefficients; points – data of this paper, dashed line – Monte-Carlo calculations.

determined, since the hole mobility μ_p (and, apparently, the hole diffusion coefficient D_p) does not depend on the field[11], and the dependence of $\mu_n(E)$ on E is well established[5]. The field dependence of $D_{n\perp}(E)$ (Fig. 6b, points), obtained in this way, is in good agreement with that calculated by Monte-Carlo methods[5]. The fact that the obtained maximum value of $D_{n\perp}$ is considerably smaller than the corresponding coefficient $D_{n\parallel}$ for n-GaAs[11], confirms that the previously obtained strong dependence of $D_{n\parallel}(E)$ on E is mainly determined by intervalley diffusion via intervalley hot electron scattering (the intervalley component of $D_{n\perp}$, being transverse to the field orientation, is equal to zero[5]).

In conclusion, these investigations confirm that the transient grating technique serves as a simple but powerful tool for the contactless measurement of hot carrier kinetic coefficients. Mono- and bipolar diffusion in strong electric fields at different free carrier concentrations can be investigated. Up to now the technique of HC ambipolar diffusion coefficient measurements presented is the only one of its kind.

REFERENCES

1. H. E. Eichler, P. Günter, and D. W. Pohl, Laser-induced Dynamic Gratings, (Springer-Verlag, Berlin-Heidelberg, 1986).
2. Special Issue on Dynamic Gratings and 4-wave Mixing, ed. H. J. Eichler, IEEE Journ. Quant. Electr., QE-22, No. 8, (1986).
3. S. A. Akhmanov and N. I. Koroteev, Metody Nelinejnoj Optiki i Spektroskopii Rassejanija Sveta (Moscow, Nauka, 1981).

4. Optical Phase Conjugation, ed. R. A. Fisher (Academic Press, New York, 1983).

5. V. Bareikis, A. Matulionis, J. Požela, et al. Hot Carrier Diffusion, ed. J. Pozela (Mokslas, Vilnius, 1981).

6. A. Miller, D. A. B. Miller, and S. D. Smith, Adv. Phys. 30, 679 (1981).

7. J. Vaitkus, R. Pranaitis, K. and Jarašiūnas, Kvantovaja Elektronika, 13, 1868 (1986).

8. J. K. Požela, Transport parameters from Microwave conductivity and noise measurements. In: Topics in Appl. Phys., vol. 58, ed. L. Reggiani (Springer-Verlag, Berlin-Heidelberg, 1985) pp. 113–147.

9. J. Vaitkus, L. Subačius, K. Jarašiūnas, Soviet Physics Collections, vol. 25, No. 4 (Allerton Press, Inc., New York, 1985) pp. 75–80.

10. J. Vaitkus, L. Subačius, K. Jarašiūnas, Pis´ma JETP 45, 154 (1987).

11. V. L. Dalal, Appl. Phys Lett. 16, 489 (1970).

PULSED HOLOGRAPHIC DIAGNOSTICS OF AN ELECTRON—HOLE PLASMA IN
SEMICONDUCTORS

A. A. Bugayev

A. F. Ioffe Physico—Technical Institute, Academy of Sciences
of the USSR, 142092 Leningrad, USSR

ABSTRACT

The nonlinear interaction of an intense laser pulse with a semi—
conductor at a photon energy below the bandgap is determined by a multi—
phonon interband transition of the free carriers which are created at
this transition[1] In this case the nonlinear contribution to the
dielectric permeability of the medium will be complex. Because of that
the accurate analysis of this interaction requires a determination of the
amplitude—phase distortions of the exciting pulse.

This paper presents the results of a study of nonlinear intractions
under conditions of induced absorption (IA) in Si and GaAs obtained by
the methods of pulsed holographic interferometry at the wavelength
$\lambda = 1.056$ μm. Holograms were recorded on films of vanadium dioxide[2].

Experiments have shown that self—interaction of the pumping pulse in
Si and GaAs at IA acquires a predominantly amplitude nature that is
caused by the production of an absorbing screen of free carriers. The
production of an absorbing screen is confirmed by numerical analysis of
the nonlinear system of equations describing the absorption of the
pumping pulse as a result of multiphonon absorption. The data obtained
confirm the existence of two—photon indirect absorption ($\hbar\omega = 1.17$ eV) in
Si at room temperature[1].

The results of interference measurements of the spatial distribution
of the electron—hole plasma density in the bulk of the semiconductor at
various levels of excitation are presented. A calculation is made of the
contribution of the real and imaginary parts of the dielectric
permeability in the build—up of the interaction volume.

INTRODUCTION

Excitation of silicon by a powerful light pulse with a quantum of
energy at about the bandgap is accompanied by a nonlinear response of the
medium whose value and nature determine the practical interest in its
utilization for the spatial—temporal transformation of the exciting
pulse[1,2]. The most significant contribution to the nonlinear response
is provided by the generation of free carriers which induce changes in
both the real ε' and imaginary ε'' parts of the nonlinear dielectric

constant of the medium $\varepsilon = \varepsilon´ + i\varepsilon''$. For this reason it will be of great interest to study the process of absorption of the exciting pulse in the bulk of the silicon since this process is the result of the joint action of both nonlinear absorption and refraction.

In this paper we present the part of our studies devoted to the development and application of methods of picosecond holographic diagnostics for the investigation of nonlinear absorption in semiconductors. To study the excited silicon volume we carried out the experiments according to the excite-and-probe technique[3]. In this method we have used the principle of holographic recording of the information[4]. In the general case this technique permits one to obtain an amplitude-phase image of a controlled volume for a known temporal delay between the exciting and probing pulses. The experimental layout is shown in Fig. 1. In this experiment we have used a single pulse selected from a train whose duration was 35 ps. This pulse passed

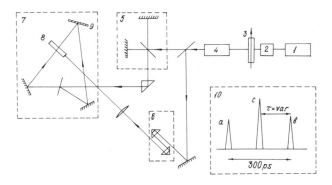

Fig. 1. Experimental layout of picosecond holographic diagnostics. 1- generator of picosecond pulses; 2- switch-off system of single pulse; 3- cell with saturating absorber; 4- amplifier; 5- pulse doubling scheme; 6- variable time delay; 7- scheme of holographic recording, 8- sample; 9- film of VO_2; 10- temporal sequence of experiment. a,b- registered pulses; c- exciting pulse.

through an additional cell with a saturable absorber and was amplified up to an energy of 5mJ. Further, the pulse was divided into two, one of which excited the sample and the other-- due to the doubling pulse scheme-- was used for recording the hologram by the method of double exposure[4]. A doubling scheme based on the Michelson interferometer has been employed to provide the recording of a double exposure hologram by two picosecond pulses, which are identical in quality and energy. Between the probing and exciting channels we have introduced the variable temporal delay that in conjunction with the doubling pulse scheme determined the temporal measurement sequence as indicated in Fig. 1. Thus, the experiment allowed us to produce the temporal diagnostics of amplitude-phase variations of a controlled silicon volume in which absorption was induced by the exciting pulse. As a sample we have used a plate of silicon of 2 mm thickness. Excitation has been carried out through the lateral surface using a long focal-length (F = 100 cm) cylindrical lens. The recording of the hologram has been performed on the VO_2 film whose holographic characteristics are presented in[5,6]. During the recording and reconstruction of the hologram the temperature of the VO_2 film corresponded to the middle of the temperature hysteresis loop[5].

At the first stage of the experiment during one exposure we were able to register the temporal sequence of the development of the nonlinear response ε_n of the silicon volume during the absorption process of the exciting pulse. With this doubling scheme we excluded the first reference pulse (a - in Fig. 1) by covering one of the mirrors in the scheme. Reconstructed images ($\lambda = 10.63\mu m$) of the silicon volume corresponding to the various instants of excitation time are shown in Fig. 2. It is seen that the absorption of the exciting pulse resulted in the production in the volume of a dark cone ($\tau = -20ps$), next a system of these cones ($\tau = 0ps$), after which a dark cone reappears ($\tau = +20ps$). It is also seen that the registered length of the nonlinear interaction reaches approximately 1 mm.

The analysis of the images obtained shows that in the range of temporal delay (-20,+20)ps the nonlinear response of the medium, averaged over the duration of the pulse, is recorded on the hologram. In this case the reconstructed image represents interference of the sequence of instantaneous values of the complex amplitude of the subject pulse. This situation is similar to the holographic interferometry of vibrating surfaces[7] where it has been shown[8] that an averaged interferogram (for instance Fig. 2b) approximately corresponds to the double exposure

Fig. 2. Temporal sequence during absorption of an exciting pulse in the silicon volume. $\tau = -20$ ps (a); $\tau = 0$ ps (b); $\tau = + 20$ ps (c). Image contains a linear scale: one division equals 0.1 mm.

interferogram of excited ($\tau = 0ps$) and nonexcited states of the volume. Further increase of the temporal delay results in the complete disappearance of the interferogram that is produced due to averaging of the nonstationary response. We have determined the development of the phase shift ϕ of the nonlinear interaction volume in similar experiments which were also carried out using the double exposure method (two pulses on the output of the doubling scheme). The measured value of ϕ has been determined by the number of interference fringes over the nonlinear interaction length. In this case we have obtained the sequence of values of $\phi(\tau)$ that describes the stepwise nature of the nonlinear response: $\pi(-20ps)$; $4\pi(0ps)$; $7\pi(+20ps)$; $8\pi(+200ps)$. Thus, the nonlinear response has a phase nature that varies over the whole period of the exciting pulse absorption. The data obtained correspond qualitatively to the mechanism of nonlinearity caused by the generation of free carriers[9] whose temporal response neglecting recombination and diffusion[10] is proportional to $\sim \int_{-\infty}^{t} I(t)dt$. (Here, $I(t)$ is the intensity of the exciting pulse.)

The method of picosecond holographic interferometry enables us to determine directly the value of the nonlinear refractive index Δn caused by the generation of carriers. In this case it is of interest to measure the maximum value of Δn that could be reached in the absence of the melting of the silicon surface. For measurements we have used the scheme in Fig. 1 according to which the exciting pulse propagates at a small angle ($\sim 7^{\circ}$) with respect to the probing pulse. As a sample we have used a silicon plate with a thickness of $d = 200$ μm. The energy density of the exciting pulse that has been focussed by the spherical lens ($F = 100$ cm) was $1J/cm^2$ and this energy density was slightly lower than that corresponding to the melting of silicon, $E_m = 1.5J/cm^2$ [11]. A silicon interferogram obtained with a time delay of 60 ps after excitation of the sample is presented in Fig. 3. It is worth mentioning that the difference of the phases $\Delta\phi = 8\pi$ between the exciting and nonexciting regions of the silicon is the maximum possible since a further increase in the excitation energy would result in surface melting. This value of $\Delta\phi$ corresponds to the change of refractive index $\Delta n = \Delta\phi \cdot \lambda/2\pi d = 2\cdot10^{-2}$. The values of $\Delta\phi$ and Δn are averaged over the thickness of the specimen. Using the Drude relation $\Delta n = -e^2\Delta N/2\varepsilon_o n_o m^* \omega^2$ one can determine the concentration of free carriers ΔN averaged over the thickness. Assuming $\omega = 1.9\cdot10^{15}s^{-1}$; $n_o = 3.5$; $m^* = 0.15\ m_o$ (12,13); $m_o = 10^{-27}g$; $e = 4.8 \cdot10^{-10}$ CGS; $\varepsilon_o = 1/4\pi$; we find $\Delta N = 1.3\cdot10^{21}\Delta n$. Hence, for $\Delta n = 2\cdot10^{-2}$ the value of $\Delta N = 2.6\cdot10^{19}cm^{-3}$.

It is known[9] that the distribution of the free carrier concentration over the thickness z is described by the ratio $N(z) = N(0)/(1 + \frac{1}{2}\ \sigma N(0)z)$ (where $\sigma = 5\cdot10^{-1}\ cm^2$ is the free carrier cross section). Averaging $N(z)$ over the thickness we find for the concentration of carriers on the surface $N(0) = 2(\exp(d\sigma\Delta N/2)-1)/\sigma d$. Since the value of the averaged concentration equals $\Delta N = 2.6\cdot10^{19}\ cm^{-3}$, we obtain $N(0) = 5.3\cdot10^{19}\ cm^{-3}$. On the other hand, the carrier concentration on the surface $N(0)$ could be determined using the ratio $N(0) = \alpha(1-R)E(0)/h\omega$ where $E(0)$ is the energy density of the exciting pulse, R is the reflection coefficient, and $\alpha = 10\ cm^{-1}$ is the coefficient of linear absorption. Substituting the value $E(0) = 1J/cm^2$ we find $N(0) = 4 \cdot 10^{19}\ cm^{-3}$, which agrees well with the experimental results of Fig. 3.

Thus the values of Δn and $N(0)$ obtained seem to be the upper limits that could be reached for nondestructive excitation of the silicon by a picosecond pulse with a quantum of energy of 1.17 eV and pulse duration 35 ps.

It is clear that the spatial shape of the volume in which the absorption of the exciting pulse takes place depends on nonlinear absorption and refraction. Analysis of this shape could be found from the interferogram of the exciting volume. We have carried out experiments in which the spatial distribution of the exciting pulse energy was extremely inhomogeneous. For this purpose we have placed a thin nontransparent screen (half-plane) on the lateral surface of the

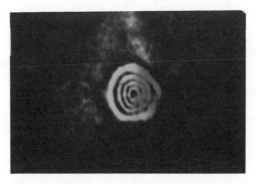

Fig. 3. Double exposure interferogram of the excited silicon region (τ = + 60 ps). The diameter of the interference pattern is 400 μm.

Fig. 4. Double exposure silicon interferogram during spatially inhomogeneous excitation of the volume (τ = + 60 ps).

sample. This screen masked half of the exciting area. In this case the characteristic feature of the exciting volume will be the presence of a very sharp boundary (light-shadow) that has been oriented on the normal to the lateral surface. For silicon the influence of nonlinear refraction in the process of self-interaction would result in the declination of this boundary towards the region of the geometric shadow screen. An interferogram of the excited volume corresponding to a time-delay of

447

60 ps after pulse interaction is shown in Fig. 4. It is seen that the shape of the volume is determined only by the process of nonlinear absorption of the given pulse. Hence the influence of nonlinear refraction on the formation of the nonlinear interaction volume in the silicon could be neglected.

In conclusion we find that the synthesis of the methods of pico-second temporal diagnostics and the principles of holographic recording of the information significantly broadens the experimental potential for the study of nonlinear interactions in solids.

REFERENCES

1. S. A. Akhmanov, Usp. Fiz. Nauk 149, 126 (1986).
2. R. K. Jain, Optical Engineering 21, 199 (1982).
3. E. P. Ippen and C. V. Shank, in Ultrashort Light Pulses, ed. by S. L. Shapiro (Springer-Verlag, New York, 1977).
4. R. J. Collier, C. B. Burckhardt, and L. N. Lin, Optical Holography, (Academic Press, New York and London, 1971).
5. A. A. Bugayev, F. A. Chudnovski, and B. P. Zakharchenya, in Semiconductor Physics, ed. V. M. Tuchkevich. (Consultants Bureau, New York and London, 1986), p. 265.
6. A. A. Bugayev, F. A. Chudnovski, and B. P. Zakharchenya, Phase Transition and its Application (Nauka, Leningrad, 1979).
7. R. L. Powell and K. A. Stetson, J. Opt. Soc. Am. 55, 1593 (1965).
8. M. A. Monahan and K. Bromley, J. Acoust. Soc. Am. 44, 1225 (1968).
9. K. B. Svantesson, J. Phys. D12, 425 (1979).
10. M. Combescot, J. Luminescence 30, 1 (1985).
11. A. L. Smipl, T. F. Boggess, and S. C. Moss, J. Luminescence 30, 272 (1985).
12. Guo-Zhen Yang and N. Bloembergen, IEEE Transaction in Quant. Electr. QE-22, 195 (1986).
13. L. A. Lompre, J. M. Lin, H. Kurz, and N. Bloembergen, Appl. Phys. Lett. 44, 3 (1984).

For USA-USSR Symposium on Laser Optics of Condensed Matter

GaAs ETALONS AND WAVEGUIDES:

BULK VERSUS MULTIPLE-QUANTUM-WELL MATERIAL

H. M. Gibbs, G. Khitrova, S. Koch, N. Peyghambarian, D. Sarid,
A. Chavez-Pirson, W. Gibbons, A. Jeffery, K. Komatsu,[†] Y. H. Lee,
D. Hendricks,[††] J. Morhange,[*] S. H. Park, and M. Warren

Optical Sciences Center
University of Arizona
Tucson, Arizona 85721

A. C. Gossard[**] and W. Wiegmann

AT&T Bell Labs
Murray Hill, New Jersey 07974

M. Sugimoto

NEC Opto-Electronics Research Laboratories
Kanagawa 213, Japan

Abstract

Nonlinear absorption spectra of bulk GaAs and 299-, 152-, and 76-Å GaAs/AlGaAs multiple-quantum-well (MQW) samples are reported.   Nonlinear index spectra are calculated by Kramers-Kronig transformations of the absorption spectra. It is concluded that the principal nonlinear refractive mechanisms for optical bistability in bulk GaAs etalons are band filling and reduction of the Coulomb enhancement of continuum states; exciton saturation and band-gap renormalization occur at lower carrier densities, but their contributions are of opposite sign and almost cancel.   Exciton saturation nonlinearities dominate for bistability in MQW samples with <100-Å wells.

Optical bistability is reported in a strip-loaded waveguide formed by reactive-ion etching; the guiding layer consisted of 60 periods of 100-Å GaAs and 100-Å $Al_{0.28}Ga_{0.72}As$.   The power for bistability is about the same in the etalon and in the waveguide. The 100-fold increase in length of the nonlinear material from the etalon to the waveguide requires a greater detuning to keep the absorption acceptably low, resulting in lower carrier density and a weaker nonlinear refractive effect per unit length.   The cleaved ends of the waveguide were uncoated.

Present addresses:
[†]NEC Corporation, 1-1 Miyazaki 4-Chome, Miyamae-Ku Kawasaki, Kanagawa 213, Japan
[††]1058 E. 17th Street, Salt Lake City, Utah, USA 84105
[*]Laboratoire de Physique des Solides, Universite P. et M. Curie, 4 Place Jussieu, 75230 Paris Cedex 05, France
[**]University of California, Santa Barbara, California, USA 93106

This article is restricted to a discussion of optical nonlinearities and logic operations using GaAs materials. For completeness we give reference to our other work related to this conference. ZnS interference filters, first demonstrated to be optically bistable by Karpushko and Sinitsyn,[1] have been used as AND and OR gates to demonstrate one-bit addition by symbolic substitution[2] and to recognize three-spot patterns in the shape of a V or Γ in a 2×9 input array.[3] These simple demonstrations of nonlinear decision-making[4] devices to parallel optical computing illustrate logic operations, cascading, and a small amount of parallelism. Optical instabilities in sodium vapor using a single feedback mirror[5] exhibit many transverse effects that must be related to the hybrid-system transverse instabilities reported at this conference by Worontsov and Shmalhausen.

In GaAs we have shown that the AC Stark shift of the exciton can be used to shift a Fabry-Perot etalon peak, situated just below the exciton, to perform an optical NOR gate; its response is determined by the pump pulse's duration.[6] However, this gate does not have gain; i.e., the energy of the pulse being controlled is less than the pump-pulse energy, unlike the NOR gate based on above-bandgap absorption of the input pulses.[7]

But the response time of the latter gate is determined by the lifetime of the carriers, typically several nanoseconds,[7] but as low as $\cong 50$ ps in 0.3-μm-thick GaAs with no AlGaAs windows.[8] This two-wavelength operation complicates cascading because the output is at a different wavelength from the inputs. One-wavelength bistable or transistor operation[4] avoids this complication, but gain with picosecond pulses has not been shown experimentally. Here we focus attention on bistable operation just below the band edge using 100-ns to 1-μs pulses, short enough to avoid thermal effects and long enough that the carriers are in thermodynamic equilibrium under the action of the pump light and relaxation processes.

The band-edge optical nonlinearity measurements[9] were motivated by the great similarity[10] between optical bistability (switch-up powers and hysteresis loops) using bulk or MQW etalons. The photoluminescence from a 299-Å MQW platelet excited by 514.5-nm 0.8-μs light served as a broad-band probe source, so that absorption data could be taken simultaneously over the band edge; see Figure 1a. The 821-nm, 1-μs pump pulses were focused to about a 15-μm-diameter spot, larger than the probe. These data are compared with the Banyai-Koch[11] plasma theory (shown in Figure 1c), which is based on the Haug and Schmidt-Rink[12] many-body theory of band-edge nonlinearities of direct-gap semiconductors. Included in the theory are band filling and plasma screening of the Coulomb interaction, leading to three other mechanisms, namely bandgap renormalization, exciton saturation, and reduction of the enhancement of the continuum states (which, of course, makes the band edge square rather than parabolic). Because of the long duration of the pulses and the resultant quasi-equilibrium in the system of electronic excitations, one can calculate the corresponding intensity dependence of the index of refraction by making a Kramers-Kronig calculation:

$$\Delta n = \frac{\hbar c}{\pi} P \int_{E_1}^{E_2} \frac{\Delta\alpha(E')}{E'^2 - E^2} dE' .$$

The Kramers-Kronig transformations of the nonlinear absorption data of Figure 1a are shown in Figure 1b, and they are consistent with direct measurements.[13] The clear similarity between the data and the theory gives us the confidence to identify the principal mechanisms leading to bistability; in fact, one can see a clear discrepancy between the data and theory if any one of exciton bound states or band filling or bandgap renormalization is neglected. At low intensities, exciton saturation and bandgap renormalization dominate, but their index contributions have opposite signs and very nearly cancel. The multiple-quantum-well data are handled similarly, but the theory is not so well developed for the realistic case of quantum wells with a finite thickness, which are neither three dimensional or two dimensional. The empirical technique of Chemla and Miller[14] was used, in which the band-edge absorption is fitted to the sum of Gaussians for the heavy- and light-hole excitonic transitions and for the band-to-band transitions; see Figure 2. The maximum index change at each intensity is shown in Figure 3 for bulk GaAs and for MQW with 299-Å, 152-Å, and 76-Å wells. The index change is seen to increase with narrower wells. However, the pump absorption was different for the various samples; Figure 4 shows the maximum change in index per carrier, Δn/N, versus the carrier concentration N. Clearly this ratio is almost the same in all four samples as long as one assumes that the carrier lifetime is the same (20 ns) in all four samples. But even if the carrier lifetimes turn out to be unequal, these results point to why bistability looks so similar in bulk and MQW etalons. The best operating point must be close to the wavelength at which the linear absorption αL equals the mirror transmissivity T=1-R. Then if the absorption is the same in two samples, so will be the carrier density and consequently the index changes and bistability. However, we stress that the mechanisms are different (at room temperature): excitonic in MQW and a combination of band filling and reduction of Coulomb enhancement in bulk GaAs.

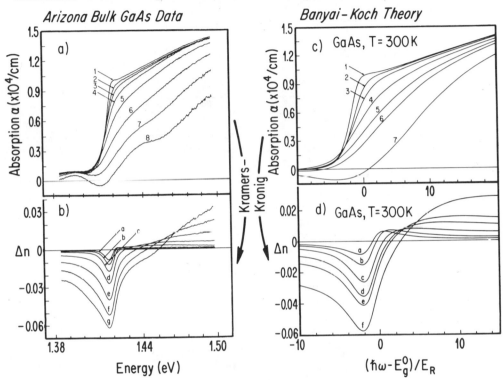

Figure 1. *Room-temperature bulk GaAs optical nonlinearities: experiment and theory. (a) Experimental absorption spectra for different excitation intensities I (mW): 1) 0; 2) 0.2; 3) 0.5; 4) 1.3; 5) 3.2; 6) 8; 7) 20; 8) 50 on a 15-μm-diameter spot. (b) Nonlinear refractive index changes corresponding to the measured absorption spectra. The curves (a-g) in Fig. 1(b) are obtained by the Kramers-Kronig transformation of the corresponding experimental data (2-8) in Fig. 1(a). (c) Calculated absorption spectra for different electron-hole pair densities N (cm⁻³): 1) 10¹⁵; 2) 8×10¹⁶; 3) 2×10¹⁷; 4) 5×10¹⁷; 5) 8×10¹⁷; 6) 10¹⁸; 7) 1.5×10¹⁸. E⁰_g = 1.433 eV and E_R = 4.2 meV. (d) Calculated nonlinear refractive index changes. The curves (a-f) in Fig. 1(d) are obtained from the curves (2-7) in Fig. 1(c), respectively.*

Clearly much of the motivation for understanding and optimizing GaAs nonlinearities and bistability is the hope that someday arrays of such devices will play a major role in parallel computing, associative memories, learning machines, etc. Reactive ion etching of arrays is relatively easy.[15] For example, 10⁴ NOR gates on 4 cm² operated at 1 GHz and requiring 40 pJ per bit operation, would require 400 W of average laser power and produce 100 W/cm² of heat. But the system would perform 10¹³ bit operations per second -- more than a CRAY.

In many applications in optical communications, interconnects, multiplexing, encryption, etc., the data are handled sequentially in a pipeline fashion. In such cases waveguide devices could be used easily, assuming performance superior to etalons. In quest of such, we have used reactive ion etching to fabricate strip-loaded waveguides in which electronic optical bistability has been seen.[16] Of course, this is not the first observation of bistability in a waveguide. Thermal bistability has been seen using prism coupling into ZnS and ZnSe slabs,[17,18] using either increasing absorption or nonlinear

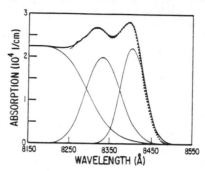

Figure 2. 76-Å MQW linear absorption and fit to two Gaussians, simulating the heavy-hole and light-hole excitons, together with a broadened two-dimensional continuum (Sommerfeld factor), to simulate the band-to-band transitions.

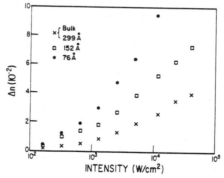

Figure 3. Maximum change in refractive index Δn as a function of input light intensity I.

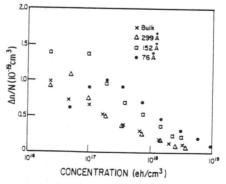

Figure 4. Maximum change in index per carrier versus the electron-hole concentration.

452

dispersion in a Fabry-Perot-like GaAs MQW slab,[19] and using increasing absorption in a ZnSe slab.[20] Hybrid self-electro-optic-effect-device bistability was seen in a GaAs MQW slab.[21] By far the closest to our work is that in a dispersive Fabry-Perot-like GaAs MQW channel guide in which a Si_3N_4 strip on top strains the MQW at its edges, resulting in guiding.[22] Our guide has yielded better bistability (wider loops and sharper switching) and should be easier to model and to fabricate. The sample was grown at NEC using a GaAs substrate, 3-μm-thick $Al_{0.3}Ga_{0.7}As$ layer to reduce losses into the substrate, then 60 periods of 100-Å GaAs and 100-Å $Al_{0.28}Ga_{0.72}As$, and finally 1-μm-thick $Al_{0.13}Ga_{0.87}As$ on top. A 3-μm-wide strip was etched about 0.9 μm into the top layer, resulting in lateral guiding in the MQW layer so that wall roughness in the top strip is less significant. A typical guide is shown in Figure 5. Optical bistability in a 200-μm-long guide with 300-ns input pulses is shown in Figure 6. Thermal effects have been ruled out by checking that the laser frequency is above the Fabry-Perot peak. With longer pulses ($\cong 8$ μs), heating in the upper branch results in switch-down at a higher power than switch-up. Multiple bistability, consisting of two bistability loops, shows that a 2π phase shift has been achieved in this waveguide, compared with about 0.5π in etalons.

Figure 5. Electron micrograph of GaAs/AlGaAs strip-loaded waveguide.

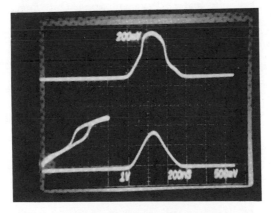

Figure 6. Experimental input-output behavior of a bistable MQW waveguide operating at 867 nm. The upper trace is the output pulse and the lower trace is the input. The trace in the lower left is a plot of output versus input. Input power: 130 mW; output power: 15 mW; estimated power coupled into waveguide: 60 mW; band-edge wavelength: 839 nm; $\alpha L = 1.15$, including both guiding and absorptive losses.

However, in some respects the waveguide bistability is disappointing. The usual argument is that a waveguide keeps the light at a high intensity over a longer interaction length so that the input power can be reduced. This has shown to be the case for many nonlinear effects in transparent materials in which the absorption length α^{-1} greatly exceeds the Rayleigh length of a tightly focused beam in free space. But that is not the situation in GaAs etalons where one operates as close to the band edge as the finesse will allow ($\alpha L \cong T$). Increasing the interaction length by 100 or more must be accompanied by increased detuning from the band edge to preserve $\alpha L \cong T$. The similarity in operating powers between MQW GaAs bistability in etalons and waveguides suggests that the nonlinear refraction decreases to compensate for the increased interaction length. On a more optimistic note, this first attempt used only cleaved-end faces ($R \cong 0.3$) and has not been optimized. Ultimately, one hopes for picosecond[6] response times in a variety of waveguide devices such as directional couplers, Mach–Zehnder interferometers, etc. The devices reported here are a first step in a material compatible with integrated electronics and opto-electronics.

In summary, bulk GaAs and MQW etalons exhibit similar optical bistability powers and loops, but band filling and reduction of the Coulomb enhancement of continuum states are the nonlinear refraction mechanisms in bulk GaAs and exciton saturation effects in MQW. The tunability of the band edge with well thickness in a MQW is very useful for diode laser operation. Bistability has been seen in a MQW strip-loaded waveguide, but the power required is similar to that for etalon bistability.

Acknowledgments

We gratefully acknowledge support from the AFOSR, ARO, NSF, DARPA/RADC, SDI, and the Optical Circuitry Cooperative.

References

1. F. V. Karpushko and G. V. Sinitsyn, *Appl. Phys. B*, **28**:137 (1982); and *J. Appl. Spectrosc. USSR*, **29**:1323 (1978).
2. M. T. Tsao, L. Wang, R. Jin, R. W. Sprague, G. Gigioli, H.-M Kulcke, Y. D. Li, H. M. Chou, H. M. Gibbs, and N. Peyghambarian, *Opt. Eng.*, **26**:41(1987).
   L. Wang, H. M. Chou, H. M. Gibbs, G. C. Gigioli, G. Khitrova, H.-M. Kulcke, R. Jin, H. A. Macleod, N. Peyghambarian, R. W. Sprague, and M. T. Tsao, *Soc. Photo-Opt. Instrum. Eng. O-E LASE '87*.
3. H. M. Gibbs, G. Khitrova, L. Wang, et al.: OSA Second Topical Meeting on Optical Computing, Lake Tahoe, March 1987.
4. H. M. Gibbs and N. Peyghambarian, Optical and Hybrid Computing, *Soc. Photo-Opt. Instrum. Eng.*, **634**:142 (1986).
   H. M. Gibbs, "Optical Bistability: Controlling Light with Light," Academic, New York (1985).
5. H. M. Gibbs, M. W. Derstine, K. Tai, J. F. Valley, J. V. Moloney, F. A. Hopf, M. Le Berre, E. Ressayre, and A. Tallet, *in:* "Optical Instabilities," R. W. Boyd, M. G. Raymer, and L. M. Narducci, eds., Cambridge University, Cambridge (1986).
   G. Giusfredi, J. F. Valley, R. Pon, G. Khitrova, H. M. Gibbs: Int'l. Workshop on Instabilities, Dynamics, and Chaos in Nonlinear Optical Systems, Lucca, Italy, July 1987.
6. D. Hulin, A. Mysyrowicz, A. Antonetti, A. Migus, W. T. Masselink, H. Morkoc, H. M. Gibbs, and N. Peyghambarian, *Appl. Phys. Lett.*, **49**:749 (1986).
   A. Mysyrowicz, D. Hulin, A. Migus, A. Antonetti, H. M. Gibbs, N. Peyghambarian, and H. Morkoc: OSA Topical Meeting on Picosecond Electronics and Opto-electronics, Lake Tahoe, January 1987.
7. J. L. Jewell, Y. H. Lee, M. Warren, H. M. Gibbs, N. Peyghambarian, A. C. Gossard, and W. Wiegmann, *Appl. Phys. Lett.*, **46**:918 (1985).
8. Y. H. Lee, H. M. Gibbs, J. L. Jewell, J. F. Duffy, T. Venkatesan, A. C. Gossard, W. Wiegmann, and J. H. English, *Appl. Phys. Lett.*, **49**:486 (1986).

9. Y. H. Lee, A. Chavez-Pirson, S. W. Koch, H. M. Gibbs, S. H. Park, J. Morhange, A. Jeffery, N. Peyghambarian, L. Banyai, A. C. Gossard, and W. Wiegmann, *Phys. Rev. Lett.*, **57**:2446 (1986).

10. S. Ovadia, H. M. Gibbs, J. L. Jewell, and N. Peyghambarian, *Opt. Eng.*, **24**:565 (1985).

11. L. Banyai and S. Koch, *Z. Phys. B*, **63**:283 (1986).

12. H. Haug and S. Schmitt-Rink, *Prog. Quantum Electron.*, **9**:3 (1984).
    H. Haug, *in:* "Nonlinear Optical Properties of Semiconductors," H. Haug, ed., Academic, New York (1987).

13. Y. H. Lee, A. Chavez-Pirson, B. K. Rhee, H. M. Gibbs, A. C. Gossard, and W. Wiegmann, *Appl. Phys. Lett.*, **49**:1505 (1986).

14. D. S. Chemla and D. A. B. Miller, *J. Opt. Soc. Am. B*, **2**:1155 (1985).
    D. S. Chemla, D. A. B. Miller, P. W. Smith, A. C. Gossard, and W. Wiegmann, *IEEE J. Quantum Electron.*, **QE-20**:265 (1984).

15. T. Venkatesan, B. Wilkens, M. Warren, Y. H. Lee, G. Olbright, H. M. Gibbs, N. Peyghambarian, J. S. Smith, and A. Yariv, *Appl. Phys. Lett.*, **48**:145 (1986).
    Y. H. Lee, M. Warren, G. R. Olbright, H. M. Gibbs, N. Peyghambarian, T. Venkatesan, J. S. Smith, and A. Yariv, *Appl. Phys. Lett.*, **48**:754 (1986).

16. M. Warren, W. Gibbons, K. Komatsu, D. Sarid, D. Hendricks, H. Gibbs, and M. Sugimoto: IQEC Baltimore (1987), Post-deadline PD9.

17. W. Lukosz, P. Pirani, and V. Briguet, *in:* "Optical Bistability III", H. M. Gibbs, P. Mandel, N. Peyghambarian, and S. D. Smith, eds., Springer-Verlag, Berlin (1986).

18. G. Assanto, B. Svensson, D. Kuchibhatla, U. J. Gibson, C. T. Seaton, and G. I. Stegeman, *Opt. Lett.*, **11**:644 (1986).

19. A. C. Walker, J. S. Aitchinson, S. Ritchie, and P. M. Rodgers, *Electron. Lett.*, **22**:366 (1986).

20. B. Y. Kim, Elsa M. Garmire, N. Shibata, and S. Zembutsu, CLEO, Baltimore, (1987).

21. J. S. Weiner, D. A. B. Miller, D. S. Chemla, T. C. Damen, C. A. Burrus, T. H. Wood, A. C. Gossard, and W. Weigmann, *Appl. Phys. Lett.*, **47**:1148 (1985).

22. P. Li Kam Wa, P. N. Robson, J. P. R. David, G. Hill, P. Mistry, M. A. Pate, and J. S. Roberts, *Electron. Lett.*, **22**:1129 (1986).

RESONANT NONLINEARITIES AND OPTICAL BISTABILITY IN SEMICONDUCTORS

V. S. Dneprovskii, A. I. Furtichev, V. I. Klimov, E. V.
Nazvanova, D. K. Okorokov, and U. F. Vandishev

Department of Physics, Moscow State University, 119899
Moscow, USSR

ABSTRACT

Changes in the absorptive and refractive index caused by absorption
of laser light in CdS and GaSe semiconductors ("dynamic" nonlinearities
which may lead to optical bistability) were investigated. Nonlinear
effects may be very large (strong nonlinearities) due to the resonant
enhancement near the sharp band edge. In this case optical properties
can be varied significantly by creating small numbers of interacting free
carriers, free and bound excitons, biexcitons, etc. The interplay of
optical nonlinearity and feedback leads to optical bistability (OB) – the
existence of two stable output states of an optical system for the same
input parameters. OB allows one to reproduce the operational modes of
electronic devices (amplification, switching, memory, logical operations)
by purely optical means.

I. ABSORPTIVE BISTABILITY IN CdS AT LOW PUMPING INTENSITIES

The sharp decrease of transmission in the vicinity of the absorption
edge of CdS (80K) and absorptive resonatorless OB was observed at an
intensity of the Argon laser pumping pulses lower than 10 kW/cm^2 (with a
pulse duration of $\tau < 10$ µs). Analysis of the luminescence spectra
showed that at such manner of excitation the fast change of the
absorption is due to the process of exciton-exciton interaction.

Optical properties of semiconductors may be significantly changed in
the case of resonant excitation of excitons. Density-dependent collision
broadening of the free-exciton resonance[1,2] leads to a nonlinear
increase of absorption in the vicinity of the absorption edge. Resonatorless OB based on this increasing nonlinear absorption has been shown
to exist[3-5]. The intrinsic feedback arises from the dynamic relation
between the generation and recombination rates of the particles. The
resonatorless absorptive OB was observed in cooled CdS samples at rather
high excitation intensities (~ 1 MW/cm^2). The induced absorption was
attributed to the formation of an electron-hole plasma and renormalization of the energy gap at high density of the carriers.

The CDS platelet (3-5 µm) cooled to the liquid nitrogen temperature
was resonantly excited (Fig. 1) by an Argon laser (λ_p = 488 nm, $\vec{E} \parallel c$, c-
the optical axis of the crystal). The pulses of tunable duration

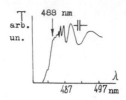

Fig. 1.   The transmission spectrum of CdS (80K, $\vec{E} \| c$)

(7 µs– 10 ms) were created with the help of an electrooptical modulator. The diameter of the focused beam on the surface of the sample was about 30 µm.   The shape of the input and the output pulses and luminescence spectra from the sample were simultaneously measured at different values of excitation intensity.   Oscillograms of the input (intensity S) and the transmitted (S_T) pulses and the hysteresis loop relating them are presented in Fig. 2 for $\tau = 7$ µs.   The fast switching of the bistable system (the switching on and off times are not longer than the time-resolution of the applied photoreceivers – 1-2 µs) is due to the increasing absorption in the vicinity of the absorption edge of CdS (80K).

The increasing absorption may arise due to the following processes: 1. the collison broadening of the free-exciton resonance; 2. band gap shrinkage; 3. excitation-induced heating of the sample.   The features of the luminescence spectra (Fig. 3) allow us to assume that the process of exciton-exciton interaction dominates.   The P-band of luminescence arises

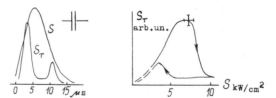

Fig. 2.   The oscillograms of the input (intensity S) and output (S_T) pulses; the hysteresis loop $S_T(S)$.

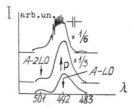

Fig. 3.   The luminescence spectra of CdS (80K), S = 4,9,15 kW/cm^2).

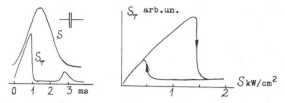

Fig. 4. The oscillograms of the input and output pulses; the hysteresis loop $S_T(S)$.

at the values of S used for excitation of the bistable mode of transmission. The spectral position of the P-band (the red shift from the free exciton energy is about the binding energy of the exciton), the superlinear dependence of its intensity upon the pumping intensity S, the difference of the shape of the P-band from the shape of the phonon-replica of the A-exciton (A-LO-band appears at low intensities and has a high energy shoulder), and the appearance of stimulated emission in the vicinity of the red shoulder of the P-band allows us to explain it by the process of exciton-exciton interaction: ex + ex = hν + (e + h) (e + h is the electron-hole pair). The estimated density of excitons for switching of the bistable system was about, or lower than, 10^{16} cm^{-3}. Thus, the nonlinear increasing of the absorption cannot be explained by the renor-malization of the energy gap. To switch the system one needs to "shift" the absorption edge (the "shift" must be about 10-13 meV, Fig. 1). Such a shift may be achieved in CdS at a carrier-density of n ⩾ 10^{17} cm^{-3} (6). The observed absorptive bistability cannot be explained by laser induced heating of CdS. The duration of the pumping pulse (7 μs) was rather short to create thermal bistability. The measured time of the sample's cooling was 50-100 μs. The red shift of the A-2LO-band (the heating of the crystal) was not observed at the intensities of excitation used (Fig. 3). The thermal bistability (nonlinearity arising from the thermal shift of the gap) could be clearly seen when the duration of the input pulse was about some milliseconds (Fig. 4).

It is possible to see the competition and coexistence of two types of induced absorption (a "fast" one, which arises due to the exciton-exciton interaction, and a "slow" one due to the heating of the sample) by increasing the input pulse duration. The fast fall of the trans-mission (Fig. 5a) is probably due to the exciton-exciton interaction in the sample heated to a higher temperature than that presented in Fig. 5b (at lower S the crystal is heated to a higher temperature for longer time

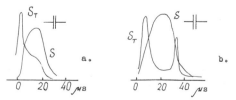

Fig. 5. The oscillograms of the input and output pulses. a. S = 2 kW/cm^2, b. S = 6 kW/cm^2.

periods until the moment of "excitonic" switching). The absorption edge of CdS shifts due to the heating of the sample, and at λ_p = 488 nm feedback arises (due to the S-type dependence of the transmission upon the density of excitons). At higher S (Fig. 5b) the switching-off time is faster (!) than the switching-on time. This may be explained by the coexistence of thermal and "excitonic" effects during the switching-on process and by fast "excitonic" switching off (the crystal is heated and the absorption spectrum is tuned to the region where the feedback for the "excitonic" switching-off exists).

II. STRONG NONLINEARITIES AND BISTABILITY IN GaSe

The layered crystals GaSe, GaSSe are convenient for the creation of OB devices[7]. These semiconductors have strong nonlinearities which exist in the case of resonant excitation of excitons and interband transitions[8,9]. It is rather easy to prepare a Fabry-Perot resonator from a layered crystal by cleaving along the planes of the layers. The atomic structure of GaSe consists of fourfold layers of Ga and Se which are held together by Van der Waals bonds to form a three-dimensional crystal. So GaSe shows a large mechanical anisotropy, which accounts for its easy cleavage. It is possible to create OB devices from GaSe operating even at room temperature[11,12] by using resonant excitation of excitons (excitonic absorption can be easily seen at room temperature because of the rather high (20 meV) exciton binding energy; and the surfaces of the cleaved samples are of high quality). GaSe has a suitable absorption ($\sim 10^3$ cm^{-3}) in the vicinity of the absorption edge. Absorption in the $\vec{E} \perp c$ (the c-axis is normal to the layer plane) geometry is only weakly allowed by spin-oribit-coupling effects. Such absorption makes it possible to operate with resonators of about 10 μm length. Finally, the spectral position of the absorption edge of GaSe (0.58–0.6 μm) lies within the widespread range of dye-laser frequencies. OB was observed in a GaSe etalon in the case of resonant excitation of direct excitons by 100–200 ns laser pulses[13]. The input intensity was about 100–300 kw/cm^2. The bistability was primarily dispersive with the nonlinear refractive index arising from light-induced changes in exciton absorption. The wavelength of the laser (λ) was just below the exciton peak $\lambda_{ex} (\lambda > \lambda_{ex}; \lambda-\lambda_{ex} \leqslant \Delta\lambda$, where $\Delta\lambda$ is the resonance half-width at half-maximum).

In Ref. 14 OB was discovered at a much lower input intensity (0.5–2 kW/cm^2). The wavelength of the dye-laser satisfied the inequality $\lambda > \lambda_{ex} (\lambda-\lambda_{ex} > \Delta\lambda)$. In this case OB could be explained by two simul-
ceeding processes: the decreasing of the refractive index
the interference spectrum, and the increasing of absorption
laser wavelength.

the paper the origin of the above-mentioned non-
zed. Excite-and-probe transmission measurements
Se were performed. Rhodamine dye-lasers with
second harmonic of a Nd:YAG-laser
(τ = 7 ns) were utilized for the
ns and free carriers. A

synchronously pumped cell with the appropriate dye was used for creating a probe beam. The samples of 2–10 μm thickness were cleaved from an ε–GaSe monocrystal. The changes in the transmission spectra of GaSe at different intensities of the pumping pulse (τ = 130 ns) in the case of resonant excitation of excitons may be seen in Fig. 6. The bleaching of the exciton absorption, the broadening of the exciton line and the increase of absorption at the red shoulder become noticeable at intensities above 50 kW/cm^2. Nearly the same bleaching and broadening of the exciton absorption line was observed in the case of interband excitation of free carriers (Fig. 7).

The luminescence spectra of GaSe (80K) show (Figs. 8, 9) that simultaneously with the A_Γ-line (the luminescence of free direct excitons, λ_{ex} = 590 nm) the P_Γ-band arises at high excitation. The excitonic nature of the P_Γ-band is confirmed by the spectra (Fig. 9) which were obtained at different energies of the exciting photon. These spectra are almost identical – they have practically the same shape and spectral position of the bands. The intensity of the A_Γ-line was $I_{ex} \sim S^{1\pm0.1}$ at S = 10-100 kW/cm^2 (S is the intensity of the pump beam) and $I_{ex} \sim S^{0.5\pm0.1}$ at S > 100 kW/cm^2, (Fig. 10). Such a dependence indicates the change of the linear recombination by the process of bimolecular

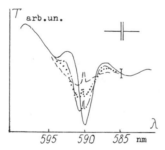

Fig. 6. The transmission spectra of GaSe (resonant excitation of excitons, S = 0, 80, 120, 220 kW/cm^2).

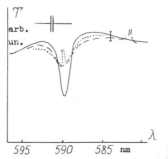

Fig. 7. The transmission spectra of GaSe (S = 200 kW/cm^2).

recombination. The red shift of the P_Γ-band relative to the A_T-line is about the binding energy of the direct exciton. The intensity of the P_Γ-band $I_{P_\Gamma} \sim I_{ex}^2$ (Fig. 10) at all values of pumping intensities S used. Thus $I_P^\Gamma \sim N_{ex}^2$ (the intensity of the exciton luminescence line $I_{ex} \sim N_{ex}$, where N_{ex} is the density of the direct excitons). The P_Γ-band may be attributed to the process of exciton-exciton interaction: $ex^\Gamma + ex^\Gamma = h\nu^\Gamma + (e^\Gamma + h)$, where ex^Γ is the direct exciton at the Γ-point of the Brillouin zone, $h\nu^\Gamma$ is the energy of the emitted photon, and $(e^\Gamma + h)$ is the electron-hole pair.

The observed bleaching and broadening of the exciton absorption line in GaSe (80K) with increasing resonant excitation (Figs. 6 and 7) seems to arise mainly due to the process of exciton-exciton interaction.

The application of shorter pulses ($\tau = 7$ ns) allows one to obtain the transmission spectra at different moments after interband excitation (an optical delay line was used) and luminescence spectra at higher

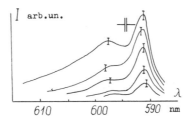

Fig. 8. The luminescence spectra of GaSe (80K), interband excitation S = 20, 75, 160, 250, 300 kW/cm^2.

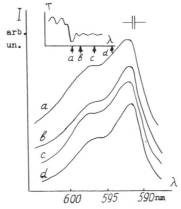

Fig. 9. The luminescence spectra of GaSe (80K) at different energies of the exciting photon, S = 200 kW/cm^2.

462

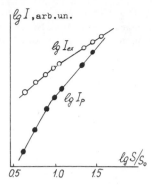

Fig. 10. The intensities of the A_Γ-line (I_{ex}) and the P_Γ-band (I_p) of GaSe as functions of the intensity S of the pump beam.

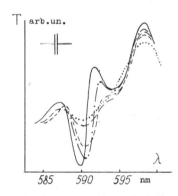

Fig. 11. The transmission spectra of GaSe (80K) at different moments after interband excitation (Dt = 7, 12, 18, 30 ns).

excitation (Figs. 11, 12). The analysis of the features of the restoration of transmission showed that two relaxation times "fast" (< 1 ns) and "slow" (> 30 ns) exist. A new band P_M arises. The spectral position of this band, the quadratic dpendence of its intensity upon the density of direct excitons, the stimulated radiation at high excitation, and the existence of the "slow" restoration time of absorption allows us to attribute the P_M-band to the process of interaction between direct and indirect excitons: $ex^\Gamma + ex_M = h\nu^M + (e_M + h)$, where ex_M and e_M are the indirect exciton and electron at the M-point of the Brillouin zone.

Other processes explaining the change of the spectrum at the absorption edge of a GaSe crystal may develop: Mott transition, self-screening of excitons and the screening of excitons by free carriers, the

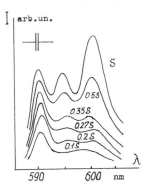

Fig. 12. The luminescence spectra of GaSe (80K), interband excitation, τ = 7 ns, S = 500 kW/cm.

gap shrinkage due to the renormalization at high density of the carriers, and thermal effects. The results of the performed experiments show that the exciton-exciton interaction is one of the dominating processes.

In summary, strong nonlinearities and different types of optical bistability may exist in CdS and GaSe due to the effective exciton-exciton interaction.

REFERENCES

1. H. Kalt, V. G. Lyssenko, R. Renner, and C. Klingshirn, Sol. St. Commun. 9, 675 (1984).
2. C. Clingshirn, K. Bohnert, H. Kalt, V. G. Syssenko, and K. Kempf, J. Lum. 30, 188 (1985).
3. H. Rossmann, F. Henneberger, J. Voigt, phys. stat. sol. (b) 115, K63 (1983).
4. F. Henneberger, phys. stat. sol. (b) 137, 371 (1986).
5. K. Bohnert, F. Fidorra, and C. Klingshirn, Z. Phys. B 57, 263 (1984).
6. M. Rosler and R. Zimmerman, phys. stat. sol. (b) 83, 85 (1977).
7. V. S. Dneprovskii, Izv. Akad. Nauk SSSR, Ser. Fiz. 50, 661 (1986); Uspekhi Fiz. Nauk 145, 149 (1985).
8. A. M. Bakiev, U. V. Vandishev, G. S. Volkov, V. S. Dneprovskii, Z. D. Kovaluk, A. R. Lesiv, S. V. Savinov, and A. I. Furtichev, Fiz. Tverd. Tela 28, 1035 (1986).
9. V. S. Dneprovskii, V. D. Egorov, D. S. Khechinashvili, and H. X. Nguyen, phys. stat. sol. (b) 138, K39 (1986).
10. G. P. Golubev, V. S. Dneprovskii, Z. D. Kovaluk, and V. A. Stadnik, Fiz. Tverd. Tela 27, 432 (1985).
11. A. M. Bakiev, V. S. Dneprovskii, Z. D. Kovaluk, and V. A. Stadnik, Pis´ma Zh. Exper. Teor Fiz. , 38, 493 (1983).
12. A. M. Bakiev, V. S. Dneprovskii, G. S. Volkov, and Z. D. Kovaluk, Zh. Tekhn. Fiz. 55, 1177 (1985).

13. A. M. Bakiev, V. S. Dneprovskii, Z. D. Kovaluk, and V. A. Stadnik, Doklad. Akad. Nauk SSSR <u>271</u>, 611 (1983).
14. G. P. Golubev, V. S. Dneprovskii, E. A. Kisilev, Z. D. Kovaluk, and

MONOLITHIC MICRORESONATOR ARRAYS OF OPTICAL SWITCHES

S. L. McCall
AT&T Bell Laboratories
Murray Hill, New Jersey 07974

J. L. Jewell
AT&T Bell Laboratories
Holmdel, New Jersey 07733

Part of the purpose of this paper is to describe some very small optical logic gates that have been recently constructed and demonstrated. The distance between gate centers in an array is as small as 3 μm and 4$\times$8 arrays have shown essentially uniform response. This work[1] is due to five people. This work is part of an effort to make optical logic gates that are competitive with or superior to semiconductor electronic gates. First some history and background is presented so the reader may place this work in perspective and be aware of motivations.

When optical bistability was first demonstrated,[2] it was suggested by many that possibly optical logic could be useful, if good gates could be constructed. Suggestions of this sort had been made before that time and Keyes and Armstrong[3] in particular investigated whether there were physical limitations that might favor one type of logic gate and system over another. The reasoning they employed at that time led to the conclusion that optical gates were inherently more power consumptive at the fastest speeds. This reasoning is now known to be incomplete and, as will be seen later, there is no clear choice between optical and semiconductor electronic logic based on power consumption alone. It must be emphasized that this sort of evaluation is a multi-dimensional subject and a single consideration is not enough. For example, based on this aspect alone, superconductor logic would be superior because the energy gap in superconductors is smaller than the gap in semiconductors such as GaAs and will ultimately consume less power. It may be that superconducting logic will eventually be preferred for computers, or maybe it won't. In addition to devices with many logic gates, devices with one or a few gates are of interest, for example in time division multiplexing in optical communications. There speed of operation takes a more important but not a singular role. In the largest computer today, the telephone system, optics is already operating in fiber data transmission both over long distances and in some specialized switching machines.

Before believing that there is unquestionably a unlimited future for optical logic, let us examine the past of computer technology and use it to extrapolate into the future. For about thirty years, this technology has advanced about a factor two

467

each year as measured by the amount of computing one can do per unit currency. Year after year engineers have bypassed or surmounted barriers described by others as impossible to overcome. Recently parallelism to a large degree, described before as belonging only to the province of optics, has entered electronic computer systems. Computing rates of 10^{10} floating point operations per second, i.e. about 10^{13} bit operations per second, have been demonstrated. If we extrapolate another 13 years at this rate of a factor two per year, we find about 10^{17} bit operations per second about the year 2000, which compares with the computing power of the optic nerve of advanced mammals. What sort of task might such a machine do? We can try to imagine a somewhat specialized machine designed to predict the weather. This machine could be constructed of many processing units with an interconnection topology of a spherical shell representing the earth's atmosphere and surface. Each processor would calculate conditions for a small area of the earths surface and for a limited depth of the atmosphere. Data communication would mainly be necessary between nearby processors. Code for the various processors would be almost identical. One can imagine that with such a machine, reliability of prediction would be limited by the quantity and quality of input data about the weather at a moment in the future.

Clearly, to speak of the potential capabilities of optical processing and to compare that with present day electronics is not a relevant comparison. Comparison should be made with accurate or at least believable estimates of future electronic capabilities. Here we focus on a single aspect of such comparisons, namely heat dissipation. Of various "limits" and "barriers", one that must be considered is heat, because it is a major factor and places the same limitation on any technology. This consideration immediately leads to the question of the required energy to perform a logic function.

We consider a length of nonlinear material an absorption length long, and estimate the energy required to bleach an atomic resonance. The light can be focused to an area about a square wavelength. An atomic cross section can be about one square wavelength, so about one atom is necessary in the active volume and about one photon is required to bleach it. This argument was employed by McCall and Gibbs[4] in a discussion of the limits of optical bistable devices and they pointed out that although such a single or few atom device could in principle be constructed according to classical considerations, it would spontaneously switch and be unreliable. They pointed out that a larger number of atoms were needed to avoid spontaneous switching and suggested 1000 quanta or the energy of 1000 photons as a limit. That limit, the statistical limit, is closer[5] to 300 for an error rate $\approx 10^{-17}$. This limit also applies to electronic switches and to superconducting switches. In both cases, about 300 electrons are needed at a gate input to switch a gate reliably. Placing each electron at the gate input requires about one electron volt of energy for semiconductor electronic transistors and about 0.01 electron volt for superconducting switches. In all cases one needs about 300 quanta for reliable switching. Superconducting quanta at 0.01 electron volts are smaller than 1 electron volt so that less energy is required for superconducting gates than for semiconductor or optical gates switching at the statistical limit.

As Keyes and Armstrong implicitly pointed out, when fast switching is considered, an atomic resonance cross section must scale downward with increasing speed because of physical limitations, for example the oscillator sum strength rule. Using the Na D line as an example, and 300 quanta as a limit, the lifetime of 30 nsec can only be reduced to 100 psec before the total resonant cross section of 300 such atoms drops below about a square wavelength. Optical bistability provides an example, and Fork[6] and Smith[7] pointed out use of a resonator lowers energy requirements, in essence the cross section is increased by a factor approximately the inverse mirror transmission. Finesse values of about 30 are typical in optical switches today, so the time of 100 psec is reduced to about 10 psec. However, 10 psec corresponding to a cycle time of about 30 psec is still too long a time for favorable comparison with future electronic devices. Also, Na atoms are not obviously attractive for constructing optical devices.

Atoms in free space are not necessarily the best nonlinear materials for this application. Firstly, a molecular transition may involve several electrons, thus increasing the oscillator strength. In semiconductors, the cooperative effect of several electrons can show up in a effective mass which is less than the mass of a free electron. Secondly the saturation mechanism can be a different physical mechanism. For example, the creation of excitons by photon absorption is a different process than the creation of excited atoms from ground state atoms. In one case two particles, a photon and a ground state atom, are present in the initial state, whereas in the exciton creation case only one particle, a photon, is present in the initial state. As an example, we choose a material familiar to us, GaAs, the particular sample of reference 8. The measured absorption curve shows an exciton feature with about 5 Å full width at one-half maximum. This width corresponds to an uncertainity limited intensity decay time of 0.68 psec. The peak absorptivity is $4.8 \times 10^{-4} cm^{-1}$ The number of absorbed photons required to saturate a volume an absorption length long and a square wavelength in the material in cross section may be estimated to be 933, the sample volume divided by an exciton volume. The exciton is assumed to be a sphere of radius 140 Å. A sample with a slightly narrower line of 3.5 Å corresponding to a 1 psec decay using about 622 quanta will be a low temperature GaAs example.

We conclude the statistical limit of about 300 quanta can be reached in low temperature GaAs with an additional factor about 2 from resonator enhancement at high switching speeds if the light can be confined to very small regions. As explained above this same limit applies to room temperature electronic devices. About one year ago, experimental semiconductor chips with large numbers of transistors were using 5-50 fJ (roughly 10^{10} electrons) per transistor operation.

In addition, other innovations might be possible. Advocates of optical processing might point out that energy which is dissipated need not be dissipated in the devices themselves if efficient fluorescence or luminescence carries the energy away. This would allow larger optical device packing. Semiconductor advocates surely can present ideas favorable to semiconductor electronic logic. We focus on the single issue of energy requirements because we believe it to be one of the most important issues and because to try to do too much at once will lead to less done.

On a GaAs substrate quarter-wave alternate layers of AlAs and GaAs were deposited using molecular beam epitaxy techniques. Nine and one-half periods of such layers provided an interference mirror as described by van der Ziel and Iligems.[9] Then approximately 1.6 μm of GaAs were deposited followed by another mirror made of 7 periods of AlAs and GaAs. This monolithically grown sample yielded high quality bistable loops and optical logic etalon performance which was uniform over small (fractions of a square mm) areas. Gradients of AlAs and GaAs thickness were deliberately included in the design, because of uncertainties in growth rate, because of uncertainties in design thicknesses, and because different operations required different material thicknesses. More detail is given in reference 10.

The recovery time is limited by the carrier recombination time in the bulk material in such planar designs. To shorten recovery times without degrading the sharpness of the band edge feature, surface recombination has been used to reduce cycle times from several nanoseconds to about 70 psec.[11] With AlAs layers present that does not happen because carriers do not recombine on AlAs-GaAs interfaces. Furthermore, diffraction of light with a finesse of about thirty spreads the transverse optical beam inside the cavity to about 5-10 μm diameter. Carriers also diffuse this distance or farther. Both effects increase energy requirements, and introduce cross-talk.

The sample of reference 10 was etched using ion beam assisted etching techniques with an 5:2 Al:Cl$_2$ mixture at 8×10^{-4} Torr leaving an array of microresonators of various shapes and diameters, some of which are shown in Figure 1. With this geometry, carriers move to the surface of the "posts" and recombine allowing cycle times less than 150 psec as shown in Figure 2. Logic operation with 5:1 contrast has been observed with as little as 600 fJ in the control pulses. Arrays 4×8 in size show essentially uniform response. Posts 1.5 μm diameter separated by 3 μm center-to-center showed these performances with very little crosstalk. The operating wavelengths were typically 0.88 μm for the probe and 0.85 μm for the control pulses. Input pulse lengths were 5-10 psec.

Prior to etching, NOR gate operation at 5:1 contrast required 20 pJ control energy. After etching 1.5 pJ was required,[1] and after further work 600 pJ was achieved.

We compare these numbers with limits and expectations. It should be emphasized that experimental conditions are not perfect. For example, the full width (zero-to-zero) of the Airy disk for the lens used was very small, about 2.65 μm, but the wrong size for matching to 1.5 μm diameter posts.

Let us compare the 600 fJ result with the 300 quanta (0.07 fJ at 1.42 electron volts). The sample diameter of 1.5 μm can eventually be decreased to about 1/4 μm , about one wavelength in the material. This should reduce 600 fJ by a factor 36 to 17 fJ, where 36 is the ratio of cross sectional areas. We also anticipate conservatively a factor 4 due to design improvements

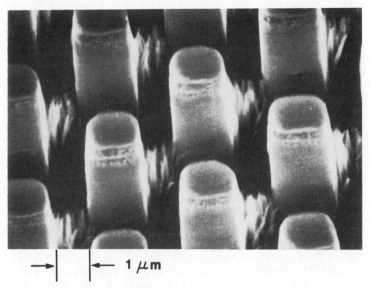

Fig. 1. Small section of the array showing ~2 μm devices used in the experiments. The slight bulges about 1 μm below the tops mark the positions of the 1.6 μm GaAs sections between mirrors.

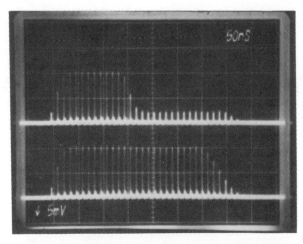

Fig. 2. NOR gate operation. The upper part shows the response of a ~1.75 μm post to 1.5 pJ input control pulses. The left side is the output with no input control pulses. The right side is the output with input control pulses. The lower part is the same as the upper except probe pulses were delayed 150 psec.

of various kinds and a factor 2 due to better coupling of the control beam. The 600 fJ then scales to about 2 fJ, still 30 times the statistical limit.

Room temperature GaAs, however, has different material parameters than low temperature GaAs. The exciton line width is about 30Å instead of 3.5 Å, yielding a factor 8.6. Furthermore, the nonlinear mechanism is not nearly as effective at room temperature because excitons quickly ionize creating carriers. Carriers screen the coulomb interaction, reducing the exciton feature and band edge coloumb enhancement, thus reducing the refractive index just below the gap. At the same time, band edge renormalization increases the refractive index just below the gap. The two contributions unfortunately largely cancel. We attribute a factor 10 to the remaining effect to obtain a net effect of a factor 86 potentional improvement by operating at low tempeatures. This may be compared with the factor 30 above.

We conclude at this time that room temperature GaAs will eventually allow switching energies of about 2 fJ. This energy is about 30 times the statistical limit. The 2 fJ figure may decline through a series of small improvements in GaAs, with a new idea, or through use of a better material. Potential better material candidates include narrow quantum well material superior to that available today, one-dimensional systems such as polydiacetylene, quantum dots (semiconductor of $\sim$50Å size) or special large molecules with very strong optical transitions.

The 150 psec cycle or recovery time is determined by the time required for carriers to diffuse to the surface of the 1.5 μm diameter "posts". Diameter reduction to 1/4 μm, the wavelength in the material, would reduce this time by a factor 36, the square of the two diameters if diffusive behavior is still dominant yielding about 5 psec cycle time.

In conclusion, a big step in the direction of creation of fast optical gates which work at the statistical limit has been made. Although not easy but possible reduction in diameter to a wavelength in the material should reduce the 600 fJ and 150 psec parameters to 17 fJ and 5 psec through simple scaling. Other improvements might reduce 17 fJ to 2 fJ. The 2 fJ energy is still about a factor 30 larger than the statistical limit. Other improvements may or may not be enough to reach the statistical and another material may be required.

REFERENCES

1. J. L. Jewell, A. Scherer, S. L. McCall, A. C. Gossard, and J. H. English, Appl. Phys. Letters 51, 94 (1987).

2. S. L. McCall, H. M. Gibbs, G. G. Churchill, and T. N. C. Venkatesan, Bull. Am. Phys. Soc. 20, 636 (1975); H. M. Gibbs, S. L. McCall, and T. N. C. Venkatesan, Phys. Rev. Letters 36, 1135 (1976).

3. R. W. Keyes, and J. A. Armstrong, Appl. Optics 8, 2549 (1969)

4. S.L. McCall and H.M. Gibbs, in "Optical Bistability", C.M. Bowden, M. Ciftan, and H.R. Robl, eds. (Plenum, New York, 1981), p. 1.

5. S. L. McCall, S.Ovadia, H. M. Gibbs, F. A. Hopf, and D. L. Kaplan, IEEE J. Quant. Elect. QE—21, 1441 (1985).

6. R. L. Fork, Phys. Rev. A 26, 2049 (1982).

7. P. W. Smith, Bell Syst. Tech. J. 61 1975 (1982).

8. H. M. Gibbs, A. C. Gossard, S. L. McCall, A. Passner, and W. Weigman, Solid State Comm. 30, 271 (1979).

9. J. P. van der Ziel and M. Ilegems, Appl. Opt. 14, 2627 (1975).

10. S. L. McCall, A. C. Gossard, J. H. English, J. L. Jewell, and J. F. Duffy, CLEO '86 Tech. Dig. (Optical Society of America, San Francisco, CA), paper FK3.

11. Y. H. Lee, H. M. Gibbs, J. L. Jewell, J. F. Duffy, T. Venkatesan, A. C. Gossard, W. Weigman, and J. H. English, Appl. Physics Lett. 49, 486 (1986).

OPTICAL BISTABILITY IN VACUUM-DEPOSITED

SEMICONDUCTOR FABRY-PEROT INTERFEROMETERS

S. P. Apanasevich F. V. Karpushko and G. V. Sinitsyn
Institute of Physics, BSSR Academy of Sciences
220602 Minsk, USSR

The interest in direct "light-light" transformation is provided by predicted and, in some cases, realized possibilities of subpicosecond switching in optical bistability (OB), for certain types of dispersive nonlinearity. This has no analogy in electronic systems. Moreover, the devices of purely optical switching and all-optical processing, even those which have no extreme properties, have some advantages such as the absence of connective conductors, insensitivity to noise, the ability to carry out simultaneous analysis of two-dimensional information signals, etc.

The first work which reported the OB-properties and possible applications of passive nonlinear Fabry-Perot interferometers appeared in 1969-71.[1,2,3] But it is only in 1975 that McCall, Gibbs and Venkatesan demonstrated experimentally an OB-device using as the nonlinear medium sodium vapor with a highly saturable transition at an excitation close to the resonance.[4,5] This successful OB-experiment showed that for nonlinear devices to operate in the OB-regime, a good dispersive nonlinearity was required.

Two of the co-authors of the present report observed in 1976, together with other researchers,[6,7] strong optical nonlinearity in multilayer semiconductor structures, i.e. thin-film semiconductor Fabry-Perot interferometers (TFS FPI). The TFS FPI is schematically represented in Fig. 1a. Such interferometers are made by a simple method of vacuum evaporation. The interaction of TSF FPI with laser radiation is largely explained by dispersive nonlinearity, which is very high in this case. Reversible changes in the refractive index of the TFS FPI internal layer amount to $\sim 10^{-2}$ at incident radiation densities of several kW/cm^2. In 1977, in such systems on ZnS optial bistability was reached.[8] We believe it was the first OB-experiment with semiconductor devices. After that we demonstrated the OB-regimes on the ZnSe, MgF$_2$ and Na$_3$AlF$_6$ systems. Similar experiments were also done by other authors (see, for example,[9,10]).

There is reason to discuss the investigations on OB and the conditions of it in the TFS FPI. The purpose of the work in this field was to study both the nonlinear interaction of lasr beams with TFS FPI and the possibility of practical realization and application of OB-devices.

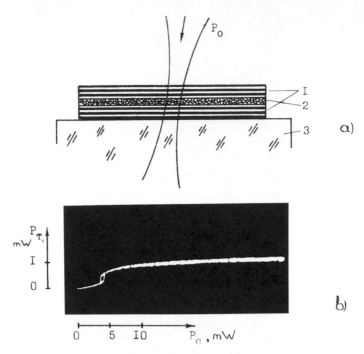

Fig. 1. Diagram (a) of the TFS FPI having the structure of an interference filter (1 - multilayer dielectric mirrors, 2 - internal semiconductor layer, 3 - substrate) and a typical hysteresis loop (b) for an optimized TFS FPI on ZnSe (P_O and P_T - input and output powers of the light beam, respectively.

Our later investigations showed that strong nonlinear response to laser irradiation is common for the broad class of thin-film interference structures based on different semiconductor (ZnS, ZnSe, CdS) and dielectric (MgF_2, LiF, Na_3AlF_6, SiO, LuF_3, YbF_3, BaF_2) materials. Even the films of such media as erbium, lutecium, and ytterbium fluorites, and cryolite which do not exhibit any optical nonlinearity when used as crystals have a strong nonlinear response. This fact points to the universality of the nonlinear interaction and its association with the specific properties of thin vacuum-deposited films. First of all, it became clear to us that it was necessary to take into consideration the structural peculiarities of such films. It has been found by electron microscopic and X-ray methods that the investigated films are polycrystal media with typical sizes of individual microcrystals of about 300-400 $\overset{\circ}{A}$.[11,12] An admixture of an amorphous component is present. As a result, the optical properties of the thin vacuum-deposited films differ from the source materials. The developed surface of the granular structure of the film, the boundaries between individual grains, and other peculiarities affect the polycrystal film response to the laser irradiation as compared to that in the initial monocrystal medium. The changes in the TFS FPI transmission peak and, therefore, in the internal layer refractive index are temperature-dependent and exhibit hysteresis during the heating-cooling cycle. This is shown in Fig. 2, which exactly corresponds to Fig. 7 from our reprint.[13] So, prolonged heating leads to an irreversible change in the TFS FPI transmission spectrum due to water evaporation from the pores, recrystallization, and other processes. After 5-6 cycles, which were carried out for a period of several hours the above irreversability disappears. The values of thermal changes in the optical thickness of the investigated layers obtained during such experiments are

476

essentially higher than those for the monocrystals of the source materials, and are identical with the refractive index changes observed under laser irradiation. Thus, the light-induced changes in the TFS FPI are interpreted as having a thermal origin. We understood the importance of the thermal processes in explaining the optical nonlinearity of the TFS FPI, in 1981-82.[13] However, we continue our efforts to find the conditions in which electron-origin nonlinearities may be observed, because it is very important for practical applications of the OB-devices. For instance, far from the edge of the fundamental absorption band, in about the middle of the forbidden enegy band of the ZnS- and ZnSe- films, additional electron processes such as luminescence have been observed.[14,15] These processes are absent in the initial ZnS- and ZnSe- monocrystals and are the result of the nonequilibrium population of the local energy states including the surface states which arise in the forbidden energy band because of the structure disorder of the polycrystal film.

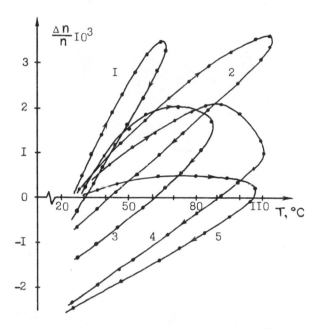

Fig. 2. Changes in the refractive index of spacers at quasistationary heating-cooling. 1 - ZnSe; 2 - Na$_3$AlF$_6$; 3 - ZnS: 4 - ZnSe (another technique); 5 - MgF$_2$.

The indisputable quality of the OB-devices based on TFS FPI is their ability to continuously operate at room temperature. The OB element can remain in one of its steady states for several hours until the switching pulse occurs. In this way, the utilization has been experimentally demonstrated of OB thin-film interferometers as devices for reversible writing and storage of digital optical signals. The typical values of the laser power required to reach the bistable response are about 15 mW (or ~ 800 W/cm^2). By optimizing the initial parameters of the ZnSe OB device the value of the switching-on power of ~ 4 mW has been obtained (see Fig. 1b). The space distribution of both the laser beam and the OB system properties leads to new effects which accompany the transition from one steady state to another of the OB device interacting with a spatially non-uniform light beam. Space hysteresis or the hysteresis of the beam profile is one of such effects.[16] This hysteresis phenomenon is accompanied by switching autowaves at whose boundaries the OB system is in two different steady states. The physical cause of similar transverse effects in the OB is the diffusion of both the medium parameters and the light field (diffraction). The first experimental observation of the hysteresis of the transverse distribution form of a limited light beam passing through the OB-device has been made in our papers.[17,18] In the case of the Gaussian input beam the switching-on proceeds as follows. As the on-axis intensity of the light beam becomes equal to the threshold value, the switching-on of the on-axis zone occurs. Then the disturbance from this zone begins to propagate to the edges of the light spot. This is an autowave process called the switching wave. Fig. 3a shows three random phase of the switching-on wave. The switching-on wave front rapidly reaches its spatially steady state at an intensity value close to the switching-off threshold but a bit higher. In the input power continues to increase the size of the switching-on zone also increases in the quasistationary manner. The switching-off process is somewhat different. The point is that the switching-off wave starts from the initial switching-on distribution of the beam intensity with the amplitude slightly exceeding the plane wave switching-off value of the intensity.

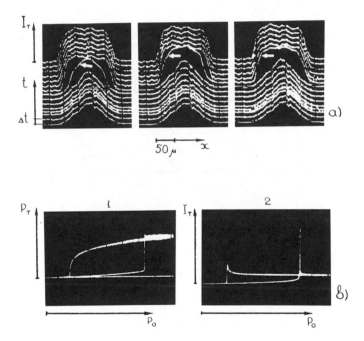

Fig. 3. Cross-sections (a) of the output beam profile $I_T(x)$ at time intervals $\Delta t = 40$ ms (arrows point to random phases of the switching-on wave) and the hysteresis loop (b) of integral output power $P_T(1)$ and on-axis output intensity $I_T(2)$. P_o - integral input power.

Thus, there occurs a relatively narrow interval of input intensities inside the hysteresis loop when the sign of the switching wave is "-" unlike the case where the sign of the switching wave is "+". Fig. 4 shows both the stationary switch-off wave (a) and the optical control of the switching wave near steady state (b). Of course, the switching autowaves substantially change the differential kinetics of the OB device transition from one steady state to another (see Fig. 3b) which shows both the integral and the on-axis hysteresis loops for the same Gaussian input

beam.)   These experimental results are in good agreement with the
theoretical results of Rozanov et al..[16,19,20]   So, omitting some details,
the hysteresis of the transverse distribution form of the spatially non-
uniform light beam interacting with the distributed OB system is as
follows: According to the prehistory of the OB system the output beam has
two different types of intensity profile at the same input.   The practical
importance of the transverse effect kinetics in the OB of the thin-film
interference structures is determined by the possibility of using these
effects to transport the information signal in the plane of the same
etalon on which the OB elements are formed.

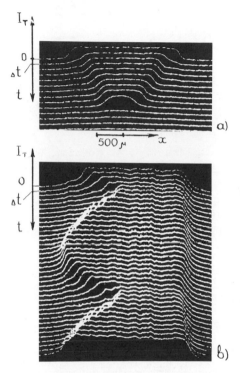

Fig. 4.   The kinetics of the transverse distribution of the output beam
          intensity $I_T(x)$ after a decrease in the uniformly distributed
          input power from 7 mW to 6.5 mW, Δt = 40 ms (a).   The moving
          kinetics of the switching wave by modulation of the input power
          in the range from 6 to 7 mW according to the triangle law with a
          period of 880 ms (b).

     One of the important parameters of the OB device is its speed of
response.   When studying the properties of the optical flip-flop based on
the TFS FPI, recommendations have been worked out to increase the
response speed of the OB devices.   These recommendations take account of
both the thermal origin of the nonlinearity and the beam profile
hysteresis.   In this way a time of switching-on of about 500 ns was
reached.[17]   The complete set of optical logic devices for all-optical
computing based on the ZnS and ZnSe TFS FPI was realized experimentally.[21]
On the same basis an optical transistor with gain factor of 20 dB/act, a
regenerative oscillator, and, what is more, an optically controlled
processor elements have been demonstrated.

We believe that of the several optically bistable devices which claim to be the basic components in elaborating and modelling all-optical computer architectures, thin-film interference structures made from semiconductor materials by vacuum deposition may be considered as playing one of the leading roles. Practical interest in these structures can be attributed to the technological ease of their fabrication, the micron size of an individual OB element, the low holding power, the ability of operating continuously for many hours at room temperatures, the possibility of forming two-dimensional integro-optical schemes on etalons with high uniformity over areas of ~ 10 cm², the free choice of controlling beam wavelengths in a wide spectral range (0.4 - 1.1 μm).

References

1. A. Szoke, V. Daneu, J. Goldhar, N. A. Kurnit, Appl. Phys. Lett. 15:376 (1969).
2. J. W. Austin, L. G. DeShazer, J. Opt. Soc. Amer. 61:650 (1971).
3. E. J. Spiller, Opt. Soc. Amer. 61:669 (1971).
4. S. L. McCall, H. M. Gibbs, T. N. C. Venkatesan, J. Opt. Soc. Amer. 65:1184 (1975).
5. H. M. Gibbs, S. L. McCall, T. N. C. Venkatesan, Phys. Rev. Lett. 36:1135 (1976).
6. F. V. Karpushko, A. S. Kireev, I. A. Morozov, G. V. Sinitsyn, N. V. Stryjenok, Abstracts of the VIII Conference on KiNo. Tbilisi 1:78 (1976).
7. F. V. Karpushko, A. S. Kireev, I. A. Morozov, G. V. Sinitsyn, N. V. Stryjenok, Zh. Prikl. Spektrosk. 26:269 (1977).
8. F. V. Karpushko, G. V. Sinitsyn, Zh. Prikl. Spektrosk. 29:820 (1978).
9. D. A. Weinberger, H. M. Gibbs, C. F. Li, M. C. Rushford, JOSA 72:1769 (1982).
10. S. D. Smith, J. G. H. Mathew, M. R. Taghizadeh, A. C. Walkner, B. S. Wherrett, A. Hendry, Opt. Communs. 51:357 (1984).
11. S. P. Apanasevich, O. V. Goncharova, F. V. Karpushko, G. V. Sinitsyn, Izv. AN SSSE, ser. fiz. 47:963 (1983).
12. O. V. Goncharova, S. A. Porukevich, Abstract of the VIII Conference of Young Researchers on Spectroscopy and Quantum Electronics, Vilnius p. 121 (1987).
13. S. P. Apanasevich, F. V. Karpushko, G. V. Sinitsyn, reprint of the Institute of Physics, BSSR Academy of Sciences, No. 265, Minsk (1982).
14. O. V. Goncharova, F. V. Karpushko, G. V. Sinitsyn, Fiz. Tekhn. Poluprovodnikov 20:1750 (1986).
15. S. P. Apanasevich, O. V. Goncharova, F. V. Karpushko, G. V. Sinitsyn, Zh. Prikl Spektrosk. 47:2:272 (1987).
16. N. N. Rozanov, V. E. Semenov, Optika i Spektrosk. 48:108 (1980).
17. S. P. Apanasevich, F. V. Karpushko, G. V. Sinitsyn, Kvant. Elektron. 11:1274 (1984).
18. S. P. Apanasevich, F. V. Karpushko, G. V. Sinitsyn, Kvant. Elektron. 12:387 (1985).
19. N. N. Rozanov, V. E. Semenov, V. V. Khodova, Kvant. Elektron. 9:354 (1982).
20. N. N. Rozanov, Proc. of the Vavilov GOI 59:193 (1985).
21. G. V. Sinitsyn, Kvant. Elektron. 14:3:523 (1987).

OPTICAL BISTABILITY WITHOUT OPTICAL FEEDBACK AND

ABSORPTION-RELATED NONLINEARITIES

Elsa Garmire

Center for Laser Studies
University of Southern California
Los Angeles, CA   90089-1112

ABSTRACT

We consider several means for obtaining optical
bistability without specific optical feedback. We show recent
results on absorption-induced optical bistability in single-
crystal ZnSe waveguides. We also describe some recent results
on the absorption saturation in GaAs/GaAlAs multiple quantum
well materials grown by Metalorganic Chemical Deposition.

INTRODUCTION

For the purpose of this paper, we define optical
bistability as any optical device or system with two (or more)
optical outputs for one input, depending on the history of
excitation. The first optically bistable devices were two
lasers pointed at each other [1] which led to the classic
investigation of bistable lasers for optical computing [2].
More recently emphasis has focused on the nonlinear Fabry-Perot
etalon [3]. One example of the research in this area is the
study of bistability in reflection performed at a wavelength of
3 um in InAs with an HF laser [4]. A typical bistability curve
generated in this experiment is shown in Fig. 1. Thresholds as
low as 3 mW (I_{th} = 75 W/cm^2) were measured.

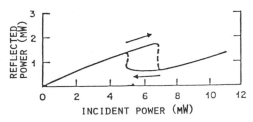

Fig. 1. Experimental measurement of optical bistability in
InAs, shown by plotting reflected power as a function of
incident power, from a Nonlinear Fabry-Perot [After ref. 4].

In the near infrared and visible, the threshold for bistability becomes considerably higher than in the 3 - 5 μm wavelength region, suggesting that perhaps the nonlinear Fabry-Perot devices may be more useful in the longer wavelength regions. It is useful, therefore, to explore some of the techniques other than the nonlinear Fabry-Perot which may be used to generate optical bistability and switching.

HYBRID BISTABLE DEVICES

In 1978 Garmire and Marburger introduced the idea that any optical device whose transmission is a function of a variable (such as voltage) may be made bistable by feeding some of the output back as a voltage across the device [5]. This can be seen by considering the experimental arrangement of Fig. 2. Since the voltage across the modulator is proportional to the light power output I_o (with the possible addition of a bias voltage, V_B) the following expression relates the incident power I_{in} to the output power through the transmission T:

$$I_o / I_{in} = T(aI_o + V_B),\qquad\qquad(1)$$

where a is the proportionality constant which relates the output power and the applied voltage. The factor a contains any system amplification.

The solution to Eq. 1 can be obtained by plotting the right hand side and left hand side on the same graph, both as a function of output power (represented by the applied voltage), as shown in Fig. 2b. The solid line in Fig. 2b represents the experimental transmission as a function of voltage. Under certain conditions of input power (slopes of the straight line within the two dashed curves of Fig. 2b), there are more than one intersection. The intersection of opposite slopes is stable, but intersections of the same slope lines are unstable against small perturbations. Thus, this describes a bistable system. Its advantage is that amplification may be introduced electronically, improving the sensitivity. Figure 2c shows the experimental result for the $LiNbO_3$ electro-optic waveguide modulator with the transmission shown in Fig. 2b, using only a small HeNe laser [6]. Hybrid devices can be constructed with no external voltage applied [7] and recently have been suggested in GaAs multiple quantum wells (MQW) [8].

INTRINSIC OPTICAL BISTABILITY

Under some conditions, no external feedback is required and the bistability may be intrinsic to the medium. This cannot occur if the medium response depends uniquely on the electric field. Rather, it requires the expression of the electric field within the medium in terms of the medium response through nonlinear constituitive relations. This is analogous to ferromagnetism in which the magnetization defines the magnetic field and hysteresis results. Domains describe the local magnetization and reflect the long range forces within the medium.

An example which we have considered is nonlinear molecular vibrations [9], in which the polarization can be described by atomic displacement vectors x, which obey Duffing's anharmonic oscillator equation:

$$P = P_O + Nex, \tag{2}$$

with $\qquad \ddot{x} + i\dot{x}/\tau + \omega_O{}^2\chi - ax^3 = -eE/m,$

where N is the molecular density, ω_O the resonant frequency, τ the lifetime of the state, E the optical field and a is the anharmonic term in the molecular potential.

In steady state

$$P(A - a|P|^2) = -eE/m, \tag{3}$$

where $\qquad A = \omega_O{}^2 - \omega^2 - i\omega/\tau.$

We see that in this system the electric field is a function of the polarization and that the polarization cannot be expressed as a Taylor series in the electric field.

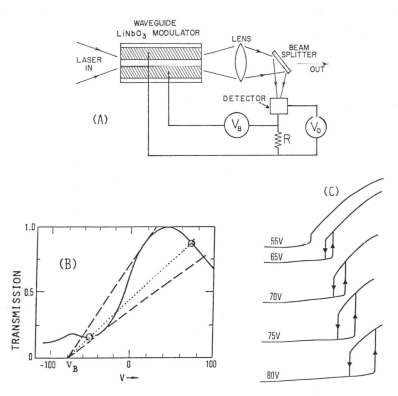

Figure 2. Hybrid electro/optical bistable device. A) Experimental setup, with I_{in} incident on a modulator whose transmission is a function of voltage. Some portion of the output light is detected and fed back through amplifier A as a voltage onto the modulator. B) Transmisison of the experimental modulator as a function of voltage. Regions of bistability lie between two lines (of slope ~ $1/I_{in}$ originating at V_B which intersect the transmission in more than one point. C) Experimental transfer curves (output vs. input) for several bias voltages [After Ref. 6].

Re-expressing in terms of a normalized intensity I

$$|P|^2[|A|^2 - 2a(ReA)|P|^2 + a|P|^4] = I.$$ (4)

This equation is cubic in $|P|^2$ and is plotted in Fig. 3. The maximum and minimum in the curve of I vs. $|P|^2$ can be calculated by setting the derivative, $dI/d|P|^2$ equal to zero. When $|A| \sim ReA$, we find $I_{MAX} = 4A^3/27aB$, and $I_{MIN} = 0$. This means that the intensity required to see bistability, I_{MAX}, is inversely proportional to the amount of anharmonicity. A sufficiently anharmonic system has not yet been investigated, and there are no reports of experimental observations of this effect.

LOCAL FIELD CORRECTION

One non-local effect which has been explored theoretically is the local field correction [10]. The Lorentz correction introduces the concept of the local field, E_{LOC}, in terms of an applied field, E_{APPL}, through

$$E_{LOC} = \frac{E_{APPL}}{f[\epsilon_{CAV}/\epsilon_B - 1] + 1},$$ (5)

where ϵ_{CAV} is the dielectric constant inside the cavity and ϵ_B is the dielectric constant of the background medium outside the cavity. The factor f is chosen by geometry. When the geometry is a sphere, f = 1/3. When a platelet, such as a portion of a multiple quantum well (MQW), f = 1.

When a nonlinear medium is placed within the cavity, $\epsilon_{CAV} = \epsilon_C + \Delta\epsilon_{NL}$ and

$$\Delta\epsilon_{NL} = n_2 I_{LOC} = \frac{n_2 I_{APPL}}{[f(\epsilon_C/\epsilon_B - 1) + \Delta\epsilon_{NL}/\epsilon_B + 1]^2}.$$ (6)

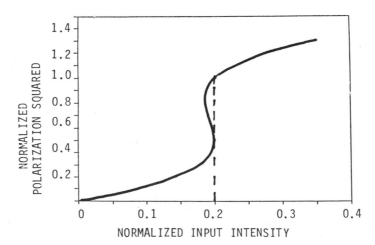

Fig. 3. Bistability in the square of the polarization as a function of intensity for a nonlinear molecular vibration. All units are normalized; in eq. 4, A = 1 and a = 0.8. Threshold for bistability is a normalized input intensity I = 0.2.

This equation gives the applied intensity as a cubic function of the nonlinear dielectric constant, $\Delta\epsilon_{NL}$:

$$I_{APPL} = X(A-X)^2 \qquad (7)$$

where $\qquad A = f\epsilon_C/\epsilon_B + 1 - f.$ $\qquad X = -\Delta\epsilon_{NL}f/\epsilon_B,$

and $\qquad I = -n_2 I_{APPL} A/\epsilon_B.$

The form of this equation is shown in Fig. 4. Simple differentiation shows that the turning point in intensity is given by $I_{MAX} = A/3$, resulting in

$$n_2 I_{MAX} = 4\epsilon_C^3/27\epsilon_B^2 f.$$

If $f \sim 1/3$, $\epsilon_C \sim \epsilon_B$, then $n_2 I_{MAX} \sim \epsilon_C/2 \sim 1.$

Thus the nonlinearity required to see the local field correction must lead to $\Delta n = n_2 I \sim 1$, which is much too large to be of practical significance.

ROLE OF SPATIAL DISPERSION

The existence of non-local material response, which leads to the concept of spatial dispersion, arises from the transport of excitation. We can define a nonlinear spatial dispersion relation:

$$P_{NL}(r) = X_{NL}(r)E(r) = \int X_{NL}(r,r')E(r')dr'.$$

That is, optical excitation at a point r' must be able to induce an optical nonlinearity at point r. A number of potential systems may be explored, some of which have already been shown to exhibit bistability. These include effects at surfaces and interaction with polaritons, ferroelectrics, orientational forces such as seen in liquid crystals and other large molecules. Of course, materials at their phase

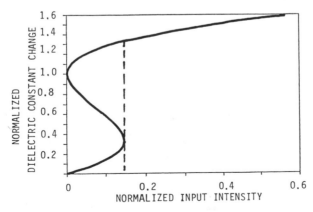

Fig. 4. Bistability in the dielectric constant change due to the local field correction inside a cavity. The threshold I_C requires a nonlinear refractive index change ~ 1, before this effect is important.

485

transition threshold may exhibit optically bistable effects. Other transport mechanisms include carrier excitation in semiconductors; charge transfer such as in the photo-refractive effect; and excitation diffusion, including radiation trapping. The effect most recently studied experimentally for the observation of optical bistability is thermally induced.

BISTABILITY DUE TO INCREASING ABSORPTION

Bistability due to increasing absorption has been seen in a number of semiconductors at the band edge [11]. This effect can be understood by defining the field through a temperature-dependent absorption, $a(T)$. The temperature increase T from ambient, T_o, is proportional to the incident light power, P, so that

$$I = \frac{\Delta T}{g[1-e^{a(T_o+\Delta T)L}]},\qquad (8)$$

where g is a constant which depends on thermal conductivity and geometry. To understand the bistability, write

$$a(T) = a_o + a_1(\Delta T)\qquad (9)$$

where a_o is the absorption at ambient and $a_1(\Delta T)$ is the functional dependence of absorption on temperature. This equation can be inverted to obtain a functional form:

$$\Delta T = f(a-a_o).\qquad (10)$$

This means the temperature increase is a function of the absorption increase. Inserting this into Eq. (8) gives

$$I = f(a-a_o)/g[1-e^{-aL}].\qquad (11)$$

If f is a suitably nonlinear function of a, bistability results.

As an example, consider the temperature dependence of absorption at the band edge of a semiconductor, which is usually exponential:

$$a(T) = a_o + a_1 \exp(\Delta T/T_c)\qquad (12)$$

or

$$f(a-a_o) = T_c \ln[(a-a_o)/a_1].$$

Through Eqs. (11) and (12), the functional form of a vs. intensity is shown in Fig. 5 as the solid line. It can be seen that for a given intensity, two stable values of a result. The dashed line is included in order to demonstrate bistability in the output light as a function of input light. The simultaneous intersection of the two curves corresponds to the following condition:

$$I_{out} = I_{in}[1-a(I_{out})L]$$

so we write

$$1 - I_{out}/I_{in} = a(I_{out})L.\qquad (13)$$

That is, the absorption is determined by the amount of light leaving the slab, and the left hand side of the equation is a straight line with slope proportional to $1/I_{in}$. This equation should be valid for a thin slab.

We have performed experiments to demonstrate this bistability in ZnSe single crystal waveguides fabricated by MOVPE on GaAs substrates [12]. The observed optical switching and hysteresis, introduced by a focused argon laser, are shown in Fig. 7. The threshold inside the guide is 30 mW, with a 10 μsec time constant.

SATURABLE ABSORPTION AT THE BANDEDGE IN GaAs/GaAlAs MQW

The interest in nonlinearities in semiconductors at their band edge has led us to investigate multiple quantum wells of GaAs/GaAlAs fabricated by Metal-Organic Chemical Vapor Deposition [13]. We have measured the spectral dependence of nonlinear absorption as shown in Fig. 8, obtaining results similar to those seen in MQW's fabricated by molecular beam epitaxy [14]. From careful measurements of the saturation as a function of intensity over several orders of magnitude, we have been able to fit the data by adding with two roughly equal contributions, one for exciton bleaching and one for band filling. The two saturable absorptions have saturation intensities which differ by more than an order of magnitude. Their values for the 100 A MQW's are:

$$a = \frac{a_0/2}{1+I/250} + \frac{a_0/2}{1 + I/9500} \quad , \quad \text{where I is intensity in W/cm}^2.$$

Measurements are in progress for wells of varying thickness and will be published soon.

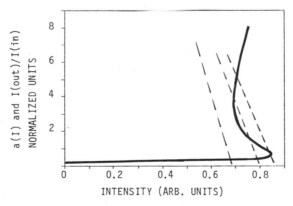

Figure 5. Calculations of absorption as a function of intensity (solid line), as well as straight lines of slope $1/I_{in}$ (dashed line). The intersections of these two curves determines the operating points. The existence of two such points indicates optical bistability.

The work reported here was supported in part by ARO, AFOSR, ONR AND NSF. The author acknowledges the many students and colleagues who were collaborators in much of this work. All are authors in the referenced papers.

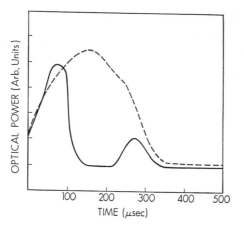 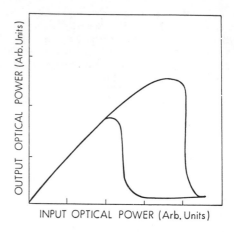

Figure 6. Experimental results for a pulse of Argon laser light focussed into a ZnSe single crystal waveguide. A) Input and output as a function of time. B) Output vs. input, demonstrating optical bistability [Ref. 12].

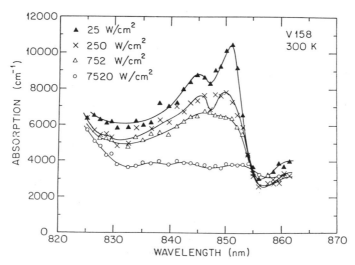

Figure 7. Curve of absorption as a function of wavelength for a GaAs/GaAlAs multiple quantum well [13]. The top curve corresponds to low intensity results. Successively lower curves occur when the input power is successively increased, indicating saturable absorption [Ref. 13].

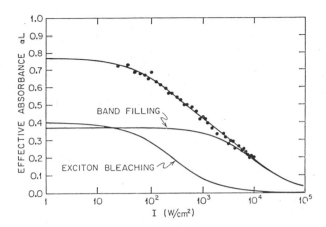

Figure 8. Absorption as a function of intensity, for a wavelength on the exciton resonance. The experimental curve is fit by two saturable absorptions, which are shown as the two lower curves [Unpublished data accumulated by M. Kawase, A. Kost and C. H. Lee].

REFERENCES

1. G. J. Lasher and A. B. Fowler, "Mutually quenched injection lasers as bistable devices", IBM J. Res. and Dev. **8**, 471 (1964) A. B. Fowler, "Quenching of GaAs injection lasers" Appl. Phys. Lett. **3**, 1, 1963. G. J. Lasher, "Analysis of Proposed Bistable Injection Lasers" Solid State Electronics **7**, 707, 1964

2. R. W. Keyes and J. A. Armstrong, Appl Opt. **8**, 2549 (1969)

3. See, for example, the special issue of Journal of Quantum Electronics on Optical Bistability, edited by Garmire, October, 1985.

4. C. D. Poole and E. Garmire, "Bandgap Optical Nonlinearities in InAs and their Use in Optical Bistability", IEEE J. Quantum Electr. **QE-21**, 1370 (1985)

5. E. Garmire, J. H. Marburger and S. D. Allen, "Coherent Mirrorless Bistable Optical Devices", Appl. Phys. Lett. **32**, 320 (1978)

6. E. Garmire, J. H. Marburger, S. D. Allen, C. Verber, Opt. Lett. **3**, 69 (1978)

7. P. W. Smith, I. P. Kaminow, P. J. Maloney, L. W. Stulz, Appl. Phys. Lett. **34** 62 (1979)

8. D. A. B. Miller, D. S. Chemla, T. C. Damen, T. H. Wood, C. A. Burrus, A. C. Gossard, W. Wiegmann, IEEE J. Quantum Electr. **QE-21**, 1462 (1985)

9.   J. A. Goldstone and E. Garmire, "Macroscopic Manifestations of Microscopic Bistability", Phys. Rev. Lett., $\underline{53}$, 910 (1984)

10.   I. Abram and A. Maruani, Phys. Rev. $\underline{B26}$, 4759 (1982); F. A. Hopf, C. M. Bowden, W. H. Louisell, Phys. . $\underline{A29}$, 2591 (1984);   K. M. Leung, Phys. Rev. $\underline{A33}$, 2461 (1986);   D. S. Chemla and D. A. B. Miller, Opt. Lett. $\underline{11}$, 522 (1986).

11.   J. Hajto, I. Janossy, Philos. Mag. $\underline{B47}$, 347 (1983);   M. R. Taghizadeh, I. Janossy and S. D. Smith, Appl. Phys. Lett. $\underline{46}$, 331 (1985); K. Bohnert, H. Kalt and C. Klingshirn, Appl. Phys. Lett. $\underline{43}$, 1088 (1983); H. Rossmann, F. Henneberger and J. Voight, Phys. Stat. Sol. $\underline{B115}$, K63 (1983);   M. Dagenais and W. F. Sharfin, Appl. Phys. Lett. $\underline{45}$, 210 (1984);   D. A. B. Miller, A. C. Gossard and W. Wiegmann, Opt. Lett. $\underline{9}$, 162 (1984).

12.   B. G. Kim, E. Garmire, N. Shibata and Z. Zembutsu, Appl. Phys. Lett., August, 1987

13. H. C. Lee, A. Hariz, P. D. Dapkus, A. Kost, M. Kawase and E. Garmire, Appl. Phys. Lett. $\underline{150}$, 1182 (1987)

14.   Miller et. al. Appl. Phys. lett. $\underline{41}$, 679 (1982)

OPTICAL FREEDERICKSZ TRANSITIONS AND ASSOCIATED EFFECTS IN LIQUID CRYSTALS

Y. R. Shen

Department of Physics
University of California
Berkeley, California 94720 USA

Liquid crystals have extraordinary nonlinear optical properties.[1] The nonlinearity arises from a high degree of electron delocalization in the molecules, the large anisotropy of the molecular structure, and the strong correlation of molecular motion under the influence of light. Propagation of laser beams in liquid crystals can lead to very unusual, but interesting, nonlinear optical effects.[2] The effects are often so strong that perturbation theories are no longer applicable. In this respect, liquid crystals as nonlinear optical media are rather unique for studies of some very highly nonlinear optical phenomena. Among these, the optical-field-induced orientational structural transitions, usually known as the optical Freedericksz transitions, are probably the most interesting. Thus, in view of the page limitation, we shall focus our discussion on the optical Freedericksz transitions in this paper.

Let us begin with a brief introduction on the orientational nonlinearity of liquid crystals.[1,2] The optical properties of liquid crystals can be easily modified by a dc field through reorientation of the highly anisotropic molecules. This is particularly true for the mesophases owing to the very strong correlation among molecules. The situation is closely analogous to the spins in a ferromagnetic phase. Typically, a dc field of E ~ 100 v/cm or H ~ 0.1 T is sufficient to induce a significant molecular reorientation and lead to a refractive index change, Δn, as large as 0.01 to 0.1. For molecular reorientation, an optical field is equivalent to a dc field if there is no permanent dipole on the molecules. Since a field of 200 v/cm should correspond to a beam intensity of 100 W/cm^2, we expect that a significant Δn can be readily induced in a liquid crystal by a CW laser beam.

If the applied field is in a direction perpendicular to the molecular alignment, the molecular reorientation can only happen when the field is above a certain threshold value. This is generally known as the Freedericksz transition. As a structural transition, it possesses all the characteristic behavior of a phase transition. The dc Freedericksz transitions have been well studied in a variety of liquid crystal films.[4] More recently, it has been demonstrated that the optical Freedericksz transitions can also be readily observed in a homeotropically aligned nematic film with a CW laser beam.[3]

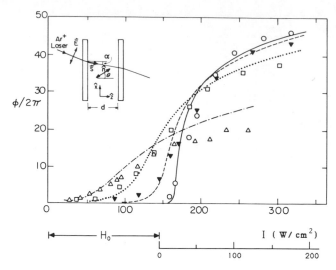

Fig. 1 Experimental points and theoretical curves for the laser-induced birefringence at different incidence angles α: circle and solid curve, α = 0°; solid triangles and dashed curve, α = 3°; squares and dotted curve, α = 11°; open triangles and dot-dashed curve, α = 30°. Inset shows the experimental geometry. The lower absissa denotes how a dc bias field H_0 along $\hat{x}$ helps reorient the molecules.

Figure 1 shows the results of an experiment on optical Freedericksz transition in a 4-cyano-4'-pentylbiphenyl (5CB) film with a linearly polarized CW Ar⁺ laser beam. The induced phase shift Δφ in the figure is given by $\Delta\phi = \int_{-1/2d}^{1/2d} (\omega/c)\Delta n\,dz$, where Δn as a function of z arises from the induced molecular reorientation. It is seen that for the case of a normally incident beam, there indeed exists a threshold for molecular reorientaiton. The transition is of the second order and is characterized by the critical slowing-down behavior near the threshold.[3] Theoretical calculations based on the formalism of free energy minimization agree well with the experiment. We note that Δn and Δφ are highly nonlinear functions of the laser intensity, not obtainable by perturbation calculations.

The threshold of the optical Freedericksz transition can be greatly reduced with the help of a dc bias field. As shown in Fig. 1, a proper dc bias field E_0 along the direction of the linear polarization can reduce the threshold to ~ 10 W/cm². A 20 W/cm² beam can then induce a phase shift Δφ ~ 40 π rad or Δn ~ 0.1. This corresponds to an extremely large optical nonlinearity. As a consequence, many highly nonlinear optical effects which are difficult to realize in other media can be readily observed with relatively weak CW laser beams.[2] These include[1] multi-order wave mixing, multiple loops in optical bistability, self-focusing and self-phase modulation, and optical transistor action.[5] They make nonlinear optics in lqiuid crystals unique and interesting.

What would happen if instead, the dc bias bield is applied along the molecular alignment? Clearly, the threshold for the optical Freedericksz transition in a homeotropic film should increase. This is because the dc field makes the molecular reorientation by the optical field more

492

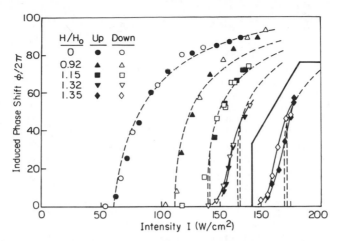

Fig. 2　Phase retardation $\phi/2\pi$ vs intensity for fixed magnetic field
strengths.　Open symbols were measured with increasing intensity;
solid symbols, with decreasing intensity.　Solid lines were drawn
to aid visualization of the data.　Broken lines are the
theoretical fits with $k_{11} = 9.00 \times 10^{-7}$ dyn, $k_{33} = 9.51 \times 10^{-7}$
dyn, $n_\perp = 1.54$, $n_\parallel = 1.73$, $\lambda = 514.5$ nm, and $d = 380$ μm.　The H/H_0
values are determined from the fit for each curve.　The
experimental H/H_0 values are 0, 0.92, 1.13, 1.20, and 1.26

difficult.　Experimental results, presented in Fig. 2, are in good
agreement with this prediction.[6]　One could say that the dc field has made
the anisotropy of the liquid crystal medium effectively larger.　In a
number of recent publications,[7] it has been shown that if the medium has a
sufficiently large anisotropy, the optical Freedericksz transition can
become first-order.　Unlike the dc case, the torque exerted on the
molecules by the optical field is proportional to $(1 - u \sin^2\theta)^{-3/2}\sin\theta$
$\cos\theta$, where θ is the molecular tilt angle from the surface normal, and $u =$
$1 - n_\perp^2/n_\parallel^2$, with $n_\perp$ and $n_\parallel$ being the refractive indices perpendicular and
parallel to the molecular alignment, respectively.　The $(1 - u \sin^2\theta)^{-3/2}$
term acts as a positive feedback in the molecular reorientation, that is,
the torque on the molecules increases with the increase of θ as long as θ
is not too large.　This positive feedback becomes larger with larger
optical anisotropy.　If it is sufficiently strong, then a first-order
transition will result.　The fact that a bias field can enhance the
anisotropy suggests that by increasing the bias field, we should be able
to change the optical Freedericksz transition from second-order to
first-order.[8]　Figure 2 shows that this is indeed the case.[6]　When the
bias magnetic field H becomes larger than the tricritical field H_0 (~ 210
Oe), a hysteresis loop appears in the induced phase shift, which is
characteristic of a first-order transition.　The results agree fairly well
with the theoretical calculations.　We note that the hysteresis loop
actually manifests a mirrorless intrinsic optical bistability behavior.[9]

Another important difference between dc and optical Freedericksz
transitions lies in the fact that the optical beam polarization is not
necessarily linear and can vary continuously in a medium.　Consequently,
optical Freedericksz transitions are much more colorful in their behavior
and can lead to many interesting phenomena.　As an example, we consider
here the case of a circularly polarized input beam normally incident on a
homeotropically aligned nematic film.

It can be easily shown that the threshold for the optical Freedericksz transition increases by a factor of 2 if the beam polarization changes from linear to circular.[10] Above the thresold, the circularly polarized light would reorient the molecules by tilting them away from the surface normal and spreading them randomly in the azimuthal plane. Minimization of elastic energy, however, requires the molecules to be tilted in a single direction. In reality, the conflict is resolved by the existence of some residual anisotropy in the system which defines the direction of the molecular tilt. As a result of this molecular reorientation, the medium appears birefringent to the input beam, and the beam evolves into elliptical polarization in the medium. If the beam intensity is now reduced to a value below the transition threshold for circularly polarized light, we should expect the molecular orientation to remain in a distorted state, because the transition threshold for elliptically polarized light is lower. This will last until the elliptically polarized beam in the medium is no longer capable of sustaining the molecular reorientation, at which point the molecules will return precipitously to the homeotropical alignment. The above qualitative picture has actually been observed in our recent experiment.[11] As seen in Fig. 3, with increasing I, the ellipticity of the output beam (which is an indication of the molecular reorientation in the nematic film) jumps suddenly at $I = I_{th}$ and then, as I decreases, it drops precipitously at $I = 0.88\ I_{th}$. The hysteresis loop here is another illustration of an intrinsic optical bistable behavior.

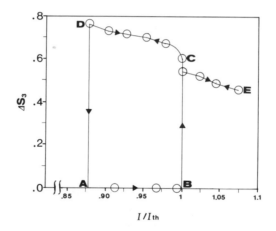

Fig. 3 Ellipticity change ΔS_3 of the laser beam traversing the sample as a function of the normalized light intensity. The branch CD is characterized by a stable uniform rotation of the output polarization ellipse

Changing of a circularly polarized input beam into an elliptically polarized output beam means that the optical field has deposited a finite amount of angular momentum in the medium. This transfer of angular momentum results in the exertion of a torque on the medium and causes the liquid crystal molecules and hence the birefringent axis to precess around the surface normal. The output polarization ellipse, following the molecular precession, should rotate with the same velocity. The rotational angular velocity can be calculated simply from the rate of the angular momentum transfer to the medium, and can be related directly to the ellipticity of the output beam. Such an induced polarization rotation has actually been observed in our experiment.[11] The measured angular velocity in the range $0.88\ I_{th} < I < I_{th}$ agrees quite well with the value calculated from the observed ellipticity of the output beam, as shown in Fig. 4. For $I > I_{th}$, the molecular orientation tends to break into a combined motion of precession and nutation. The nutation may be the result of a larger average birefringnece induced in the medium. The overall dynamics then becomes so complex that a complete theory is yet to be found to explain the results quantitatively.

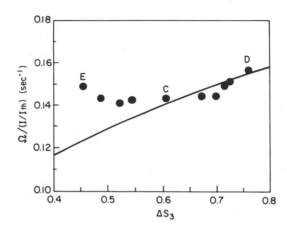

Fig. 4 Angular velocity Ω (normalized against I/I_{th}) of the output polarization ellipse as a function of the ellipticity change ΔS_3 across the sample. The continuous lines obtained from theory

The angular momentum transfer should be accompanied by an energy transfer from the beam to the medium. Since the medium is transparent, this can happen only if part of the beam is down-shifted in frequency. Indeed, the rotation of the polarization ellipse with an angular velocity Ω means that the two circularly polarized components of the elliptical polarization have different frequencies ω and ω' with $\omega - \omega' = 2\Omega$. We can therefore regard the induced polarization rotation as a stimulated light scattering process in which a new frequency component at ω' is generated.[11] Unlike other stimulated scattering processes, hwoever, the frequency shift, 2Ω, in the present case is not simply a characteristic constant of the medium, but depends on the incoming laser beam intensity.

In conclusion, we have seen how the study of optical Freedericksz transitions can lead to many interesting results. So far, only optical Freedericksz transitions with a normally incident beam in homeotropically aligned nematic films have been carefully investigated. Those with other beam geometries or molecular alignment are yet to be explored. Interesting results can be anticipated in general in the nonlinear dynamic behavior of the system resulting from the unusually strong laser-matter interaction.

The work described in ths paper came from the collaborative effort with S. D. Durbin, S. M. Arakelian, A. J. Karn, H. L. Ong, E. Santamato, A. J. Karn, H. L. Ong, E. Santamato, B. Baino, M. Romagnoli, and M. Settembre. This work was supported by the National Science Foundation, Solid State Chemistry, Grant DMR84-14053.

References

1. See, for example, Y. R. Shen, in Optical Bistability III, eds. H. M.

Gibbs et al. (Springer-Verlag, Berlin, 1986); I. C. Khoo and Y. R. Shen, Optical Eng. $\underline{24}$, 579 (1985).

2. N. V. Tabiryan, A. V. Sukhov, and B. Ya. Zel'dovich, Mol. Cryst. Liq. Cryst. $\underline{136}$, 1 (1986).

3. A. S. Zolot'ko, V. F. Kitaeva, N. Kroo, N. N. Sobolev, and L. Chillag, Pis'ma Zh. Eksp. Teor. Fiz. $\underline{32}$, 170 (1980) [JETP Lett. $\underline{32}$, 158 (1980)]; B. Ya. Zel'dovich, N. F. Pilipetskii, A. V. Sukhov, and N. V. Tabiryan, Pis'ma Zh. Eksp. Teor. Fiz. $\underline{32}$, 287 (1980) [JETP Lett. $\underline{32}$, 263 (1980)]; S. D. Durbin, S. M. Arakelian, and Y. R. Shen, Phys. Rev. Lett. $\underline{47}$, 1411 (1981).

4. See, for example, P. Sheng, in Introduction to Liquid Crystals, eds. E. B. Priestley et al. (Plenum, New York, 1975), p.103.

5. E. Santamato, A. Sasso, R. Bruzzese, and Y. R. Shen, Optics Lett. $\underline{11}$, 452 (1986).

6. A. J. Karn, S. M. Arakelian, Y. R. Shen, and H. L. Ong, Phys. Rev. Lett. $\underline{57}$, 448 (1986).

7. B. Ya. Zel'dovich, N. V. Tabiryan, and Yu S. Chilingaryan, Zh. Eksp. Teor. Fiz. $\underline{81}$, 72 (1980) [JETP $\underline{54}$, 32 (1981)]; H. L. Ong, Phys. Rev. A $\underline{28}$, 2392 (1983).

8. S. R. Nersisyan and N. V. Tabiryan, Opt. Spektrosk. $\underline{55}$, 782 (1983) [Opt. Spectros. (USSR) $\underline{55}$, 469 (1983)] and Mol. Cryst. Liq. Cryst. $\underline{116}$, 111 (1984); H. L. Ong, Phys. Rev. A $\underline{31}$, 3450 (1985) and Appl. Phys. Lett. $\underline{46}$, 822 (1985).

9. J. A. Goldstone and E. Garmire, Phys. Rev. Lett. $\underline{53}$, 910 (1984).

10. B. Ya. Zel'dovich and N. V. Tabiryan, Zh. Eksp. Teor. Fiz. $\underline{82}$, 1126 (1982) [JETP $\underline{55}$, 99 (1982)].

11. E. Santamato, B. Diano, M. Romagnoli, M. Settembre, and Y. R. Shen, Phys. Rev. Lett. $\underline{57}$, 2423 (1986).

LIGHT-INDUCED SELF-EXCITATION OF 3-D ORIENTATIONAL GRATINGS IN LIQUID CRYSTALS

T. V. Galstyan, E. A. Nemkova, A. V. Sukhov, and B. Ya. Zel'dovich

Institute of Applied Mechanics, USSR Academy of Sciences, Moscow, USSR

ABSTRACT

Transient excitation of spatially periodic 3-D director reorientation in a planarly aligned nematic sample via the interference field of ordinary (o) and extraordinary (e) waves is treated both theoretically and experimentally. The dependence of all the parameters of the process upon a single space-time argument is demonstrated. Stimulated scattering, four-wave mixing, and transient energy transfer between both waves are treated in detail. The influence of the heating of the medium due to absorption on the processes mentioned above is also studied.

BASIC EQUATIONS

The geometry of the interaction of light waves in nematic liquid crystals to be discussed in this paper is presented in Fig. 1. Let two waves, namely an ordinary (o) wave $\vec{E}_o = (\vec{e}_y \cos\delta + \vec{e}_z \sin\delta) \times E_o(z,t)\exp(i\vec{k}_\perp \cdot \vec{r} - i\omega t)$ and an extraordinary (e) wave $\vec{E}_e = \vec{e}_x E_e(z,t)\exp(i\vec{k}_\parallel \cdot \vec{r} - i\omega t)$, propagate in a planarly aligned nematic sample normal to its undisturbed optical axis, the angles between their wave vectors $|\vec{k}_{\parallel,\perp}| = \frac{2\pi}{\lambda} n_{\parallel,\perp}$ and the director being equal ($\delta \ll 1$). Here $n_{\parallel,\perp}$ stands for the principal values of the refractive index, and λ for the wavelength in free space. Interference between $\vec{E}_e$ and $\vec{E}_o$ leads to a spatially periodic orientational deformation $\delta\vec{n} = \theta_R(\vec{e}_y \cos\delta + \vec{e}_z \sin\delta)$, whose wave vector is $\vec{q} = \vec{k}_\parallel - \vec{k}_\perp$ (the grating optical nonlinearity (GRON), see[1]). It should be pointed out that this orientational grating is a sufficiently three-dimensional one due to the large difference between $n_{\parallel,\perp}$ creating a rather large q_z component. Scattering of both waves by this grating leads to their volume self-diffraction. The system of equations for the reorientation mentioned above, containing shortened wave equations for $\vec{E}_{o,e}$ and a dynamic equation for θ_R, obtained by traditional variational methods, looks as follows:

$$\frac{\partial}{\partial z} E_e = \frac{i\varepsilon_a \pi \theta_R}{\lambda n_\parallel \cos\delta} E_o \exp(i\vec{q}\cdot\vec{r})$$

$$\frac{\partial}{\partial z} E_o = \frac{i\varepsilon_a \pi \theta_R}{\lambda n_\perp \cos\delta} E_e \exp(-i\vec{q}\cdot\vec{r}) \qquad (1)$$

$$\eta \frac{\partial}{\partial t} \theta_R + K_1 \Delta\theta_R = \frac{\varepsilon_a}{16\pi} \left(E_o E_e^* \exp(i\vec{q}\cdot\vec{r}) + c.c. \right).$$

Here $\varepsilon_a = n_\parallel^2 - n_\perp^2$ stands for the dielectric constant anisotropy, K_1 for Frank's constant, and η (poise) for the orientational viscosity of the nematic. Let us emphasize the fact that the first two equations do not contain intramodulational terms, for neither of the waves alone can create reorientation in the geometry discussed in an approximation linear in θ_R. Thus, their self-focusing is impossible. The solution of such a system in steady state gives a value for θ_R proportional to $|\vec{q}|^{-2}$, i.e. significantly dependent upon the beam intersection angle 2δ. Such nonlocality can give rise to valuable reconstruction quality by reduction of speckle-field holograms due to orientational nonlinearity. Nevertheless, if the pulse duration $\tau_p \ll \tau = \eta/(k_1 |\vec{q}|^2)$, i.e. if the reorientation is sufficiently transient, the second term on the left hand side of the third of Eqs. (1) can be neglected. Thus, the medium's response becomes essentially local. Below we shall deal with such a duration range, corresponding numerically to $\tau_p < 1$ ms for typical values of K_1, η, and δ. Now if we introduce the complex quantity $\theta_R = \theta \exp(i\vec{q}\cdot\vec{r})$ + c.c., omit the terms in the equations for $\vec{E}$ which do not satisfy Bragg's conditions, and put $\cos\delta = 1$ within the accuracy of the approximation used, we can introduce $A = E_e n_\parallel^{1/2}$, $B = E_o n_\perp^{1/2}$, $q = \frac{\varepsilon_a^2}{16\lambda\eta n_\parallel n_\perp}$, and $\sigma = \frac{\pi\varepsilon_a}{\lambda(n_\parallel n_\perp)^{1/2}} \theta$, thus obtaining the following system:

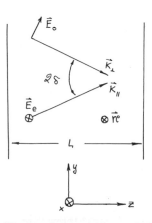

Fig. 1. The geometry of the interaction of light waves in a nematic.

$$\frac{\partial}{\partial z} A = i\sigma B; \quad \frac{\partial}{\partial z} B = i\sigma^* A; \quad \frac{\partial}{\partial t} \sigma = qAB^*. \tag{2}$$

A remarkable property of this system is that it can be readily rewritten in total derivatives (see Ref. 2). In fact, assuming that A and B at the entrance plane differ only by a constant phase multiplier, having the same temporal envelope, $A(z,t) = \alpha(z,t) E(t)$, $B(z,t) = \beta(z,t) E(t)$, we can introduce a new argument $y = qz\int_0^t |E(t')|^2 dt'$, put $\sigma = \frac{y}{z} M$, and obtain the following:

$$d\alpha/dy = iM\beta \qquad \alpha(0) = \alpha_o ; \qquad |\alpha_o|^2 + |\beta_o|^2 = 1$$

$$d\beta/dy = iM^*\alpha \qquad \beta(0) = \beta_o$$

$$M(0) = \alpha_o \beta_o^* \tag{3}$$

$$dM/dy = \frac{1}{y} [\alpha\beta^* - M] \qquad \left(\frac{dM}{dy}\right)_o = \frac{i\alpha_o\beta_o^*}{2} [|\beta_o|^2 - |\alpha_o|^2].$$

Here an additional condition for dM/dy is required due to the singularity of the third equation at $y = 0$. The system (3) provides the basic equations for the description of processes caused by GRON in the transient case.

FOUR-WAVE MIXING AND PHASE CONJUGATION

It is natural to start the treatment of system (3) with the simplest case, when one of the waves (e.g. E_e, as it was in our experiments) is significantly more intense than the other, and is assumed not to be depleted ($\alpha \equiv 1$), y being small. The expression for $M(y)$ under these assumptions is given by the boundary conditions and looks as follows:

$$M(y) = M_o + \left(\frac{dM}{dy}\right)_o y = \beta_o^* + i \frac{1}{2} \beta_o^* y \approx \beta_o^*. \tag{4}$$

If we now have one more (reference) wave counterpropagating with E_e, $E_e^- = \alpha_-(y)E(t)n_\parallel^{-1/2}$, $\alpha_-(y) = $ const, its scattering by the grating gives rise to the wave $E_o^- = \beta_-(y)E(t)n_\perp^{-1/2}$, which is phase conjugated to E_o. Within Born's approximation the following equation for the four-wave mixing (FWM) "reflectivity" can be deduced:

$$R^{NL} = \frac{W_o^-(y)}{W_o(y)} = \frac{|\beta_-|^2}{|\beta_o|^2} = |\alpha_-|^2 y^2. \tag{5}$$

Here W_o, W_o^- represent the power densities of the corresponding waves. Thus the current reflectivity does not depend upon the intersection

499

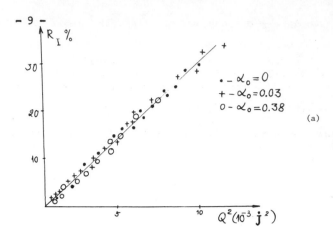

(a)

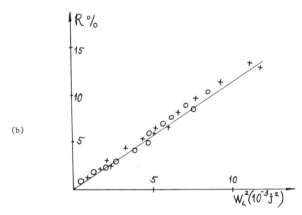

(b)

Fig. 2. (a) The dependence of $R_I\%$ on the current value of the pulse energy $Q^2(t)$ (for various intersection angles $2\delta_o$). (b) The dependence of the energetic reflectivity R_Q on $Q^2(\infty)$.

angle 2δ, enabling one to achieve phase conjugation. Such FWM with phase conjugation was observed in Ref. 3 (a detailed description of the experiment is given in Ref. 3).

The main results are the following: with the increase of the pulse energy Q a signal E_O arises that depends nonlinearly on Q. Measurements of the dependence of $R^{NL}(t)$ on the current value of the pulse energy $Q^2(t)$ were performed for various intersection angles $2\delta_o$ (Fig. 2a). One can see that independently of the total pulse energy and angle $2\delta_o$ all the experimental points correspond to one common straight line, in agreement with expression (4). Current values of $R^{NL} \approx 30$ per cent were achieved. It should be pointed out that in the steady-state case R_o^{NL} for $2\delta_o$ = 0.38 rad must be 25 times smaller than for δ_o = 0.

The dependence of the energetic reflectivity R_Q on $Q^2(\infty)$ (see Fig. 2b) also appeared to be linear, which is in agreement with the theoretical results under the additional assumption of identical pulse shapes for the different pulses, which was satisfied in our experiment. Values of $R_Q \approx 10$ per cent were achieved. The locality of the transient orientational nonlinearity allowed us to observe phase conjugation of the $\overset{\circ}{E}$ wave with a finite angular spectrum. For this purpose a phase plate was introduced into the $\overset{\circ}{E}$ beam, spoiling its divergence from $\theta_o = 6\cdot10^{-4}$ rad to $3\cdot10^{-3}$ rad. The energetic reflectivity was not changed by this introduction, and the angular divergence of the $\overset{\circ}{E}$ wave was practically θ_o after penetrating the phase plate backwards Thus we conclude that phase conjugation took place.

FORWARD-DIRECTED STIMULATED SCATTERING OF LIGHT

Let us consider now the similar case of an e-type non-exhausted pump, but without the assumption that y << 1; let β_o be small (spontaneous scattering of the e-wave by thermal orientational fluctuations). The system of equations (3) is now reduced to the following:

$$\frac{d\beta^*}{dy} = -iM \qquad \beta(0) = \beta_o$$

$$M(0) = \beta_o^* \tag{6}$$

$$\frac{dM}{dy} = \frac{1}{y}\left[\beta^* - M\right] \qquad \left(\frac{dM}{dy}\right)_o = -\frac{i\beta_o^*}{2}.$$

The results again do not depend upon δ. The asymptotic form of the solution of (6) for y >> 1 has the following form:

$$|\beta(y)|^2 = |\beta_o|^2 \frac{1}{4\pi\sqrt{y}} \exp(2\sqrt{2}y). \tag{7}$$

Thus we have obtained exponential-space-time gain for a weak signal, the so-called transient forward-directed stimulated scattering (SS) of the e-wave into the o-wave. In what follows the weak pre-exponential factor will be neglected. Such SS was observed in Ref. 4. Forward-directed SS was observed in a 70 μm thick 5CB planar sample. It was induced by e-polarized free-running ruby laser pulses of about 80 μs duration. An estimate of SS relaxation times gives $\tau \approx 5$ms, SS thus having a sufficiently transient nature. Results of measurements of the temporal evolution of the SS signal are presented in Fig. 3 as the dependence of $\ln(W_o/W_e)$ on the current value $Q^{1/2}(t)$. The readily seen linearity of this dependence demonstrates good agreement of the results with the theoretical calculations. For a quantitative comparison of this coefficient with the theoretical one, the transverse size of the laser beam inside the sample was measured to be a = 90 μm (FWHM). The experimental value of this coefficient appeared to be 1.2 times smaller than the theoretical one, evaluated from the expression

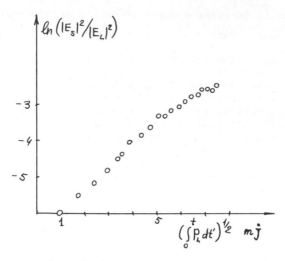

Fig. 3. The temporal evolution measurements of the SS signal (as the dependence of $\ln(W_o/W_e)$ upon the current value of $q^{1/2}(t)$).

$\frac{400\varepsilon_a}{a}\left(\frac{L}{3cn_\parallel n_\perp \lambda\eta}\right)^{1/2}$ $(mj^{-1/2})$, which also confirms the theory. The experimental value of the spontaneous scattering level ν was $\nu \approx 5 \times 10^{-3}$. The estimate for this level using standard formulae for spontaneous scattering in nematics (see e.g. Ref. 5) for an angular aperture of the recording system of $\theta_{reg} \approx 0.15$ gives $\nu \approx 0.42$. The measurements of the angular spectra of the pump beam and SS signal gave the following main results. The angular divergence of the SS signal was at least 4 times greater than that of the pump, being larger than the angular aperture of the recording system. The angular spectrum of the SS signal is of speckle-type, which is natural for transient scattering with a pulse duration significantly shorter than the dephasing time of the spontaneous fluctuations. Thus, in good agreement with theory, the SS discussed appeared to depend upon the single spatial-temporal variable y.

TRANSIENT ENERGY TRANSFER BETWEEN WAVES WITH AN ARBITRARY RATIO OF THEIR INTENSITIES

Let us consider finally the general case of the system (3), i.e. arbitrary α, β, and y. Two remarks should made. First, since the solution of (3) in the general case is rather complicated, only the collinear interaction of two plane waves is considered. Second, as large values of y require rather high intensities, the heating of the medium via light absorption may also be signficant, which leads to the phase modulation of the waves due to the large values

of $\frac{\partial n_{\parallel,\perp}}{\partial T}$ $(\sim(3\div10)\cdot10^{-4} K^{-1})$ appropriate to nematics. This phase modulation affects the energy transport significantly. Fortunately, if the typical pulse duration is $\tau_p \sim 1$ ms and the sample thickness is ~ 100 μm, the heat relaxation time is about 10 ms. Hence the medium heating during the pulse also possesses a transient behavior, and thus depends only on the same variable y as the orientational perturbation. Let the medium possess a weak polarization that is independent of the

502

absorption κ (cm^{-1}) (perhaps of impurity origin). Then Eqs. (3) with
thermal phase modulation taken into account are rewritten as follows:

$$d\alpha/dy = iM\beta + i\gamma_\parallel \alpha \qquad \alpha(0) = \alpha_o$$

$$\beta(0) = \beta_o$$

(8)

$$d\beta/dy = iM^*\alpha + i\gamma_\perp\beta \qquad M(0) = \alpha_o\beta_o^*$$

$$dM/dy = \frac{1}{y}\left[\alpha\beta^* - M\right] \qquad \left(\frac{dM}{dy}\right)_o = \frac{i\alpha_o\beta_o}{2}\left[|\beta_o|^2 - |\alpha_o|^2 + \gamma_\parallel - \gamma_\perp\right].$$

Here $\gamma_{\parallel,\perp} = \dfrac{4n_\perp n_\parallel \kappa c \eta}{\rho c_p \varepsilon_a^2}\left(\frac{\partial}{\partial T}n_{\parallel,\perp}\right)_p$, c stands for the speed of light
and ρ and C_p are the density and heat capacity of the nematic, respect-
ively. Numerical estimates[6] show that $\gamma_\parallel < 0$, $\gamma_\perp > 0$, and
$|\gamma_{\parallel,\perp}| \sim 1$ if $\kappa \gtrsim 10^{-2} \div 10^{-3}$ cm^{-1}. Equations (8) have a conservation
law $|\alpha|^2 + |\beta|^2 \cong 1$. Describing the evolution of 6 real variables, Eqs.
(8) can be readily reduced to the description of only 3 real variables.
In fact, by rewriting $V = |\beta|^2 - |\alpha|^2$, $(C+iD)\alpha\beta^* = M$, and omitting the
non-essential phases of α and β, one can obtain a system of equations
for real V, C, and D:

$$dV/dy = (1 - V^2)D$$

$$dC/dy = 2CDV + (1-C)/y + \gamma_a D$$

$$dD/dy = V(D^2 - C^2) - D/y - \gamma_a C.$$

(9)

These equations have an equilibrium solution if $|\gamma_a| < 1$, namely
$D \equiv 0$, $C \equiv 1$, $V = -\gamma_a$. Thus, if $V_o = -\gamma_a$, energy transfer does not
take place. We treat the stability of such an equilibrium solution by
giving small initial perturbations $V = w - \gamma_a$, $D = 0(w)$, $C = 1 + 0(w^2)$. Then
the linearized equations (9) give the following:

$$\frac{d^2w}{dy^2} + \frac{1}{y}\frac{dw}{dy} + (1 - \gamma_a^2)w = 0.$$

(10)

The solution of (10) is Bessel's function of zero order,

$$w(y) = w_o J_o\left[\frac{y}{(1-\gamma_a^2)^{1/2}}\right].$$

(11)

Thus we obtain two main conclusions:

First, light absorption by the medium leads to unidirectional energy
transfer from the e-wave to the o-wave (the equilibrium intensity ratio
is not equal to 1).

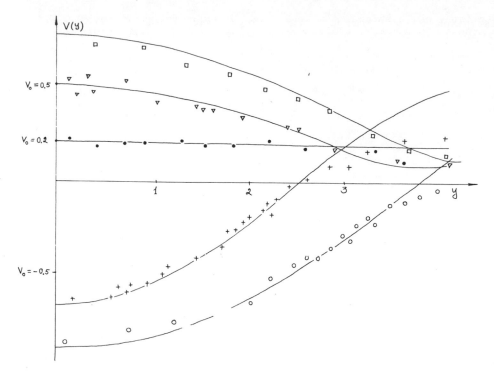

Fig. 4. The results of a numerical solution of (9) (solid curves) for $\gamma_a = -0.2$ and various V_o, and the corresponding experimental points.

Second, the normalized intensity difference V approaches its equilibrium value with the increase of y not monotonically but in an oscillatory manner. This statement is confirmed once more by the numerical solution of (9) for V_o drastically different from $-\gamma_a$, for which (10) is not valid.

The situation discussed was treated experimentally for the excitation of a grating in a 5CB sample by a free-running ruby laser pulse, polarized linearly at some angle to the sample optical axis. The transmitted beam was divided by a calcite prism into E_o and E_e components, whose intensity difference I_Δ was measured by photodiodes with the help of a differential input oscilloscope. Time envelopes of the exciting pulse I_Σ and the current value $Q(t)$ were also recorded. Measurements of the dependence of $V = I_\Delta/I_\Sigma$ on y were carried out. For $V = +0.2$, $V(y)$ appeared to degrade to $V(y) = V_o$, hence the experimental value for γ_a was $\gamma_a = -0.2$.

In Fig. 4 results of a numerical solution of (9) (solid curves) for $\gamma_a = -0.2$ and various V_o, and the corresponding experimental points are presented. Here the unknown value of η was used as a free parameter. It appeared that the best coincidence was achieved if η is put equal to 1.35 poise, which is quite realistic for a nematic near the melting point ($\sim 20^\circ$ C in our case). Experimentally, the $V(y)$ curves and $V = +0.2$ line intersected, confirming qualitatively the theoretical prediction of oscillations in $V(y)$. Using the experimental value $\gamma_a = -0.2$, an estimate for κ was obtained, $\kappa \approx 2 \cdot 10^{-3}$ cm^{-1}. Such negligibly small light absorption in nematics could never be measured by any direct method.

Thus, in this paper the dependence of all parameters of transient GRON on the single variable y is shown theoretically and confirmed experimentally.

REFERENCES

1. N. V. Tabiryan and B. Ya. Zel'dovich, Molecular Crystals and Liquid Crystals 62, 237 (1981).
2. B. Ya. Zel'dovich and E. A. Nemkova, Kratkiye soobscheniya po fisike, N1, pp. 21-23, Moscow, 1987.
3. B. Ya. Zel'dovich, N. F. Pilipetsky, and A. V. Sukhov, Pis'ma v Zh. Eksp. Teor. Fiz. 43, 122 (1986).
4. B. Ya Zel'dovich, S. K. Merzlikin, N. F. Pilipetsky, and A. V. Sukhov, Pis'ma v Zh. Eksp. Teor. Fiz. 41, 418 (1985).
5. P. de Gennes, The Physics of Liquid Crystals. (Clarendon Press, Oxford, 1974).
6. I. C. Khoo and R. Normandin, IEEE J. Quant. Electron. 21, 329 (1985).

ROTATORY INSTABILITY OF THE SPATIAL STRUCTURE OF LIGHT FIELDS IN

NONLINEAR MEDIA WITH TWO-DIMENSIONAL FEEDBACK

M. A. Vorontsov, V. Ju. Ivanov, and V. I. Shmalhauzen

Moscow State University, Physics Faculty, Moscow 119899, USSR

ABSTRACT

Optical bistability, instabilities, and chaos are usually observed as processes developing with time. (The only exception is the so-called transverse optical bistability.)

The three-dimensional nature of optical fields enables us to propose radically new nonlinear phenomena in systems with suitably organized feedback.

In this paper a new class of nonlinear optical systems is presented and experimentally tested - coherent systems with two-dimensonal feedback. We discuss the conditions of initiation of spatial instabilities, spatial bistability, optical self-organizing processes, and optical turbulence, which develops in space and time like hydrodynamic turbulence.

These phenomena were oberved in a hybrid system with a liquid crystal. New possibilities of applications in physics and optical computing provided by systems with two-dimensional feedback are noted.

INTRODUCTION

Nonlinear interactions in optics are usually considered as evolution processes of the initial parameters of light fields, which proceed in some local area during a restricted time interval. Usually, they are problems with initial conditions (the Cauchy problem).

Of special interest in nonlinear optics is the solution of the vast class of boundary value problems: the dynamics of nonlinear resonators, systems with optical feedback, adaptive nonlinear systems, etc.[1,2]. Linear problems of this kind are rather thoroughly investigated[3]. Systems with a strong nonlinear response can be analyzed only in the framework of some approximations which allow the "decoupling" of the separate space and time variables (the approximation of geometrical optics, setting field, the nonaberration approximation, etc.).

The analysis of dynamic processes in nonlinear Fabry-Perot interferometers is a classical example of such an approach[1,4]. Even a one-dimensional approximation to such a system exhibits nontrivial behavior: optical bistability, multistability, chaos[5].

The next stage of investigations in this direction is connected with taking into account spatial effects. Diffusion of the particles of the nonlinear medium or diffraction of the wave from the mirror aperture necessitates simultaneous analysis of the spatial-temporal evolution of a light beam. The space-spread nature of the field dominates the system dynamics (it is already impossible to factorize space and time variables). Observed spatial effects, such as transverse bistability and switching waves, are examples of complicated dynamic phenomena that exist in systems of this type[6-8].

Besides the coupling of variables by diffusion mentioned above, the use of different spatial transformations of the light field in nonlinear systems opens up much better prospects. One of the simplest examples is field conversion with the help of the spherical mirrors of a resonator, resulting in all points of the beam diameter being linked. In contrast to the diffusion-type processes the space-field transformation leads to a nonlocal coupling of points. Diffraction or diffusion of nonlinear-medium particles contributes to an additional local coupling.

The first experimental investigations of such systems showed a variety of dynamic processes: formation of structures, spatial stochastization of the light field, different types of autowaves, optical turbulence, etc.[9,10]. Some types of the phenomena enumerated were observed earlier in noncoherent TV optical systems with two-dimensional feedback[11].

In coherent nonlinear systems the possibilities of spatial light-beam transformations are essentially wider: not only conversions of intensity, but also of phase and polarization of coherent light waves can be used here. Accordingly, the dynamics of such systems is more interesting and complicated.

Apart from the purely scientific interest connected with the development of ideas of nonlinear optical processes, the abovementioned phenomena are of great importance in view of investigations of spatial instabilities in diverse loop systems of wave front conjugation[12], adaptive optical systems operating under conditions of nonlinear distortions[2], nonlinear resonators[1], etc.

It can be mentioned that spatial instability of the light field is also observed in other problems of light-beam propagation in nonlinear media, i.e. actually in problems with initial conditions. Nevertheless, the influence of nonlinearity in boundary value problems of nonlinear optics is displayed to a much greater extent, and results in essentially different dynamical processes in the system.

This circumstance can be easily understood, if one deals with the standard method for analyzing boundary value problems in optics. These problems are usually solved by means of a successive approximation method, i.e. they, in fact, are reduced to problems with initial conditions (for example: the use of equivalent lens systems for calculating the field in a laser resonator[3]).

This approach often has a realistic physical basis, and reflects a process of field stabilization in the system. The peculiarity consists in the fact that during the stabilization process the same part of the nonlinear medium affects a spreading light wave, accumulating over and over again all the previous interactions. The field transformation brings in an additional complicated tie between the nonlinear response of different parts of the medium. Strictly speaking, field stabilization in nonlinear systems may not be achieved, so the question concerning the correlation between the stabilization process and concrete nonlinear boundary value dynamical problems remain unsolved.

Nonlinear optical systems with two-dimensional feedback are, apparently the simplest objects being investigated in this area. Wave propagation along one direction can be comparatively easily realized in such systems. This fact essentially simplifies the analysis and enables the creation of an adequate mathematical model of the process, which is not always possible in systems as complicated as a nonlinear resonator. At the same time, the space-spread nature of the nonlinear interaction is clearly displayed in systems with two-dimensional feedback.

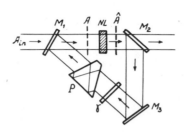

Fig. 1. Nonlinear ring resonator with the feedback field rotated: M_1, M_2 - Mirrors, NL - thin layer of nonlinear medium, p - turning prism, γ - element with controlling absorption coefficient.

NONLINEAR RING RESONATOR WITH ROTATORY SHIFT

Dynamical effects occurring in a nonlinear optical system with two-dimensional feedback with the feedback field rotatively shifted, are analyzed in this article. The optical scheme of the system is shown in Fig. 1. It represents a nonlinear passive ring resonator. Let us denote the complex amplitude of the field directly in front of the nonlinear medium layer as $A(\vec{r},t)$ ($z = 0$, $\vec{r} = (r,\theta)$ is the radius-vector in the plane perpendicular to the direction of wave propagation Oz). The complex amplitude of the field $\hat{A}(\vec{r},t)$ in the plane $z = \ell$ is connected with $A(\vec{r},t)$ by the relation

$$\hat{A}(\vec{r},t) = A(\vec{r},t)\exp\left[-g\ell/2+ik\ell n\right], \tag{1}$$

where g is the absorption coefficient of the medium layer, ℓ is the length of the layer, $n=n(\vec{r},t)$ is the refractive index of the medium, and $k = 2\pi/\lambda$ is the wave number.

We will describe the nonlinear response of the medium with the help of a Debye relaxation equation (medium with a Kerr nonlinearity):

$$\tau\frac{dn}{dt} + n = n_o + n_2|A|^2, \tag{2}$$

where n_o is the undisturbed value of the refractive index.

We assume the delay time in the feedback circuit to be negligible in comparison with the characteristic relaxation time of the nonlinearity τ.

At the first stage of the investigation we will confine ourselves to an analysis of light-wave propagation processes in the framework of a geometrical optics approximation, assuming the system length L to be much

509

shorter than the characteristic length of diffraction blooming for an input setting diameter of the beam a_o, $(L \ll ka_o^2)$.

A Dove prism is placed in the feedback loop (Fig. 1). By means of the prism the light beam can be turned around an optical axis through an arbitrary angle Δ. Thus, the simplest field transformation is realized in the nonlinear system feedback.

Taking into account the rotatory shift of the field, the following equation can be written for the complex amplitude $A(\vec{r},t)$:

$$A(r,\theta,t) = (1-R)^{1/2} A_{in}(r) + \gamma R \hat{A}(r,\theta-\Delta,t). \tag{3}$$

Here, R is a reflection coefficient of the resonator mirrors M_1 and M_2, γ is the coefficient of the field attenuation in the feedback circuit, and $A_{in}(r)$ is the complex amplitude of the incident field. Substitution of (1) into (3) results in a recurrence equation for the definition of the field $A(\vec{r},t)$:

$$A(r,\theta,t) = (1-R)^{1/2} A_{in}(r) + B\exp\left[ik\ell n(r,\theta-\Delta,t)\right]A(r,\theta-\Delta,t),$$

$$\tag{4}$$

where $B = \gamma R \exp(-g\ell/2)$.

This equation can be rewritten in the equivalent form:

$$A(r,\theta,t) = (1-R)^{1/2} A_{in} + Be^{ik\ell n(r,\theta-\Delta,t)}\left[(1-R)^{1/2} A_{in} + \right.$$

$$\left. + Be^{ik\ell n(r,\theta-2\Delta,t)}A(r,\theta-2\Delta,t)\right]. \tag{5}$$

We will take into consideration only a single passage of the wave through the resonator, assuming $B \ll 1$. In this case Eq. (5) can be limited to the two first addends:

$$A(r,\theta,t) = (1-R)^{1/2} A_{in}\left[1+B\exp(ik\ell n(r,\theta-\Delta,t))\right].$$

An approximate expression for the intensity $|A|^2$ is

$$I(R,\theta,t) = (1-R)I_{in}\left[1+2B\cos\phi(r,\theta-\Delta,t)\right], \tag{6}$$

where $\phi = k\ell n$ is the phase shift of the wave in the nonlinear medium, and $I_{in} = |A_{in}|^2$.

By multiplying (2) by $k\ell$ and using the expression (6) one can derive an equation for the nonlinear phase modulation dynamics:

$$\tau \partial u/dt + u = K\cos\left[u(r,\theta-\Delta,t)+\phi_o\right], \tag{7}$$

where $u(r,\theta,t) = \phi(r,\theta,t)-\phi_o$, $\phi_o = k\ell\left[n_o+n_2(1-R)I_{in}\right]$, and $K = 2n_2k\ell(1-R)B$ is a parameter whose value determines the strength of the nonlinear effects in the system. In what follows we will refer to K as an amplification factor of the feedback circuit by analogy with the theory of an automatic regulation system. In a passive system the coefficient K can be altered by changing, for example, energy absorption in the feedback (the coefficient γ). It is assumed that although $B \ll 1$, $K \sim 1$ owing to the strong nonlinearity of the medium. The phase shift

ϕ is determined by the intensity of the input field and can be changed with the help of external (noncoherent) light illuminating the nonlinear medium.

The effects of diffusion in a nonlinear medium and (in the first approximation) the diffraction of light in the system can be taken into consideration phenomenologically. For this purpose one can add to the right hand side of Eq. (7) a transverse-coordinates Laplacian operator:

$$\tau \partial u/\partial t + u = d\Delta_\perp u + K\cos[u(r,\theta-\Delta,t)+\phi_o],\tag{8}$$

where d is an effective diffusion coefficient, and $\Delta_\perp = \dfrac{1}{r}\dfrac{\partial}{\partial r}\; r\; \dfrac{\partial}{\partial r} +$

$+ \dfrac{1}{r^2}\dfrac{\partial^2}{\partial\theta^2}$ is the Laplacian operator.

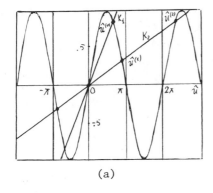

(a)

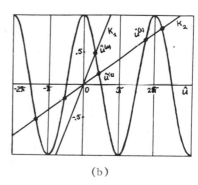

(b)

Fig. 2. Graphical solution of Eq. (18) for different values of K:
a) $\phi_o = \pi/2$, b) $\phi_o = 0$; $\hat{u}^{(o)}$, $\hat{u}^{(1)}$, $\hat{u}^{(2)}$ - stationary states,
$(\hat{K}_1 > \hat{K}_2)$.

STABILITY ANALYSIS IN THE FRAMEWORK OF THE LINEAR APPROXIMATION

An equation for stationary solutions $\hat{u} = \hat{u}(r,\theta)$ can be obtained from (8), if the time derivative is set equal to zero:

$$\hat{u} = d\Delta_\perp\hat{u} + K\cos[\hat{u}(r,\theta-\Delta)+\phi_o].\tag{9}$$

Let us investigate the stability of the stationary solutions. $u_1(r,\theta,t)$ means a small deviation from the stationary solution, i.e. $u = \hat{u}+u_1$. Substitution of this expression into (8), accurate within second-order terms in u_1, leads to the equation:

$$\tau \partial u_1/\partial t + u_1 = d\Delta_\perp u_1 + \hat{K}u_1(r,\theta-\Delta,t),$$

$$\hat{K} = -K\sin(\hat{u}+\phi_o).\tag{10}$$

511

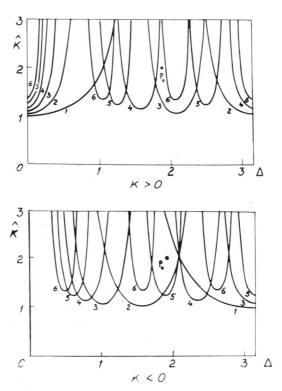

Fig. 3. Boundaries of the steady areas for waves with different spatial periods: a) $\phi_o = \pi/2$, $\hat{K} > 0$), b) $\phi_o = 0$, $\hat{K} < 0$).

We will consider a one-dimensional model, assuming that the light beam in the system has the form of a thin ring with radius a_0. In this case instead of (10) the following equation is obtained:

$$\frac{\partial u_1}{\partial t} + u_1 = \hat{d}\frac{\partial^2 u_1}{\partial \theta^2} + \hat{K}u_1(\theta-\Delta,t), \quad (\hat{d} = d/a_0^2), \tag{11}$$

with periodic boundary conditions:

$$u_1(\theta,t) = u_1(\theta+2\pi,t), \quad \frac{\partial u_1}{\partial \theta}(\theta,t) = \frac{\partial u_1}{\partial \theta}(\theta+2\pi,t). \tag{12}$$

The time t in Eq. (11) is normalized to τ. Let the solution of this equation have the form of a travelling wave:

$$u_1(\theta,t) = A\exp[pt+i\kappa\theta]. \tag{13}$$

Substituting Eq. (13) int Eq. (11) we get:

$$p = -(1+\hat{d}\kappa^2) + \hat{K}\exp(i\kappa\Delta). \tag{14}$$

From the boundary conditions (12), a discretization of the spatial frequencies emerges: $\kappa = \kappa_n = n$, (n = 1,2,3,...). Accordingly, in place of (14) one can write

$$p_n = -(1+\hat{d}n^2) + \hat{K}\exp(in\Delta), \quad n = 1,2,... \tag{15}$$

Writing p_n in the form $p_n = \delta_n + i\omega_n$, from (15) we get:

$$\delta_n = -(1+\hat{d}n^2) + \hat{K}\cos(n\Delta), \tag{16a}$$

$$\omega_n = \hat{K}\sin(n\Delta). \tag{16b}$$

Thus when $\delta_n > 0$ the instability has the form of a rotating wave $u_1(\theta,t) = A\exp(\delta_n t)\exp[i\omega_n t+in\theta]$. The parameter n determines the number of tops (petals) in the period 2π, ω_n is an angular frequency of the developed rotational instability, and δ_n is a relaxation decrement. When the value of n is fixed, the boundaries of the stable areas can be determined from the condition $\delta_n = 0$, i.e.

$$\hat{K}_n = \frac{1+dn^2}{\cos(n\Delta)} \quad \text{or} \quad -\hat{K}\sin(\hat{u}+\phi_0) = (1+\hat{d}n^2)\cos^{-1}(n\Delta). \tag{17}$$

Let us investigate the stability of the spatially-homogeneous stationary solutions $\hat{u}(\theta) = $ const. From (9) we get:

$$\hat{u}/K = \cos(\hat{u}+\phi_0). \tag{18}$$

Examples of graphical solutions of this equation when $\phi_0 = \pi/2$ and $\phi_0 = 0$ are shown in Fig. 2. There are a number of diverse stationary solutions for large values of K. The boundaries of the steady areas for the waves with different spatial periods in the case of $\phi_0 = \pi/2$ (a) and $\phi_0 = 0$ (b) ($\hat{d} = 0.01$) are given in Fig. 3. The boundaries of

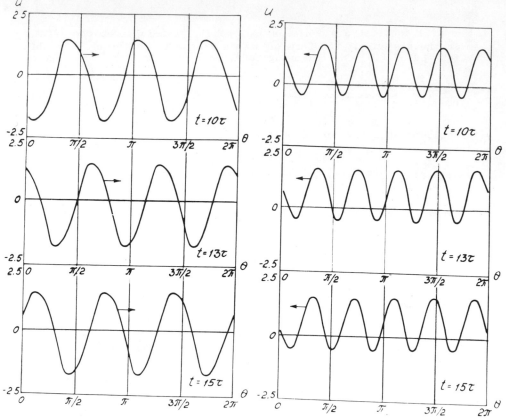

Fig. 4. Transformation of the wave space structure when the parameter ϕ_o is changed ($|\hat{K}| = 2$, $\hat{d} = 0.01$, $\Delta = 1.91$ rad):
a) $\phi_o = \pi/2$, b) $\phi_o = 0$; (point p_o in Fig. 3).

stability were determined in accordance with formulas (17). These areas correspond to the stationary solutions denoted in Fig. 2 by $\hat{u}^{(1)}$.

Let us analyze the main features of rotatory instability. When $|\hat{K}| \leqslant 1 + \hat{d}$ the system is stable. If $|\hat{K}| > 1 + \hat{d}$, multipetal rotating structures may occur in the system. When the shifts are equal to $\Delta_n^m = \frac{2\pi m}{n}$ (m is an integer), the structures are immovable ($\omega_n = 0$). An angular shift relative to the points Δ_n^m results in the structures being rotated with the angular speed determined by (16b). As Δ increases the speed of rotation also increases at first and then the wave changes its spatial period.

The same effects can be observed in the system if the constant phase shift ϕ_o is changed. Functions $u = u(\theta)$ at different moments of time t, obtained by means of a numerical solution of the one-dimensional equation (8), are represented in Fig. 4a. The parameters of the system were the following: $|\hat{K}| = 2$, $\hat{d} = 0.01$, $\Delta = 1.91$ rad, $\phi_o = \frac{\pi}{2}$. When $|\hat{K}| \lesssim 2$ the angular speed of the rotation structure is accurately described by

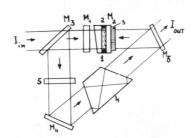

Fig. 5. Scheme of the experimental device: M_1, M_2 – Fizeau interferometer mirrors, M3 – beamsplitter, M4, M5 – rotating mirrors in
the feedback circuit, 1 – optically controlling transparency, 2 –
liquid crystal layer, 3 – photoconductor, 4 – Dove prism, 5 –
polarizer (amplification factor of the feedback circuit alteration).

(16b). On the whole, according to the results of numerical simulations
the linear theory gives an accurate prediction of the appearance and
motion of the structures.

When the constant phase shift ϕ_0 is altered (Fig. 4b) the system
switches to a state with a different spatial structure. (The other
parameters of the system are the same as in Fig. 4a.) As a result, the
spatial period of the wave and the direction and speed of its rotation
are changed.

For the experimental investigation of the rotatory instability in
nonlinear systems with two-dimensional feedback, a structure with a
photoconductor-dielectric mirror-liquid crystal layer was used[13]. The
scheme of the device is shown in Fig. 5.

Modulation of the liquid-crystal-layer refractive index was
accomplished by illuminating the photoconductor with controlling light.
The resistance of the photoconductor was changed proportionally to the
intensity of the controlling light. As a consequence, the voltage
applied to the thin ($\sim 10\mu m$) layer of liquid crystal was spatially
modulated according to the intensity distribution in the feedback
circuit. This, in turn, caused the modulation of the liquid-crystal-
layer refractive index.

The input field is reflected by the dielectric mirror; since it does
not reach the photoconductor, it has practically no influence upon the
refractive index change.

Equation (8), written for the system with purely optical feedback,
can also be used for describing the hybrid system. In this case the
diffusion part of the equation is mainly connected with charge diffusion
in the photoconductor.

Rotation of the field in the feedback circuit was accomplished by
the Dove prism. The structures obtained with different rotation angles
Δ are presented in Fig. 6. All of them are usually moving. The spatial
period of the structures and the speed and direction of its rotation are
in good agreement with the theory discussed above. The one-dimensional
model being considered by no means describes the complicated spatial
configuration of the structures shown in Fig. 6. The two-dimensional
transverse problem (8) must be solved. The multipetal structures in Fig.
6(b,c) differ from each other by the effective diffusion coefficient (in
case (b) this coefficient is larger). If the amplification factor (light

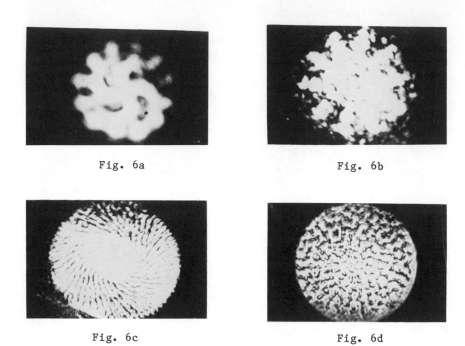

Fig. 6a   Fig. 6b

Fig. 6c   Fig. 6d

Fig. 6. Experimental results : an examples of the rotatory
instability.

intensity in the feedback circuit) is large enough, the regular structures are destroyed and stochastization of the field occurs (Fig. 6d).

In conclusion we want to emphasize that strong nonlinear self-actions in systems with two-dimensional feedback cause the generation of light fields that have no analogs in linear optics. A noteworthy example in this sense is the rotatory instability (rotating fields) considered in this article.

We are particularly grateful to Dr. S. A. Akhmanov for fruitful discussions of the results.

REFERENCES

1. H. M. Gibbs, Optical Bistability: Controlling Light with Light (Academic Press, New York City, 1985).
2. M. A. Vorontsov and V. I. Shmal´gauzen, Principles of Adaptive Optics (Nauka, Moscow, 1985).
3. Yu. A. Anan´ev, Optical Resonators and the Problem of the Divergence of Laser Emission (Nauka, Moscow, 1979).
4. N. N. Rozanov, Zh. Eksp. Teor. Fiz. 20, 96 (1981).
5. K. Ikeda, J. de Physique C1, 183 (1983).
6. N. N. Rozanov and G. V. Khodova, Kvant. Electron. 13, 368 (1986).
7. Yu. A. Balkarei, V. A. Grigor´yants, and Yu. A. Rzhanov, Kvant. Electron. 14, 128 (1987).
8 L. A. Lugiato and R. Lefever, Phys. Rev. Lett. 58, 2209 (1987).
9. S. A. Akhmanov, M. A. Vorontsov, and V. I. Shmal´gauzen, Preprint No. 33 from the Physical Faculty of Moscow State University (1986).
10. S. A. Akhmanov, M. A. Vorontsov, and V. I. Shmal´gauzen, Proceedings of the IXth Vavilov Conference, Novosibirsk, 1987.
11. G. Ferrano and G. Hausler, Optical Engineering 19, 442 (1980).
12. M. Cronin-Golomb, B. Fischer, J. O. White, and A. Yariv, IEEE QE-20, 12 (1984).
13. S. S. Ignatocyan, V. P. Simonov, and B. M. Stepanov, Opto-Mechanical Industry, issue no. 10, 7 (1986).

INTRINSIC OPTICAL MULTISTABILITY AND INSTABILITIES IN LIQUID CRYSTALS

S. M. Arakelian, Yu. S. Chilingaryan, R. B. Alaverdyan, and
A. S. Karayan

Yerevan State University, Yerevan, Armenia, 375049

ABSTRACT

The effects of intrinsic optical multistability and temporal
instabilities during wave interactions in an inhomogeneous anisotropic
medium with threshold nonlinearity are discussed.

1. INTRODUCTION

Optical bistability (multistability) and dynamic instabilities in
wave phenomena without any external feedback are the subjects of intense
study at the present time[1]. The multivalued regimes arise for these
cases because the propagating light waves induce dynamic gratings of the
refractive index inside the highly nonlinear medium. Liquid crystals
(LC) are unique objects for these nonlinear distributed-feedback systems
in two respects.

First, the very large nonlinearity of LC leads to real laser-induced
structural phase transitions without any temperature variations of the
substance; different spatially modulated structures arise in the
medium[2]. This is intrinsic optical multistability which is due to the
physical properties of the developed nonlinear phenomena; the feedback
arises from a nonlocal nonlinear response of the medium to the laser
field, because of the elastic forces[3].

Second, in LC, due to the anisotropy of the medium, two waves of
different polarization travelling through the medium create laser-induced
gratings along the length (d) of the sample (z-direction). An energy
interchange occurs between the two polarization components in the laser-
distorted inhomogeneous anisotropic medium with spatial modulation of the
optical axis $\vec{n}(z)$ – nonadiabatic deformations[4]. A description of these
processes on the basis of their analogy to two coupled (orthogonal)
oscillators is useful.

In the case of threshold reorientation of nematic LC (NLC) – a
homogeneous orientation of the anisotropic molecules from the
initial $(\vec{n}_o)$ direction – at oblique incidence (angle α) of the light
field $\vec{E}$ on the sample $(\vec{E} \perp \vec{n}_o, \vec{n}_o \parallel z)$, the characteristic dependences are
shown in Fig. 1. Figure 1b demonstrates that for large α (large I) the
system can switch into any of the possible states following one of the
branches (which in particular, is an occurrence) if α decreases.

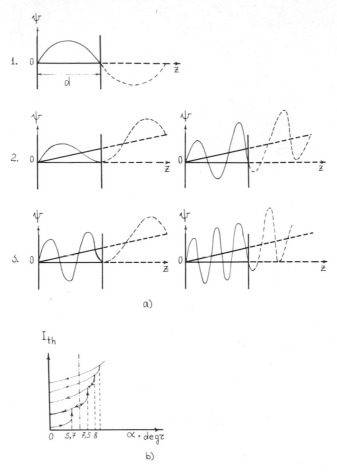

Fig. 1. a) The structure of light-induced distortions in NLC (the threshold reorientation, oblique incidence of o-wave, MBBA, d = 100 μm) for different α. 1: $\alpha = 0, \psi \sim \sin (\pi z/d)$; 2 : $\alpha = \alpha_1 = 5.7°$ (first jump), $\psi \sim [\sin(\frac{3}{2} \frac{\pi z}{d}) + z/d]$; 3: $\alpha = \alpha_2 \approx 7.5°$ (second jump), $\psi \sim \lfloor \sin(\frac{5}{2} \frac{\pi z}{d}) + \frac{z}{d} \rfloor$; ψ is the reorientation angle. The curves on the left and right hand sides determine the modulated structures on the upper and lower branches at the I_{th} jumps, respectively. b) the multistability of $I_{th}(\alpha)$. The vertical dash-dot line shows the different states of the system at a fixed value of α when I is varied. (These effects can be observed if a static magnetic field $\vec{H} \| \vec{E}$ is used – nonlinear coupling of $\vec{E}$ and $\vec{H}$.)

We confine ourselves here to the comment that the expression describing the relationship between the amplitudes of waves with ortho-gonal polarizations includes not only a term corresponding to the phase retardation ΦNL but also a term containing the change in the amplitude parameters. The characteristic spatial scale for this energy interchange in NLC is determined by the value of $1/vA_i^2$, where A_i^2 is the intensity of each component of the light-field ($i=1,2$), $v = q_z^{-1} \varepsilon_\alpha /16\pi K$, q_z is the z-component of the difference in wave vectors $\vec{q} = \vec{K}_e - \vec{K}_o$ of the e- and o-waves, while $\varepsilon_\alpha = \varepsilon_\| - \varepsilon_\perp$ and K are the optical anisotropy and elastic parameter of the NLC, respectively[4]. In cholesteric LC (CLC) the periodic energy redistribution between the waves leads to the so-called pendulum beatings (over space)[2]. For both of these cases the continuous output can be replaced by strong oscillations in the outgoing waves, and eventually by chaos (cf. Ref. 8).

There are many reasons for two different polarizations arising in LC[4]. We shall discuss a few of them below.

2. Different schemes of dynamic self-diffraction of light in aniso-tropic media have been realized experimentally, and pulsations in time have been obtained (for CW input), both in NLC (two linear orthogonal polarizations of the light propagating through the medium) and in CLC (two circular polarizations in the medium). We used a mixture of CLC and NLC.

In the first case, the oscillations of the ring pattern arising for the passing laser beam because of the oscillations of the angle ψ of molecular reorientation (from the initial orientation $\vec{n}_o \| z$)[5] *have been observed in the following experiments: (a) the excitation of non-adiabatic deformations in threshold reorientation of NLC for oblique incidence; (b) reorientation in hybrid aligned NLC; (c) two counterpropa-gating coherent waves with different linear polarizations in a NLC at normal incidence; (d) elliptically polarized light at normal incidence; and (e) two waves at oblique incidence to each other (but symmetric with respect to the initial orientation of the NLC) with different linear polarizations in the plane of incidence.

In CLC the oscillations of intensity for the passing radiation were obtained (f) during self-interaction under the Bragg reflection condition for the light and (g) for oblique incidence (o- or e-wave) in a homeo-tropic ($\vec{n}_o \| z$) aligned sample.

The dependences of the oscillation period τ on the intensity I of the incident light in cases (a) and (b) were measured[7]. An increase of τ with respect to I was obtained in the (a)-geometry; no oscillations occurred at larger values of I – the reorientation was steady-state.

If I continued to increase a regime of instability also arose. In contrast to (a) for the case (b) a decreas of τ with respect to I was obtained; at $I \sim 5kW/cm^2$ the oscillations are replaced by a steady-state (saturation of reorientation).

In the (c)-geometry the oscillations arise for both waves, but in antiphase[7]. If the intensities of the waves are equal ($I_1 = I_2$) the oscillations are damped out in about ten periods. They are largest and last longest when the waves ($\vec{E}_1 \perp \vec{E}_2$) have orthogonal polarizations. When the angle between $\vec{E}_1$ and $\vec{E}_2$ decreases the oscillations are weaker, and for $\vec{E}_1 \| \vec{E}_2$ the reorientation is stable. Different regimes of regenerative pulsations are obtained when the relative intensity of the waves is varied.

---

*In addition to these intensity oscillations others, due to polarization rotation, have been observed in (6) for circularly polarized input light.

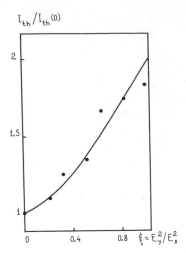

Fig. 2. I_{th} vs. polarization ellipticity $\xi \approx E_y^2/E_x^2$ at normal incidence of the light (NLC 5CB, d = 200 μm). $I_{th}(0)$ is the value for linear (along x) polarization ($\xi = 0$).

In the (d)-geometry the reorientation threshold intensity I_{th} increases if the polarization of the input waves is changed from linear (along x, E_x-component) to elliptical (an E_y-component arises) (Fig. 2). If E_y increases (the reorientation due to E_x still exists) there is a value of I, I_{yo}, above which oscillations take place. If $E_x \sim E_y$ they arise at $I_{yo} \sim I_{th}$ and disappear after a few oscillation periods. When $I_y \gg I_x$ the steady-state picture of threshold reorientation is reconstructed.

In the (e)-experiment, in contrast to geometric factors, a new effect, viz. a dramatic increase, rather than a decrease, of the nonlinear phase retardation ΦNL of the probe beam due to reorientation of the molecules was observed when the second field $\vec{E}_2$ was switched on, but the reorientation due to the first field $\vec{E}_1$ had already been saturated (in steady-state). This effect has a threshold for the sum field $\vec{E}_1+\vec{E}_2$.

In the (f)-case the initial shift (due to temperature) from the Bragg-resonance condition (the thermal nonlinearity of laser heating in a CLC-NLC-dye mixture was used) determined the oscillations of intensity of the passing light, both attenuated and non-attenuated.

The results for the (g)-experiment are shown in Fig. 3.

3. Qualitative explanations of the above effects can be given by taking into account the energy interchange and competitions between waves of different polarizations passing through the nonlinear medium.

In fact, e.g. in the (a)-case, I_{th} exists for the o-wave only, but an e-wave is also generated due to reorientation and energy transfer (because of transient processes as well as the induced shift gratings due to the dependence of the grating period on $I^{(4)}$) to the e-wave from the o-wave; this can decrease the intensity of the o-wave below the threshold

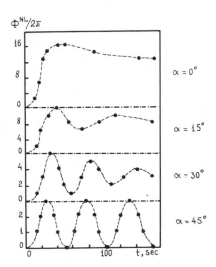

Fig. 3. The oscillations in the field at oblique incidence of an e-wave ($I = 2kW/cm^2$) for different angles α; homeotropic aligned CLC (a mixture of NLC 5CB and a chiral addition (0.017% by weight), which causes $\vec{n}$ to rotate out of the $\vec{E}\vec{k}$-plane because of reorientation, as a result of which two waves of orthogonal polarizations propagate through the medium).

value. Then reorientation no longer occurs in the LC, and the molecules
return to the initial unperturbed state. In this case the e-wave disappears, the intensity of the o-wave again exceeds I_{th}, and the entire
cycle repeats – i.e., oscillations occur. This explanation works near
the reorientation threshold only, which is in agreement with the experimental situation, where the oscillations arise exactly near I_{th}. In the
(b)-, (f)-, and (g)-experiments a second component of the polarization
also arises as a result of the action of the laser radiation on the LC,
and the oscillations have the same explanation as in the (a)-case. The
period of oscillations for all of these cases is determined by the
efficiency of energy transfer between the components of the
polarizations.

An important point to mention is that the development of oscillations in antiphase for two waves in the (d)-geometry strongly supports
the conjecture about the importance of energy transfer (for co-
propagating, but not for counterpropagating, waves for which the phase-
retardation effects -- because of the interference of the waves, but not
direct energy interchange -- play the principal role).

The phase-retardation effects (laser-induced ΦNL, which gives the
variation of polarization on passing through the nonlinear anisotropic
medium wave) can also lead to oscillations. In fact, linear polarization
of the light ($\xi = 0$) with intensity $I > I_{th}$ on the input plane become
elliptical ($\xi \neq 0$) inside the medium (because of anisotropy and ΦNL);
according to Fig. 2 ((d)-geometry) the value of I_{th} (ξ) increases, so the
fixed I can be smaller than $I_{th}(\xi)$. (This means that because of
reorientation the effective anisotropy of the medium changes and I_{th}
increases.) Thus, we have a similar case as in the explanation of the
oscillations in the (a)-geometry.

This approach has a quite general meaning -- if under laser radiation any parameter which determines I_{th} changes (e.g. spot size or
curvature of the laser beam, temperature, etc.) this can lead to
oscillations in the way described above.

The results of the (e)-experiment can be explained by direct
calculations of reorientation for this case, but also taking into account
the higher orders of perturbation theory in the reorientation angle ψ for
the high field. This effect of reorientation enhancement in crossed
polarized light is very important if $I_1 \approx I_2$; a necessary approximation
for that is the retention of terms $\sim \psi$ in the material equations of
motion of NLC-molecules. These terms have the same sign for both fields,
and the reorientation is proportional to $I_1 + I_2$. (We carried out a
separate experiment which showed that during the reorientation $\vec{n}$ does not
rotate out of the $\vec{E}_1\vec{E}_2$-plane).

4. The exact solution of this problem meets with serious difficulties even for numerical calculations. Here we discuss a few of the
results of an approximate analysis carried out by the procedure described
in Ref. 4.

The light-induced reorientation of $\vec{n}$ in NLC is characterized by two
angles: ψ and ϕ. We select them in the following way: $\vec{n} = (\sin\psi\cos\phi,$
$\sin\psi,\sin\phi,\cos\psi)$. That means that the angle ψ determines the deviation
of $\vec{n}$ from the initial homogeneous molecular orientation $\vec{n} \parallel z$ and leads to
oscillations of the ring pattern. The angle ϕ determines the azimuthal
rotation of $\vec{n}$ in the xy-plane $\perp z$ (oscillations of the polarization).

The first oscillations arise if the condition $\partial\phi/\partial z \neq 0$ is satisfied. If it is not ($\partial\phi/\partial z = 0$) this reduces to the relations $\partial\psi/\partial t = 0$,

$\partial\phi/\partial t$ = const for $t \to \infty$, and only oscillations of the polarizations develop.

In the approximation when $\psi \ll 1$ and the phase-retardation difference between the e- and o-waves

$$\Delta g \equiv \frac{\omega}{c} \int_0^z [\varepsilon_\perp^{1/2} / (1 - \frac{\varepsilon_a}{\varepsilon_\parallel} \sin^2\psi)]dz - \varepsilon_\perp^{1/2} z \ll 1,$$

where ω and c are the frequency and speed of light, energy transfer between the two different components of polarization exists for elliptical but not for circularly polarized input light. In the last case the solution is obtained in the form $\phi(z,t)$ = const $\cos \{\Omega\ell n|\cos \pi z/d| + \frac{\Omega\pi^2}{d^2} \frac{2K}{\gamma} t\}$; the frequency of rotation of the polarization is determined by the expression $\frac{\Omega\pi^2}{d^2} \frac{2K}{\gamma}$, where γ is the viscosity, and Ω is a constant. This reduces to a temporal dependence for ψ also:

$$\psi(t) = \text{const}^\cdot \exp\{(\frac{\varepsilon_a\varepsilon_\perp}{8\pi\varepsilon_\parallel} E^2 - K(\frac{\pi}{d})^2 - \frac{2K}{\pi} I_1\text{const})t +$$

$$+ \frac{d^2\gamma}{2\Omega\pi^3} \text{const}[\sin(\frac{\Omega\pi}{d^2} \frac{4K}{\gamma} t + I_2) - \sin I_2]\},$$

where $I_{1,2}$ are the numerical values of integrated relations.

Thus, in this approximation the time-periodic regimes can arise due to an effective variation of the anisotropy caused by the rotation of ϕ without direct energy interchange between the components of the polarizations*. But the value of Ω which is determined by the field (E^2) is not defined.

In the next approximation ($\sim\psi^2$) the energy transfer occurs for both elliptical and circularly-polarized light. Again, in the simplest case of circular polarization, the rotation of ϕ ($\phi = \Omega t$) leads to oscillations of the e-component of the light-wave; this leads to periodic solutions for ψ as well -- oscillations of the ring-pattern arise.

In conclusion we point out that the analysis of instabilities, oscillations, and chaotic regimes during laser-induced reorientation of LC is possible by using the traditional language of four-wave mixing spectroscopy in nonlinear optics[8]. Then the dynamic (shifted) grating is created in the medium by stimulated scattering of the laser radiation. The high energy interchange, which leads to the instabilities, e.g. in the (c)-geometry for the simplest case of two waves with the same polarization, takes place if the frequencies of the waves are not equivalent (ω and $\omega^\cdot$) and have a non-zero shift $\Omega \equiv \omega-\omega^\cdot$ which satisfies a phase-matching condition $\Omega = 1/\tau$, where τ is the nonlinear response time of the medium[8]. The energy transfers to the wave with the lower frequency; this is the process of parametric oscillation with distributed feedback. The difference between ω and $\omega^\cdot$ in the limit of radiation band width can arise due to different processes of dissipation of energy during the interaction of the laser field with the medium.

---

* The coupling of different waves inside the medium is realized by phase terms.

REFERENCES

1. S. A. Akhmanov, Usp. Fiz. Nauk, 149 361 (1986).
2. S. M. Arakelian and Yu. S. Chilingarian, Nonlinear Optics of Liquid Crystals, (Nauka, Moscow, 1984).
3. S. M. Arakelian, Usp. Fiz. Nauk 153 No. 3 (1987).
4. S. M. Arakelian and Yu. S. Chilingarian, IEEE J. of Quantum Electron. QE-22, 1276 (1986).
5. A. D. Zolot´ko, V. F. Kitaeva, and N. Kroo et al., Zh. Eksp. Teor. Fiz. 87, 859 (1984); 88, 1514 (1985).
6. E. Santamato, B. Daino, M. Romagnoli, M. Settemore, and Y. R. Shen, Phys. Rev. Lett. 57, 2423 (1986).
7. R. B. Alaverdian, S. M. Arakelian, and Yu. S. Chilingarian, Pis´ma v. Zh. Eksp. Teor. Fiz. 42,366 (1985); Pis´ma v Zh. Tech. Fiz. 13, 119 (1987).
8. Y. Silberberg and I. Bar Joseph, Phys. Rev. Lett. 48, 1541 (1982).

CONCLUDING REMARKS

Herman Z. Cummins

City College of the City University of New York

Having now reached the conclusion of this third USA-USSR Binational Symposium on Laser Optics of Condensed Matter, the 22 American participants are about to begin a week of post-symposium visits to universities and institutions in various locations in the Soviet Union. We take this opportunity to thank the Soviet Organizing and Program Committee for the outstanding work they have done in putting together this superb meeting, and to thank the many Soviet participants and their colleagues whose efforts on our behalf have made this week in Leningrad so rewarding and enjoyable.

Twice the time that passed between the first Symposium (Moscow 1975) and the second (New York 1979) has passed between the second and third Symposia, owing in large part to the deterioration of relations in general and scientific exchange in particular between the USA and the USSR. This meeting takes place at a time when relations are improving and we hope that future Symposia will occur more frequently, close to the two year schedule originally planned when the series began.

In the twelve years since the first Symposium many new developments have occurred in this field, and the range of subjects included in the program has been broadened to reflect this growth. For example, the dramatic decrease in length of laser pulses down to a few femtoseconds has opened new areas of laser spectroscopy, as discussed by C.V. Shank and E.P. Ippen, while the range of problems studied with picosecond pulses has also expanded as described by J. Aaviksoo and R.M. Hochstrasser. New applications of laser spectroscopy to surface physics were discussed by S.A. Akhmanov and G. Stegeman. Traditional laser Raman and luminescence techniques have also been extended to problems such as investigating the symmetry of isolated molecules in host crystals (L. Rebane) and properties of excitons in mixed semiconductor crystals (S.A. Permogorov).

Among the new areas discussed at this meeting were quantum wells and heterostructures (D.S. Chemla, M.V. Klein, M.D. Sturge), localization of photons (P.A. Fleury, J.L. Birman) and excitons (M.D. Sturge, S.A. Permogorov), and nonlinear optics of liquid crystals (Y.R. Shen, B.Ya Zel'dovich). The program also included presentations in areas of laser optics not included in the previous Symposia which significantly enriched this meeting. These included phase conjugation (J. Feinberg), optical bistability (V.S. Dneprovskii, E. Garmire, V.M. Agranovich), optical computing (H.M. Gibbs) and new applications of photorefractive materials (K.K. Rebane, R.W. Hellwarth, S.G. Odovlov).

The Symposium was also enhanced by outstanding visual materials in-
cluding M.A. Worontsov's film of instabilities in a liquid crystal, R.K.
Chang's photographs of stimulated Raman scattering in liquid droplets and J.
Feinberg's phase conjugation films and slide of a conjugated cat!

A point of continuity between this meeting and the first Symposium in
1975 may be of historical interest. J.P. Wolfe has described elegant new
studies of exciton transport in Cu_2O, a material studied extensively here at
the Ioffe Institute in Leninghrad by E.F. Gross and his coworkers and discussed in our previous Symposia. His method of producing an excitonic drift
velocity, production of inhomogenous stress by a small stylus pressed against
the crystal, resembles the method by which he produced electron hole drops
in germanium. A photograph of the experiment made with light produced by
recombination was shown in the first Symposium and appears on the jacket of
the proceedings: "Theory of Light Scattering in Condensed Matter", B. Bendow,
J.L. Birman and V.M. Agranovich, editors, Plenum Press, New York (1975).

S. A. Akhmanov
Moscow State University

All of us may say now that these were a really beautiful five days in
Leningrad. Our Soviet-American Symposium became an important event in the
life of the international scientific community as well as in the cooperation
between our two countries.

Many excellent papers were presented at the Symposium and the discussions also provided a lot of interesting information. Speaking today about
the success of our Symposium, I cannot ignore one of its very important components; the remarkable scientific and cultural traditions of Leningrad and
the whole atmosphere of this great city undoubtedly contributed a lot to the
success of our Symposium.

This year was marked by tremendous achievements in physics. The disco-
very of high-temperature superconductivity and the supernova observation
became real scientific sensations. Nevertheless, brilliant achievements presented at our Symposium allow us to say that 1987 is also the year of laser
superspectroscopy and laser superoptics of condensed matter.

Actually, taking into account only a comparitively narrow field - the
spectroscopy of light scattering in solids, which was the main subject of our
Symposia in 1975 and 1979, we may declare excellent achievements in this
field during the last few years. I have in mind the pico- and femtosecond
spectroscopy of spontaneous and coherent Raman scattering, investigatins of
scattering on superlattices and quantum wells, the research of the photon
localization on time and space effects in strongly scattered matter, stimulated scattering in optical fibers and scattering on the interface and the
defects in deformed crystals.

It has also become apparent that the new physical ideas and new experimental techniques have had a dramatic impact on the optics of condensed
matter. I would like to emphasize several problems and groups of questions
discussed at our seminar (I have to apologize for being biased) in which these
new changes were obvious.

I think we will start with ultrafast (pico- and femtosecond) optical
phenomena. Very short light pulses (we have heard here about the spectroscopy with pulse duration of about 6.10^{-15} s) allow one to apply direct time
measurement techniques (exciting-probing) to a wide range of elementary excitations in liquids, semiconductors and metals. A lot of data on the relaxation processes obtained by these methods were presented at our meetings.

When considering the relaxation processes in semiconductors we cannot disregard the progress made in the techniques based on hot photoluminescence and four-wave mixing spectra - I believe there is a lot of promise in the combination of direct and "indirect" methods, the time resolution 10^{-14} s has become quite available now.

Also very promising are the experimental investigations of femtosecond molecular dynamics in the condensed phase. With the mathematical simulation of molecular dynamics on big computers the methods discussed above are laying the foundations for a new branch of condensed matter physics and chemistry. Finally, as we were convinced here, the nonstationary wave problems - precursors, nonstationary diffraction on 3-dimensional lattices (one could achieve picosecond pulses in the x-rays region now) are still of great interest to the researcher.

"Laser and surface" is another field that should be mentioned here. Even the investigation of linear optical phenomena on rough surfaces keep supplying important practical information. Conditions of strong interaction of difracted waves, abnormal light absorption by rough surfaces and sharp surface amplification of local fields have been revealed. Optical nonlinearity obviously made the phenomena on the surface more complicated.

The new data about laser-modulated phase transitions on the surface are very interesting. It is noteworthy that optical methods, especially those based on nonlinear light reflection, provide very special information about them. Discussions at the seminar have demonstrated that nonlinear methods of surface transformation diagnostics (especially ultrafast) have serious advantages in comparison with difraction of slow electrons and x-ray methods.

Many new interesting facts were revealed during the discussions on nonlinear optics, nonlinear and laser materials. We can claim now that the Symposium has elucidated the present day situation in the investigation of strong nonlinear effects, such as optical bistability, instability and chaos (optical turbulence). 3-dimensional character of optical field opens up wide opportunities for the investigation and application of strong nonlinear wave interactions and selfinfluence. The rapid development of this research has been to a great extent determined by the progress made in the investigation and production of new nonlinear materials with very high (sometimes "gigantic") nonlinear susceptibilities such as semiconductors with 2-dimensional structures, photorefractive materials and liquid crystals.

Undoubtedly, the papers presented at our seminar are clear evidence of the highest level of the investigations in these rapidly developing branches of science. Of special interest are the studies on nonlinear optics of such "nontraditional" objects as drops, clusters, fractals, etc. Of all the practical applications of condensed matter laser optics discussed at the Symposium (we have just heard here about the new important results in the development of a new generation of solid-state lasers and lasers based on semiconductors heterostructures) I would like to emphasize the problems of physics and technology of nonlinear optical computers. Semicondutor nonlinear microresonators may become elements of high-speed optical digital processors (we have heard here about the project of an optical processor with 10^{13} operations per second - more than can be obtained on the Cray computer). Is this demonstration sufficient to judge about the rapid implementation of nonlinear optics into digital computers technology? We heard here different opinions on this subject. We have also seen at the Symposium that optical methods employing lasers invade energetically different branches of computer physics and technology.

The technique of stable spectral hole-burning in inhomogeneously broadened lines may lay the foundation for new systems of optical memory - spectral memory.

The phenomenon of spatial bistability and multistability, the techniques of nonlinear wave structures generation in nonlinear optical systems with 2-dimensional feedback may be the base for the new type of optical computers - nonlinear analogue computers. Such systems, as biocomputers, intensively discussed now, will operate with nonlinear samples, and not with numbers.

We hope that in the near future we will witness new remarkable results and discoveries in laser optics of condensed matter. The "hot spots" here are: strong nonlinear excitation, laser-induced phase transitions of different types (including pico- and femtosecond time scale), femtosecond molecular dynamics, linear and nonlinear optics of new objects (clusters, fractals, quantum wells, quantum points, etc.), and the physics of nonlinear optical computers.

Undoubtedly, optical and especially nonlinear optical methods will aid in understanding the high-temperature superconductivity phenomenon and, of course, the new laser methods developing now will strongly affect the development of condensed matter physics (such as the technique of pico- and femtosecond electron clots and ultrashort x-ray pulses, based on the ultrashort laser pulses.)

Predictions are certainly a risky business. However I am sure you will agree that we are now working in a rapidly developing field with an unending flow of new ideas, new experimental methods, new people, new scientific groups.

Finally, I should like to thank all the participants on behalf of the Program Committee and express cordial gratitude to our hosts - Leningrad physicists, for the excellent organization of the Symposium.

I do hope we will meet in two years at the 4th Binational USSR-USA Symposium on Laser Optics of Condensed Matter.

Thank you for your attention.

Karl K. Rebane
Institute of Physics, Academy of Sciences, Estonian SSR

We are closing the Third USSR-USA Symposium on Laser Optics of Condensed Matter. We have had and discussed a number of interesting presentations, both oral and posters.

We shall study the papers to be published and recall the discussions we had here during the Symposium. The Proceedings are to be published both in Russian in this country and in English in the US. There are good prospects that the Proceedings will be in print quite soon, around the end of this year.

The general feeling of the participants and certainly my own opinion is that the Symposium was a success. The topics were up-to-date. Several modern problems of lasers and their applications were presented by the leading scientists of today's field of physics and also chemistry, biology and computer science. The discussions were active and well-pointed.

The level of a scientific meeting is determined by the body of its participants. The US delegation to the Third Symposium was representative and of high qualification, including both the scientists who have been well known for many years already and the physicists of a younger generation who have started to make important contributions to lasers and their applications.

On behalf of the Soviet side of the permanent Organizing Committee of the series of our Symposia I would like to thank Professors of the City University of New York Joseph Birman and Herman Cummins for their well pointed and expert efforts to create such a powerful team of physicists and bring it to the Third Symposium in Leningrad.

The participants unanimously share the opinion that the Leningrad Organizing Committee headed by Academician Vladimir Maximovitch Tuchkevich and Vice-chairman Alexander Alexandrovitch Kaplyanskii performed smoothly - the Symposium was well organized both in scientific and organizational aspects.

On behalf of all the participants many thanks to them and their assistants (helpers, aids). Many thanks to Ija Pavlovna Ipatova and to Lyudmila Alexejevna Burejeva, key figures of the Seminar in Moscow.

Of major importance for the successful work of the sessions was the expert translation. For that many thanks from all of us to our colleagues - physicists Georgi Vladimirovitch Skrebtsov and Arseni Borisovitch Berezin.

Well, tonight we'll meet in a friendly dinner. Tomorrow our American colleagues will start on trips aimed at visiting laboratories in different towns. I wish them a successful journey in this country and a measure of benefit from new and old contacts with Soviet scientists. As to our Soviet colleagues who do not live in Leningrad - a happy journey home!

Thank you for your attention.

USA PARTICIPANTS

J. L. Birman	City College of the City University of New York
E. Burstein	University of Pennsylvania
R. K. Chang	Yale University
D. S. Chemla	AT&T Bell Laboratories, Holmdel
H. Z. Cummins	City College of the City University of New York
J. Feinberg	University of Southern California
P. A. Fleury	AT&T Bell Laboratories, Murray Hill
E. Garmire	University of Southern California
H. M. Gibbs	University of Arizona
R. W. Hellwarth	University of Southern California
R. M. Hochstrasser	University of Pennsylvania
E. P. Ippen	Massachusetts Institute of Technology
M. V. Klein	University of Illinois
M. Lax	City College of the City University of New York
A. A. Maradudin	University of California, Irvine
S. L. McCall	AT&T Bell Laboratories, Murray Hill
J. F. Scott	University of Colorado
C. V. Shank	AT&T Bell Laboratories, Holmdel
Y. R. Shen	University of California, Berkeley
G. I. Stegeman	University of Arizona
M. D. Sturge	Dartmouth College
J. P. Wolfe	University of Illinois

USSR PARTICIPANTS

J.Yu. Aaviksoo	Institute of Physics, Tartu, Estonian SSR
G. I. Abutalibov	Institute of Physics, Baku, Azerbaijan SSR
V. M. Agranovich	Institute of Spectroscopy, Troitsk
A. S. Agabekyan	Yerevan State University, Georgian SSR
V. N. Ageev	A.F. Ioffe Physico-Technical Institute, Leningrad
S. A. Akhmanov	Moscow State University
Z. I. Alferov	A.F. Ioffe Physico-Technical Institute, Leningrad
A. A. Anik'iev	Physico-Technical Institute, Dushanbe, Tajik SSR
S. M. Arakelian	Yerevan State University
A. G. Aronov	Institute of Nuclear Physics, Leningrad
B. M. Ashkinadze	A.F. Ioffe Physico-Technical Institute, Leningrad
V. M. Asnin	A.F. Ioffe Physico-Technical Institute, Leningrad
S. N. Bagaev	Institute of Thermophysics, Novosibirsk
L. A. Bureyeva	Scientific Council on Spectroscopy, Moscow
A. A. Bugayev	A.F. Ioffe Physico-Technical Institute, Leningrad
V. P. Chebotaev	Institute of Thermophysics, Novosibirsk
Yu.S. Chilingaryan	Yerevan State University, Armenian SSR
A. A. Danilov	Institute of General Physics, Moscow
V. S. Dneprovskii	Moscow State University
A. L. Efros	A.F. Ioffe Physico-Technical Institute, Leningrad
S. E. Egorov	Institute of Spectroscopy, Troitsk
A. I. Ekimov	S.I. Vavilov Optical Institute, Leningrad
M. D. Galanin	P.N. Lebedev Physical Institute, Moscow
Yu. Galperin	A.F. Ioffe Physico-Technical Institute, Leningrad

D. Z. Garbuzov	A.F. Ioffe Physico-Technical Institute, Leningrad
A. A. Gorokhovskii	Institute of Physics, Tartu, Estonian SSR
S. A. Gurevich	A.F. Ioffe Physico-Technical Institute, Leningrad
V. L. Gurevich	A.F. Ioffe Physico-Technical Institute, Leningrad
V. V. Hizhnyakov	Institute of Physics, Tartu, Estonian SSR
I. P. Ipatova	A.F. Ioffe Physico-Technical Institute. Leningrad
V. I. Ivanov-Omskii	A.F. Ioffe Physico-Technical Institute, Leningrad
E. L. Ivchenko	A.F. Ioffe Physico-Technical Institute, Leningrad
Yu.M. Kagan	Kurchatov Atomic Energy Institute, Moscow
A. A. Kaplyanskii	A.F. Ioffe Physico-Technical Institute, Leningrad
F. V. Karpushko	Institute of Physics, Minsk, Byelorussian SSR
R. V. Katilyus	A.F. Ioffe Physico-Technical Institute, Leningrad
V. B. Khalfin	A.F. Ioffe Physico-Technical Institute, Leningrad
P. S. Kop'ev	A.F. Ioffe Physico-Technical Institute, Leningrad
O. V. Konstantinov	A.F. Ioffe Physico-Technical Institute, Leningrad
N. I. Koroteev	Moscow State University, Moscow
V. A. Kosobukin	A.F. Ioffe Physico-Technical Institute, Leningrad
V.Yu. Kovalchuk	A.F. Ioffe Physico-Technical Institute, Leningrad
N. M. Kreines	Institute of Physical Problems, Moscow
A. P. Levanyuk	Institute of Crystallography, Moscow
I. B. Levinson	Institute of Microelectronics, Chernogolovka
I. V. Lerner	Institute of Spectroscopy, Troitsk
T. A. Leskova	Institute of Spectroscopy, Troitsk
V. V. Lemanov	A.F. Ioffe Physico-Technical Institute, Leningrad
T. I. Maksomova	A.F. Ioffe Physico-Technical Institute, Leningrad
Yu.F. Markov	A.F. Ioffe Physico-Technical Institute, Leningrad
D. N. Mirlin	A.F. Ioffe Physico-Technical Institute, Leningrad
S. A. Moskalenko	Institute of Physics, Kishinev, Moldavian SSR
S. G. Odoulov	Institute of Physics, Kiev, Ukrainian SSR
Yu.A. Ossipyan	Inst. of Solid State Physics, Chernogolovka
V. I. Perel	A.F. Ioffe Physico-Technical Institute, Leningrad
S. A. Permogorov	A.F. Ioffe Physico-Technical Institute, Leningrad
R. I. Personov	Institute of Spectroscopy, Troitsk
M. P. Petrov	A.F. Ioffe Physico-Technical Institute, Leningrad
R. V. Pisarev	A.F. Ioffe Physico-Technical Institute, Leningrad
Yu. N. Polivanov	Institute of General Physics, Moscow
E. I. Rashba	L.D. Landau Inst. of Theor. Phys., Chenogolovka
B. S. Razbirin	A.F. Ioffe Physico-Technical Institute, Leningrad
A. K. Rebane	Institute of Physics, Tartu, Estonian SSR
K. K. Rebane	Institute of Physics, Tartu, Estonian SSR
L. A. Rebane	Inst. of Chem. & Bio. Physics, Tallin, Estonian SSR
A. N. Reznitsky	A.F. Ioffe Physico-Technical Institute, Leningrad
A. A. Rogachev	A.F. Ioffe Physico-Technical Institute, Leningrad
B. S. Ryvkin	A.F. Ioffe Physico-Technical Institute, Leningrad
V. I. Safarov	A.F. Ioffe Physico-Technical Institute, Leningrad
R. P. Seysyan	A.F. Ioffe Physico-Technical Institute, Leningrad
B. I. Shklovskii	A.F. Ioffe Physico-Technical Institute, Leningrad
V. V. Shuvalov	Moscow State University
S. G. Tikhodeev	P.N. Lebedev Physical Institute, Moscow
V. B. Timofeev	Institute of Solid State Physics, Chernogolovka
E. D. Trifonov	A.I. Herzen Pedagogical Institute, Leningrad
I. N. Uraltzev	A.F. Ioffe Physico-Technical Institute, Leningrad
Y. Y. Vaitkus	Vilnius State University
V. I. Vladimirov	A.F. Ioffe Physico-Technical Institute, Leningrad
M. A. Voronstov	Moscow State University
I. D. Yaroshetskii	A.F. Ioffe Physico-Technical Institute, Leningrad
B. P. Zakharchenya	A.F. Ioffe Physico-Technical Institute, Leningrad
B.Ya. Zel'dovich	Institute of Applied Mechanics, Moscow

Energy (continued)
  electron, quasi-elastic scattering, 248
  ionization, field desorption, 191
  microcrystal polariton, 205
  liquid crystals, 523, 525
  quasi-elastic light scattering, 248
  nonequilibrium electron temperatures and, 17
  single-particle light scattering in crystals, 245-246
  superconducting vs. optical switches, 468
  surface polariton, 149, 151, 152-154
  switching, 469-470, 472
  2-D electron density of states, 61-64
Energy relaxaton, see Relaxation
Energy spectrum
  direct gap materials, 245
  Fermi-fluid, Coulomb gaps in, 64-69
  microcrystal hole, quantization of, 203-204
  one-electron, gap values in, 61
  photoluminescence spectroscopy, 19-22
  polaritons, incoherent, 331
  2-D electrons, 62-64
Energy transfer
  LC three-D gratings, light-induced self-excitation, 502-505
  between molecules, 194
  optical Freedericksz transitions, 495
  quasistationary distribution of excitons, 319
  ruby photocurrents, 374-380
Energy transformation coefficients, 149, 151
Enhanced coherent backscattering, 223
Entropy fluctuations
  central mode scattering, barium sodium niobate, 266
  and scattering intensity, 252-253
Epitaxial crystal growth, 79
Epitaxial structures, 162-164
Equal-pulse correlation, 11
Erbium fluorite, 476
Etalons, 481
  bulk vs. quantum well material, 449-454
  Fab+ry-Perot, 481
Etching, switches, 470
Ethylene glycol, 13
Europium chalcogenides, 287-292
Evaporation, films, 279-280

E-wave, 501, 503, 521
Exchange interactions, 73, 74
Excitation, see Excitons; Polaritons
Excitation energy, bleaching at, 8
Excitation power densities, 90
Excitation profiles, RRS (REP), 297-302
Excitation-profile spectra, 173
Excite-and-probe measurements, 444, 460
Excited carrier density, 13
Excited states (see also Excitons; Polaritons, surface)
  magnon scattering, 287-292
  in phonon branches, 165
  Raman scattering, 169-171
  3-D gratings in liquid crystals, 497-505
  vibrational relaxation of crystal impurities, 53-59
Exciton binding energy, 460
Exciton-exciton interactions (see also Bistability; Nonlinearities)
  bistability in CdS at low pumping intensity, 457-460
  GaSe, 460, 461, 462, 464
  in quantum wells, 82
Excitonic polaritons
  bottleneck region, 315-319
  in dispersive media, 303-313
  surface polaritons, 149, 154
Exciton-polariton spectroscopy, 303-313
Excitons (see also Polaritons)
  bistability in CdS at low pumping intensity, 457-460
  bleaching, 487
  bound, 355-362
    absorption spectrum, 356-357
    electron spin resonance, 362
    magnetic splitting, 360-362
    Raman scattering, 357
    $TlGaS_2$ luminescence, 358-362
  carrier relaxation in quantum wells, 95-100
  dielectric, 154
  in dispersive media, 303-313
    precursors, 312-313
    reflectivity, 308, 309
  GaAs nonlinearities, 450, 451
  GaSe, 460, 461, 462
  localized, in semiconductor solid solutions, 347-353
  microcrystal, 205, 206
    electron-hole interactions, 202-203
    impurity luminescence, 205-206
    size quantization, 201-202
  quantum statistics, 337-345

Ring pattern, 521
Ring phase conjugator, 400, 401, 406, 407, 408, 409
Ring resonator, 509–512
Rosette, dislocation, 211, 212, 213
Rotary instability, 507–518
Rotary shift, 509–512
Rotational processes, in liquids, 41–52
Rotator, Faraday, 406
Rough surfaces, optical interactions at, 127–136
  deep lamellar grating, 130–132
  diagnostics, 157–168
  isolated ridge or groove, 128–130
  laser-induced, 128, 171
  random grating, Goos Hänchen effect, 132–133
  random metal surface, 134–136
Ruby lasers, conductivity of, 373–380
  microscopic nature of phenomenon, 377–380
  photocurrent kinetics, 374–375
  photocurrent spectral response, 376–377
Rutile, 279

Salol, 251, 252
Saturation spectroscopy, 31–38
Scaling coefficient, 299–300
Scaling theory of localization, 219
Scattering
  classical photons, randomly scattered, 238–241
  disordered dielectric, classical photon localization in, 237–242
  femtosecond spectroscopy, 8
    carrier-carrier, 14
    electron-phonon, 17
    hot electrons, 11
    metals, 16–17
  hot electron, 441
  intra- and interalley, 20
  lamellar grating and, 131, 132
  LC 3-D gratings, light-induced self-excitation, 501–502
  light propagation through disordered media, 217–222
  surface polariton excitation, 128, 129–130
Second derivative spectra, 204, 205
Second harmonic generation
  anisotropy, 158, 159
  dipole contribution, 163
  Si surface deposited films, 162–164
  surface changes, 158
  surface modification of silicon, 159–161

Second order nonlinearity, 164
Self-diffraction, 436, 437, 439, 521
Self-excitation, 3-D gratings in liquid crystals, 497–505
Self-screening, excitons, 464
Semiconductor crystals, see Crystals
Semiconductors
  bistability in, dynamic nonlinearities and, 457–464
  carrier heating in, transient grating techniques, 435–441
  femtosecond spectroscopy, 5–9, 11–14
  Franz-Keldysh effect, 190
  photoluminescence spectroscopy, energy spectrum and relaxation times, 19–22
  plasmas
    cooling of, 25–29
    pulsed holographic diagnostics, 443–448
  quantum wells, see Quantum wells
  surface structure transformations, 157–168
Semi-linear open cavity, 393
Separate-confinement heterostructure lasers, 103–110
s-functions, 21–22
Shear viscosity, 43
Short period superlattices, 89–85
Sideband, phonon, 73, 422
Side alleys, 20–21
Sigma model, nonlinear, 219
Signal wave, 396
Silicon
  Franz-Keldysh effect, 191
  induced absorption in, 443–448
  thermal oxidation, 162–164
Silicon dioxide
  single-frequency diode lasers, 113–118
  thermal oxidation of Si, 162–164
  thin-film interference devices, 476
Silicon film deposition, 158
Silicon lattice dynamics, 164–168
Silicon melting, 158
Silicon-MOS structure, 61–62
Silver
  grating, 131
  phthalocyanine adsorbate, 175–177
Single-beam phase conjugator, 399, 400, 401
Single-bend conjugators, 400, 403
Single-carrier model, 381–382
Single crystals
  hot carrier diffusion, 438
  light scattering, 245–250
  zero-phonon lines, 422
Single-crystal waveguides, 281
Single micron-size droplets, 193–197